WITHDRAWN
LVC BISHOP LIBRA

NONLINEAR PROGRAMMING

ANALYSIS AND METHODS

Prentice-Hall
Series in Automatic Computation

AHO, ed., *Currents in the Theory of Computing*
AHO AND ULLMAN, *The Theory of Parsing, Translation, and Compiling*, Volume I: *Parsing;* Volume II: *Compiling*
ANDREE, *Computer Programming: Techniques, Analysis, and Mathematics*
ANSELONE, *Collectively Compact Operator Approximation Theory and Applications to Integral Equations*
AVRIEL, *Nonlinear Programming: Analysis and Methods*
BENNETT, JR., *Scientific and Engineering Problem-Solving with the Computer*
BLAAUW, *Digital System Implementation*
BLUMENTHAL, *Management Information Systems*
BRENT, *Algorithms for Minimization without Derivatives*
BRINCH HANSEN, *Operating System Principles*
BRZOZOWSKI AND YOELI, *Digital Networks*
COFFMAN AND DENNING, *Operating Systems Theory*
CRESS, et al., *FORTRAN IV with WATFOR and WATFIV*
DAHLQUIST, BJÖRCK, AND ANDERSON, *Numerical Methods*
DANIEL, *The Approximate Minimization of Functionals*
DEO, *Graph Theory with Applications to Engineering and Computer Science*
DESMONDE, *Computers and Their Uses*, 2nd ed.
DIJKSTRA, *A Discipline of Programming*
DRUMMOND, *Evaluation and Measurement Techniques for Digital Computer Systems*
ECKHOUSE, *Minicomputer Systems: Organization and Programming (PDP-11)*
FIKE, *Computer Evaluation of Mathematical Functions*
FIKE, *PL/1 for Scientific Programmers*
FORSYTHE AND MOLER, *Computer Solution of Linear Algebraic Systems*
GEAR, *Numerical Initial Value Problems in Ordinary Differential Equations*
GILL, *Applied Algebra for the Computer Sciences*
GORDON, *System Simulation*
GRISWOLD, *String and List Processing in SNOBOL4: Techniques and Applications*
HANSEN, *A Table of Series and Products*
HARTMANIS AND STEARNS, *Algebraic Structure Theory of Sequential Machines*
HILBURN AND JULICH, *Microcomputers/Microprocessors: Hardware, Software, and Applications*
JACOBY, et al., *Iterative Methods for Nonlinear Optimization Problems*
JOHNSON, *System Structure in Data, Programs, and Computers*
KIVIAT, et al., *The SIMSCRIPT II Programming Language*
LAWSON AND HANSON, *Solving Least Squares Problems*
LORIN, *Parallelism in Hardware and Software: Real and Apparent Concurrency*
LOUDEN AND LEDIN, *Programming the IBM 1130*, 2nd ed.
MARTIN, *Computer Data-Base Organization*
MARTIN, *Design of Man-Computer Dialogues*
MARTIN, *Design of Real-Time Computer Systems*
MARTIN, *Future Developments in Telecommunications*
MARTIN, *Principles of Data-Base Management*
MARTIN, *Programming Real-Time Computing Systems*

MARTIN, *Security, Accuracy, and Privacy in Computer Systems*
MARTIN, *Systems Analysis for Data Transmission*
MARTIN, *Telecommunications and the Computer*
MARTIN, *Teleprocessing Network Organization*
MARTIN AND NORMAN, *The Computerized Society*
MCKEEMAN, et al., *A Compiler Generator*
MEYERS, *Time-Sharing Computation in the Social Sciences*
MINSKY, *Computation: Finite and Infinite Machines*
NIEVERGELT, et al., *Computer Approaches to Mathematical Problems*
PLANE AND MCMILLAN, *Discrete Optimization: Integer Programming and Network Analysis for Management Decisions*
POLIVKA AND PAKIN, *APL: The Language and Its Usage*
PRITSKER AND KIVIAT, *Simulation with GASP II: A FORTRAN-based Simulation Language*
PYLYSHYN, ed., *Perspectives on the Computer Revolution*
RICH, *Internal Sorting Methods Illustrated with PL/1 Programs*
RUDD, *Assembly Language Programming and the IBM 360 and 370 Computers*
SACKMAN AND CITRENBAUM, eds., *On-Line Planning: Towards Creative Problem-Solving*
SALTON, ed., *The SMART Retrieval System: Experiments in Automatic Document Processing*
SAMMET, *Programming Languages: History and Fundamentals*
SCHAEFER, *A Mathematical Theory of Global Program Optimization*
SCHULTZ, *Spline Analysis*
SCHWARZ, et al., *Numerical Analysis of Symmetric Matrices*
SHAH, *Engineering Simulation Using Small Scientific Computers*
SHAW, *The Logical Design of Operating Systems*
SHERMAN, *Techniques in Computer Programming*
SIMON AND SIKLOSSY, eds. *Representation and Meaning: Experiments with Information Processing Systems*
STERBENZ, *Floating-Point Computation*
STOUTEMYER, *PL/1 Programming for Engineering and Science*
STRANG AND FIX, *An Analysis of the Finite Element Method*
STROUD, *Approximate Calculation of Multiple Integrals*
TANENBAUM, *Structured Computer Organization*
TAVISS, ed., *The Computer Impact*
UHR, *Pattern Recognition, Learning, and Thought: Computer-Programmed Models of Higher Mental Processes*
VAN TASSEL, *Computer Security Management*
VARGA, *Matrix Iterative Analysis*
WAITE, *Implementing Software for Non-Numeric Application*
WILKINSON, *Rounding Errors in Algebraic Processes*
WIRTH, *Algorithms + Data Structures = Programs*
WIRTH, *Systematic Programming: An Introduction*
YEH, ed., *Applied Computation Theory: Analysis, Design, Modeling*

NONLINEAR PROGRAMMING

ANALYSIS AND METHODS

MORDECAI AVRIEL

Technion—Israel Institute of Technology
Haifa, Israel

PRENTICE-HALL, INC.

ENGLEWOOD CLIFFS, NEW JERSEY

Library of Congress Cataloging in Publication Data

AVRIEL, M.
Nonlinear programming: Analysis and Methods.

Bibliography: p.
Includes index.
1. Nonlinear programming. I. Title.
T57.8.A9 519.7′6 75–45324
ISBN 0–13–623603–0

©1976 by Prentice-Hall, Inc., Englewood Cliffs, N. J.

All rights reserved. No part of this book may be reproduced in any form, by mimeograph or any other means, without permission in writing from the publisher.

10 9 8 7 6 5 4 3 2

Printed in the United States of America

PRENTICE-HALL INTERNATIONAL, INC., *London*
PRENTICE-HALL OF AUSTRALIA PTY. LIMITED, *Sydney*
PRENTICE-HALL OF CANADA, LTD., *Toronto*
PRENTICE-HALL OF INDIA PRIVATE LIMITED, *New Delhi*
PRENTICE-HALL OF JAPAN, INC., *Tokyo*
PRENTICE-HALL OF SOUTH-EAST ASIA PRIVATE LIMITED, *Singapore*

298044

T
57.8
.A9

To the Memory of My Parents

CONTENTS

PART II
METHODS

8 ONE-DIMENSIONAL OPTIMIZATION

9 MULTIDIMENSIONAL UNCONSTRAINED OPTIMIZATION WITHOUT DERIVATIVES: EMPIRICAL AND CONJUGATE DIRECTION METHODS

10 SECOND DERIVATIVE, STEEPEST DESCENT AND CONJUGATE GRADIENT METHODS

11 VARIABLE METRIC ALGORITHMS

12 PENALTY FUNCTION METHODS

13 SOLUTION OF CONSTRAINED PROBLEMS BY EXTENSIONS OF UNCONSTRAINED OPTIMIZATION TECHNIQUES

14 APPROXIMATION-TYPE ALGORITHMS

PREFACE

Optimization is the science of selecting the best of many possible decisions in a complex real-life environment. The subject of this book is nonlinear optimization or, as it is more commonly called, nonlinear programming, a discipline playing an increasingly important role in such diverse fields as operations research and management science, engineering, economics, system analysis, and computer science.

The systematic approach to decision-making generally involves three closely interrelated stages. The object of the first stage is to develop a mathematical model representing the decision problem under consideration. The decision-maker forms the model first by identifying the variables; then collecting the relevant data; formulating the objective function to be optimized—that is, maximized or minimized; and arranging the variables and data into a set of mathematical relations, such as equations and inequalities, called constraints. The second stage continues the process with an analysis of the mathematical model and selection of an appropriate numerical technique for finding the optimal solution. If one or more of the constraints, or the objective function of the model, are nonlinear, the optimization problem is a nonlinear program. Analysis of a nonlinear program provides valuable insight into the structure of the problem and answers questions about the existence and characterization of feasible and optimal decisions. Numerical solution techniques of nonlinear programs depend on the nature of the objective and constraint functions and the structure of the optimization problem. Based on the analysis, the decision-maker recognizes what existing numerical solution techniques, if any, can yield a sought optimum. The third stage consists of finding an optimal solution, in most cases on a computer. In obtaining the solution, the decision-maker uses either an available computer code or develops a new code, implementing the techniques chosen in the second stage. The process is complete with the subsequent reinterpretation of the numerical

solution in terms of the decisions to be made, and finally the evaluation of the practical aspects of convergence, efficiency, and error analysis of the solution technique used.

Since accurate representation of real-world situations often results in mathematical models involving nonlinear functions, the analysis of nonlinear programs and the development of efficient numerical algorithms have become essential subject matters for study by students and practitioners of various disciplines involving optimal decision-making. This book offers a comprehensive overview of nonlinear programming, presenting the reader the state of the art of the previously mentioned second stage of the decision-making process.

The book consists of two parts. Part I, *Analysis*, first derives necessary and sufficient optimality conditions for mathematical programs. Next we give an introduction to convex sets and functions and derive the theory of convex programming. Nonlinear convex duality is then developed using the modern approach of conjugate functions. Generalized convexity and the analysis of selected nonlinear programs close the first part. Part II, *Methods*, deals with numerical solution techniques for nonlinear optimization problems. A discussion of unconstrained optimization methods is presented first, including chapters on nonderivative, conjugate direction and variable metric techniques. Penalty function methods, extensions of unconstrained techniques, and approximation-type methods are among the most widely used or recently developed promising algorithms for constrained nonlinear programs discussed in this part.

It is almost impossible to cover in a single volume every aspect of nonlinear programming. Consequently, I have either left out or merely mentioned certain important subjects, such as the theory of complementarity, discrete optimization and stochastic programming. Similarly, I have omitted certain computational details of the numerical techniques, such as the question of roundoff errors, ill-conditioning, and computer implementation. I assume that the interested reader can find these subjects in more specialized texts.

The material in this book has served as a basis for teaching nonlinear optimization for advanced undergraduate and graduate students at the Technion–Israel Institute of Technology, Tel-Aviv University, and the University of British Columbia. The book has great flexibility. The instructor may present each part separately in a standard one-semester course and in any order since cross-reference from one part of the book to the other is minimal. Another combination is to cover the material of the two parts in parallel, in a one-year course on nonlinear programming. In this integrated approach, which I have used several times, the instructor may teach Chapters 1, 2, and 3 from Part I and Chapter 8, 9, 10 and 11 from Part II in the first term and the rest of the book in the second term.

The mathematical background required is differential calculus (partial deratives, series expansions), elementary linear algebra (matrix and vector operations), and real analysis (sets, limits and sequences, continuity). In addition, a basic familiarity with the notion of presenting results by theorems and proofs is helpful. The text assumes no previous experience in optimization; in particular, it requires no previous knowledge of linear programming.

The book aims toward readers from a variety of disciplines. It is recommended that the readers of this book supplement the exercises appearing at the end of the chapters by formulating nonlinear optimization problems arising from their fields of interests. In a formal class, the instructor may assign term projects along these lines which the students can subsequently present and discuss in class. In addition, the reader can obtain most valuable insight into the characteristics of the numerical techniques by writing computer codes for the simpler techniques appearing in the book or by using available codes written by others. Thus, some knowledge of computer programming is useful but not necessary.

It is a real pleasure to acknowledge the persons who have influenced my thinking and whose professional interests stimulated mine. I am indebted to Professor D. J. Wilde of Stanford University who first introduced me to optimization and to Dr. A. C. Williams of the Mobil Oil Corporation who taught me more of this subject.

I am grateful to Professor W. Resnick of the Technion for his continued interest and encouragement in writing this book and to Professors J. H. Dreze and G. deGhellinck of the University of Louvain, who provided me with excellent facilities and intellectual stimulation for writing the drafts of the first chapters.

During the five years of writing this book I greatly benefited from discussions with my former students, E. U. Choo, Y. Menidor, and I. Zang, whose many comments on the manuscript were incorporated in the final version. I also wish to thank Professors S. Schaible, R. A. Tapia, R. J-B. Wets, and D. J. Wilde for their valuable suggestions. Special thanks go to Mrs. Ora Naor and Mrs. Aviva Rappaport for their excellent typing of the manuscript. Last but not least, I am indebted to my wife, Hava, and children, Dorith, Ron and Tal, for their patience and understanding during the many lost evenings and weekends.

MORDECAI AVRIEL

1 INTRODUCTION

Mathematical programming is a branch of optimization theory in which a single-valued **objective function** f of n real variables $x_1, \ldots, x_n$ is minimized (or maximized), possibly subject to a finite number of **constraints**, which are written as inequalities or equations. Generally we can define a mathematical program of, say, minimization as

$$\text{(MP)} \qquad \min f(x) \tag{1.1}$$

subject to

$$g_i(x) \geq 0, \qquad i = 1, \ldots, m \tag{1.2}$$

$$h_j(x) = 0, \qquad j = 1, \ldots, p \tag{1.3}$$

where x denotes the column vector whose components are $x_1, \ldots, x_n$. In other words, (MP) is the problem of finding a vector x^* that satisfies (1.2) and (1.3) and such that $f(x)$ has a minimal—that is, optimal value. If one or more of the functions appearing in (MP) are nonlinear in x, we call it a **nonlinear program**, in contrast to a linear program, where all these functions must be linear. The study of some basic aspects of nonlinear programming is the subject of this book.

Nonlinear programming problems arise in such various disciplines as engineering, economics, business administration, physical sciences, and mathematics, or in any other area where decisions (in a broad sense) must be taken in some complex (or conflicting) situation that can be represented by a mathematical model. In order to illustrate some types of nonlinear programs, a few examples are presented below.

NONLINEAR CURVE FITTING

Suppose that in some scientific research, say in biology or physics, a certain phenomenon f is measured in the laboratory as a function of time. Also suppose that we are given a mathematical model of the phenomenon, and from the model we know that the value of f is assumed to vary with time t as

$$f(t) = x_1 + x_2 \exp(-x_3 t). \tag{1.4}$$

The purpose of the laboratory experiments is to find the unknown parameters x_1, x_2, and x_3 by measuring values of f at times $t^1, t^2, \ldots, t^M$. The decision-making process involves assigning values to the parameters, and it is reasonable to ask for those values of x_1, x_2, and x_3 that are optimal in some sense. For example, we can seek optimal values of the parameters in the least-squares sense—that is, those values for which the sum of squares of the experimental deviations from the theoretical curve is minimized. Formally, we have the nonlinear program

$$\min F(x_1, x_2, x_3) = \sum_{i=1}^{M} [f(t^i) - x_1 - x_2 \exp(-x_3 t^i)]^2. \tag{1.5}$$

Note that this is an unconstrained program that, if solved, may yield unacceptable values of the parameters. To avoid such a situation, we can impose restrictions in the form of constraints. For example, the parameter x_3 can be restricted to have a nonnegative value—that is,

$$x_3 \geq 0. \tag{1.6}$$

Also suppose that, for the particular phenomenon under consideration, the mathematical model proposed can be acceptable only if the parameters are so chosen that at $t = 0$ we have $f(0) = 1$. Hence we must add a constraint

$$x_1 + x_2 = 1. \tag{1.7}$$

Solving (1.5), subject to (1.6) and (1.7), is then a constrained nonlinear programming problem having a nonlinear objective function with linear inequality and equality constraints.

LOCATION PROBLEM

Suppose now that the location of a supply center to serve m customers having fixed spatial locations in a city must be selected. The commodity to

be supplied from the center may be electricity, water, milk, or some other goods. The criterion for selecting the location of the supply center is to minimize some distance function from the center to the customers. It may happen, for example, that we are interested in minimizing the **maximal** distance from the center to any particular customer. Since the supply of goods in this city must be along perpendicular lines (e.g., streets), the appropriate distance function is the so-called **rectangular** distance. Stating it as a mathematical model, let (x_1, x_2) denote the unknown location (coordinates) of the supply center and let (a^i, b^i) be the given location of customer i. Our problem is then

$$\min_{x_1, x_2} \{\max_{1 \leq i \leq m} [|a^i - x_1| + |b^i - x_2|]\}, \tag{1.8}$$

where the preceding formulation means that, first, for every possible value of (x_1, x_2) we must find that index i that maximizes the rectangular distance given between the square brackets, and, second, among all those maximal distances, depending on (x_1, x_2), we must find the smallest one. Again, if every location (x_1, x_2) is acceptable, then our problem is an unconstrained one. Sometimes, however, it is advantageous to simplify some expressions at the expense of adding extra variables and constraints. For example, define a new variable x_0 by

$$x_0 = \max_{1 \leq i \leq m} [|a^i - x_1| + |b^i - x_2|] \tag{1.9}$$

or

$$x_0 \geq |a^i - x_1| + |b^i - x_2|, \qquad i = 1, \ldots, m. \tag{1.10}$$

We obtain, consequently, a nonlinear program in three variables x_0, x_1, x_2:

$$\min f(x) = x_0 \tag{1.11}$$

subject to

$$g_i(x) = x_0 - |a^i - x_1| - |b^i - x_2| \geq 0, \qquad i = 1, \ldots, m. \tag{1.12}$$

The reader can easily verify that problems (1.8) and (1.11) to (1.12) are equivalent in the sense that (x_1^*, x_2^*) is an optimal solution of (1.8) if and only if (x_0^*, x_1^*, x_2^*) is an optimal solution of (1.11) and (1.12) with

$$x_0^* = |a^k - x_1^*| + |b^k - x_2^*| \tag{1.13}$$

for some k, $1 \leq k \leq m$. The reader can also show that, by introducing more variables, (1.8) can be transformed into a linear program.

PROCESS DESIGN

Consider the problem of manufacturing a given quantity F_B gram moles per hour of chemical product B from a feed consisting of an aqueous solution of reactant A, in a continuous stirred tank (backmix) reactor. The chemical reaction is

$$2A \longrightarrow B \tag{1.14}$$

with an empirical rate equation, established in the laboratory and based on unit volume of reacting fluid,

$$-\frac{dC_A}{dt} = 8.4(C_A)^2 = 8.4[C_A^0(1 - x_A)]^2 \qquad \left(\frac{\text{g mole}}{\text{liter-hr}}\right), \tag{1.15}$$

where C_A = concentration of A in the reactor (g mole/liter)
C_A^0 = concentration of A in the feed (g mole/liter)
t = time (hours)
x_A = conversion, fraction of reactant converted into product.

Suppose that the feed solution is available at a continuous range of concentrations of A and that its unit cost p_A is given by the relation

$$p_A = 4(C_A^0)^{1.4} \qquad (\$/\text{liter}). \tag{1.16}$$

The operating cost of the continuous stirred tank reactor (CSTR) is given by

$$p_{\text{CSTR}} = 0.75(V)^{0.6} \qquad (\$/\text{hr}), \tag{1.17}$$

where V (liter) is the volume of the reactor. Assume that the product B can be sold at a price of 10 \$/g mole. Our problem is to determine the rate of feed solution F_A^0 (liter/hr), its concentration C_A^0, the volume of the reactor V, and the conversion x_A for optimum operation—that is, for maximizing total profit per hour—given by

$$p_T = 10F_B - p_A F_A^0 - p_{\text{CSTR}} \qquad (\$/\text{hr}). \tag{1.18}$$

Material balance around the reactor yields

$$F_A^0 C_A^0 = F_A^0 C_A^0(1 - x_A) - \left(\frac{dC_A}{dt}\right)V. \tag{1.19}$$

From (1.14) we get

$$\tfrac{1}{2}F_A^0 C_A^0 x_A = F_B \tag{1.20}$$

and from (1.15) and (1.19) we obtain

$$8.4C_A^0(1 - x_A)^2V - F_A^0x_A = 0. \tag{1.21}$$

Our design problem then becomes

$$\max p_T = 5F_A^0C_A^0x_A - 4(C_A^0)^{1.4}F_A^0 - 0.75(V)^{0.6} \tag{1.22}$$

subject to constraint (1.21). This is a nonlinear program in variables C_A^0, F_A^0, V, x_A in which both the objective and the constraint functions are nonlinear. Note that here the objective function is maximized.

Throughout the text we shall mainly be concerned with nonlinear programs in which the objective function is minimized. This does not represent any restriction, since every problem of the type $\max f(x)$ can be equivalently analyzed and solved by considering $\min \bar{f}(x)$, where $\bar{f}(x) = -f(x)$.

Finally, a few words on notation and terminology used in later chapters. No special symbols will be used to denote vectors. The dimension of a vector, if not specifically mentioned, should always be clear from the formula in which it appears. All vectors are assumed to be column vectors. Components of a vector will be denoted by subscripts; thus $x_1, x_2, \ldots, x_n$ are the components of the n vector x. Superscripts on vectors will be used in order to distinguish between different vectors; thus $x^1, x^2, \ldots, x^m$ are meant to be m different vectors. In order to avoid confusion, exponents on real numbers will be used together with parentheses; that is, $(a)^2$ is the square of the number a. The notation x^T will be used to indicate a row vector—that is, the transpose of a column vector. The real line, that is, the set of all real numbers, is denoted by R and the n-dimensional real Euclidean space by R^n. Vectors $x \in R^n$ are also frequently referred to as points in R^n. Specific points in R^n will frequently be written in terms of their coordinates—for example, $x^0 = (1, -2)$. The notation $x \geq 0$ means that every component of x is nonnegative; thus if $x \in R^n$, then 0 is also an n-dimensional vector, each component of which is zero. The notation $x \neq 0$ means that at least one component of x is different from zero.

Only matrices having real elements will be used. Similar to vectors, no special symbols will be used for matrices (although capital letters will generally be employed). The dimension of a matrix will also be uniquely determined from the formula in which it appears. If A is an $m \times n$ matrix (m rows and n columns), then A^T is the transpose of A, having n rows and m columns. The inverse of a matrix A will be denoted by A^{-1}—that is, $AA^{-1} = A^{-1}A = I$, where I is the identity (unit) matrix.

Norm of a vector $x \in R^n$ is defined by

$$||x|| = [(x_1)^2 + (x_2)^2 + \cdots + (x_n)^2]^{1/2}. \tag{1.23}$$

We shall frequently use the notion of neighborhoods. A set

$$N_\delta(x^0) = \{x : x \in R^n, \ ||x - x^0|| < \delta\}, \tag{1.24}$$

where δ is a positive number is called a (spherical) neighborhood of the point x^0.

Matrix norms will be used in a few places in the text and then they will be the norms induced by vectors. Formally, if A is an $m \times n$ matrix and x is an n vector, then

$$||A|| = \sup \left\{\frac{||Ax||}{||x||} : ||x|| \neq 0\right\} = \sup \{||Ax|| : ||x|| = 1\}. \tag{1.25}$$

Functions will be always single valued; and as we shall see later, they can sometimes take on the values $+\infty$ or $-\infty$. In the few places where a more advanced mathematical concept is needed, we either define it in the text or, if such a definition would require extensive background material, we sacrifice completeness and refer the reader to appropriate references. Those readers unfamiliar with elementary linear algebra, real analysis, or topology may wish to consult introductory textbooks on these subjects, such as Apostol [1], Bartle [2], Hall and Spencer [3], and Noble [4].

REFERENCES

1. APOSTOL, T. M., *Mathematical Analysis*, 2nd ed., Addison-Wesley Publishing Co., Reading, Mass., 1974.

2. BARTLE, R. G., *The Elements of Real Analysis*, 2nd ed., John Wiley & Sons, New York, 1976.

3. HALL, D. W., and G. L. SPENCER, *Elementary Topology*, John Wiley & Sons, New York, 1955.

4. NOBLE, B., *Applied Linear Algebra*, Prentice-Hall, Englewood Cliffs, N.J., 1969.

PART I
ANALYSIS

The solution of a nonlinear programming problem, if not specified otherwise, consists of finding an optimal solution vector x^* and not all optimal solutions that may exist. Recognizing an optimal x^* and studying its properties form the central theme of the first part of this book, which deals with some analytic aspects of nonlinear programming problems. We shall see that if a vector x is a candidate for an optimal solution, it must satisfy certain necessary conditions of optimality. Unfortunately, however, there may be vectors other than the optimal ones that also satisfy these conditions. Consequently, necessary conditions are primarily useful in the negative sense: if a vector x does not satisfy them, it cannot be an optimal solution. To verify optimality, we may, therefore, look for sufficient conditions of optimality that, if satisfied together with the necessary ones, give a clear indication of the nature of a particular solution vector under consideration. These two types of optimality conditions constitute the first subject to be discussed in some detail. In particular, the classical method of Lagrange multipliers is extended to optimization problems with inequality constraints. In a general nonlinear program it may happen that vectors satisfying one type of optimality conditions do not satisfy the other one. Programs in which necessary optimality conditions are also sufficient are important, since a solution of a system of equations and inequalities representing, say, necessary conditions of optimality in a program under consideration is ensured as the sought optimum of that program.

A class of nonlinear programs, called convex programs and involving convex and concave functions in a certain configuration, possesses this nice property. Such convex programs are especially convenient to analyze. Associated with each such program there exist so-called dual programs, which, similar to linear programming, have some interesting theoretical properties.

Duality relations will be derived with the aid of a modern approach based on conjugate functions.

Convex programs belong to that class of nonlinear programs in which every local minimum is global. Since there are many nonconvex real-life problems, we shall also be concerned with the important question of finding more general nonconvex functions and programs for which this local-global optimum property holds. The analysis part of this text will conclude with a few selected nonlinear programs in which some of the theoretical results will be illustrated.

2 CLASSICAL OPTIMIZATION—UNCONSTRAINED AND EQUALITY CONSTRAINED PROBLEMS

The problem of finding extrema—that is, minima or maxima of real-valued functions—plays a central role in mathematical optimization. We begin the topic of extrema with the simplest case of unconstrained problems and then proceed to the subject of minima and maxima in the presence of constraints, expressed as equations. Here we shall treat the classical Lagrange multiplier theory and some necessary and sufficient conditions for extrema of differentiable functions. Treatment of these topics goes back a few centuries, hence the name "classical." In later chapters we shall discuss optimization problems in which the constraints are expressed as inequalities. All the remarkable results obtained for such problems can be classified as "modern" because they are a consequence of intensified interest in inequality constrained problems during the last two to three decades. All the "classical" results can be considered as special cases of the more general "modern" theory. We chose to present the classical results first because they can serve as a bridge between the material presented in most first- and second-year university courses of calculus or real analysis and the more advanced subject of mathematical programming. In addition, the classical theory is simpler than the modern theory in the sense that results, such as necessary and sufficient conditions for extrema, are not obscured by the more complicated requirements in the case of inequality constraints.

2.1 UNCONSTRAINED EXTREMA

Consider a real-valued function f with domain D in R^n. Then f is said to have a **local minimum** at a point $x^* \in D$ if there exists a real number $\delta > 0$ such that

$$f(x) \geq f(x^*) \tag{2.1}$$

for all $x \in D$ satisfying $\|x - x^*\| < \delta$. We define a **local maximum** in a similar way but with the sense of the inequality in (2.1) reversed. If the inequality (2.1) is replaced by a strict inequality

$$f(x) > f(x^*), \qquad x \in D, \quad x \neq x^* \tag{2.2}$$

we have a **strict** local minimum; and if the sense of the inequality in (2.2) is reversed, we have a strict local maximum. The function f has a **global minimum** (strict global minimum) at $x^* \in D$ if (2.1) [or (2.2)] holds for all $x \in D$. A similar definition holds for a **global maximum** (strict global maximum). An **extremum** is either a minimum or a maximum. Not every real function has an extremum; for example, a nonzero linear function has no extremum on R^n. It is clear from these definitions that every global minimum (maximum) of f in D is also a local minimum (maximum). The converse of this statement is, in general, false, and the reader can easily demonstrate it by examples. In later chapters we shall discuss functions, such as convex functions, that, however, have the remarkable property that every local minimum is also a global minimum.

Let $x \in D \subset R^n$ be a point where the real function f is differentiable. Recall that if a real-valued function f is differentiable at an interior point $x \in D$, then its first partial derivatives exist at x. If, in addition, the partial derivatives are continuous at x, then f is said to be continuously differentiable at x. Similarly, if f is twice differentiable at $x \in D$, then the second partial derivatives exist there. And if they are continuous at x, then f is said to be twice continuously differentiable at x. We define the **gradient** of f at x as the vector $\nabla f(x)$, given by

$$\nabla f(x) = \left(\frac{\partial f(x)}{\partial x_1}, \quad \ldots, \quad \frac{\partial f(x)}{\partial x_n}\right)^T. \tag{2.3}$$

Similarly, if f is twice differentiable at x, we define the **Hessian matrix** of f at x as the $n \times n$ symmetric matrix $\nabla^2 f(x)$, given by

$$\nabla^2 f(x) = \left[\frac{\partial^2 f(x)}{\partial x_i \partial x_j}\right], \qquad i, j = 1, \ldots, n. \tag{2.4}$$

In this section we discuss necessary and sufficient conditions for extrema of functions without constraints. We start by stating the following well-known result.

Theorem 2.1 (Necessary Condition)

Let x^* be an interior point of D at which f has a local minimum or local maximum. If f is differentiable at x^*, then

$$\nabla f(x^*) = 0. \tag{2.5}$$

This theorem will be restated and proved as part of Theorem 2.3.

Now we turn to sufficient conditions for a local extremum.

Theorem 2.2 (Sufficient Conditions)

Let x^* be an interior point of D at which f is twice continuously differentiable. If

$$\nabla f(x^*) = 0 \tag{2.6}$$

and

$$z^T \nabla^2 f(x^*) z > 0 \tag{2.7}$$

for all nonzero vectors z, then f has a local minimum at x^*. If the sense of the inequality in (2.7) *is reversed, then f has a local maximum at x^*. Moreover, the extrema are strict local extrema.*

This theorem can be proved by using the Taylor expansion of f and is left for the reader.

In both theorems we are utilizing the behavior of the function at x^*, the extremum. If, however, we can investigate the behavior of the function in some neighborhood of the extremum in question, we have a result that provides additional conditions for a local extremum.

Theorem 2.3

Let x^* be an interior point of D and assume that f is twice continuously differentiable on D. It is necessary for a local minimum of f at x^* that

$$\nabla f(x^*) = 0 \tag{2.8}$$

and

$$z^T \nabla^2 f(x^*) z \geq 0 \tag{2.9}$$

for all z. Sufficient conditions for a local minimum are that (2.8) *holds and that for every x in some neighborhood $N_\delta(x^*)$ and for every $z \in R^n$, we have*

$$z^T \nabla^2 f(x) z \geq 0. \tag{2.10}$$

If the sense of the inequalities in (2.9) *and* (2.10) *is reversed, the theorem applies to a local maximum.*

Proof. Suppose that f has a local minimum at x^*. Then

$$f(x) \geq f(x^*) \tag{2.11}$$

for all x in some neighborhood $N_\delta(x^*) \subset D$.

We can write every $x \in N_\delta(x^*)$ as $x = x^* + \theta y$, where θ is a real number and y is a vector such that $\|y\| = 1$. Hence

$$f(x^* + \theta y) \geq f(x^*) \tag{2.12}$$

for sufficiently small $|\theta|$.

For such a y, we define F by $F(\theta) = f(x^* + \theta y)$. Then (2.12) becomes

$$F(\theta) \geq F(0) \tag{2.13}$$

for all θ such that $|\theta| < \delta$.

From the Mean Value Theorem [1], we have

$$F(\theta) = F(0) + \nabla F(\lambda\theta)\theta, \tag{2.14}$$

where λ is a number between 0 and 1.

If $\nabla F(0) > 0$, then, by the continuity assumptions, there exists an $\epsilon > 0$ such that

$$\nabla F(\lambda\theta) > 0 \tag{2.15}$$

for all λ between 0 and 1 and for all θ satisfying $|\theta| < \epsilon$. Hence we can find a $\theta < 0$ such that $|\theta| < \delta$ and

$$F(0) > F(\theta), \tag{2.16}$$

a contradiction. Assuming that $\nabla F(0) < 0$ would lead to a similar contradiction. Thus

$$\nabla F(0) = y^T \nabla f(x^*) = 0. \tag{2.17}$$

But y is an arbitrary nonzero vector. Hence we must have

$$\nabla f(x^*) = 0. \tag{2.18}$$

Turning now to the second-order conditions, we have, by Taylor's theorem,

$$F(\theta) = F(0) + \nabla F(0)\theta + \tfrac{1}{2}\nabla^2 F(\lambda\theta)(\theta)^2, \qquad 1 > \lambda > 0. \tag{2.19}$$

If $\nabla^2 F(0) < 0$, then, by continuity, there exists an $\epsilon' > 0$ such that

$$\nabla^2 F(\lambda\theta) < 0 \tag{2.20}$$

for all λ between 0 and 1 and for all θ satisfying $|\theta| < \epsilon'$.

Since $\nabla F(0) = 0$, (2.20) would imply for such a θ that

$$F(\theta) < F(0), \tag{2.21}$$

a contradiction. Consequently,

$$\nabla^2 F(0) = y^T \nabla^2 f(x^*) y \geq 0. \tag{2.22}$$

Since this inequality holds for all y, subject to the arbitrary restriction on the norm of y, it must hold for all vectors z. This completes the proof of the first part of the theorem.

For the second part, suppose that (2.8) and (2.10) hold but that x^* is not a local minimum. Then there exists a $w \in N_\delta(x^*)$ such that $f(x^*) > f(w)$. Let $w = x^* + \theta y$, where $\|y\| = 1$ and $\theta > 0$. By Taylor's theorem,

$$f(w) = f(x^*) + \theta y^T \nabla f(x^*) + \tfrac{1}{2}(\theta)^2 y^T \nabla^2 f(x^* + \lambda\theta y) y, \tag{2.23}$$

where $1 > \lambda > 0$. Our assumptions lead then to

$$y^T \nabla^2 f(x^* + \lambda\theta y) y < 0, \tag{2.24}$$

contradicting (2.10), since $x^* + \lambda\theta y \in N_\delta(x^*)$. The proof for a local maximum is similar. ■

Theorem 2.2 provides sufficient conditions for a strict local extremum of f at x^*, based on the behavior of the function at that point. We shall show that it is easy to find examples of extrema for which these sufficient conditions are not satisfied. In Theorem 2.3 we have sufficient conditions for a local (not necessarily strict) extremum based on the behavior of f in a neighborhood of x^*. Finally, we present sufficient conditions for a strict local extremum, also based on a neighborhood of x^*.

Theorem 2.4

Let x^* be an interior point of D and assume that f is twice continuously differentiable. If

$$\nabla f(x^*) = 0 \tag{2.25}$$

and

$$z^T \nabla^2 f(x) z > 0 \tag{2.26}$$

for any $x \neq x^*$ in a neighborhood of x^* and for any nonzero z, then f has a strict local minimum at x^*. Reversing the sense of the inequality in (2.26) results in sufficient conditions for a strict local maximum.

The proof of this theorem is similar to preceding ones and is left as an exercise for the reader.

We can illustrate the foregoing theorems by a simple example.

Example 2.1.1

Let $f(x) = (x)^{2p}$, where p is a positive integer, and let D be the whole real line. The gradient ∇f is given by $\nabla f(x) = 2p(x)^{2p-1}$. At $x = 0$, the gradient vanishes; that is, the origin satisfies the necessary conditions for a minimum or a maximum as stated in Theorem 2.1.

The Hessian, $\nabla^2 f$, is given by

$$\nabla^2 f(x) = (2p - 1)(2p)(x)^{2p-2}. \tag{2.27}$$

For $p = 1$, $\nabla^2 f(0) = 2$; that is, the sufficient conditions for a strict local minimum (Theorem 2.2) are satisfied.

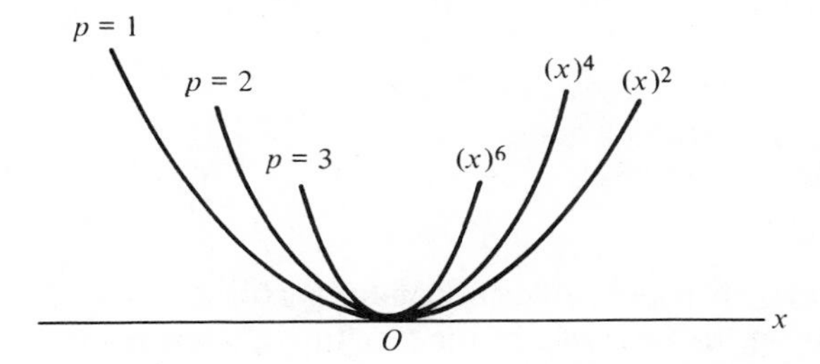

Fig. 2.1 The functions x^{2p} in the neighborhood of the origin.

If, however, we take $p > 1$, then $\nabla^2 f(0) = 0$ and the sufficient conditions of Theorem 2.2 are not satisfied; yet f has a minimum at the origin, as can be seen in Figure 2.1. On the other hand, by taking any neighborhood of the origin, the reader can verify that all the conditions for a local minimum (both the necessary and the sufficient) of Theorem 2.3 are satisfied. Moreover, by Theorem 2.4, we conclude that the minimum at $x^* = 0$ is a strict minimum. In fact, the origin is actually a strict global minimum of f. ∎

The preceding theorems contain second-order conditions that involve the nature of functions, called **quadratic forms**, given by

$$z^T H(c)z = \sum_{i=1}^{n} \sum_{j=1}^{n} h_{ij}(c)z_i z_j, \tag{2.28}$$

where $H = [h_{ij}]$ is a real symmetric matrix. In order to investigate the sign of such functions or, equivalently, the definiteness of the matrix $H(c)$, we

can compute the determinants $d_k(c)$:

$$d_k(c) = \det \begin{bmatrix} h_{11}(c), & \dots, & h_{1k}(c) \\ \cdot & & \cdot \\ \cdot & & \cdot \\ \cdot & & \cdot \\ h_{k1}(c), & \dots, & h_{kk}(c) \end{bmatrix}, \quad k = 1, \dots, n. \tag{2.29}$$

If $d_k(c) > 0$ for $k = 1, \dots, n$, the quadratic form $z^T H(c) z$ is positive for all nonzero z and $H(c)$ is positive definite. If, however, $d_k(c)$ has the sign of $(-1)^k$ for $k = 1, \dots, n$—that is, the values of $d_k(c)$ are alternately negative and positive—the quadratic form given in (2.28) is negative for all nonzero z and $H(c)$ is negative definite. These checks are useful if we are interested in the behavior of f at a certain **point** c, as in the sufficient conditions of Theorem 2.2, which involve the matrix $\nabla^2 f(c)$. They become, however, quite impractical if we must determine the sign of the quadratic form in a **neighborhood** of c, as in Theorem 2.3.

2.2 EQUALITY CONSTRAINED EXTREMA AND THE METHOD OF LAGRANGE

In the previous section we discussed necessary and sufficient conditions for the existence of extrema of functions in the interior of their domain and without additional subsidiary conditions. In this section we study extremum problems in which the object is to find minima or maxima of a real-valued function in a specified region contained in the domain of the function, such that the admissible region is described by a finite set of equations, called **constraints.**

Consider, then, the problem of finding the minimum (or maximum) of a real-valued function f with domain $D \subset R^n$, subject to the constraints

$$g_i(x) = 0, \qquad i = 1, \dots, m \quad (m < n) \tag{2.30}$$

where each of the g_i is a real-valued function defined on D. The assumption that the number of constraint equations is less than the number of variables will simplify subsequent discussions. The problem is, therefore, to find an extremum of f in the region determined by the equations in (2.30). The first and most intuitive method of solution of such a problem involves the elimination of m variables from the problem by using the equations in (2.30). The conditions for such an elimination will be stated later in the Implicit Function Theorem, and the proof can be found in most advanced calculus textbooks (see, e.g., [1, 8]). This theorem assumes the differentiability of the functions g_i and that the $n \times m$ Jacobian matrix $[\partial g_i/\partial x_j]$ has rank m. The

actual solution of the constraint equations for m variables in terms of the remaining $n - m$ can often prove a difficult, if not impossible task. For this reason, and since this method reduces a problem with equality constraints to an equivalent unconstrained one, discussed in the previous section, we do not pursue it further.

Another method, also based on the idea of transforming a constrained problem into an unconstrained one, was proposed by Lagrange. Many recent results in mathematical programming are actually a direct extension and generalization of Lagrange's method, mainly to problems with inequality constraints.

Before presenting a result that will provide a direction in which we must proceed in order to transform an equality constrained problem into an equivalent unconstrained problem, we state the well-known Implicit Function Theorem.

Theorem 2.5 (Implicit Function Theorem)

Suppose that ϕ_i are real-valued functions defined on D and continuously differentiable on an open set $D^1 \subset D \subset R^{m+p}$, where $p > 0$ and $\phi_i(x^0, y^0) = 0$ for $i = 1, \ldots, m$ and $(x^0, y^0) \in D^1$. Assume that the Jacobian matrix $[\partial\phi_i(x^0, y^0)/\partial x_j]$ has rank m. Then there exists a neighborhood $N_\delta(x^0, y^0) \subset D^1$, an open set $D^2 \subset R^p$ containing y^0 and real-valued functions ψ_k, $k = 1, \ldots, m$, continuously differentiable on D^2, such that the following conditions are satisfied:

$$x_k^0 = \psi_k(y^0), \qquad k = 1, \ldots, m. \tag{2.31}$$

For every $y \in D^2$, we have

$$\phi_i(\psi(y), y) = 0, \qquad i = 1, \ldots, m \tag{2.32}$$

where $\psi(y) = (\psi_1(y), \ldots, \psi_m(y))$ and for all $(x, y) \in N_\delta(x^0, y^0)$ the Jacobian matrix $[\partial\phi_i(x, y)/\partial x_j]$ has rank m. Furthermore, for $y \in D^2$, the partial derivatives of $\psi_k(y)$ are the solutions of the set of linear equations

$$\sum_{k=1}^{m} \frac{\partial\phi_i(\psi(y), y)}{\partial x_k} \frac{\partial\psi_k(y)}{\partial y_j} = -\frac{\partial\phi_i(\psi(y), y)}{\partial y_j}, \qquad i = 1, \ldots, m. \tag{2.33}$$

Before introducing the method of Lagrange, we present the following result.

Theorem 2.6

Let f and g_i, $i = 1, \ldots, m$, be real-valued functions on $D \subset R^n$ and continuously differentiable on a neighborhood $N_\epsilon(x^*) \subset D$. Suppose that x^*

is a local minimum or local maximum of f for all points x in $N_\epsilon(x^*)$ that also satisfy

$$g_i(x) = 0, \qquad i = 1, \ldots, m. \tag{2.34}$$

Also assume that the Jacobian matrix of $g_i(x^*)$ has rank m. Under these hypotheses the gradient of f at x^* is a linear combination of the gradients of g_i at x^*; that is, there exist real numbers λ_i^* such that

$$\nabla f(x^*) = \sum_{i=1}^{m} \lambda_i^* \nabla g_i(x^*). \tag{2.35}$$

Proof. By suitable rearrangement and relabeling of rows, we can always assume that the $m \times m$ matrix, formed by taking the first m rows of the Jacobian $[\partial g_i(x^*)/\partial x_j]$, is nonsingular. The set of linear equations

$$\sum_{i=1}^{m} \frac{\partial g_i(x^*)}{\partial x_j} \lambda_i = \frac{\partial f(x^*)}{\partial x_j}, \qquad j = 1, \ldots, m \tag{2.36}$$

has a unique solution for the λ_i, denoted λ_i^*. Let $\hat{x} = (x_{m+1}, \ldots, x_n)$. Then, by applying Theorem 2.5 to (2.34) at x^*, there exist real functions $h_j(\hat{x})$ and an open set $\hat{D} \subset R^{n-m}$ containing x^*, such that

$$x_j^* = h_j(\hat{x}^*), \qquad j = 1, \ldots, m \tag{2.37}$$

and

$$f(x^*) = f(h_1(\hat{x}^*), \ldots, h_m(\hat{x}^*), x_{m+1}^*, \ldots, x_n^*). \tag{2.38}$$

As a result of the last expression, it follows from Theorem 2.1 that the first partial derivatives of f with respect to $x_{m+1}, \ldots, x_n$ must vanish at x^*. Thus

$$\frac{\partial f(x^*)}{\partial x_j} = \sum_{k=1}^{m} \frac{\partial f(x^*)}{\partial x_k} \frac{\partial h_k(\hat{x}^*)}{\partial x_j} + \frac{\partial f(x^*)}{\partial x_j} = 0, \qquad j = m+1, \ldots, n. \tag{2.39}$$

From (2.33) we have for every $j = m + 1, \ldots, n$

$$\sum_{k=1}^{m} \frac{\partial g_i(x^*)}{\partial x_k} \frac{\partial h_k(\hat{x}^*)}{\partial x_j} = -\frac{\partial g_i(x^*)}{\partial x_j}, \qquad i = 1, \ldots, m. \tag{2.40}$$

Multiplying each of the equations in (2.40) by λ_i^* and adding up, we get

$$\sum_{i=1}^{m} \sum_{k=1}^{m} \lambda_i^* \frac{\partial g_i(x^*)}{\partial x_k} \frac{\partial h_k(\hat{x}^*)}{\partial x_j} + \lambda_i^* \frac{\partial g_i(x^*)}{\partial x_j} = 0, \qquad j = m+1, \ldots, n. \tag{2.41}$$

Subtracting (2.41) from (2.39) and rearranging yield

$$\sum_{k=1}^{m}\left[\frac{\partial f(x^*)}{\partial x_k} - \sum_{i=1}^{m} \lambda_i^* \frac{\partial g_i(x^*)}{\partial x_k}\right]\frac{\partial h_k(\hat{x}^*)}{\partial x_j} + \frac{\partial f(x^*)}{\partial x_j} - \sum_{i=1}^{m} \lambda_i^* \frac{\partial g_i(x^*)}{\partial x_j} = 0,$$
$$j = m+1, \ldots, n. \tag{2.42}$$

But the expression in the brackets is zero by (2.36), and so

$$\frac{\partial f(x^*)}{\partial x_j} - \sum_{i=1}^{m} \lambda_i^* \frac{\partial g_i(x^*)}{\partial x_j} = 0, \qquad j = m+1, \ldots, n. \tag{2.43}$$

The last expression, together with (2.36), yields the desired result. ■

The relation between the gradient of the function to be minimized or maximized and the gradients of the constraint functions at a local extremum, as expressed in the last theorem, leads to the formulation of the Lagrangian, $L(x, \lambda)$, given by

$$L(x, \lambda) = f(x) - \sum_{i=1}^{m} \lambda_i g_i(x), \tag{2.44}$$

where the λ_i are called **Lagrange multipliers.**

Lagrange's method consists of transforming an equality constrained extremum problem into a problem of finding a stationary point of the Lagrangian. This can be seen by the following

Theorem 2.7

Suppose that $f, g_i, i = 1, \ldots, m$, satisfy the hypotheses of Theorem 2.6. Then there exists a vector of multipliers $\lambda^* = (\lambda_1^*, \ldots, \lambda_m^*)^T$ such that

$$\nabla L(x^*, \lambda^*) = 0. \tag{2.45}$$

Proof. Follows directly from Theorem 2.6 and the definition of L as given by (2.44). ■

Several different proofs of the last two theorems exist in the literature; see, for example [2, 6, 7]. We chose the one based on the Implicit Function Theorem, since it requires no additional background material. As indicated earlier, the theorems in this section can be regarded as special cases of some recent, more general results. In the next chapter we shall present them, too, and will indicate proofs along different lines.

Theorem 2.7 provides necessary conditions for an extremum of f with equality constraints. As in the previous section, we turn now to a discussion of sufficient conditions for such an extremum. The notation $\nabla_\xi^j \phi(\xi, \eta)$ is

used when ϕ is being differentiated j times but only with respect to ξ (for $j = 1$, the superscript is omitted). We have then

Theorem 2.8

Let $f, g_1, \ldots, g_m$ be twice continuously differentiable real-valued functions on R^n. If there exist vectors $x^* \in R^n$, $\lambda^* \in R^m$ such that

$$\nabla L(x^*, \lambda^*) = 0 \tag{2.46}$$

and for every nonzero vector $z \in R^n$ satisfying

$$z^T \nabla g_i(x^*) = 0, \qquad i = 1, \ldots, m \tag{2.47}$$

it follows that

$$z^T \nabla_x^2 L(x^*, \lambda^*) z > 0, \tag{2.48}$$

then f has a strict local minimum at x^*, subject to $g_i(x) = 0$, $i = 1, \ldots, m$. If the sense of the inequality in (2.48) is reversed, then f has a strict local maximum at x^*.

Proof [6]. Assume that x^* is not a strict local minimum. Then there exist a neighborhood $N_\delta(x^*)$ and a sequence $\{z^k\}$, $z^k \in N_\delta(x^*)$, $z^k \neq x^*$, converging to x^* such that for every $z^k \in \{z^k\}$

$$g_i(z^k) = 0, \qquad i = 1, \ldots, m \tag{2.49}$$

and

$$f(x^*) \geq f(z^k). \tag{2.50}$$

Let $z^k = x^* + \theta^k y^k$, where $\theta^k > 0$ and $\| y^k \| = 1$. The sequence $\{\theta^k, y^k\}$ has a subsequence that converges to $(0, \bar{y})$, where $\| \bar{y} \| = 1$. By the Mean Value Theorem [1], we get for each k in this subsequence

$$g_i(z^k) - g_i(x^*) = \theta^k (y^k)^T \nabla g_i(x^* + \eta_i^k \theta^k y^k) = 0, \qquad i = 1, \ldots, m \tag{2.51}$$

where η_i^k is a number between 0 and 1 and

$$f(z^k) - f(x^*) = \theta^k (y^k)^T \nabla f(x^* + \xi^k \theta^k y^k) \leq 0, \tag{2.52}$$

where ξ^k is, again, a number between 0 and 1.

Dividing (2.51) and (2.52) by θ^k and taking limits as $k \to \infty$, we get

$$(\bar{y})^T \nabla g_i(x^*) = 0, \qquad i = 1, \ldots, m \tag{2.53}$$

and

$$(\bar{y})^T \nabla f(x^*) \leq 0. \tag{2.54}$$

From Taylor's theorem we have

$$L(z^k, \lambda^*) = L(x^*, \lambda^*) + \theta^k (y^k)^T \nabla_x L(x^*, \lambda^*) + \tfrac{1}{2}(\theta^k)^2 (y^k)^T \nabla_x^2 L(x^* + \eta^k \theta^k y^k, \lambda^*) y^k, \tag{2.55}$$

where $1 > \eta^k > 0$.

By (2.44), (2.46), (2.49), and (2.50), and dividing (2.55) by $\frac{1}{2}(\theta^k)^2$, we obtain

$$(y^k)^T \nabla_x^2 L(x^* + \eta^k \theta^k y^k, \lambda^*) y^k \leq 0. \tag{2.56}$$

Letting $k \longrightarrow \infty$, we obtain from the last expression

$$(\bar{y})^T \nabla_x^2 L(x^*, \lambda^*) \bar{y} \leq 0. \tag{2.57}$$

This completes the proof, since $\bar{y} \neq 0$, and it satisfies (2.47). ■

The sufficient conditions stated in the last theorem involve determining the sign of a quadratic form, subject to linear constraints. This task can be accomplished by a result, due to Mann [5]. Let $A = [\alpha_{ij}]$ be an $n \times n$ symmetric real matrix and $B = [\beta_{ij}]$ an $n \times m$ real matrix. Denote by M_{pq} the matrix obtained from a matrix M by keeping only the elements in the first p rows and q columns.

Theorem 2.9

Suppose that $\det [B_{mm}] \neq 0$. *Then the quadratic form*

$$\sum_{i=1}^{n} \sum_{j=1}^{n} \alpha_{ij} \xi_i \xi_j \tag{2.58}$$

is positive for all nonzero ξ *satisfying*

$$\sum_{i=1}^{n} \beta_{ij} \xi_i = 0, \qquad j = 1, \ldots, m \tag{2.59}$$

if and only if

$$(-1)^m \det \begin{bmatrix} A_{pp} & B_{pm} \\ B_{pm}^T & 0 \end{bmatrix} > 0 \tag{2.60}$$

for $p = m + 1, \ldots, n$. *Similarly,* (2.58) *is negative for all nonzero* ξ *satisfying* (2.59) *if and only if*

$$(-1)^p \det \begin{bmatrix} A_{pp} & B_{pm} \\ B_{pm}^T & 0 \end{bmatrix} > 0 \tag{2.61}$$

for $p = m + 1, \ldots, n$.

The proof of this theorem can be found in [3] or [5].

Suppose now that the $n \times m$ Jacobian matrix $[\partial g_i(x^*)/\partial x_j]$ has rank m and the variables are indexed in such a way that

$$\det \begin{bmatrix} \frac{\partial g_1(x^*)}{\partial x_1}, \dots, \frac{\partial g_m(x^*)}{\partial x_1} \\ \cdots \quad\quad \cdots \\ \frac{\partial g_1(x^*)}{\partial x_m}, \dots, \frac{\partial g_m(x^*)}{\partial x_m} \end{bmatrix} \neq 0. \tag{2.62}$$

Then we have the following

Corollary 2.10

Let $f, g_1, \dots, g_m$ be twice continuously differentiable real-valued functions. If there exist vectors $x^* \in R^n$, $\lambda^* \in R^m$, such that

$$\nabla L(x^*, \lambda^*) = 0 \tag{2.63}$$

and if

$$(-1)^m \det \begin{bmatrix} \frac{\partial^2 L(x^*, \lambda^*)}{\partial x_1 \partial x_1}, \dots, \frac{\partial^2 L(x^*, \lambda^*)}{\partial x_1 \partial x_p} & \frac{\partial g_1(x^*)}{\partial x_1}, \dots, \frac{\partial g_m(x^*)}{\partial x_1} \\ \vdots \quad\quad \vdots & \vdots \quad\quad \vdots \\ \frac{\partial^2 L(x^*, \lambda^*)}{\partial x_p \partial x_1}, \dots, \frac{\partial^2 L(x^*, \lambda^*)}{\partial x_p \partial x_p} & \frac{\partial g_1(x^*)}{\partial x_p}, \dots, \frac{\partial g_m(x^*)}{\partial x_p} \\ \frac{\partial g_1(x^*)}{\partial x_1}, \quad \dots, \frac{\partial g_1(x^*)}{\partial x_p} & 0, \quad \dots, \quad 0 \\ \vdots \quad\quad \vdots & \vdots \quad\quad \vdots \\ \frac{\partial g_m(x^*)}{\partial x_1}, \quad \dots, \frac{\partial g_m(x^*)}{\partial x_p} & 0, \quad \dots, \quad 0 \end{bmatrix} > 0 \tag{2.64}$$

for $p = m + 1, \dots, n$, then f has a strict local minimum at x^*, such that

$$g_i(x^*) = 0, \qquad i = 1, \dots, m. \tag{2.65}$$

Proof. Follows directly from Theorems 2.8 and 2.9. ■

The similar result for strict maxima is obtained by changing $(-1)^m$ in (2.64) to $(-1)^p$.

Example 2.2.1

Consider the problem

$$\max f(x_1, x_2) = x_1 x_2 \tag{2.66}$$

subject to the constraint

$$g(x_1, x_2) = x_1 + x_2 - 2 = 0. \tag{2.67}$$

First we form the Lagrangian:

$$L(x, \lambda) = x_1 x_2 - \lambda(x_1 + x_2 - 2). \tag{2.68}$$

Next we set $\nabla L(x^*, \lambda^*) = 0$:

$$\frac{\partial L(x^*, \lambda^*)}{\partial x_1} = x_2^* - \lambda^* = 0 \tag{2.69}$$

$$\frac{\partial L(x^*, \lambda^*)}{\partial x_2} = x_1^* - \lambda^* = 0 \tag{2.70}$$

$$\frac{\partial L(x^*, \lambda^*)}{\partial \lambda} = -x_1^* - x_2^* + 2 = 0. \tag{2.71}$$

Solution of the last three equations yields

$$x_1^* = x_2^* = \lambda^* = 1. \tag{2.72}$$

The point $(x^*, \lambda^*) = (1, 1, 1)$ therefore satisfies the necessary conditions for a maximum as stated in Theorem 2.7.

The linear dependence between ∇f and ∇g at the optimum, as asserted by Theorem 2.6, is clearly illustrated in Figure 2.2, where $\nabla f(x^*)$ is actually equal to $\nabla g(x^*)$.

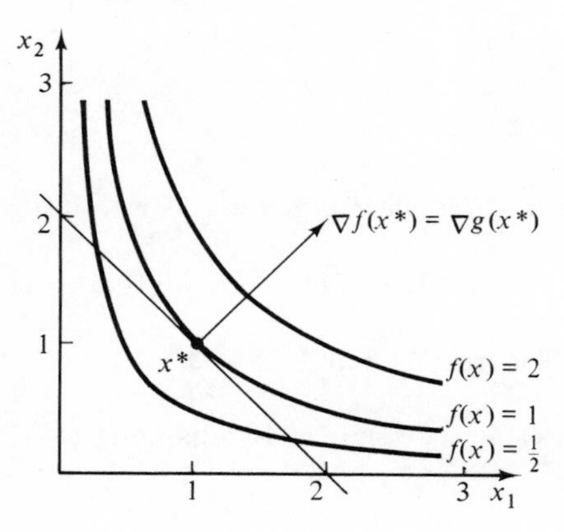

Fig. 2.2 Constrained maximum.

Turning to the sufficient conditions, we compute $\nabla_x^2 L(x^*, \lambda^*)$:

$$\frac{\partial^2 L(x^* \lambda^*)}{\partial x_1 \partial x_1} = 0, \quad \frac{\partial^2 L(x^*, \lambda^*)}{\partial x_1 \partial x_2} = 1, \quad \frac{\partial^2 L(x^*, \lambda^*)}{\partial x_2 \partial x_2} = 0. \tag{2.73}$$

Hence

$$z^T \nabla_x^2 L(x^*, \lambda^*) z = (z_1, z_2) \begin{bmatrix} 0 & 1 \\ 1 & 0 \end{bmatrix} \begin{pmatrix} z_1 \\ z_2 \end{pmatrix} = 2 z_1 z_2 \tag{2.74}$$

and, by Theorem 2.8, we must determine the sign of $2z_1 z_2$ for all $z \neq 0$ that satisfy $z^T \nabla g(x^*) = 0$.

Since

$$\frac{\partial g(x^*)}{\partial x_1} = 1 \quad \text{and} \quad \frac{\partial g(x^*)}{\partial x_2} = 1, \tag{2.75}$$

the last condition is equivalent to $z_1 + z_2 = 0$. Substituting into (2.74), we get

$$z^T \nabla_x^2 L(x^*, \lambda^*) z = -2(z_1)^2 < 0. \tag{2.76}$$

Thus (1, 1) is a strict local maximum.

Finally, we can check the sufficient condition given by Corollary 2.10. Here $p = 2$ and

$$(-1)^2 \det \begin{bmatrix} 0 & 1 & 1 \\ 1 & 0 & 1 \\ 1 & 1 & 0 \end{bmatrix} = 2 > 0, \tag{2.77}$$

thereby verifying our previous conclusions. ▮

Second-order necessary conditions, similar to those of Theorem 2.3, were derived [4, 6] for more general constrained extremum problems. In cases of problems with equality constraints, these conditions are an almost straightforward generalization of the second-order necessary and sufficient conditions for an unconstrained problem.

In Theorem 2.6 we assumed that the Jacobian matrix $[\partial g_i(x^*)/\partial x_j]$ has rank m ($<n$), equal to the number of constraint equations. We conclude this chapter by stating a slight generalization of Theorem 2.6 that does not require conditions on the rank of the Jacobian.

Theorem 2.11

Let f and g_i, $i = 1, \ldots, m$, be continuously differentiable real-valued functions with domain $D \subset R^n$. If x^* is a local minimum or local maximum of

f for all points x in a neighborhood of x^* satisfying

$$g_i(x) = 0, \qquad i = 1, \ldots, m \tag{2.78}$$

then there exist $m + 1$ real numbers $\lambda_0^*, \lambda_1^*, \ldots, \lambda_m^*$, not all zero, such that

$$\lambda_0^* \nabla f(x^*) - \sum_{i=1}^{m} \lambda_i^* \nabla g_i(x^*) = 0. \tag{2.79}$$

This theorem can be considered a corollary of more general results dealing with extrema in the presence of equality as well as inequality constraints, to be presented in the next chapter. From these results we can conclude that if x^* is a local extremum, then the rank of the $(m + 1) \times n$ **augmented Jacobian matrix** $[\partial f(x^*)/\partial x_j, \partial g_i(x^*)/\partial x_j]$ is less than $(m + 1)$. Moreover, it can be shown that if the rank of the preceding matrix is equal to the rank of the Jacobian $[\partial g_i(x^*)/\partial x_j]$, then $\lambda_0^* \neq 0$ and we can normalize to get $\lambda_0^* = 1$. If, however, the rank of the augmented Jacobian is greater than the rank of the Jacobian $[\partial g_i(x^*)/\partial x_j]$, then λ_0^* must vanish. This situation can occur, for example, in the case of a feasible set containing a single point.

Example 2.2.2

$$\min f(x) = x \tag{2.80}$$

subject to

$$g_1(x) = (x)^2 = 0. \tag{2.81}$$

The feasible set contains the single point $x = 0$. At this point, the rank of the Jacobian $[dg_1/dx]$ is zero, whereas the rank of the augmented Jacobian $[df/dx, dg_1/dx]$ is one. From (2.79) we get

$$\lambda_0^* - \lambda_1^* \cdot 0 = 0. \tag{2.82}$$

That is, $\lambda_0^* = 0$. ∎

EXERCISES

2.A. Prove Theorem 2.2.

2.B. Find points satisfying necessary conditions for extrema of the function

$$f(x) = \frac{x_1 + x_2}{3 + (x_1)^2 + (x_2)^2 + x_1 x_2}. \tag{2.83}$$

Try to establish the nature of these points by checking sufficient conditions.

2.C. Investigate whether the origin in R^3 is an extremum of the function

$$f(x) = \alpha(x_1)^2 e^{x_2} + (x_2)^2 e^{x_3} + (x_3)^2 e^{x_1}, \tag{2.84}$$

where α is a parameter.

2.D. Find minima of the function

$$f(x) = [(x_2)^2 - x_1]^2 \tag{2.85}$$

among all the points satisfying necessary conditions for an extremum.

2.E. Prove Theorem 2.4.

2.F. Given the system of linear equations

$$Ax = b, \tag{2.86}$$

where A is an $m \times n$ real matrix, b is an m vector, and $x \in R^n$ is unknown. Assume that $m < n$ and that the rank of A is m. Find an explicit solution in terms of A and b to that solution x^* of (2.86) for which $\frac{1}{2}x^T x$ is minimal. You can check your result by showing that $x^* = (41, -9, -1)^T$ is indeed such a solution of

$$2x_1 - x_2 + 5x_3 = 86 \tag{2.87}$$

$$x_1 \qquad - 2x_3 = 43. \tag{2.88}$$

2.G. The necessary condition (2.35) need not hold if the gradients of $g_i(x^*)$ are linearly dependent. Illustrate this fact by a numerical example.

2.H. Find the shortest distance from the point $\bar{x} = (1, 0)$ to the curve

$$4x_1 - (x_2)^2 = 0 \tag{2.89}$$

by (a) direct elimination of one of the variables and (b) by the method of Lagrange. Did you obtain the same answer in both cases?

REFERENCES

1. Bartle, R. G., *The Elements of Real Analysis*, 2nd ed., John Wiley & Sons, New York, 1976.

2. Beltrami, E. J., "A Constructive Proof of the Kuhn-Tucker Multiplier Rule," *J. Math. Anal. & Appl.*, **26**, 297–306 (1967).

3. Debreu, G., "Definite and Semidefinite Quadratic Forms," *Econometrica*, **20**, 295–300 (1952).

4. Fiacco, A. V., "Second Order Sufficient Conditions for Weak and Strict Constrained Minima," *SIAM J. Appl. Math.*, **16**, 105–108 (1968).

5. MANN, H. B., "Quadratic Forms with Linear Constraints," *Amer. Math. Monthly*, **50**, 430–433 (1943).

6. McCORMICK, G. P., "Second Order Conditions for Constrained Minima," *SIAM J. Appl. Math.*, **15**, 641–652 (1967).

7. PHIPPS, C. G., "Maxima and Minima Under Restraint," *Amer. Math. Monthly*, **59**, 230–235 (1952).

8. RUDIN, W., *Principles of Mathematical Analysis*, 2nd ed., McGraw-Hill Book Co., New York, 1964.

3 OPTIMALITY CONDITIONS FOR CONSTRAINED EXTREMA

The problems treated in the last chapter were limited to unconstrained or equality constrained problems. In this chapter we begin discussing mathematical programming problems with inequality as well as equality constraints. As noted, the introduction of inequalities to optimization problems marks the end of the "classical" era of optimization and the beginning of the "modern" theory of mathematical programming. Inequality constraints are seldom strict inequalities; they can be satisfied either as equations or as strict inequalities. This feature of inequalities causes some complications in the analytic treatment of optimality conditions, but they are more than compensated for by the immensely rich class of problems that can be formulated by the use of inequalities in the constraints.

Several approaches to the question of optimality conditions can be found in the literature, each approach being characterized by the assumptions made on the functions involved. The optimality conditions stated by most of these approaches are related in one way or another to the concept of the Lagrangian, which can be most conveniently treated, without loss of much generality, if the objective and constraint functions are differentiable (or twice differentiable), an assumption we use throughout most of this chapter. Relaxing the differentiability assumptions usually leads to optimality conditions that can best be expressed as optimality conditions on some other problem, such as finding a saddlepoint of the Lagrangian or solving a so-called dual program.

3.1 FIRST-ORDER NECESSARY CONDITIONS FOR INEQUALITY CONSTRAINED EXTREMA

We begin deriving first-order necessary conditions for inequality and equality constrained extremum problems involving only first derivatives by stating the most general mathematical program to be discussed in this chapter:

$$\text{(P)} \qquad \min f(x) \tag{3.1}$$

subject to the constraints

$$g_i(x) \geq 0, \qquad i = 1, \ldots, m \tag{3.2}$$

$$h_j(x) = 0, \qquad j = 1, \ldots, p. \tag{3.3}$$

The functions $f, g_1, \ldots, g_m, h_1, \ldots, h_p$ are assumed to be defined and differentiable on some open set $D \subset R^n$. Let $X \subset D$ denote the **feasible set for problem (P)**—that is, the set of all $x \in D$ satisfying (3.2) and (3.3). Members of the feasible set are called **feasible points**. As before, let $N_\delta(x^0)$ be a spherical neighborhood of the point x^0 with radius δ.

A point $x^* \in X$ is said to be a **local minimum** of problem (P), or a **local solution** of (P), if there exists a positive number $\hat{\delta}$ such that

$$f(x) \geq f(x^*) \tag{3.4}$$

for all $x \in X \cap N_{\hat{\delta}}(x^*)$. If (3.4) holds for all $x \in X$, then x^* is said to be a **global minimum** (global solution) of problem (P).

Every point x in a neighborhood of x^* can be written as $x^* + z$, where z is a nonzero vector if and only if $x \neq x^*$. A vector $z \neq 0$ is called a **feasible direction vector** from x^* if there exists a number $\delta_1 > 0$ such that $(x^* + \theta z) \in X \cap N_{\delta_1}(x^*)$ for all $0 \leq \theta < \delta_1/||z||$ (see Figure 3.1).

Feasible direction vectors are important in many numerical optimization algorithms. Momentarily, we are interested in them for the simple reason that if x^* is a local solution of problem (P) and z is a feasible direction vector, then we must have $f(x^* + \theta z) \geq f(x^*)$ for sufficiently small positive θ. Let us characterize the feasible direction vectors in terms of the constraint functions g_i and h_j.

Define

$$I(x^*) = \{i : g_i(x^*) = 0\} \tag{3.5}$$

and suppose that $z^T \nabla g_k(x^*) < 0$ for some $k \in I(x^*)$ and z, a feasible direc-

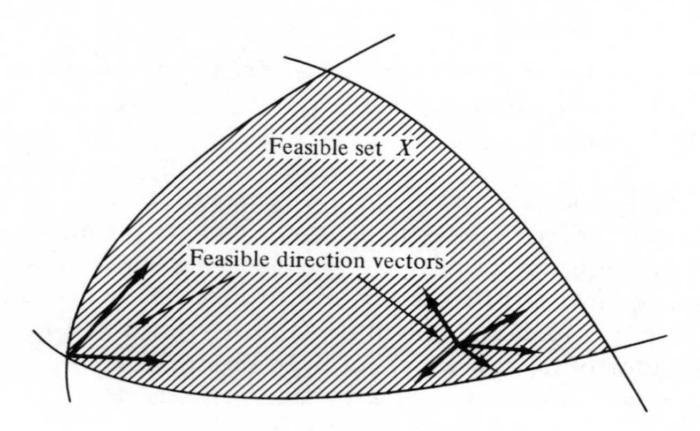

Fig. 3.1 Feasible direction vectors.

tion vector from x^*. By the differentiability assumption, we can write

$$g_k(x^* + \theta z) = g_k(x^*) + \theta z^T \nabla g_k(x^*) + \theta \epsilon_k(\theta), \tag{3.6}$$

where $\epsilon_k(\theta)$ tends to zero as $\theta \longrightarrow 0$. If θ is small enough, then $z^T \nabla g_k(x^*) + \epsilon_k(\theta) < 0$; and since $g_k(x^*) = 0$, we obtain $g_k(x^* + \theta z) < 0$ for all sufficiently small $\theta > 0$, contradicting the fact that z is a feasible direction vector from x^*. Hence we must have $z^T \nabla g_i(x^*) \geq 0$ for all $i \in I(x^*)$. Similar reasoning can be applied to show that, for a feasible direction vector, we must also have $z^T \nabla h_j(x^*) = 0$ for $j = 1, \ldots, p$. Define now

$$Z^1(x^*) = \{z : z^T \nabla g_i(x^*) \geq 0,\ i \in I(x^*),\ z^T \nabla h_j(x^*) = 0,\ j = 1, \ldots, p\}. \tag{3.7}$$

From the foregoing discussion it follows that if z is a feasible direction vector from x^*, then $z \in Z^1(x^*)$. A set $K \subset R^n$ is called a **cone** if $x \in K$ implies $\alpha x \in K$ for every nonnegative number α. The set $Z^1(x^*)$ is clearly a cone. It is also called the **linearizing cone** of X at x^* [1, 17], since it is generated by linearizing the constraint functions at x^*. Let us define $Z^2(x^*)$, another "linearizing" set that will be needed later:

$$Z^2(x^*) = \{z : z^T \nabla f(x^*) < 0\}. \tag{3.8}$$

If $z \in Z^2(x^*)$, it can be shown that there exists a point $x = x^* + \theta z$, sufficiently close to x^*, such that $f(x^*) > f(x)$.

The following lemma, due to Minkowski and Farkas [13] and bearing the latter's name, is needed in the sequel. The proof of the lemma is given in the next chapter.

Lemma 3.1 (Farkas Lemma)

Let A be a given $m \times n$ real matrix and b a given n vector. The inequality $b^T y \geq 0$ holds for all vectors y satisfying $Ay \geq 0$ if and only if there exists an m vector $\rho \geq 0$ such that $A^T \rho = b$.

An illustration of the Farkas lemma is given in Figure 3.2 for a 3×2 matrix A. The vectors A_1, A_2, A_3 are the row vectors of the matrix A. Consider the set Y consisting of all vectors y that make an acute angle with every row vector of A. The Farkas lemma then states that b makes an acute angle with every $y \in Y$ if and only if b can be expressed as a nonnegative linear combination of the row vectors of A. In Figure 3.2, b^1 is a vector that satisfies these conditions, whereas b^2 is a vector that does not.

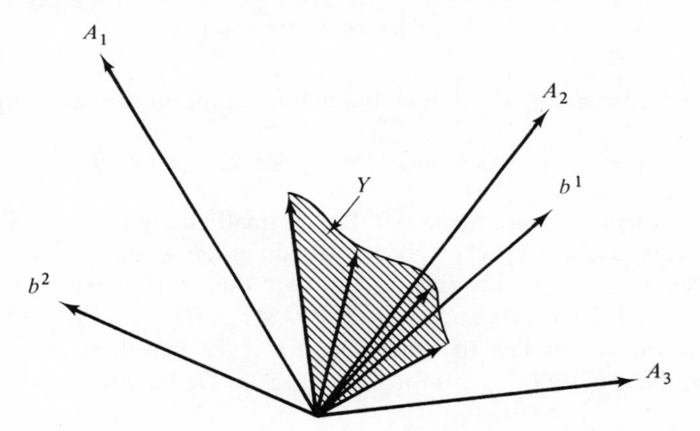

Fig. 3.2 Interpretation of the Farkas lemma for a 3×2 matrix A.

As in the previous chapter, we now define the Lagrangian associated with problem (P) as

$$L(x, \lambda, \mu) = f(x) - \sum_{i=1}^{m} \lambda_i g_i(x) - \sum_{j=1}^{p} \mu_j h_j(x), \tag{3.9}$$

and we can prove the following

Theorem 3.2

Suppose that $x^0 \in X$. Then $Z^1(x^0) \cap Z^2(x^0) = \emptyset$ if and only if there exist vectors λ^0, μ^0 such that

$$\nabla_x L(x^0, \lambda^0, \mu^0) = \nabla f(x^0) - \sum_{i=1}^{m} \lambda_i^0 \nabla g_i(x^0) - \sum_{j=1}^{p} \mu_j^0 \nabla h_j(x^0) = 0 \tag{3.10}$$

$$\lambda_i^0 g_i(x^0) = 0, \qquad i = 1, \ldots, m \tag{3.11}$$

$$\lambda^0 \geq 0. \tag{3.12}$$

Proof. The set $Z^1(x^0)$ is never empty, since the origin always belongs to it; and $Z^1(x^0) \cap Z^2(x^0)$ is empty if and only if for every z satisfying

$$z^T \nabla g_i(x^0) \geq 0, \qquad i \in I(x^0) \tag{3.13}$$

$$z^T \nabla h_j(x^0) = 0, \qquad j = 1, \ldots, p \tag{3.14}$$

we have

$$z^T \nabla f(x^0) \geq 0. \tag{3.15}$$

We can write (3.14) as two inequalities

$$z^T \nabla h_j(x^0) \geq 0, \qquad j = 1, \ldots, p \tag{3.16}$$

$$z^T[-\nabla h_j(x^0)] \geq 0, \qquad j = 1, \ldots, p. \tag{3.17}$$

It follows from Lemma 3.1 that (3.15) holds for all vectors z satisfying (3.13), (3.16), and (3.17) if and only if there exist vectors $\lambda^0 \geq 0$, $\mu^1 \geq 0$, $\mu^2 \geq 0$ such that

$$\nabla f(x^0) = \sum_{i \in I(x^0)} \lambda_i^0 \nabla g_i(x^0) + \sum_{j=1}^{p} (\mu_j^1 - \mu_j^2) \nabla h_j(x^0). \tag{3.18}$$

Letting $\lambda_i^0 = 0$ for $i \notin I(x^0)$, $\mu^0 = \mu^1 - \mu^2$, we conclude that $Z^1(x^0) \cap Z^2(x^0)$ is empty if and only if (3.10) to (3.12) hold. ∎

Conditions (3.10) to (3.12) are, of course, natural candidates to become the desired extension of the necessary conditions given in Theorem 2.6 for a solution of an equality constrained problem. They can, indeed, become necessary conditions for optimality in the general mathematical programming problem (P) if we could guarantee that the set $Z^1(x^*) \cap Z^2(x^*)$ is empty at x^*, a local solution of (P).

The interested reader may try at this point to formulate simple mathematical programs with equality and inequality constraints, obtain their solutions, and construct the sets $Z^1(x)$ and $Z^2(x)$ at the solution point. For most problems he will find that $Z^1(x) \cap Z^2(x)$ is indeed empty at the solution of the problem, and, therefore, the Lagrangian conditions (3.10) to (3.12) hold at that point. Such, however, is not always the case, as can be seen from the following example taken from [25].

Example 3.1.1

Consider the constraints in R^2 (see Figure 3.3):

$$g_1(x) = (1 - x_1)^3 - x_2 \geq 0 \tag{3.19}$$

$$g_2(x) = x_1 \geq 0 \tag{3.20}$$

$$g_3(x) = x_2 \geq 0. \tag{3.21}$$

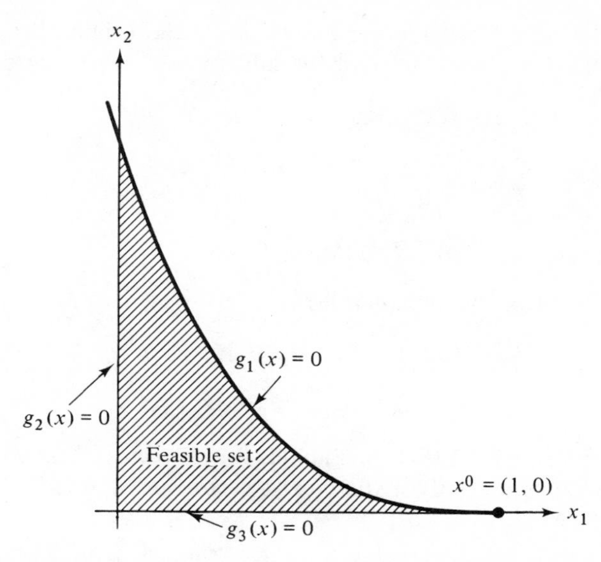

Fig. 3.3 An example of constraints and an optimal point without associated Lagrange multipliers.

The point $x^0 = (1, 0)$ is feasible, and we can easily verify that

$$Z^1(x^0) = \{(z_1, z_2) : z_2 = 0\}. \tag{3.22}$$

Letting $f(x) = -x_1$, we can see that x^0 is a solution of the problem

$$\min -x_1 \tag{3.23}$$

subject to constraints (3.19) to (3.21). At x^0,

$$Z^2(x^0) = \{(z_1, z_2) : z_1 > 0\} \tag{3.24}$$

and $Z^1(x^0) \cap Z^2(x^0)$ is nonempty. Hence there exists no λ^0 satisfying (3.10) to (3.12). ∎

It is possible to derive weak necessary conditions for optimality without requiring the set $Z^1(x) \cap Z^2(x)$ to be empty at the solution, by introducing a multiplier for the objective function in the definition of the Lagrangian function associated with problem (P). Let the **weak Lagrangian** $\tilde{L}$ be defined by

$$\tilde{L}(x, \lambda, \mu) = \lambda_0 f(x) - \sum_{i=1}^{m} \lambda_i g_i(x) - \sum_{j=1}^{p} \mu_j h_j(x). \tag{3.25}$$

In his pioneering work in 1948 Fritz John [22] stated and proved necessary conditions for inequality constrained mathematical programs (without

equality constraints) via the weak Lagrangian and assuming only continuous first partial derivatives of the functions involved. John's conditions were later extended by Mangasarian and Fromovitz to problems with equality and inequality constraints—that is, to problem (P), presented at the beginning of this section. We shall state these conditions in a moment and present a proof for the somewhat simpler case of having only inequality constraints. The complete proof for inequalities and equations can be found in [27]. First we need the following result, called a "theorem of the alternative."

Theorem 3.3

Let $\tilde{A}$ be an $m \times n$ real matrix. Then either there exists an n vector x such that

$$\tilde{A}x < 0 \tag{3.26}$$

or there exists a nonzero m vector u such that

$$u^T\tilde{A} = 0, \qquad u \geq 0 \tag{3.27}$$

but never both.

Proof. Suppose that there exist x and u such that both (3.26) and (3.27) are satisfied. Then we have $u^T\tilde{A}x < 0$ and $u^T\tilde{A}x = 0$ simultaneously, a contradiction. Suppose now that there exists no x satisfying (3.26). This means that we cannot find a negative number w satisfying

$$\tilde{A}_i x = \sum_{j=1}^{n} \tilde{a}_{ij}x_j \leq w, \qquad i = 1, \ldots, m \tag{3.28}$$

for every $x \in R^n$. Letting $y = (w, x)^T$, $b = (1, 0, \ldots, 0)^T \in R^{n+1}$, $A = [e^T, -\tilde{A}]$, where $e = (1, \ldots, 1)^T \in R,^m$ and invoking the previously stated Farkas lemma, we conclude that there exists an m vector $u \geq 0$ such that

$$\sum_{i=1}^{m} u_i = 1 \tag{3.29}$$

$$\sum_{i=1}^{m} u_i a_{ij} = 0, \qquad j = 1, \ldots, n. \tag{3.30}$$

Hence u solves (3.27). ■

For additional theorems of the alternative, the reader is referred to Mangasarian [26]. Necessary conditions based on the weak Lagrangian defined above will be stated now. In the proof, however, we assume that no equality constraints $h_j(x) = 0$ are present in the problem. The next theorem reduces, then, to the original one of Fritz John [22].

Theorem 3.4

Suppose that $f, g_1, \ldots, g_m, h_1, \ldots, h_p$ are continuously differentiable on an open set containing X. If x^* is a solution of problem (P), *then there exist vectors $\lambda^* = (\lambda_0^*, \lambda_1^*, \ldots, \lambda_m^*)^T$ and $\mu^* = (\mu_1^*, \ldots, \mu_p^*)^T$ such that*

$$\nabla_x \tilde{L}(x^*, \lambda^*, \mu^*) = \lambda_0^* \nabla f(x^*) - \sum_{i=1}^{m} \lambda_i^* \nabla g_i(x^*) - \sum_{j=1}^{p} \mu_j^* \nabla h_j(x^*) = 0 \quad (3.31)$$

$$\lambda_i^* g_i(x^*) = 0, \qquad i = 1, \ldots, m \quad (3.32)$$

$$(\lambda^*, \mu^*) \neq 0, \qquad \lambda^* \geq 0. \quad (3.33)$$

Proof. We shall consider necessary conditions for the solution x^* of the problem

$$\min f(x) \quad (3.34)$$

subject to

$$g_i(x) \geq 0, \qquad i = 1, \ldots, m. \quad (3.35)$$

The conditions are the existence of a vector λ^* such that

$$\lambda_0^* \nabla f(x^*) - \sum_{i=1}^{m} \lambda_i^* \nabla g_i(x^*) = 0 \quad (3.36)$$

$$\lambda_i^* g_i(x^*) = 0, \qquad i = 1, \ldots, m \quad (3.37)$$

$$\lambda^* \neq 0, \qquad \lambda^* \geq 0. \quad (3.38)$$

If $g_i(x^*) > 0$ for all i, then $I(x^*) = \emptyset$. Choose $\lambda_0^* = 1$, $\lambda_1^* = \lambda_2^* = \cdots = \lambda_m^* = 0$, and (3.36) to (3.38) hold with $\nabla f(x^*) = 0$.

Suppose now that $I(x^*) \neq \emptyset$. Then for every z satisfying

$$z^T \nabla g_i(x^*) > 0, \qquad i \in I(x^*) \quad (3.39)$$

we cannot have

$$z^T \nabla f(x^*) < 0. \quad (3.40)$$

This result follows from the previously seen fact that if there exists a z satisfying (3.39), then we can find a sufficiently small δ such that the point $x = x^* + \theta z$ satisfies

$$g_i(x) > 0, \qquad i = 1, \ldots, m \quad (3.41)$$

for all $0 < \theta < \delta$; that is, x is feasible. If (3.40) also holds, then

$$f(x) < f(x^*), \quad (3.42)$$

contradicting that x^* is a minimum. Thus the system of inequalities (3.39) and (3.40) has no solution. By Theorem 3.3, there exists a nonzero vector $\lambda^* \geq 0$ such that

$$\lambda_0^* \nabla f(x^*) + \sum_{i \in I(x^*)} \lambda_i^*[-\nabla g_i(x^*)] = 0. \tag{3.43}$$

Letting $\lambda_i^* = 0$ for $i \notin I(x^*)$, we get from (3.43), after rearrangement,

$$\lambda_0^* \nabla f(x^*) - \sum_{i=1}^{m} \lambda_i^* \nabla g_i(x^*) = 0 \tag{3.44}$$

and clearly

$$\lambda_i^* g_i(x^*) = 0, \qquad i = 1, \ldots, m. \tag{3.45}$$

■

Conditions (3.31) to (3.33) of Theorem 3.4 become conditions (3.10) to (3.12) of Theorem 3.2 if λ_0^* is positive. Conversely, the conditions of Theorem 3.2 trivially imply those of Theorem 3.4 with $\lambda_0^* = 1$.

Example 3.1.2

Consider again the minimization problem discussed in Example 3.1.1. Letting $\lambda_0^* = 0$, $\lambda_1^* = 1$, $\lambda_2^* = 0$, and $\lambda_3^* = 1$, we observe that Fritz John's necessary conditions are satisfied at $x^* = (1, 0)$. ■

This example also illustrates the main weakness in Theorem 3.4: substituting $\lambda_0^* = 0$, conditions (3.31) to (3.33) are, in fact, satisfied at (1, 0) for any differentiable objective function f whether it has a local minimum at that point or not.

The reader may have wondered why our basic mathematical programming problem has both equality and inequality constraints when it is easy to transform such a problem into an equivalent one involving either equality or inequality constraints only. Suppose that we have the inequality $g(x) \geq 0$ in a problem. Letting y be an additional variable, we can write the equivalent equality

$$g(x) - (y)^2 = 0. \tag{3.46}$$

Conversely, suppose that we have an equality constraint $h(x) = 0$. It can be replaced by two inequalities

$$h(x) \geq 0 \tag{3.47}$$

$$h(x) \leq 0. \tag{3.48}$$

Thus equality-inequality type problems can be transformed into equivalent problems having constraints of only one type, at the expense of increasing

the number of variables or the number of constraints. Such transformations, while advantageous in some cases, can also cause considerable weakening of certain results, as was pointed out by Mangasarian and Fromovitz [27] in the case of the Fritz John conditions.

Suppose that we write each equality constraint in problem (P) as

$$h_j(x) = g_{m+j}(x) \geq 0, \qquad j = 1, \ldots, p \tag{3.49}$$

$$-h_j(x) = g_{m+p+j}(x) \geq 0, \qquad j = 1, \ldots, p. \tag{3.50}$$

Hence we transform problem (P) into forms (3.34) and (3.35). Choosing $\lambda_0^* = \cdots = \lambda_m^* = 0$, $\lambda_{m+1}^* = \cdots = \lambda_{m+p}^* = \lambda_{m+p+1}^* = \cdots = \lambda_{m+2p}^* = 1$, we find that conditions (3.36) to (3.38) are, in fact, satisfied for every feasible $x \in R^n$, not necessarily the optimal ones.

In 1951 Kuhn and Tucker, in their fundamental paper [25], presented necessary conditions for inequality constrained mathematical programs that are stronger than those of John. Restricting the constraint functions by a regularity condition, called **constraint qualification**, they were able to ensure that the multiplier λ_0 is indeed positive, and thus the extended version of John's necessary conditions becomes equivalent to conditions (3.10) to (3.12). The type of restrictions imposed on the constraint functions to ensure the existence of Lagrange multipliers at the solution of a mathematical program has been the subject of intensive research effort. In addition to the original constraint qualification of Kuhn and Tucker, several other regularity conditions were proposed. We mention here the works of Abadie [1], Arrow, Hurwicz, and Uzawa [2], Beltrami [6], Canon, Cullum, and Polak [10], Evans [12], Gould and Tolle [17], Guignard [19], Karlin [23], Mangasarian and Fromovitz [27], Slater [35], and Varaiya [36]. In subsequent discussions we shall generally follow the works of Abadie, Gould and Tolle, and Varaiya. For extensive discussions of various constraint qualifications, the reader is referred to surveys by Bazaraa, Goode and Shetty [5] and Gould and Tolle [18].

Actually, we should have called these regularity conditions "first-order" constraint qualifications in order to distinguish them from some other constraint qualifications that we shall mention in the next section in connection with second-order necessary conditions. Where no confusion may arise, we shall, however, refer to them simply as constraint qualifications.

We begin our discussion of constraint qualifications by introducing the notion of the cone of tangents to a nonempty set $A \subset R^n$ at the point $x \in A$. Denote by $\tilde{S}(A, x)$ the intersection of all closed cones containing the set $\{a - x: a \in A\}$. Then the **closed cone of tangents** of the set A at x, denoted

by $S(A, x)$, is defined as

$$S(A, x) = \bigcap_{k=1}^{\infty} \tilde{S}(A \cap N_{1/k}(x), x), \tag{3.51}$$

where $N_{1/k}(x)$ is a spherical neighborhood of x with radius $1/k$ and k is a natural number. The following lemma characterizes $S(A, x)$.

Lemma 3.5

A vector z is contained in $S(A, x)$ if and only if there exists a sequence of vectors $\{x^k\} \subset A$ converging to x and a sequence of nonnegative numbers $\{\alpha^k\}$ such that the sequence $\{\alpha^k(x^k - x)\}$ converges to z.

Proof [7]. Suppose that $z \in S(A, x)$. Then $z \in \tilde{S}(A \cap N_{1/k}(x), x)$ for every $k = 1, 2, \ldots$, and, by definition,

$$\tilde{S}(A \cap N_{1/k}(x), x) = \text{cl}\,\{\alpha(y - x) : \alpha \geq 0, y \in A \cap N_{1/k}(x)\}, \quad k = 1, 2, \ldots \tag{3.52}$$

where cl means the closure operation of sets in R^n. Choose any sequence of positive numbers ϵ^k such that $\{\epsilon^k\}$ converges to 0, and consider the vectors $z(\epsilon^k) \in \{\alpha(y - x) : \alpha \geq 0, y \in A \cap N_{1/k}(x)\}$ such that

$$||z(\epsilon^k) - z|| < \epsilon^k. \tag{3.53}$$

Then by (3.52), they can be written as

$$z(\epsilon^k) = \alpha(\epsilon^k)(y(\epsilon^k) - x), \quad \alpha(\epsilon^k) \geq 0, \quad y(\epsilon^k) \in A \cap N_{1/k}(x). \tag{3.54}$$

Letting $k = 1, 2, \ldots$, we generate a sequence of vectors $y(\epsilon^1), y(\epsilon^2), \ldots$ that is contained in A and converges to x and a sequence of nonnegative numbers $\alpha(\epsilon^1), \alpha(\epsilon^2), \ldots$ such that, by (3.53) and (3.54), the sequence $\{\alpha(\epsilon^k)(y(\epsilon^k) - x)\}$ converges to z. Conversely, suppose that there exist a sequence of vectors $\{x^k\} \subset A$ converging to x and a sequence of nonnegative numbers $\{\alpha^k\}$ such that $\{\alpha^k(x^k - x)\}$ converges to z. Let p be any natural number. Then there exists a natural number K such that $k \geq K$ implies $x^k \in A \cap N_{1/p}(x)$, or

$$\alpha^k(x^k - x) \in \tilde{S}(A \cap N_{1/p}(x)), \qquad k \geq K \tag{3.55}$$

and since $\tilde{S}$ is closed,

$$z \in \tilde{S}(A \cap N_{1/p}(x)). \tag{3.56}$$

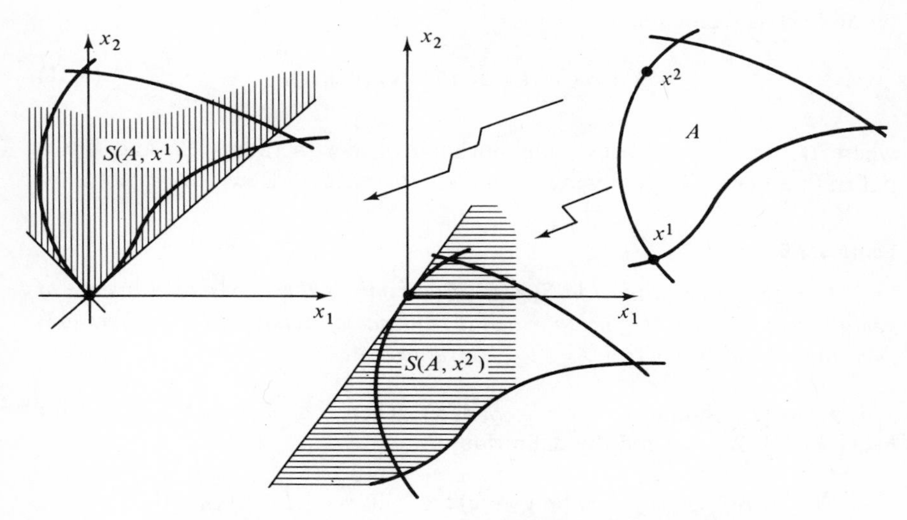

Fig. 3.4 An arbitrary set A and cones of its tangents.

Because the last expression holds for any natural number p, it follows that

$$z \in \bigcap_{p=1}^{\infty} \tilde{S}(A \cap N_{1/p}(x)) = S(A, x). \tag{3.57}$$

■

With the aid of the last lemma, it is possible to give alternative descriptions of $S(A, x)$, the cone of tangents of a nonempty set A at a point x. First we observe that the vector $w = 0$ is always in $S(A, x)$ for every A and x. Let w be a unit vector—that is, a vector with $||w|| = 1$—and suppose that there exists a sequence of points $\{x^k\} \subset A$, converging to x, $x^k \neq x$, and

$$\lim_{k \to \infty} \frac{x^k - x}{||x^k - x||} = w. \tag{3.58}$$

We can say that the sequence of vectors $\{x^k\}$ converges to x in the **direction** of w. The cone of tangents of the set A at x contains, then, all the vectors that are nonnegative multiples of the w obtained above. Accordingly, $w \in S(A, x)$ implies that there exists a sequence $\{x^k\} \subset A$ converging to x in the direction of w. Still another way of looking at cones of tangents is as follows: Translate the set A by subtracting x from each of its elements and let $\{x^k\}$ be a sequence in the translated set, $x^k \neq 0$, converging to the origin. Construct a sequence of half-lines from the origin and passing through x^k. These half-lines tend to a half-line that will be a member of $S(A, x)$. The union of all the half-lines formed by taking all such sequences will then be

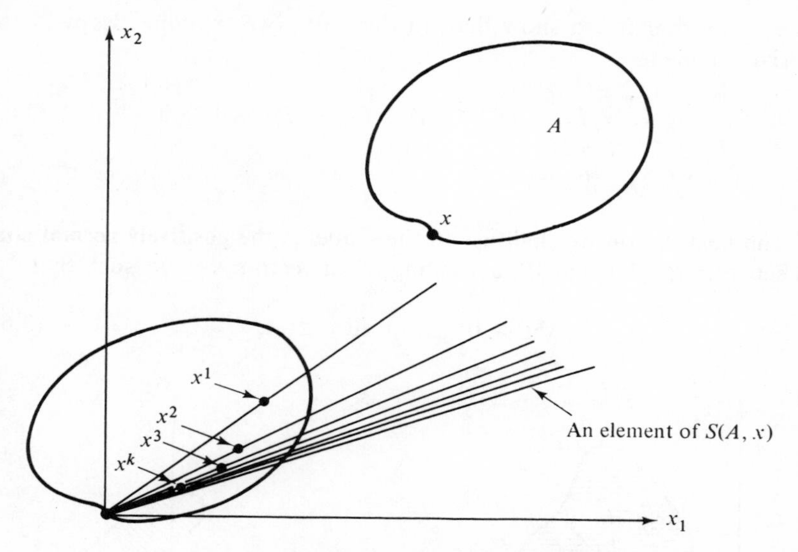

Fig. 3.5 Constructing the cone of tangents of a set A.

the cone of tangents of A at x. This construction is illustrated in Figure 3.5 and in the following example.

Example 3.1.3

Let

$$B = \{(x_1, x_2) : (x_1 - 4)^2 + (x_2 - 2)^2 \leq 1\}. \tag{3.59}$$

That is, B is a closed ball with center at (4, 2) and radius 1. Let us find the cone of tangents of B at a boundary point, say $x = (4 - \frac{\sqrt{3}}{2}, \frac{3}{2})$. First we translate B by subtracting x from each of its members, thus obtaining the ball

$$B^1 = \{(x_1, x_2) : (x_1 - \tfrac{\sqrt{3}}{2})^2 + (x_2 - \tfrac{1}{2})^2 \leq 1\}. \tag{3.60}$$

By taking sequences of points $\{x^k\}$ on the boundary of B^1, converging to the origin, we generate sequences of half-lines converging to a line that is actually the ordinary tangent line to the curve defined by the boundary of B^1 at the origin. This line satisfies

$$\tfrac{\sqrt{3}}{2}x_1 + \tfrac{1}{2}x_2 = 0. \tag{3.61}$$

Repeating this process for all sequences in the interior of B^1 converging to the origin, we get the cone of tangents of B at x as

$$S(B, x) = \{(x_1, x_2) : \tfrac{\sqrt{3}}{2}x_1 + \tfrac{1}{2}x_2 \geq 0\}. \tag{3.62}$$

It is an easy exercise to show that, in this case, $S(B, x)$ coincides with the linearizing cone to

$$g(x_1, x_2) = -(x_1 - 4)^2 - (x_2 - 2)^2 + 1 \geq 0 \tag{3.63}$$

at $x = (4 - \frac{\sqrt{3}}{2}, \frac{3}{2})$. ▮

The next notion we shall use in the sequel is the **positively normal** cone to a set $A \subset R^n$, denoted A', consisting of all vectors $x \in R^n$ such that

$$x^T y \geq 0 \qquad \text{for all } y \in A. \tag{3.64}$$

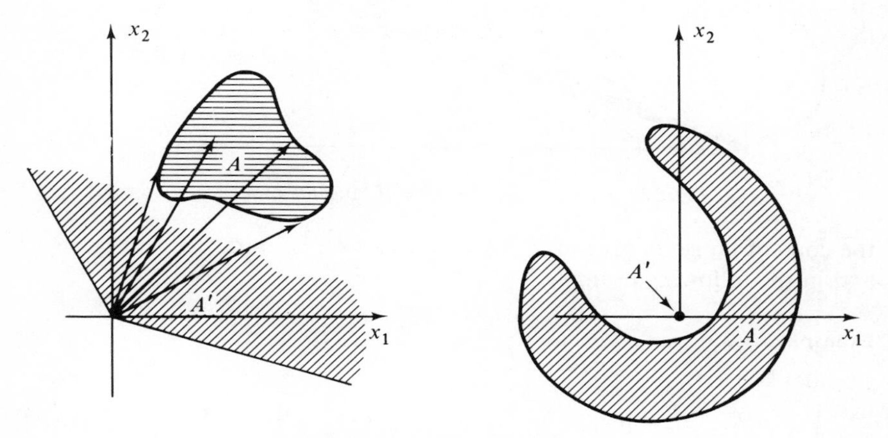

Fig. 3.6 Positively normal cones.

We use the name positively normal cone to distinguish it from the "negatively" normal, or **polar**, cone of a set defined as above with the sense of the inequality in (3.64) reversed. An important property of normal cones, which will be useful later, is the following. Given two sets $A_1 \subset R^n$, $A_2 \subset R^n$, then

$$A_1 \subset A_2 \quad \text{implies} \quad A'_2 \subset A'_1. \tag{3.65}$$

Cones of tangents and positively normal cones play a central role in establishing strong optimality conditions. We start bringing them together by the following

Lemma 3.6

Suppose that $x^0 \in X$. The set $Z^1(x^0) \cap Z^2(x^0)$ is empty if and only if

$$\nabla f(x^0) \in (Z^1(x^0))'. \tag{3.66}$$

Proof. The set $Z^1(x^0) \cap Z^2(x^0)$ is empty if and only if for every $z \in Z^1(x^0)$ we have $z^T \nabla f(x^0) \geq 0$. It follows, then, that $\nabla f(x^0)$ is contained in the positively normal cone of $Z^1(x^0)$. ∎

Lemma 3.7

Suppose that x^0 is a solution of problem (P). *Then*

$$\nabla f(x^0) \in (S(X, x^0))'. \tag{3.67}$$

Proof. We must show that $z^T \nabla f(x^0) \geq 0$ for every $z \in S(X, x^0)$. Suppose that $z \in S(X, x^0)$. Then, by Lemma 3.5, there exist a sequence $\{x^k\} \in X$, converging to x^0 and a sequence of nonnegative numbers $\{\alpha^k\}$ such that $\{\alpha^k(x^k - x^0)\}$ converges to z. If f is differentiable at x^0, we can write

$$f(x^k) = f(x^0) + (x^k - x^0)^T \nabla f(x^0) + \epsilon \| x^k - x^0 \|, \tag{3.68}$$

where ϵ is a function that tends to zero as $k \longrightarrow \infty$. Hence

$$\alpha^k[f(x^k) - f(x^0)] = (\alpha^k(x^k - x^0))^T \nabla f(x^0) + \epsilon \| \alpha^k(x^k - x^0) \|. \tag{3.69}$$

Since $x^k \in X$ and x^0 is a local minimum, it follows that, by letting $k \longrightarrow \infty$, the term $\epsilon \| \alpha^k(x^k - x^0) \| \longrightarrow 0$, and the expression $\alpha^k[f(x^k) - f(x^0)]$ converges to a nonnegative limit. Thus

$$\lim_{k \to \infty} (\alpha^k(x^k - x^0))^T \nabla f(x^0) = z^T \nabla f(x^0) \geq 0; \tag{3.70}$$

that is, $\nabla f(x^0) \in (S(X, x^0))'$. ∎

Now we can state and prove the main result of this section, a set of necessary conditions, stronger than those presented in Theorem 3.4. The conditions stated below can be viewed as a direct extension of the Kuhn-Tucker necessary conditions [25] for optimality.

Theorem 3.8 (Generalized Kuhn-Tucker Necessary Conditions)

Let x^* be a solution of problem (P) *and suppose that*

$$(Z^1(x^*))' = (S(X, x^*))'. \tag{3.71}$$

Then there exist vectors $\lambda^* = (\lambda_1^*, \ldots, \lambda_m^*)^T$ and $\mu^* = (\mu_1^*, \ldots, \mu_p^*)^T$ such that

$$\nabla f(x^*) - \sum_{i=1}^{m} \lambda_i^* \nabla g_i(x^*) - \sum_{j=1}^{p} \mu_j^* \nabla h_j(x^*) = 0 \tag{3.72}$$

$$\lambda_i^* g_i(x^*) = 0, \qquad i = 1, \ldots, m \tag{3.73}$$

$$\lambda^* \geq 0. \tag{3.74}$$

Proof. Suppose that x^* is a solution of (P). By Lemma 3.7, $\nabla f(x^*) \in (S(X, x^*))'$. If $(Z^1(x^*))' = (S(X, x^*))'$, then $\nabla f(x^*) \in (Z^1(x^*))'$. By Lemma 3.6, the set $Z^1(x^*) \cap Z^2(z^*)$ is empty; and, by Theorem 3.2, conditions (3.72) to (3.74) hold. ■

Notice that

$$Z^1(x^*) = S(X, x^*) \tag{3.75}$$

at a solution x^* of problem (P) implies the hypotheses of the last theorem. Gould and Tolle [17] have used (3.71) as a constraint qualification for a slightly more general problem than ours. In their work the additional restriction $x \in D$ is placed on the vector x, where D is an arbitrary subset of R^n. The feasible set X in their case is the intersection of D with the set of x satisfying (3.2) and (3.3). They showed that (3.71) is not only a sufficient condition for the existence of the multipliers λ^*, μ^*, such that (3.72) to (3.74) hold, but it is also **necessary** in a certain sense. The triple (g, h, D) is said to be **Lagrange regular** at x^* if and only if for every differentiable objective function f that has a local constrained minimum at x^* there exist vectors λ^*, μ^* satisfying (3.72) to (3.74). It is then shown in [17] that (g, h, D) is Lagrange regular at x^* if and only if condition (3.71) holds. Essentially, we have proved in Theorem 3.8 that (3.71) is indeed a sufficient condition for the existence of the multipliers λ^*, μ^* satisfying (3.72) to (3.74). To show that the condition is also necessary, Gould and Tolle show that for every $y \in (S(X, x^*))'$ there exists a differentiable f with a local constrained minimum such that $\nabla f(x^*) = y$. By the Lagrange regularity and Lemma 3.6, we get $y \in (Z^1(x^*))'$. Hence

$$(S(X, x^*))' \subset (Z^1(x^*))'. \tag{3.76}$$

The reader will be asked to show that for every feasible point $\hat{x}$

$$(Z^1(\hat{x}))' \subset (S(X, \hat{x}))' \tag{3.77}$$

and, therefore, equality holds.

Example 3.1.4

Consider the following nonlinear programming problem:

$$\min f(x) = x_1 \tag{3.78}$$

subject to

$$g_1(x) = 16 - (x_1 - 4)^2 - (x_2)^2 \geq 0 \tag{3.79}$$

$$h_1(x) = (x_1 - 3)^2 + (x_2 - 2)^2 - 13 = 0. \tag{3.80}$$

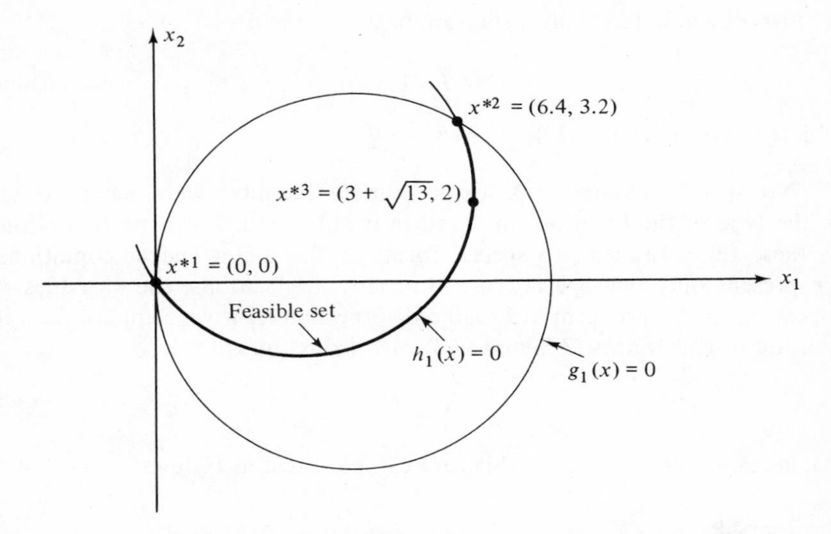

Fig. 3.7 The feasible set in Example 3.1.4 and the points satisfying the Kuhn-Tucker conditions.

It can be seen from Figure 3.7 that $f(x)$ has local minima at $x^{*1} = (0, 0)$ and at $x^{*2} = (6.4, 3.2)$. At both points, $I(x^{*1}) = I(x^{*2}) = \{1\}$. At the first point,

$$Z^1(x^{*1}) = \{(z_1, z_2) : z_1 \geq 0, z_2 = -\tfrac{3}{2}z_1\} \tag{3.81}$$

and this set is also $S(X, x^{*1})$, as can be verified by simple construction. Now

$$Z^2(x^{*1}) = \{(z_1, z_2) : z_1 < 0\}. \tag{3.82}$$

Hence $Z^1(x^{*1}) \cap Z^2(x^{*1}) = \emptyset$. The Kuhn-Tucker conditions (3.72) to (3.74) are satisfied for $\lambda_1^* = \frac{1}{8}$ and $\mu_1^* = 0$. At the second point,

$$Z^1(x^{*2}) = \{(z_1, z_2) : z_1 \geq 0, z_2 = -\tfrac{17}{6}z_1\} \tag{3.83}$$

$$Z^2(x^{*2}) = \{(z_1, z_2) : z_1 < 0\} \tag{3.84}$$

and, again, $Z^1(x^{*2}) \cap Z^1(x^{*2}) = \emptyset$. The reader can verify that at the second local minimum the required multipliers have values $\lambda_1^* = \frac{3}{40}$, $\mu_1^* = \frac{1}{5}$.

It turns out that there is another feasible point where the Kuhn-Tucker necessary conditions hold. Let $x^{*3} = (3 + \sqrt{13}, 2)$. The reader can again verify that $Z^1(x^{*2}) \cap Z^2(x^{*3}) = \emptyset$ and the corresponding multipliers are $\lambda_1^* = 0$, $\mu_1^* = \frac{\sqrt{13}}{26}$. Inspection of Figure 3.7 reveals that x^{*3} is **not** a solution

of our problem but actually a solution of the problem

$$\max f(x) = x_1 \tag{3.85}$$

subject to constraints (3.79) and (3.80). ∎

No specific assumptions, except differentiability, were made so far on the type of the functions involved in problem (P). Further assumptions on these functions lead to special forms of the Kuhn-Tucker conditions. We present only one special case. In many applications, the variables x_j appearing in the program are required to be nonnegative. Suppose that, in addition to constraints (3.2) and (3.3), we also require

$$x \geq 0. \tag{3.86}$$

The necessary conditions for this case can be stated as follows:

Theorem 3.9

Let x^* be the solution of problem (P) *with the additional constraint* (3.86) *and suppose that* (3.75) *holds. Then there exist vectors $\lambda^* = (\lambda_1^*, \ldots, \lambda_m^*)$ and $\mu^* = (\mu_1^*, \ldots, \mu_p^*)$ such that*

$$\nabla f(x^*) - \sum_{i=1}^{m} \lambda_i^* \nabla g_i(x^*) - \sum_{j=1}^{p} \mu_j^* \nabla h_j(x^*) \geq 0 \tag{3.87}$$

$$\lambda_i^* g_i(x^*) = 0, \qquad i = 1, \ldots, m, \quad \lambda^* \geq 0 \tag{3.88}$$

and

$$(x^*)^T\left[\nabla f(x^*) - \sum_{i=1}^{m} \lambda_i^* \nabla g_i(x^*) - \sum_{j=1}^{p} \mu_j^* \nabla h_j(x^*)\right] = 0. \tag{3.89}$$

The proof of this theorem is left for the reader. Note that here X, the feasible set, consists of all x satisfying (3.2), (3.3), and (3.86). Necessary conditions for the solution of problems in which f is maximized instead of being minimized as in (P), or in which some constraints have the form $g_i \leq 0$, can easily be obtained by considering the negatives of these functions; that is, by minimizing $-f(x)$ in the first case, and subject to constraints $-g_i(x) \geq 0$ in the second.

Verification of the necessary and sufficient constraint qualification (3.75) for a general nonlinear programming problem is an almost impossible task. In practice, fortunately, the constraint qualification usually holds, and it is, therefore, quite justifiable to assume the existence of the multipliers $\lambda_1^*, \ldots, \lambda_m^*, \mu_1^*, \ldots, \mu_p^*$, called **generalized Lagrange** or **Kuhn-Tucker multipliers**, satisfying the first-order conditions of Theorem 3.8. We conclude

this section by stating two other constraint qualifications that are stronger than the one presented in the sense that they imply (3.75).

The original constraint qualification of Kuhn and Tucker [25] at a point $x^0 \in X$ requires that any vector $z \in Z^1(x^0)$ be tangent to a differentiable arc contained in X; that is, for each $z \in Z^1(x^0)$ there exists a function α whose domain is the interval $[0, \epsilon] \subset R$ and whose range is in R^n, such that $\alpha(0) = x^0$, $\alpha(\theta) \in X$ for $0 \leq \theta \leq \epsilon$, α is differentiable at $\theta = 0$, and

$$\frac{d\alpha(0)}{d\theta} = \lambda z \tag{3.90}$$

for some positive number λ. Another constraint qualification was introduced by Mangasarian and Fromovitz [27]. Let g_i and h_j be differentiable and continuously differentiable, respectively, at x^0. The qualification holds if the vectors $\nabla h_j(x^0)$, $j = 1, \ldots, p$, are linearly independent and there exists a vector $z \in R^n$ such that

$$z^T \nabla g_i(x^0) < 0, \qquad i \in I(x^0) \tag{3.91}$$

$$z^T \nabla h_j(x^0) = 0, \qquad j = 1, \ldots, p. \tag{3.92}$$

Further constraint qualifications, applicable when the constraint functions are of certain types, will be presented in later chapters.

First-order necessary conditions for optimality are widely discussed in the literature. In addition to references already cited in this chapter and the preceding one, the reader can find analysis of optimality conditions in the works of Bazaraa [3], Bazaraa and Goode [4], Braswell and Marban [8, 9], Canon, Cullum, and Polak [10], Dubovitskii and Milyutin [11], Gamkrelidze [16], Halkin and Neustadt [20], Neustadt [31], Ritter [32, 33, 34], and Wilde [38]. More specialized optimality conditions, mainly for convex and generalized convex nonlinear programs, will be also discussed in subsequent chapters.

3.2 SECOND-ORDER OPTIMALITY CONDITIONS

In this section optimality conditions for problem (P) that involve second derivatives are discussed. We begin with second-order necessary conditions, which complement the Kuhn-Tucker conditions of the preceding section, and then we state and prove sufficient conditions for optimality in problem (P) that extend the corresponding results of Chapter 2.

The reader may recall that, in our derivations of first-order necessary conditions, we obtained them as a result of the fact that a certain intersec-

tion of sets was empty. The same approach could be followed here for the case of second-order conditions, and there is a very elegant derivation of this type due to Dubovitskii and Milyutin [11]. In a similar derivation by Messerli and Polak [29], some general second-order necessary conditions based on the weak Lagrangian are obtained that can be applied to problems with minimal regularity conditions. We shall, however, sacrifice generality for shorter and simpler results and proofs. The following results were obtained by McCormick [28], who also derived the sufficient optimality conditions for problem (P) that will be presented later.

In the following discussion we assume that the functions $f, g_1, \ldots, g_m, h_1, \ldots, h_p$ appearing in problem (P) are twice continuously differentiable. A second-order constraint qualification will be stated first. Let $x \in X$ and define

$$\hat{Z}^1(x) = \{z : z^T\nabla g_i(x) = 0, i \in I(x), z^T\nabla h_j(x) = 0, j = 1, \ldots, p\}. \tag{3.93}$$

The second-order constraint qualification is said to hold at $x^0 \in X$ if every nonzero $z \in \hat{Z}^1(x^0)$ is tangent to a twice differentiable arc contained in the boundary of X; that is, for each $z \in \hat{Z}^1(x^0)$ there exists a twice differentiable function α defined on $[0, \epsilon] \subset R$ with range in R^n such that $\alpha(0) = x^0$,

$$g_i(\alpha(\theta)) = 0, \quad i \in I(x^0), \quad h_j(\alpha(\theta)) = 0, \qquad j = 1, \ldots, p \tag{3.94}$$

for $0 \leq \theta \leq \epsilon$ and

$$\frac{d\alpha(0)}{d\theta} = \lambda z \tag{3.95}$$

for some positive number λ. We have then

Theorem 3.10 (Second-Order Necessary Conditions)

Let x^* be a solution of problem (P) *and suppose that there exist vectors $\lambda^* = (\lambda_1^*, \ldots, \lambda_m^*)^T$, $\mu^* = (\mu_1^*, \ldots, \mu_p^*)^T$ satisfying* (3.72) *to* (3.74). *Further suppose that the second-order constraint qualification holds at x^*. Then for $z \neq 0$ such that $z \in \hat{Z}^1(x^*)$, we have*

$$z^T\left[\nabla^2 f(x^*) - \sum_{i=1}^{m} \lambda_i^* \nabla^2 g_i(x^*) - \sum_{j=1}^{p} \mu_j^* \nabla^2 h_j(x^*)\right] z \geq 0. \tag{3.96}$$

Proof. Let $z \neq 0$ and $z \in \hat{Z}^1(x^*)$, and let $\alpha(\theta)$ be the vector-valued function assumed in the second-order constraint qualification; that is, $\alpha(0) = x^*$, $d\alpha(0)/d\theta = z$ (since $\hat{Z}^1(x^*)$ is a cone, we can assume, without loss of generality, that $\lambda = 1$). Let $d^2\alpha(0)/d\theta^2 = w$. From (3.94) and the chain

rule it follows that

$$\frac{d^2 g_i(\alpha(0))}{d\theta^2} = z^T \nabla^2 g_i(x^*) z + w^T \nabla g_i(x^*) = 0, \qquad i \in I(x^*) \tag{3.97}$$

$$\frac{d^2 h_j(\alpha(0))}{d\theta^2} = z^T \nabla^2 h_j(x^*) z + w^T \nabla h_j(x^*) = 0, \qquad j = 1, \ldots, p. \tag{3.98}$$

From (3.72) to (3.74) and the definition of $\hat{Z}^1(x^*)$, we have

$$\frac{df(\alpha(0))}{d\theta} = z^T \nabla f(x^*) = z^T \left[\sum_{i=1}^{m} \lambda_i^* \nabla g_i(x^*) + \sum_{j=1}^{p} \mu_j^* \nabla h_j(x^*) \right] = 0. \tag{3.99}$$

Since x^* is a local minimum and $df(\alpha(0))/d\theta = 0$, it follows that $d^2 f(\alpha(0))/d\theta^2 \geq 0$; that is,

$$\frac{d^2 f(\alpha(0))}{d\theta^2} = z^T \nabla^2 f(x^*) z + w^T \nabla f(x^*) \geq 0. \tag{3.100}$$

Multiplying (3.97) and (3.98) by the corresponding multipliers, subtracting from (3.100), and by (3.72) to (3.74), we obtain

$$z^T \left[\nabla^2 f(x^*) - \sum_{i=1}^{m} \lambda_i^* \nabla^2 g_i(x^*) - \sum_{j=1}^{p} \mu_j^* \nabla^2 h_j(x^*) \right] z \geq 0. \tag{3.101}$$

■

It is at least as difficult to establish a second-order constraint qualification as a first-order one. There is, however, a relatively simple situation that implies both first- and second-order constraint qualifications. If the vectors $\nabla g_i(x)$, $i \in I(x)$, $\nabla h_j(x)$, $j = 1, \ldots, p$, are linearly independent, then both types of constraint qualifications hold at $x \in X$ [28].

Example 3.2.1

McCormick [28] uses the following problem to illustrate a situation in which the second-order necessary conditions of the last theorem can be used to reduce the number of points satisfying the first-order Kuhn-Tucker conditions. Find the values of the parameter $\beta > 0$ for which the origin is a local minimum of the problem

$$\min f(x_1, x_2) = (x_1 - 1)^2 + (x_2)^2 \tag{3.102}$$

subject to

$$g_1(x_1, x_2) = -x_1 + \frac{(x_2)^2}{\beta} \geq 0. \tag{3.103}$$

The first- and second-order constraint qualifications hold (why?), and the Lagrangian is given by

$$L(x, \lambda) = (x_1 - 1)^2 + (x_2)^2 - \lambda\left[-x_1 + \frac{(x_2)^2}{\beta}\right]. \tag{3.104}$$

The Kuhn-Tucker conditions are satisfied at $x^* = (0, 0)$ by $\lambda^* = 2$ for **any** $\beta \neq 0$. Checking the second-order necessary conditions, we see that $I(x^*) = \{1\}$ and

$$\hat{Z}^1(x^*) = \{z : z \in R, z_1 = 0\}. \tag{3.105}$$

Thus any vector in $\hat{Z}^1(x^*)$ has the form $(0, z_2)^T$. Hence (3.96) becomes in this case

$$(0, z_2)\begin{bmatrix} 2 & 0 \\ 0 & 2 - \dfrac{4}{\beta} \end{bmatrix}\begin{pmatrix} 0 \\ z_2 \end{pmatrix} = \left(2 - \frac{4}{\beta}\right)(z_2)^2 \geq 0 \tag{3.106}$$

and for $\beta < 2$ the origin is clearly not a local minimum. Thus the set of values β satisfying the Kuhn-Tucker conditions has been considerably reduced. ▮

Let us turn now to sufficient conditions for optimality in problem (P). Such conditions were derived by Hestenes [21], King [24], and McCormick [28], whose results were subsequently sharpened by Fiacco [14].

Denote by $\hat{I}(x^*)$ the set of indices i for which $g_i(x^*) = 0$ and (3.72) to (3.74) are satisfied by a **positive** λ_i^*. Thus $\hat{I}(x^*)$ is a subset of $I(x^*)$. Let

$$\hat{\hat{Z}}^1(x^*) = \{z : z^T\nabla g_i(x^*) = 0, i \in \hat{I}(x^*), z^T\nabla g_i(x^*) \geq 0, i \in I(x^*), \\ z^T\nabla h_j(x^*) = 0, j = 1, \ldots, p\}. \tag{3.107}$$

Observe that $\hat{\hat{Z}}^1(x^*) \subset Z^1(x^*)$. We have, then, the following sufficient conditions, proof of which is due to McCormick [28]. These conditions are a direct generalization of Theorem 2.8.

Theorem 3.11

Let x^* be feasible for problem (P). *If there exist vectors λ^*, μ^* satisfying*

$$\nabla_x L(x^*, \lambda^*, \mu^*) = \nabla f(x^*) - \sum_{i=1}^{m} \lambda_i^* \nabla g_i(x^*) - \sum_{j=1}^{p} \mu_j^* \nabla h_j(x^*) = 0 \tag{3.108}$$

$$\lambda_i^* g_i(x^*) = 0, \qquad i = 1, \ldots, m \tag{3.109}$$

$$\lambda^* \geq 0 \tag{3.110}$$

and for every $z \neq 0$ *such that* $z \in \hat{\hat{Z}}^1(x^*)$ *it follows that*

$$z^T\left[\nabla^2 f(x^*) - \sum_{i=1}^{m} \lambda_i^* \nabla^2 g_i(x^*) - \sum_{j=1}^{p} \mu_j^* \nabla^2 h_j(x^*)\right]z > 0, \tag{3.111}$$

then x^* *is a strict local minimum of problem* (P).

Proof. Assume that (3.108) to (3.111) hold and that x^* is not a strict local minimum. Then there exists a sequence $\{z^k\}$ of feasible points $z^k \neq x^*$ converging to x^* such that for each z^k

$$f(x^*) \geq f(z^k). \tag{3.112}$$

Let $z^k = x^* + \theta^k y^k$, where $\theta^k > 0$ and $||y^k|| = 1$. Without loss of generality, assume that the sequence $\{\theta^k, y^k\}$ converges to $(0, \bar{y})$, where $||\bar{y}|| = 1$. Since the points z^k are feasible,

$$g_i(z^k) - g_i(x^*) = \theta^k (y^k)^T \nabla g_i(x^* + \eta_i^k \theta^k y^k) \geq 0, \qquad i \in I(x^*) \tag{3.113}$$

$$h_j(z^k) - h_j(x^*) = \theta^k (y^k)^T \nabla h_j(x^* + \bar{\eta}_j^k \theta^k y^k) = 0, \qquad j = 1, \dots, p \tag{3.114}$$

and from (3.112)

$$f(z^k) - f(x^*) = \theta^k (y^k)^T \nabla f(x^* + \eta^k \theta^k y^k) \leq 0, \tag{3.115}$$

where $\eta^k, \eta_i^k, \bar{\eta}_j^k$ are numbers between 0 and 1. Dividing (3.113) to (3.115) by θ^k and taking limits, we get

$$(\bar{y})^T \nabla g_i(x^*) \geq 0, \qquad i \in I(x^*) \tag{3.116}$$

$$(\bar{y})^T \nabla h_j(x^*) = 0, \qquad j = 1, \dots, p \tag{3.117}$$

$$(\bar{y})^T \nabla f(x^*) \leq 0. \tag{3.118}$$

Suppose that (3.116) holds with a strict inequality for some $i \in \hat{I}(x^*)$. Then combining (3.108), (3.116), and (3.117), we obtain

$$(\bar{y})^T \nabla f(x^*) = \sum_{i=1}^{m} \lambda_i^* (\bar{y})^T \nabla g_i(x^*) + \sum_{j=1}^{p} \mu_j^* (\bar{y})^T \nabla h_j(x^*) > 0, \tag{3.119}$$

contradicting (3.118). Therefore $(\bar{y})^T \nabla g_i(x^*) = 0$ for all $i \in \hat{I}(x^*)$ and so $\bar{y} \in \hat{\hat{Z}}^1(x^*)$. From Taylor's theorem we obtain

$$\begin{aligned} g_i(z^k) = {} & g_i(x^*) + \theta^k (y^k)^T \nabla g_i(x^*) \\ & + \tfrac{1}{2}(\theta^k)^2 (y^k)^T [\nabla^2 g_i(x^* + \xi_i^k \theta^k y^k)] y^k \geq 0, \qquad i = 1, \dots, m, \end{aligned} \tag{3.120}$$

$$h_j(z^k) = h_j(x^*) + \theta^k(y^k)^T\nabla h_j(x^*) + \tfrac{1}{2}(\theta^k)^2(y^k)^T[\nabla^2 h_j(x^* + \bar{\xi}_j^k\theta^k y^k)]y^k = 0, \qquad j = 1, \ldots, p, \tag{3.121}$$

and

$$f(z^k) - f(x^*) = \theta^k(y^k)^T\nabla f(x^*) + \tfrac{1}{2}(\theta^k)^2(y^k)^T[\nabla^2 f(x^* + \xi^k\theta^k y^k)]y^k \leq 0, \tag{3.122}$$

where $\xi^k, \xi_i^k, \bar{\xi}_j^k$ are again numbers between 0 and 1. Multiplying (3.120) and (3.121) by the corresponding λ_i^* and μ_j^* and subtracting from (3.122) yield

$$\theta^k(y^k)^T\left\{\nabla f(x^*) - \sum_{i=1}^{m}\lambda_i^*\nabla g_i(x^*) - \sum_{j=1}^{p}\mu_j^*\nabla h_j(x^*)\right\} + \tfrac{1}{2}(\theta^k)^2(y^k)^T\Big[\nabla^2 f(x^* + \xi^k\theta^k y^k) - \sum_{i=1}^{m}\lambda_i^*\nabla^2 g_i(x^* + \xi_i^*\theta^k y^k) - \sum_{j=1}^{p}\mu_j^*\nabla^2 h_j(x^* + \xi_j^k\theta^k y^k)\Big]y^k \leq 0. \tag{3.123}$$

The expression in braces vanishes by (3.108). Dividing the remaining portion by $\frac{1}{2}(\theta^k)^2$ and taking limits, we obtain

$$(\bar{y})^T\Big[\nabla^2 f(x^*) - \sum_{i=1}^{m}\lambda_i^*\nabla^2 g_i(x^*) - \sum_{j=1}^{p}\mu_j^*\nabla^2 h_j(x^*)\Big]\bar{y} \leq 0. \tag{3.124}$$

Since $\bar{y}$ is nonzero and contained in $\hat{\hat{Z}}^1(x^*)$, it follows that (3.124) contradicts (3.111). ■

Note that the expression in the brackets appearing in (3.111) is the matrix of second derivatives of $L(x, \lambda, \mu)$ taken with respect to x. Fiacco's work [14] extends the last theorem to sufficient conditions for not necessarily strict minima. The interested reader is referred to [14] and [15] for further details.

Example 3.2.2

Consider again the problem presented in Example 3.1.4. We have shown that there are (at least) three points satisfying the necessary condition for optimality. Let us check the sufficient conditions. At x^{*1} we have $\hat{\hat{Z}}^1(x^{*1}) = \{0\}$, and there are no vectors $z \neq 0$ such that $z \in \hat{\hat{Z}}^1(x^{*1})$; hence the sufficient conditions of Theorem 3.11 are trivially satisfied. The reader can verify that these conditions also hold at x^{*2}. At x^{*3}, however,

$$\hat{\hat{Z}}^1(x^{*3}) = \{(z_1, z_2) : z_1 = 0\} \tag{3.125}$$

and the quadratic form appearing in (3.111) is $(-1\sqrt{13})z^Tz$, which is negative for all $z \neq 0$. Thus x^{*3} does not satisfy the sufficient conditions. ▮

3.3 SADDLEPOINTS OF THE LAGRANGIAN

Still another type of optimality conditions is related to the Lagrangian and expressed in terms of saddlepoints of the latter function. Here we state some of these conditions that do not require special assumptions on the nature of the functions appearing in the nonlinear programming problem (P). It is interesting to note that the differentiability assumption used in the first two sections of this chapter can also be dropped in some of the coming results. In later chapters we shall strengthen these results by restricting ourselves to particular families of functions.

Let Φ be a real function of the two vector variables $x \in D \subset R^n$ and $y \in E \subset R^m$. Thus the domain of Φ is $D \times E$. A point $(\bar{x}, \bar{y})$ with $\bar{x} \in D$ and $\bar{y} \in E$ is said to be a **saddlepoint** of Φ if

$$\Phi(\bar{x}, y) \leq \Phi(\bar{x}, \bar{y}) \leq \Phi(x, \bar{y}) \tag{3.126}$$

for every $x \in D$ and $y \in E$.

Associated with the nonlinear program (P) there is a saddlepoint problem that can be stated as follows:

(S) *Find a point* $\bar{x} \in R^n$, $\bar{\lambda} \in R^m$, $\bar{\lambda} \geq 0$, $\bar{\mu} \in R^p$ *such that* $(\bar{x}, \bar{\lambda}, \bar{\mu})$ *is a saddlepoint of the Lagrangian*

$$L(x, \lambda, \mu) = f(x) - \sum_{i=1}^{m} \lambda_i g_i(x) - \sum_{j=1}^{p} \mu_j h_j(x). \tag{3.127}$$

That is,

$$L(\bar{x}, \lambda, \mu) \leq L(\bar{x}, \bar{\lambda}, \bar{\mu}) \leq L(x, \bar{\lambda}, \bar{\mu}) \tag{3.128}$$

for every $x \in R^n$, $\lambda \in R^m$, $\lambda \geq 0$, *and* $\mu \in R^p$.

A one-sided relation between a saddlepoint of the Lagrangian and a solution of problem (P) is given in the next result.

Theorem 3.12

If $(\bar{x}, \bar{\lambda}, \bar{\mu})$ *is a solution of problem* (S), *then* $\bar{x}$ *is a solution of problem* (P).

Proof. Suppose that $(\bar{x}, \bar{\lambda}, \bar{\mu})$ is a solution of problem (S). Then for all $x \in R^n$, $\lambda \in R^m$, $\lambda \geq 0$, and $\mu \in R^p$,

$$f(\bar{x}) - \sum_{i=1}^{m} \lambda_i g_i(\bar{x}) - \sum_{j=1}^{p} \mu_j h_j(\bar{x}) \leq f(\bar{x}) - \sum_{i=1}^{m} \bar{\lambda}_i g_i(\bar{x}) - \sum_{j=1}^{p} \bar{\mu}_j h_j(\bar{x})$$

$$\leq f(x) - \sum_{i=1}^{m} \bar{\lambda}_i g_i(x) - \sum_{j=1}^{p} \bar{\mu}_j h_j(x). \tag{3.129}$$

Rearranging the first inequality, we obtain

$$\sum_{i=1}^{m} (\bar{\lambda}_i - \lambda_i)g_i(\bar{x}) + \sum_{j=1}^{p} (\bar{\mu}_j - \mu_j)h_j(\bar{x}) \leq 0 \tag{3.130}$$

for all $\lambda \in R^m$, $\lambda \geq 0$, and $\mu \in R^p$. Suppose now that $h_k(\bar{x}) > 0$ for some k, $1 \leq k \leq p$. Letting $\lambda_i = \bar{\lambda}_i$ for $i = 1, \ldots, m$ and $\mu_j = \bar{\mu}_j$, $j \neq k$, $\mu_k = \bar{\mu}_k - 1$, we get a contradiction to (3.130). If $h_k(\bar{x}) < 0$ for some k, we can choose an appropriate μ that results in a similar contradiction. Thus $h_j(\bar{x}) = 0$ for $j = 1, \ldots, p$. Now set $\mu = \bar{\mu}$ and let $\lambda_1 = \bar{\lambda}_1 + 1$, $\lambda_i = \bar{\lambda}_i$, $i = 2, \ldots, m$. We obtain $g_1(\bar{x}) \geq 0$. Let $\lambda_2 = \bar{\lambda}_2 + 1$ and $\lambda_i = \bar{\lambda}_i$, $i = 1, \ldots, m$, $i \neq 2$. Then $g_2(\bar{x}) \geq 0$. Repeating this process for all i, we obtain $g_i(\bar{x}) \geq 0$ for $i = 1, \ldots, m$. It follows that $\bar{x}$ is feasible for problem (P). Next let $\lambda = 0$. Then by the first inequality in (3.129), we have

$$0 \leq -\sum_{i=1}^{m} \bar{\lambda}_i g_i(\bar{x}). \tag{3.131}$$

But $\bar{\lambda}_i \geq 0$ and $g_i(\bar{x}) \geq 0$ for $i = 1, \ldots, m$. Therefore

$$\sum_{i=1}^{m} \bar{\lambda}_i g_i(\bar{x}) = 0 \tag{3.132}$$

and so $\bar{\lambda}_i g_i(\bar{x}) = 0$ for all i.

Let us turn now to the second inequality in (3.129). From the preceding arguments we get

$$f(\bar{x}) \leq f(x) - \sum_{i=1}^{m} \bar{\lambda}_i g_i(x) - \sum_{j=1}^{p} \bar{\mu}_j h_j(x). \tag{3.133}$$

If x is feasible for (P), then $g_i(x) \geq 0$ and $h_j(x) = 0$. Thus

$$f(\bar{x}) \leq f(x) \tag{3.134}$$

and $\bar{x}$ is a solution of (P). ■

Note that the preceding sufficient condition for optimality in problem (P) holds whether the Lagrangian is differentiable or not. If, however, the functions f, g_i, h_j are indeed differentiable, we have the following interesting result (compare with Theorem 3.2).

Theorem 3.13

Suppose that $f, g_1, \ldots, g_m, h_1, \ldots, h_p$ are differentiable functions and $(\bar{x}, \bar{\lambda}, \bar{\mu})$ is a solution of problem (S). *Then $Z^1(\bar{x}) \cap Z^2(\bar{x}) = \emptyset$ and*

$$\nabla_x L(\bar{x}, \bar{\lambda}, \bar{\mu}) = \nabla f(\bar{x}) - \sum_{i=1}^{m} \bar{\lambda}_i \nabla g_i(\bar{x}) - \sum_{j=1}^{p} \bar{\mu}_j \nabla h_j(\bar{x}) = 0 \quad (3.135)$$

$$\bar{\lambda}_i g_i(\bar{x}) = 0, \qquad i = 1, \ldots, m \quad (3.136)$$

$$\bar{\lambda} \geq 0. \quad (3.137)$$

Proof. We saw in the proof of the previous theorem that if $(\bar{x}, \bar{\lambda}, \bar{\mu})$ solves (S), then $\bar{\lambda}_i g_i(\bar{x}) = 0$, $i = 1, \ldots, m$, and $\bar{\mu}_j h_j(\bar{x}) = 0$, $j = 1, \ldots, p$. Clearly, $\bar{\lambda}$ satisfies (3.137) by definition. By the second inequality of (3.128), it follows that

$$L(\bar{x}, \bar{\lambda}, \bar{\mu}) \leq L(\bar{x} + \alpha z, \bar{\lambda}, \bar{\mu}) \quad (3.138)$$

for every $z \in R^n$ and $\alpha > 0$.

Thus

$$0 \leq \frac{L(\bar{x} + \alpha z, \bar{\lambda}, \bar{\mu}) - L(\bar{x}, \bar{\lambda}, \bar{\mu})}{\alpha}. \quad (3.139)$$

Letting $\alpha \longrightarrow 0$, we get

$$0 \leq z^T \nabla_x L(\bar{x}, \bar{\lambda}, \bar{\mu}) \quad (3.140)$$

for every $z \in R^n$, and, consequently, (3.135) must hold. By Theorem 3.2, it follows that (3.135) to (3.137) hold if and only if $Z^1(\bar{x}) \cap Z^2(\bar{x}) = \emptyset$. ■

Note that although a saddlepoint of the Lagrangian implies that (3.135) to (3.137) hold without additional regularity conditions, an optimal solution x^* of (P) does not generally imply the existence of a pair (λ^*, μ^*) satisfying (3.135) to (3.137) unless a condition such as (3.71) is imposed on (P).

Theorem 3.13, together with Example 3.1.1, implies that the converse of Theorem 3.12 generally does not hold. We present now an even simpler example illustrating this fact.

Example 3.3.1

Suppose that we have the following program:

$$\min f(x) = x \quad (3.141)$$

subject to

$$-(x)^2 \geq 0. \quad (3.142)$$

The optimal solution is $x^* = 0$. The corresponding saddlepoint problem of the Lagrangian is to find a $\lambda^* \geq 0$ such that

$$x^* + \lambda(x^*)^2 \leq x^* + \lambda^*(x^*)^2 \leq x + \lambda^*(x)^2 \quad (3.143)$$

for every $x \in R$, or, equivalently,

$$0 \leq x + \lambda^*(x)^2. \tag{3.144}$$

Clearly, λ^* cannot vanish. But for any $\lambda^* > 0$ we can choose $x > -1/\lambda^*$ and (3.144) will not hold. Thus there exists no λ^* such that (x^*, λ^*) will be a saddlepoint. ▮

In the next chapter nonlinear programs in which the objective and constraint functions satisfy certain convexity and concavity requirements are discussed. If such programs also satisfy some constraint qualifications, then the converse of Theorem 3.12 **does** hold for them; that is, the existence of a solution to (S) is a necessary condition for optimality in (P).

Saddlepoints are closely related to the theory of games of strategy in which two players with conflicting interests oppose each other. For a given "payoff" function $\Phi(x, y)$, one player is minimizing Φ with respect to x while the other player is maximizing Φ with respect to y. This is called a min-max (or max-min) of Φ. The mathematical foundations of game theory and its application to economics were laid down by Von Neumann and are described in the classical work of Von Neumann and Morgenstern [30].

We conclude this chapter by relating saddlepoints to min-max of functions of two vector variables.

Lemma 3.14

Let Φ be a real function of the two vector variables $x \in D \subset R^n$ and $y \in E \subset R^m$. Then

$$\max_{y \in E} \min_{x \in D} \Phi(x, y) \leq \min_{x \in D} \max_{y \in E} \Phi(x, y) \tag{3.145}$$

provided that the above minima and maxima exist.

Proof. For any fixed $y \in E$ we have

$$\min_{x \in D} \Phi(x, y) \leq \Phi(x, y). \tag{3.146}$$

Similarly, for any fixed $x \in D$

$$\Phi(x, y) \leq \max_{y \in E} \Phi(x, y). \tag{3.147}$$

Hence for every $x \in D$ and $y \in E$

$$\min_{x \in D} \Phi(x, y) \leq \max_{y \in E} \Phi(x, y). \tag{3.148}$$

Consequently,

$$\max_{y \in E} \min_{x \in D} \Phi(x, y) \leq \min_{x \in D} \max_{y \in E} \Phi(x, y). \tag{3.149}$$

■

Example 3.3.2

A real matrix $A = [a_{ij}]$ can be regarded as a real-valued function Φ of two variables i, j, given by

$$\Phi(i, j) = a_{ij}. \tag{3.150}$$

Let

$$A_1 = \begin{bmatrix} 1 & -1 \\ -1 & 1 \end{bmatrix}. \tag{3.151}$$

Then

$$\max_j \min_i a_{ij} = \max(\min_i a_{i1}, \min_i a_{i2}) = \max(-1, -1) = -1 \tag{3.152}$$

and

$$\min_i \max_j a_{ij} = \min(\max_j a_{1j}, \max_j a_{2j}) = \min(1, 1) = 1. \tag{3.153}$$

Hence

$$\max_j \min_i \Phi(i, j) < \min_i \max_j \Phi(i, j) \tag{3.154}$$

and (3.145) holds as a strict inequality. On the other hand, taking

$$A_2 = \begin{bmatrix} 1 & 2 \\ -1 & 3 \end{bmatrix}, \tag{3.155}$$

the reader can easily show that

$$\max_j \min_i a_{ij} = 2 = \min_i \max_j a_{ij} \tag{3.156}$$

and (3.145) holds as an equation. ■

A necessary and sufficient condition for an equality in (3.145) is that Φ has a saddlepoint. Formally,

Theorem 3.15

Suppose that the hypotheses of Lemma 3.14 hold. Then

$$\max_{y \in E} \min_{x \in D} \Phi(x, y) = \Phi(\bar{x}, \bar{y}) = \min_{x \in D} \max_{y \in E} \Phi(x, y) \tag{3.157}$$

if and only if $(\bar{x}, \bar{y})$ *is a saddlepoint of* Φ.

Proof. Suppose that $(\bar{x}, \bar{y})$ is a saddlepoint of Φ. Then

$$\max_{y \in E} \Phi(\bar{x}, y) \leq \Phi(\bar{x}, \bar{y}) \leq \min_{x \in D} \Phi(x, \bar{y}). \tag{3.158}$$

Also,

$$\min_{x \in D} \max_{y \in E} \Phi(x, y) \leq \max_{y \in E} \Phi(\bar{x}, y) \tag{3.159}$$

$$\min_{x \in D} \Phi(x, \bar{y}) \leq \max_{y \in E} \min_{x \in D} \Phi(x, y). \tag{3.160}$$

Combining the last three relations, we obtain

$$\min_{x \in D} \max_{y \in E} \Phi(x, y) \leq \Phi(\bar{x}, \bar{y}) \leq \max_{y \in E} \min_{x \in D} \Phi(x, y). \tag{3.161}$$

Comparing (3.161) with (3.145), we conclude that (3.157) must hold. Conversely, let $(\bar{x}, \bar{y})$ satisfy

$$\max_{y \in E} \Phi(\bar{x}, y) = \min_{x \in D} \max_{y \in E} \Phi(x, y) \tag{3.162}$$

and

$$\min_{x \in D} \Phi(x, \bar{y}) = \max_{y \in E} \min_{x \in D} \Phi(x, y) \tag{3.163}$$

If (3.157) holds, then by (3.162), (3.163), and the definitions of minima and maxima, we get

$$\Phi(\bar{x}, y) \leq \max_{y \in E} \Phi(\bar{x}, y) = \Phi(\bar{x}, \bar{y}) = \min_{x \in D} \Phi(x, \bar{y}) \leq \Phi(x, \bar{y}). \tag{3.164}$$

Thus Φ has a saddlepoint at $(\bar{x}, \bar{y})$. ∎

EXERCISES

3.A. For the constraints

$$(x_1)^2 + (x_2 - 1)^2 \leq 1 \tag{3.165}$$

$$x_2 - x_1 \leq 0, \tag{3.166}$$

find an objective function $f(x_1, x_2)$ to be minimized such that

$$Z^1(x^0) \cap Z^2(x^0) = \emptyset \tag{3.167}$$

at $x^0 = (0, 0)$ and this point is **not** a local optimum.

3.B. Let A_1 and A_2 be nonempty subsets of R^n. Show that $A_1 \subset A_2$ implies $A_1' \subset A_2'$; that is, the positively normal cone of A_2 is included in that of A_1.

3.C. Prove that $(Z^1(\hat{x}))' \subset (S(X, \hat{x}))'$ for every $\hat{x}$ in the feasible set X of problem (P).

3.D. Show that if $\hat{x}$ is in the interior of the set $A \subset R^n$, then the cone of tangents of A at $\hat{x}$ is R^n. Find also $(S(A, \hat{x}))'$.

3.E. Find the cone of tangents of the hypercube

$$A = \{x : x \in R^n, 0 \leq x_j \leq 1, j = 1, \ldots, n\} \tag{3.168}$$

at the origin.

3.F. Prove the following corollary of Lemma 3.7. Suppose that x^0 is a solution of problem (P) and x^0 is in the interior of X. Then $\nabla f(x^0) = 0$.

3.G. Given the problem

$$\max 4x_1 + 2x_2 \tag{3.169}$$

subject to

$$x_1 \log x_1 - x_1 + 1 \geq 0 \tag{3.170}$$

$$x_1 \geq 0 \tag{3.171}$$

$$3x_1 - 2(x_2)^2 \geq 1, \tag{3.172}$$

show that there is no point $x^* \in R^2$ satisfying the Kuhn-Tucker necessary conditions for optimality. Is there a point satisfying the Fritz John conditions of Theorem 3.4?

3.H. It is suggested to solve the problem

$$\min (x_1)^2 + (x_2)^2 \tag{3.173}$$

subject to

$$(x_1 - 1)^3 - (x_2)^2 \geq 0 \tag{3.174}$$

by finding stationary points of the corresponding Lagrangian. Discuss the difficulties that may arise in doing so.

3.I. One of the constraint qualifications for problem (P) is as follows: The vectors $\nabla g_i(x^*)$, $i \in I(x^*)$, and $\nabla h_j(x^*)$, $j = 1, \ldots, p$, are all linearly independent. Show that this qualification implies $S(X, x^*) = Z^1(x^*)$ but not conversely. [*Hint:* Let $g(x) = (x_1)^2 + (x_2)^2 - x_3$, $h(x) = x_3$, and take $x^* = (0, 0, 0)$.]

3.J. State the generalized Kuhn-Tucker necessary conditions of optimality for the problem

$$\max f(x) \tag{3.175}$$

subject to

$$g_i(x) \geq 0, \qquad i = 1, \ldots, k \tag{3.176}$$

$$g_i(x) \leq 0, \qquad i = k+1, \ldots, m \tag{3.177}$$

$$h_j(x) = 0, \qquad j = 1, \ldots, p. \tag{3.178}$$

3.K. Prove the following result, due to J. W. Gibbs from 1876. Given the problem

$$\min f(x) = \sum_{j=1}^{n} f_j(x_j) \tag{3.179}$$

subject to

$$x_j \geq 0, \qquad j = 1, \ldots, n \tag{3.180}$$

$$\sum_{j=1}^{n} x_j = 1 \tag{3.181}$$

where f is a differentiable function. Let x^* be a solution of this problem. Then there exists a number μ^* such that

$$f'_j(x_j^*) = \mu^* \quad \text{if} \quad x_j^* > 0 \tag{3.182}$$

$$f'_j(x_j^*) \geq \mu^* \quad \text{if} \quad x_j^* = 0 \tag{3.183}$$

where the prime indicates differentiation.

3.L. Given the nonlinear program [37]

$$\min f(x) = \sum_{j=1}^{n} \frac{c_j}{x_j} \tag{3.184}$$

subject to

$$\sum_{j=1}^{n} a_j x_j = b \tag{3.185}$$

$$x \geq 0 \tag{3.186}$$

where the a_j, b, c_j are positive constants. Show that the optimal value of the objective function is given by

$$f(x^*) = \frac{\left[\sum_{j=1}^{n} (a_j c_j)^{1/2}\right]^2}{b}. \tag{3.187}$$

3.M. For the nonlinear program

$$\min f(x) = \sum_{j=1}^{n} c_j x_j \tag{3.188}$$

subject to

$$g_i(x) \geq 0, \qquad i = 1, \ldots, m, \tag{3.189}$$

where the c_j are given numbers, not all zero, and the g_i are differentiable

functions, show that if x^* is a solution of the problem, where a (first-order) constraint qualification holds, then $I(x^*) \neq \emptyset$.

3.N. Show that the vector $x^* = b/\|b\|$ satisfies the sufficient conditions for optimality in the problem

$$\max b^T x, \qquad x \in R^n \tag{3.190}$$

subject to

$$x^T x \leq 1, \tag{3.191}$$

where the vector $b \neq 0$ is given.

3.O. Given the constraints

$$1 - (x_1 - 2)^2 - (x_2 - 5)^2 \geq 0 \tag{3.192}$$

$$1 - (x_1 - 2)^2 - \left(\frac{x_2}{4}\right)^2 \geq 0 \tag{3.193}$$

$$x_1 - 2 \geq 0, \tag{3.194}$$

draw the feasible set and show that

$$(Z^1(x))' \neq (S(X, x))' \tag{3.195}$$

for any $x \in X$, but the second-order constraint qualification is satisfied.

3.P. The nonlinear programming problem

$$\max (x_2)^2 - 2x_1 - (x_1)^2 \tag{3.196}$$

subject to

$$(x_1)^2 + (x_2)^2 \leq 1 \tag{3.197}$$

was solved on a computer by a certain numerical algorithm. Depending on the starting conditions, the algorithm converged to four different points:

(1) $x_1 = 1.000 \quad x_2 = 0.000$
(2) $x_1 = -1.000 \quad x_2 = 0.000$
(3) $x_1 = -0.500 \quad x_2 = 0.866$
(4) $x_1 = -0.500 \quad x_2 = -0.866$

(a) Check whether all these points satisfy the Kuhn-Tucker necessary conditions for optimality. Draw the feasible set and determine if the first-order constraint qualification holds at these points.
(b) Assuming that the second-order constraint qualification holds at each of the foregoing points, check second-order necessary conditions for optimality.
(c) Finally, derive sufficient conditions for strict optimality in this problem and verify them at each of the points satisfying the necessary conditions.

REFERENCES

1. Abadie, J., "On the Kuhn-Tucker Theorem," in *Nonlinear Programming*, J. Abadie (Ed.), North-Holland Publishing Co., Amsterdam, 1967.
2. Arrow, K. J., L. Hurwicz, and H. Uzawa, "Constraint Qualifications in Maximization Problems," *Naval Res. Log. Quart.*, **8**, 175–191 (1961).
3. Bazaraa, M. S., "Nonlinear Programming: Nondifferentiable Functions," Ph.D. dissertation, Georgia Institute of Technology, Atlanta, 1970.
4. Bazaraa, M. S., and J. J. Goode, "Necessary Optimality Criteria in Mathematical Programming in the Presence of Differentiability," *J. Math. Anal. & Appl.*, **40**, 609–621 (1972).
5. Bazaraa, M. S., J. J. Goode, and C. M. Shetty, "Constraint Qualifications Revisited," *Management Science*, **18**, 567–573 (1972).
6. Beltrami, E. J., "A Constructive Proof of the Kuhn-Tucker Multiplier Rule," *J. Math. Anal. & Appl.*, **26**, 297–306 (1967).
7. Bergthaller, C., private communication, 1971.
8. Braswell, R. N., and J. A. Marban, "Necessary and Sufficient Conditions for the Inequality Constrained Optimization Problem Using Directional Derivatives," *Int. J. Systems Sci.*, **3**, 263–275 (1972).
9. Braswell, R. N., and J. A. Marban, "On Necessary and Sufficient Conditions in Non-Linear Programming," *Int. J. Systems Sci.*, **3**, 277–286 (1972).
10. Canon, M. D., C. D. Cullum, and E. Polak, "Constrained Minimization Problems in Finite Dimensional Spaces," *SIAM J. Control*, **4**, 528–547 (1966).
11. Dubovitskii, A. Y., and A. A. Milyutin, "Extremum Problems in the Presence of Restrictions," *USSR Comp. Math. and Math Phys.*, **5**, 3, 1–80 (1965).
12. Evans, J. P., "On Constraint Qualifications in Nonlinear Programming," *Naval Res. Log. Quart.*, **17**, 3, 281–286 (1970).
13. Farkas, J., "Uber die Theorie der Einfachen Ungleichungen," *J. für die Reine und Angew. Math.*, **124**, 1–27 (1902).
14. Fiacco, A. V., "Second Order Sufficient Conditions for Weak and Strict Constrained Minima," *SIAM J. Appl. Math.*, **16**, 105–108 (1968).
15. Fiacco, A. V., and G. P. McCormick, *Nonlinear Programming: Sequential Unconstrained Minimization Techniques*, John Wiley & Sons, New York, 1968.
16. Gamkrelidze, R. V., "Extremal Problems in Finite-Dimensional Spaces," *J. Optimization Theory and Appl.*, **1**, 173–193 (1967).
17. Gould, F. J., and J. W. Tolle, "A Necessary and Sufficient Qualification for Constrained Optimization," *SIAM J. Appl. Math.*, **20**, 164–172 (1971).
18. Gould, F. J., and J. W. Tolle, "Geometry of Optimality Conditions and Constraint Qualifications, *Math. Prog.*, **2**, 1–18 (1972).

19. GUIGNARD, M., "Generalized Kuhn-Tucker Conditions for Mathematical Programming Problems in a Banach Space," *SIAM J. Control*, **7**, 232–241 (1969).

20. HALKIN, H., and L. W. NEUSTADT, "General Necessary Conditions for Optimization Problems," *Proc. Natl. Acad. Sci.*, **56**, 1066–1071 (1966).

21. HESTENES, M. R., *Calculus of Variations and Optimal Control Theory*, John Wiley & Sons, New York, 1966.

22. JOHN, F., "Extremum Problems with Inequalities as Subsidiary Conditions," in *Studies and Essays, Courant Anniversary Volume*, K. D. Friedricks, et al. (Eds.), Interscience Publishers, New York, 1948.

23. KARLIN, S., *Mathematical Methods and Theory in Games, Programming and Economics*, Vols. I and II, Addison-Wesley Publishing Co., Reading, Mass., 1959.

24. KING, R. P., "Necessary and Sufficient Conditions for Inequality Constrained Extreme Values," *Ind. & Eng. Chem. Fund.*, **5**, 484–489 (1966).

25. KUHN, H. W., and A. W. TUCKER, "Nonlinear Programming," in *Proceedings of the Second Berkeley Symposium on Mathematical Statistics and Probability*, J. Neyman (Ed.), University of California Press, Berkeley, Calif., 1951.

26. MANGASARIAN, O. L., *Nonlinear Programming*, McGraw-Hill Book Co., New York, 1969.

27. MANGASARIAN, O. L., and S. FROMOVITZ, "The Fritz John Necessary Optimality Conditions in the Presence of Equality and Inequality Constraints," *J. Math. Anal. & Appl.*, **17**, 37–47 (1967).

28. MCCORMICK, G. P., "Second Order Conditions for Constrained Minima," *SIAM J. Appl. Math.*, **15**, 641–652 (1967).

29. MESSERLI, E. J., and E. POLAK, "On Second Order Necessary Conditions of Optimality," *SIAM J. Control*, **7**, 272–291 (1969).

30. NEUMANN, J. VON and O. MORGENSTERN, *Theory of Games and Economic Behavior*, 2nd ed., Princeton University Press, Princeton, N.J., 1947.

31. NEUSTADT, L. W., "An Abstract Variational Theory with Applications to a Broad Class of Optimization Problems. I. General Theory," *J. SIAM Control*, **4**, 505–527 (1966).

32. RITTER, K., "Optimization Theory in Linear Spaces. I," *Math. Ann.*, **182**, 189–206 (1969).

33. RITTER, K., "Optimization Theory in Linear Spaces Part II. On Systems of Linear Operator Inequalities in Partially Ordered Normed Linear Spaces," *Math. Ann.*, **183**, 169–180 (1969).

34. RITTER, K., "Optimization Theory in Linear Spaces Part III. Mathematical Programming in Partially Ordered Banach Spaces," *Math. Ann.*, **184**, 133–154 (1970).

35. SLATER, M., "Lagrange Multipliers Revisited: A Contribution to Nonlinear

Programming," Cowles Commission Discussion Paper, Math. 403, November 1950.

36. VARAIYA, P., "Nonlinear Programming in Banach Space," *SIAM J. Appl. Math.*, **15**, 284–293 (1967).

37. WHITTLE, P., *Optimization Under Constraints*, Wiley-Interscience, London, 1971.

38. WILDE, D. J., "Differential Calculus in Nonlinear Programming," *Operations Research*, **10**, 764–773 (1962).

4 CONVEX SETS AND FUNCTIONS

The mathematical programming problem introduced in the beginning of the previous chapter is probably the most general program we shall deal with. In fact, it is so general that not much more can be said about it. In this chapter we shall examine nonlinear programs in which the functions involved will have special features. It will become apparent that most of these problems are related, in one way or another, to convex sets and convex functions. This chapter forms the basis for understanding some very important aspects of these problems, such as the existence of optima, duality, the sensitivity of optimal solutions to small perturbations, and numerical algorithms for finding a solution. Here will be summarized some of the properties of convex sets and functions that are relevant to mathematical programming. This summary is by no means comprehensive. For a more complete study of convex sets and functions, from an optimization-oriented point of view, the reader should consult the lecture notes of Fenchel [8] or the later books of Rockafellar [13] and Stoer and Witzgall [15]. Many results appearing in this chapter can be also found in Berge [2], where the underlying topological concepts are thoroughly explained.

In the late 1960s the notion of convex functions has been extended in several ways in order to generalize results on extrema of convex functions, which are of primary importance in mathematical programming. Such generalizations will be presented in later chapters.

4.1 CONVEX SETS

A subset C of R^n is called a **convex set** if for any two points $x^1 \in C$, $x^2 \in C$, the line segment joining them is also in C. Here and in the following

three chapters q will always denote an ordered pair of numbers (q_1, q_2), called **weights**, with the following properties:

$$q_1 \geq 0, \quad q_2 \geq 0, \quad q_1 + q_2 = 1. \tag{4.1}$$

Then a set C is convex if $x^1 \in C$, $x^2 \in C$ and q, as defined above, also imply that $(q_1 x^1 + q_2 x^2) \in C$.

Example 4.1.1

Examples of convex sets are given below.

(a) The empty set, the set consisting of a single point $x \in R^n$, the whole space R^n are trivial examples of convex sets.

(b) The set

$$H = \{x : x \in R^n, c^T x = b\}, \tag{4.2}$$

where $c \neq 0$ is a given vector and b is a given number, is called a **hyperplane.** It is a convex set.

(c) Using the foregoing notation, we define the **closed half space**

$$H^c = \{x : x \in R^n, c^T x \geq b\} \tag{4.3}$$

and the **open half space**

$$H^o = \{x : x \in R^n, c^T x > b\} \tag{4.4}$$

as being **generated** by the hyperplane H. Note that each hyperplane also generates an opposite closed (open) half space, obtained by reversing the sense of the inequalities in (4.3) and (4.4). All these half spaces are convex sets.

(d) The **hypersphere**

$$S_\alpha(x^0) = \{x : x \in R^n, \|x - x^0\| \leq \alpha\}, \tag{4.5}$$

where $x^0 \in R^n$ is a given vector and α is a given number, is a convex set. ▮

We now present some simple algebraic and geometric properties of convex sets.

Given vectors $x^1, x^2, \ldots, x^p$ in R^n and numbers $\alpha_1, \ldots, \alpha_p$ satisfying

$$\alpha_1 \geq 0, \ldots, \alpha_p \geq 0 \qquad \sum_{i=1}^{p} \alpha_i = 1, \tag{4.6}$$

the **convex combination** of the vectors $x^1, \ldots, x^p$ is a vector x given by

$$x = \alpha_1 x^1 + \cdots + \alpha_p x^p. \tag{4.7}$$

Theorem 4.1

The set $C \subset R^n$ is convex if and only if every convex combination x of the points $x^1 \in C, \ldots, x^S \in C$ is contained in C.

Proof. Suppose that C is a convex set. The proof is by induction on S. For $S = 1$, the theorem is clearly true. Assume now that it is true for $S > 1$ and let

$$x = \alpha_1 x^1 + \cdots + \alpha_S x^S + \alpha_{S+1} x^{S+1}. \tag{4.8}$$

Without loss of generality, we can assume that $\alpha_{S+1} \neq 1$. Then we can write

$$x = (1 - \alpha_{S+1})z + \alpha_{S+1} x^{S+1}, \tag{4.9}$$

where

$$z = \frac{\alpha_1}{1 - \alpha_{S+1}} x^1 + \cdots + \frac{\alpha_S}{1 - \alpha_{S+1}} x^S. \tag{4.10}$$

Since

$$\frac{\alpha_1}{1 - \alpha_{S+1}} \geq 0, \ldots, \frac{\alpha_S}{1 - \alpha_{S+1}} \geq 0; \qquad \sum_{i=1}^{S} \frac{\alpha_i}{1 - \alpha_{S+1}} = 1, \tag{4.11}$$

it follows, by the induction hypothesis, that $z \in C$; and by the convexity of C, we have $x \in C$.

Conversely, if every convex combination of the points of C is contained in C, then, in particular, the statement is true for $S = 2$; that is, $x^1 \in C$, $x^2 \in C$, $\alpha_1 \geq 0$, $\alpha_2 \geq 0$, $\alpha_1 + \alpha_2 = 1$ imply $(\alpha_1 x^1 + \alpha_2 x^2) \in C$. Hence C is a convex set. ■

Theorem 4.2

The intersection of an arbitrary collection of convex sets is also a convex set.

Proof. Let x^1 and x^2 be points contained in the intersection. Then they are also contained in every member of the collection and so is $x = q_1 x^1 + q_2 x^2$. Hence x is contained in the intersection. ■

The intersection of all the convex sets containing an arbitrary set $A \subset R^n$ is called the **convex hull** of A and is denoted by Co (A). From Theorem 4.2 it is clear that the set Co (A) is a convex set; it is actually the smallest convex set in R^n containing A. By the **sum** of two sets $X \subset R^n$, $Y \subset R^n$, we mean the set

$$X + Y = \{x + y : x \in X, y \in Y\}. \tag{4.12}$$

Similarly, the **scalar multiple** of a set $X \subset R^n$ is defined as

$$\lambda X = \{\lambda x : x \in X\}, \qquad \lambda \in R. \tag{4.13}$$

We can now state

Theorem 4.3

If C and D are convex subsets of R^n and λ is a real number, then the sets $C + D$ and λC are also convex.

The proof of this theorem is left for the reader.

Many important results in mathematical programming can be proved by using so-called **separation theorems** of convex sets. These theorems deal with two nonempty convex subsets C_1 and C_2 in R^n for which there exists a hyperplane H such that C_1 lies on one side of H and C_2 on the opposite side. Such a hyperplane is said to separate C_1 and C_2. Formally, let S and T be nonempty subsets of R^n. Then the hyperplane H is said to **separate** S and T if S is contained in one of the closed half spaces generated by H and T is contained in the opposite closed half space. Such a hyperplane is called a **separating hyperplane**. A hyperplane **strictly separates** S and T if S is contained in one of the open half spaces generated by H and T is in the opposite open half space. It should be intuitively clear to the reader that, given two disjoint convex sets in R^n, it is possible to find a hyperplane that separates them.

We now state some theorems dealing with the existence of such separating hyperplanes. These theorems can be proved in several ways. Our proofs will use certain results from topology, in particular, results involving compact sets in R^n. For a more complete study of these and related results, the reader is referred to Berge [2]. Proofs of separation theorems along different lines can be found for example in Rockafellar [13].

Lemma 4.4

Let C be a nonempty closed convex set in R^n, not containing the origin. Then there exists a hyperplane that strictly separates C and the origin.

Proof. Let $S_\alpha(0)$ be a closed hypersphere with center at the origin; that is,

$$S_\alpha(0) = \{x \in R^n : ||x|| \leq \alpha\} \tag{4.14}$$

such that $C \cap S_\alpha(0) \neq \emptyset$. Then the set $C \cap S_\alpha(0)$ is a compact set, and the continuous function $||x||$ attains its minimum on $C \cap S_\alpha(0)$ at some $x^0 \in C$.

Note that $||x^0|| > 0$. It follows that for every $x \in S_\alpha(0) \cap C$ we have $||x|| \geq ||x^0||$, and, by the definition of $S_\alpha(0)$, $||x|| \geq ||x^0||$ for every $x \in C$

also. Next let x be any point in C. Since C is a convex set, we have $(\lambda x + (1 - \lambda)x^0) \in C$ for every $1 \geq \lambda \geq 0$ and

$$||\lambda x + (1 - \lambda)x^0||^2 \geq ||x^0||^2. \tag{4.15}$$

Hence

$$(\lambda x + (1 - \lambda)x^0)^T(\lambda x + (1 - \lambda)x^0) \geq (x^0)^T x^0, \qquad 1 \geq \lambda \geq 0 \tag{4.16}$$

or

$$(\lambda)^2(x - x^0)^T(x - x^0) + 2\lambda(x^0)^T(x - x^0) \geq 0, \qquad 1 \geq \lambda \geq 0. \tag{4.17}$$

Suppose that $(x^0)^T(x - x^0) = -\epsilon < 0$. We can choose a sufficiently small λ such that

$$2\epsilon > \lambda(x - x^0)^T(x - x^0) > 0. \tag{4.18}$$

It follows that

$$2\lambda(x^0)^T(x - x^0) < -(\lambda)^2(x - x^0)^T(x - x^0), \tag{4.19}$$

thereby contradicting (4.17). Thus we must have

$$(x^0)^T(x - x^0) \geq 0 \tag{4.20}$$

or for every $x \in C$

$$(x^0)^T x \geq (x^0)^T(x^0) > 0. \tag{4.21}$$

Let $\alpha = (x^0)^T(x^0)/2$. Then the hyperplane

$$H_C = \{x : x \in R^n, (x^0)^T x = \alpha\} \tag{4.22}$$

strictly separates C and the origin. ■

We can now prove the following

Theorem 4.5 (Strict Separation Theorem)

Let C_1 and C_2 be two disjoint nonempty closed convex sets in R^n and suppose that C_2 is compact. Then there exists a hyperplane that strictly separates them.

Proof. Since C_2 is compact, the set $C_1 - C_2 = C_1 + (-C_2)$ is closed (see, for example, [1, 2, 3]), and by Theorem 4.3 it is convex. Since C_1 and C_2 are disjoint, $C_1 - C_2$ does not contain the origin. By Lemma 4.4, there exists a hyperplane

$$H_{C_1-C_2} = \{x : x \in R^n, (x^0)^T x = \alpha\}, \tag{4.23}$$

where $x^0 \in C_1 - C_2$ minimizes the distance from $C_1 - C_2$ to the origin and $\alpha = (x^0)^T(x^0)/2$, which strictly separates $C_1 - C_2$ from the origin. For every $x \in C_1 - C_2$,

$$(x^0)^T x > \alpha > 0. \tag{4.24}$$

Letting $x = u - v$, where $u \in C_1$ and $v \in C_2$, we get

$$(u^0 - v^0)^T(u - v) > \alpha > 0. \tag{4.25}$$

Hence

$$\inf_{u \in C_1} (u^0 - v^0)^T u \geq \sup_{v \in C_2} (u^0 - v^0)^T v + \alpha > \sup_{v \in C_2} (u^0 - v^0)^T v. \tag{4.26}$$

Choose a number β such that

$$\inf_{u \in C_1} (u^0 - v^0)^T u > \beta > \sup_{v \in C_2} (u^0 - v^0)^T v. \tag{4.27}$$

Then the hyperplane $\{x : x \in R^n, (x^0)^T x = \beta\}$ strictly separates C_1 from C_2. ■

The key assumption in the previous results on strictly separating two convex sets was the condition that the sets are closed. If we do not restrict ourselves to closed convex sets, it is possible to obtain somewhat weaker results analogous to Lemma 4.4 and Theorem 4.5.

Lemma 4.6

Let C be a nonempty convex set in R^n, not containing the origin. Then there exists a hyperplane separating C and the origin.

Proof. For every $x \in C$, let

$$Y(x) = \{y : y \in R^n, y^T y = 1, y^T x \geq 0\}. \tag{4.28}$$

This set is nonempty and closed. Let $x^1, \ldots, x^k$ be any finite set of points in C.

Since C is convex, by Theorem 4.1 the set of points x, given by

$$x = \alpha_1 x^1 + \cdots + \alpha_k x^k \qquad \sum_{i=1}^{k} \alpha_i = 1, \qquad \alpha_i \geq 0, \tag{4.29}$$

is a compact set contained in C (it is the convex hull of the set of points $x^1, \ldots, x^k$). By Lemma 4.4, there exists a $y^0 \neq 0$ such that

$$(y^0)^T x^i > 0, \qquad i = 1, \ldots, k. \tag{4.30}$$

Without loss of generality, we can assume that $(y^0)^T y^0 = 1$, and then

$$\bigcap_{i=1}^{k} Y(x^i) \neq \emptyset. \tag{4.31}$$

Since the sets $Y(x)$ are closed subsets of the compact set $Y = \{y : y \in R^n, y^T y = 1\}$, it follows from the finite intersection property [2, 12] that

$$\bigcap_{x \in C} Y(x) \neq \emptyset. \tag{4.32}$$

Choose any $\hat{y} \in \bigcap_{x \in C} Y(x)$. Then $\hat{y}^T x \geq 0$ for every $x \in C$. Hence the hyperplane $\{x : x \in R^n, \hat{y}^T x = 0\}$ separates C and the origin. ■

Example 4.1.2

Consider the set

$$C = \{x : x \in R^2, x_1 > 0\}. \tag{4.33}$$

This set is a nonempty open convex set, not containing the origin. The hyperplane $\{x : x \in R^2, x_1 = 0\}$ separates C and the origin. There exists, however, no hyperplane that strictly separates them. ■

Theorem 4.7 (Separation Theorem)

Let C_1 and C_2 be two nonempty disjoint convex sets in R^n. Then there exists a hyperplane that separates them.

Proof. The set $C_1 - C_2$ is convex and does not contain the origin. By Lemma 4.6, there exists a vector $\hat{y}$ such that for every $x \in C_1 - C_2$ we have $\hat{y}^T x \geq 0$, or, equivalently, $u \in C_1$, $v \in C_2$ imply $\hat{y}^T(u - v) \geq 0$. Hence there exists a number β satisfying

$$\inf_{u \in C_1} \hat{y}^T u \geq \beta \geq \sup_{v \in C_2} \hat{y}^T v. \tag{4.34}$$

The hyperplane $\{x : x \in R^n, \hat{y}^T x = \beta\}$ then separates C_1 and C_2. ■

Figure 4.1 provides additional explanations on separation theorems for convex sets. In part (a) we have a simple case of strict separation, while in part (b) we show that open convex sets that do not contain the origin cannot always be strictly separated from the origin. The example illustrated in part (b) shows that the assumption of closed convex sets in Lemma 4.4 is crucial. Similarly, the assumption that C_2 is compact in Theorem 4.5 is important, as is shown by the sets in part (c). Finally, in part (d) we present a possible weakening of the hypotheses in Theorem 4.7. The assumption that the sets C_1 and C_2 are disjoint can actually be replaced by the assumption that the

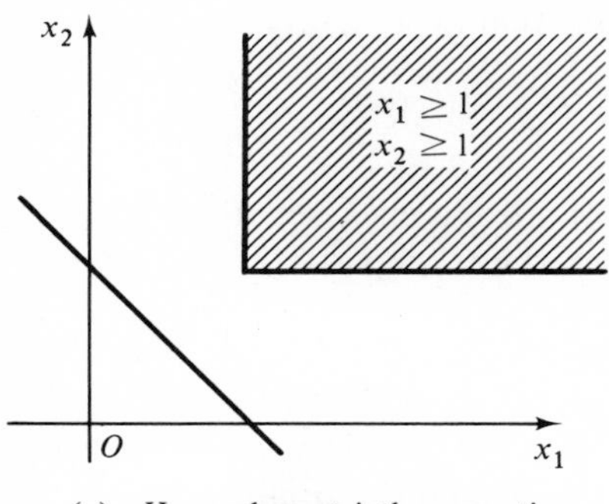

(a) Hyperplane strictly separating a closed convex set and the origin.

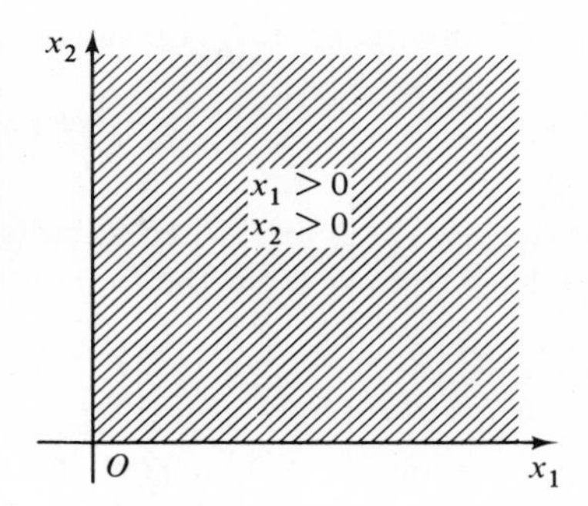

(b) Open convex set which cannot be strictly separated from the origin.

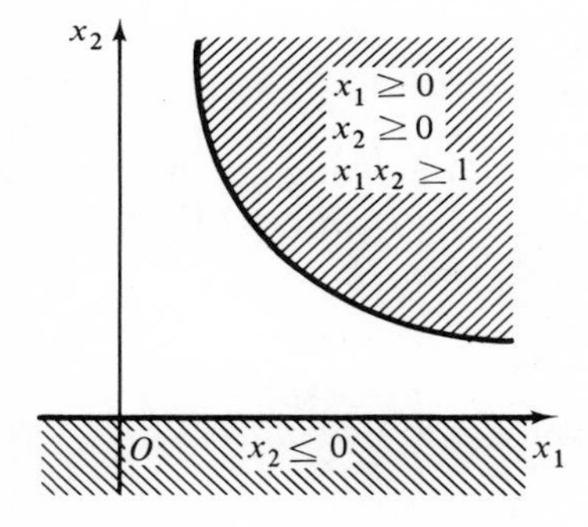

(c) Noncompact closed convex sets which cannot be strictly separated.

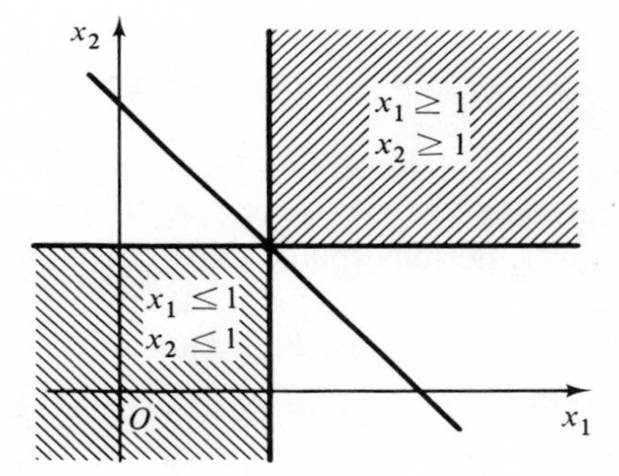

(d) Separating hyperplane for convex sets having a common boundary point.

Fig. 4.1 Separation of convex sets.

interiors of C_1 and C_2 are disjoint. Thus they may have a common boundary point.

As a first use of the preceding separation theorems, we now prove the well-known Farkas lemma, which was presented in Chapter 3. For convenience, we restate it here.

Lemma 4.8 (Farkas Lemma)

Let A be a given $m \times n$ real matrix and b a given n vector. The inequality $b^T y \geq 0$ holds for all vectors y satisfying $Ay \geq 0$ if and only if there exists an m vector $\rho \geq 0$ such that $A^T \rho = b$.

Proof. This statement is equivalent to saying that $Ay \geq 0$, $b^T y < 0$ has a solution if and only if $A^T \rho = b$ and $\rho \geq 0$ has no solution. The latter

system has no solution if and only if the nonempty closed convex sets

$$C_1 = \{x : x \in R^n, x = A^T\rho, \rho \geq 0\} \qquad C_2 = \{b\} \tag{4.35}$$

are disjoint (see Exericse 4.C). Note that the set C_2 is compact.

If these conditions hold, then by Theorem 4.5 there exist a nonzero vector $c \in R^n$ and a real number α such that $c^Tb < \alpha$. And for every $x \in C_1$ we have $c^Tx > \alpha$, or, equivalently, for every $\rho \geq 0$ we have $c^TA^T\rho > \alpha$.

Letting $\rho = 0$, we conclude that $\alpha < 0$. And by letting $\rho = (0, \ldots, \rho_j, 0, \ldots, 0)$ for $j = 1, \ldots, m$, where ρ_j is a large positive number, we obtain $c^TA^T \geq 0$. And so $Ac \geq 0$. Thus c is a solution of $Ay \geq 0$ and $b^Ty < 0$. Conversely, if $A^T\rho = b$, $\rho \geq 0$, and $Ay \geq 0$, then $b^Ty = \rho^TAy \geq 0$. ∎

The separation theorems can be used to prove similar results concerning other systems of linear equations and inequalities. An extensive list of these results can be found in Mangasarian [12]. There is considerable interest in convex sets for purposes other than mathematical programming. For a survey of additional results on convex sets, the reader is referred to Eggleston [5], Fenchel [8], Rockafellar [13], or Valentine [16].

4.2 CONVEX FUNCTIONS

The fundamentals of the theory of convex functions were laid down by Hölder [9] and Jensen [10] around the turn of the century. It is intuitive and, in most cases, also sufficient to look at convex functions from a geometrical point of view. A real function f is said to be convex provided that the chord connecting any two points of the curve $f(x)$ lies on or above the curve. The convex functions shown in Figure 4.2 illustrate the "chord above the curve" principle. A look at these drawings also reveals that convex functions are not necessarily continuous or differentiable everywhere on their domain of definition. Another important observation, which will serve as a starting point in our formal discussions on convex functions, is the close relation between convex sets and convex functions. The shaded areas above the curves

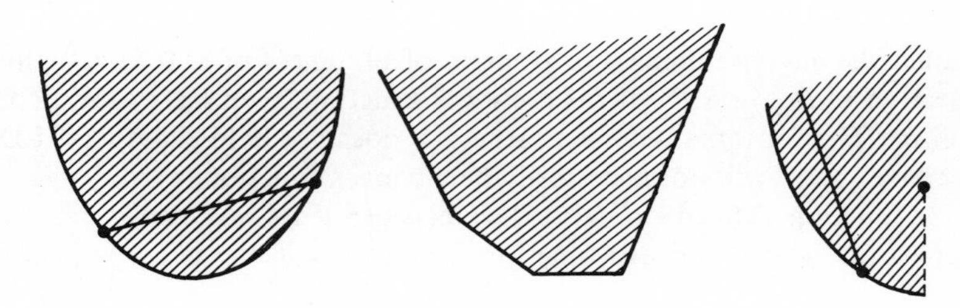

Fig. 4.2 Examples of convex functions.

in Figure 4.2 are convex sets that, in turn, can be used to define convex functions. For reasons that will become clear later, it will be convenient to study convex functions that may have infinite values.

Let f be a function defined on a subset D of R^n with values in the extended reals; that is, $f(x)$ is either a real number or it is $\pm\infty$ The subset of R^{n+1}

$$P(f) = \{(x, \alpha) : x \in D \subset R^n, \alpha \in R, f(x) \leq \alpha\} \tag{4.36}$$

is called the **epigraph** of f. We define f to be a **convex function** if $P(f)$ is a convex set. Some straightforward examples follow.

Example 4.2.1

The function $f = +\infty$ on R^n is a convex function, since $P(f) = \emptyset$ and the empty set is convex. Similarly, $f = -\infty$ on R^n is also convex, since $P(f) = R^{n+1}$.

Consider a convex function f defined on a proper subset D of R^n. Let

$$f_1(x) = \begin{cases} f(x) & \text{if } x \in D \\ +\infty & \text{if } x \notin D. \end{cases} \tag{4.37}$$

The epigraph $P(f)$ of the function f defined on D is identical to $P(f_1)$, where f_1 is a function defined on all R^n. In this way, we can always construct convex functions defined throughout R^n. In particular, let $a \in R$, $b \in R^n$ be given. Then

$$f_1(x) = \begin{cases} a & \text{if } x = b \\ +\infty & \text{if } x \neq b \end{cases} \tag{4.38}$$

is a convex function defined on R^n. ∎

As a result of Example 4.2.1, we shall assume that, unless mentioned explicitly, a convex function is defined on all R^n. The set

$$\text{ED}(f) = \{x : x \in R^n, f(x) < +\infty\} \tag{4.39}$$

is called the **effective domain** or domain of finiteness of f. It is actually the projection of $P(f)$ on R^n. If f is a convex function, then ED (f) is a convex set in R^n. The converse statement generally does not hold. That is, if ED (f) is a convex set, f will not necessarily be a convex function.

A function defined on $D \subset R^n$ is **concave** if its negative is convex. In other words, g is a concave function if and only if $-g$ is convex. The set

$$Q(g) = \{(x, \alpha) : x \in D \subset R^n, \alpha \in R, g(x) \geq \alpha\} \tag{4.40}$$

is called the **hypograph** of g and g is said to be concave if $Q(g)$ is a convex subset of R^{n+1}. In the following discussions we shall deal, in most cases, only with convex functions. The reader should be able to modify the results concerning properties of convex functions to the analogous properties for concave functions.

Allowing convex functions to take on infinite values requires some caution in arithmetic operations, such as the undefined operation $+\infty + (-\infty)$. Convex functions that have values $+\infty$ and $-\infty$ do not arise frequently in applications. We shall, therefore, be concerned mainly with **proper convex functions** defined as convex functions that nowhere have the value $-\infty$ and are not identically equal to $+\infty$. Convex functions that are not proper are called **improper**. Formally, f is a proper convex function if f is convex, $f(x) > -\infty$ for all $x \in R^n$, and ED $(f) \neq \emptyset$. For proper convex functions such that ED (f) contains more than one point, we can now derive a more familiar result, which can be taken as an alternative definition. Given $x^1 \in$ ED (f), $x^2 \in$ ED (f), if $(x^1, \alpha^1) \in P(f)$, $(x^2, \alpha^2) \in P(f)$, then for every vector of weights q we have $(q_1x^1 + q_2x^2, q_1\alpha^1 + q_2\alpha^2) \in P(f)$. That is,

$$f(q_1x^1 + q_2x^2) \leq q_1\alpha^1 + q_2\alpha^2. \tag{4.41}$$

Since the inequality in (4.41) must hold for every α^1, α^2 such that $f(x^1) \leq \alpha^1$, $f(x^2) \leq \alpha^2$, it can be modified to

$$f(q_1x^1 + q_2x^2) \leq q_1f(x^1) + q_2f(x^2). \tag{4.42}$$

The inequality in (4.42) is the "classical" definition of real convex functions, expressing the "chord above the curve" property. Generally (4.42) remains valid even if we let $f(x) = +\infty$ or $f(x) = -\infty$ for some x, provided that we adopt the arithmetic rule $0(\infty) = 0(-\infty) = 0$ and avoid the undefined operation $+\infty + (-\infty)$.

A real-valued function f defined on a convex subset C of R^n is called **strictly convex** if for any $x^1 \in C$, $x^2 \in C$, $x^1 \neq x^2$, and $1 > q_1 > 0$, $q_1 + q_2 = 1$, inequality (4.42) holds as a strict inequality. Most of the results of this chapter dealing with convex functions can be also strengthened to strictly convex functions, and the reader is invited to do so. We shall specifically mention strictly convex functions in only one or two results.

Let us examine some basic properties of convex functions.

Theorem 4.9

Let f be a proper convex function on R^n. Let $x^1, \ldots, x^S$ be points in R^n and $q_1, \ldots, q_S$ nonnegative numbers satisfying $q_1 + \cdots + q_S = 1$. Then

$$f(q_1x^1 + \cdots + q_Sx^S) \leq q_1f(x^1) + \cdots + q_Sf(x^S). \tag{4.43}$$

Proof. If $f(x^i) = +\infty$ for some i, then (4.43) trivially holds. Assume now that $f(x^i) < +\infty$ for all i. The epigraph of f is a convex set, and by Theorem 4.1 it contains every convex combination of its points. Hence $(x^1, \alpha^1) \in P(f), \ldots, (x^S, \alpha^S) \in P(f)$ imply $(q_1x^1 + \cdots + q_Sx^S,\ q_1\alpha^1 + \cdots + q_s\alpha^s) \in P(f)$; that is,

$$f(q_1x^1 + \cdots + q_Sx^S) \leq q_1\alpha^1 + \cdots + q_S\alpha^S. \tag{4.44}$$

Since the α^i satisfy $f(x^i) \leq \alpha^i$, $i = 1, \ldots, S$, the inequality in (4.43) follows. ■

Theorem 4.10

Let f be a convex function and let λ be a nonnegative number. Then λf is also a convex function. Let f and g be convex functions. Then $f + g$ is also convex, provided that the undefined operation $+\infty + (-\infty)$ is avoided.

The proof is left for the reader. The next result follows immediately from the last theorem.

Corollary 4.11

Under the hypotheses of Theorem 4.10 every linear combination $\lambda_1 f_1 + \cdots + \lambda_k f_k$ of convex functions with $\lambda_1 \geq 0, \ldots, \lambda_k \geq 0$ is also a convex function.

Given a function Ψ defined on R with values in the extended reals, it is said to be nondecreasing if for every $x^1 < x^2$ we have $\Psi(x^1) \leq \Psi(x^2)$. The following theorem is sometimes useful in identifying convex functions or in constructing new convex functions from existing ones.

Theorem 4.12

Let f be a real convex function on R^n and let Ψ be a nondecreasing proper convex function on R. Then $\Psi(f(x))$ is convex on R^n.

Proof. By (4.42), we have for every q

$$f(q_1x^1 + q_2x^2) \leq q_1f(x^1) + q_2f(x^2). \tag{4.45}$$

Since Ψ is nondecreasing,

$$\Psi(f(q_1x^1 + q_2x^2)) \leq \Psi(q_1f(x^1) + q_2f(x^2)); \tag{4.46}$$

and by the convexity of Ψ,

$$\Psi(f(q_1x^1 + q_2x^2)) \leq q_1\Psi(f(x^1)) + q_2\Psi(f(x^2)). \tag{4.47}$$

■

Note that by letting $\Psi(+\infty) = +\infty$, Theorem 4.12 can be extended to functions f on R^n that are proper convex functions.

One of the functions appearing in Figure 4.2 is a piecewise linear function. Its convexity can be proven by noting that a linear function is convex and by the following theorem.

Theorem 4.13

Let $\{f_i\}$, $i \in I$, be a finite or infinite collection of convex functions on R^n. For every $x \in R^n$, define the pointwise supremum of this collection as

$$f(x) = \sup_{i \in I} f_i(x). \tag{4.48}$$

Then f is a convex function.

Proof. The epigraphs of the f_i are convex sets. By Theorem 4.2, the intersection of the $P(f_i)$ is also a convex set. By definition,

$$\bigcap_{i \in I} P(f_i) = \{(x, \alpha) : x \in R^n, \alpha \in R, f_i(x) \leq \alpha \text{ for all } i \in I\}. \tag{4.49}$$

Thus we obtain

$$\bigcap_{i \in I} P(f_i) = \{(x, \alpha) : x \in R^n, \alpha \in R, \sup_{i \in I} f_i(x) = f(x) \leq \alpha\} \tag{4.50}$$

and f is a convex function. ■

We have seen that with every convex function f on R^n one can associate a convex set $P(f)$ in R^{n+1}. The next result deals with the converse statement—that is, constructing convex functions on R^n from convex sets in R^{n+1} by taking their "lower envelope." Note that the infimum taken over the empty set is, by convention, assumed to be $+\infty$.

Theorem 4.14

Let C be a convex set in R^{n+1} and for every $x \in R^n$ let f be defined by

$$f(x) = \inf \{\alpha : \alpha \in R, (x, \alpha) \in C\}. \tag{4.51}$$

Then f is a convex function on R^n.

Proof. We have to show that $P(f)$ is a convex set. Let (x^1, α^1) and (x^2, α^2) be any two points in $P(f)$. It follows from the definition of f that there exist numbers $\hat{\alpha}^1 \leq \alpha^1$ and $\hat{\alpha}^2 \leq \alpha^2$ such that $(x^1, \hat{\alpha}^1) \in C$ and $(x^2, \hat{\alpha}^2) \in C$. Since C is convex, by (4.51) we have

$$f(q_1 x^1 + q_2 x^2) \leq q_1 \hat{\alpha}^1 + q_2 \hat{\alpha}^2 \leq q_1 \alpha^1 + q_2 \alpha^2 \tag{4.52}$$

for every q. Thus $(q_1 x^1 + q_2 x^2, q_1 \alpha^1 + q_2 \alpha^2) \in P(f)$. ■

Example 4.2.2

Let C be the open unit circle; that is,

$$C = \{(x_1, x_2) : (x_1)^2 + (x_2)^2 < 1\}. \tag{4.53}$$

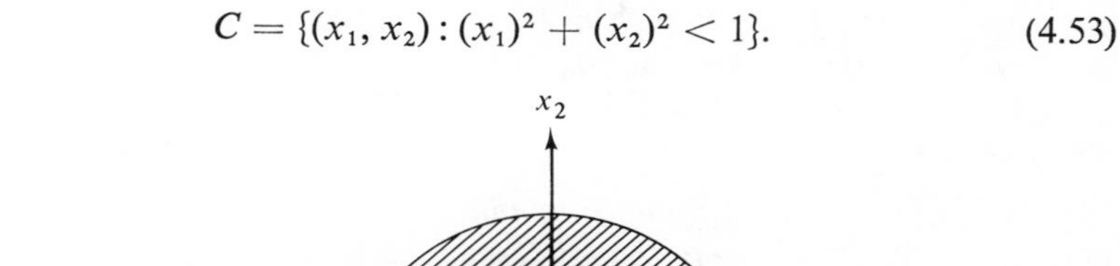

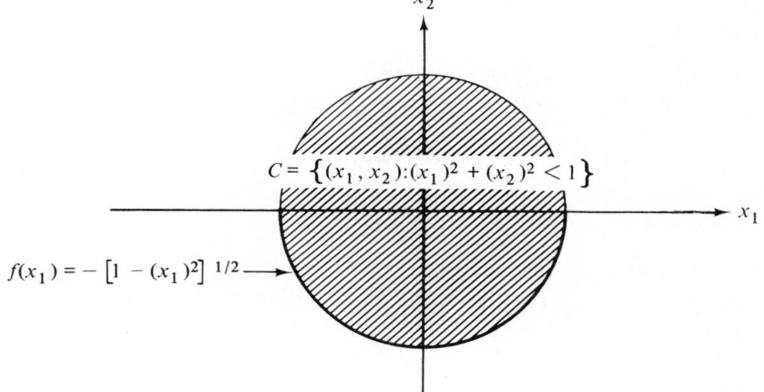

Fig. 4.3 The lower envelope of the open unit circle.

Then we construct f as the convex function defined on R by (see Figure 4.3)

$$f(x) = \begin{cases} -[1 - (x)^2]^{1/2} & \text{if } |x| < 1 \\ +\infty & \text{if } |x| \geq 1. \end{cases} \quad \blacksquare \tag{4.54}$$

The next theorem states that a function is convex on a convex set C if and only if the restriction of f to each line segment in the set C is a convex function.

Theorem 4.15

The function f is convex on R^n if and only if for every $x^1 \in R^n$, $x^2 \in R^n$ the function ϕ, defined by

$$\phi(\lambda) = f(\lambda x^1 + (1 - \lambda)x^2), \tag{4.55}$$

is convex for all $0 \leq \lambda \leq 1$.

Proof. Suppose that f is convex on R^n and let x^1, x^2 be arbitrary points in R^n. We must show that the epigraph of ϕ, given by

$$P(\phi) = \{(\lambda, \alpha) : \lambda \in R, 0 \leq \lambda \leq 1, \alpha \in R, \phi(\lambda) \leq \alpha\} \tag{4.56}$$

is a convex set. Let $(\lambda^1, \alpha^1) \in P(\phi)$, $(\lambda^2, \alpha^2) \in P(\phi)$ and let

$$z^1 = \lambda^1 x^1 + (1 - \lambda^1)x^2 \qquad z^2 = \lambda^2 x^1 + (1 - \lambda^2)x^2. \tag{4.57}$$

We have then

$$f(z^1) = \phi(\lambda^1) \leq \alpha^1 \tag{4.58}$$

$$f(z^2) = \phi(\lambda^2) \leq \alpha^2. \tag{4.59}$$

Hence $(z^1, \alpha^1) \in P(f)$, $(z^2, \alpha^2) \in P(f)$; and since $P(f)$ is a convex set, we also have $(q_1 z^1 + q_2 z^2, q_1\alpha^1 + q_2\alpha^2) \in P(f)$ for every q.

Consequently,

$$f(q_1 z^1 + q_2 z^2) \leq q_1\alpha^1 + q_2\alpha^2. \tag{4.60}$$

By (4.55) and (4.57), we get

$$f(q_1 z^1 + q_2 z^2) = \phi(q_1\lambda^1 + q_2\lambda^2) \tag{4.61}$$

and so $(q_1\lambda^1 + q_2\lambda^2, q_1\alpha^1 + q_2\alpha^2) \in P(\phi)$; that is, ϕ is convex. The proof of the converse statement is similar and is left as an exercise. ■

The reader may have noticed that we have said nothing about the continuity of convex functions. It can easily be demonstrated that convex functions can be discontinuous. Take, for example, the function f given by

$$f(x) = \begin{cases} (x)^2 - 1 & x < 1 \\ 2 & x = 1 \\ +\infty & x > 1. \end{cases} \tag{4.62}$$

(See Figure 4.4.)

This is a convex function, discontinuous on the boundary of its effective domain. Roughly speaking, discontinuities in convex functions can occur only at some boundary points of their effective domain. Some of these discontinuities can be eliminated by the closure operation for convex functions. Consider a convex function f defined on R^n and, for a given point $x^0 \in R^n$, the collection of all linear affine functions h of the form $h(x) = a^T x - b$ such that $h(x^0) \leq f(x^0)$. Let us define the **support set**

$$L(f) = \{(a, b) : a \in R^n, b \in R, a^T x - b \leq f(x) \text{ for every } x \in R^n\}. \tag{4.63}$$

Then we define the **closure of a convex function**, denoted cl f, as

$$\text{cl } f(x) = \sup_{(a,b) \in L(f)} \{a^T x - b\}. \tag{4.64}$$

Clearly, $\text{cl}\, f(x) \leq f(x)$ for every $x \in R^n$. A convex function f is said to be **closed** if $f = \text{cl}\, f$. In particular, it can be shown that a proper convex function f is closed if and only if the convex level set $\{x : x \in R^n, f(x) \leq \alpha\}$ is closed for every real number α.

Equivalently, the closure operation for proper convex functions is directly related to the closure operation for sets (the notation $\bar{C}$ is used for the closure of the set C). The epigraph of $\text{cl}\, f$ is the closure of the epigraph of f; that is, [13]

$$P(\text{cl}\ f) = \overline{P(f)}. \tag{4.65}$$

The last relation can, of course, also be used as the definition of the closure operation for proper convex functions. For improper convex functions, the closure operation has a very simple meaning. If $f(x) = +\infty$ for every $x \in R^n$, then $\text{cl}\, f = f$. If, however, $f(x) = -\infty$ for some $x \in R^n$, it follows that $L(f) = \emptyset$; and since, by convention, the supremum over the empty set is taken to be $-\infty$, we get $\text{cl}\ f(x) = -\infty$ for **every** $x \in R^n$. Consequently, the only closed improper convex functions are those that are identically $+\infty$ or $-\infty$. For proper convex functions, the notions of closedness and lower semicontinuity are identical [13]. Similarly, a proper concave function is closed if and only if it is upper semicontinuous.

Example 4.2.3

Let us apply the closure operation to the convex function defined by (4.62). Since this is a proper convex function, we simply "close" the epigraph of f as in Figure 4.4. Thus we obtain the closure of f as

$$\text{cl}\, f = \begin{cases} (x)^2 - 1 & x \leq 1 \\ +\infty & x > 1 \end{cases} \tag{4.66}$$

and

$$\text{cl}\ f(1) = 0. \tag{4.67}$$

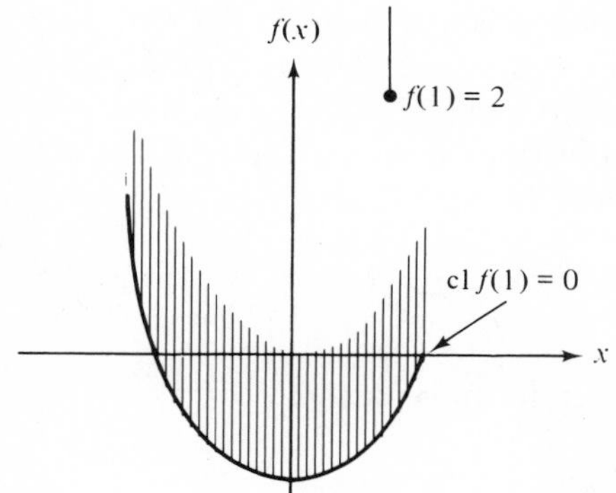

Fig. 4.4 Closure operation for a convex function.

In other words, the closure operation in this example lowered the value of f at the boundary of its effective domain until the function became continuous over ED (f). ▮

As we just saw in a simple example, the closure operation for convex functions eliminates certain discontinuities at boundary points of the effective domain. The function cl f agrees with f at every point in the interior of ED(f). The reason, as the reader might suspect, is that convex functions are continuous on the interior of their effective domains. A slightly more general result holds, but before stating it, we introduce the notion of relative interior of a nonempty convex set.

Consider a subset B of R^n with the following property. For every two points $x^1 \in B$, $x^2 \in B$ and for every value of the real number α, also $\alpha x^1 + (1 - \alpha)x^2 \in B$. A set satisfying this property is called an **affine set** [13]. Single points, lines, and hyperplanes are examples of affine sets in R^n. Given a convex set $C \subset R^n$, the intersection of all affine sets containing C is called the **affine hull** of C. The **relative interior** of C, denoted by ri C, is defined as the interior of C, viewed as a subset of its affine hull.

Example 4.2.4

Consider the convex set $C \subset R^2$ consisting of the line segment between the points $(a, 0)$ and $(b, 0)$, $a < b$; that is,

$$C = \{(x_1, 0) : a \leq x_1 \leq b\}. \tag{4.68}$$

This set has no interior if we view C as a subset of R^2 (since one cannot find an open hypersphere in R^2 contained in C). The affine hull of C is, of course, the whole x_1 (one-dimensional) axis, and the relative interior of C is given by

$$\text{ri}\, C = \{(x_1, 0) : a < x_1 < b\}, \tag{4.69}$$

which is the interior of C viewed as a subset of the x_1 axis. ▮

Note that if the convex set $C \subset R^n$ is n dimensional, such as a square in R^2 or a cube in R^3, then the affine hull of C is R^n itself, so the relative interior of C coincides with the interior of C.

The proofs of the next two theorems need certain additional concepts from topology and further properties of convex sets and functions, not directly related to the subjects of our discussions. For this reason, we shall only state them here; the interested reader may find the proofs in [13].

The first theorem deals with certain improper convex functions. Such a function either has the value $+\infty$ everywhere or it takes on the value $-\infty$ at some point of its effective domain.

Theorem 4.16

If f is an improper convex function, then $f(x) = -\infty$ for every x in the relative interior of its effective domain.

The main result we state now is a generalization of the fact that a real-valued convex function defined on an open convex set is continuous; see, for example, [2, 5].

Theorem 4.17

A convex function is continuous on the relative interior of its effective domain.

From this theorem we have immediately

Corollary 4.18

A real-valued convex function on R^n is continuous everywhere.

We conclude this section by using the separation theorems for convex sets, presented earlier, to prove some results on the existence of solutions to systems of inequalities involving convex functions. The following result is due to Fan, Glicksberg, and Hoffman [6].

Theorem 4.19

Let $f_1, \ldots, f_m$ be proper convex functions and let C be a nonempty convex set such that $C \subset \bigcap_{i=1}^{m} \mathrm{ED}(f_i)$.

Then exactly one of the alternatives holds:

(i) *There exists an $x^0 \in C$ such that*

$$f_i(x^0) < 0, \qquad i = 1, \ldots, m. \tag{4.70}$$

(ii) *There exist nonnegative numbers $\alpha_1, \ldots, \alpha_m$, not all zero, such that for every $x \in C$*

$$\sum_{i=1}^{m} \alpha_i f_i(x) \geq 0. \tag{4.71}$$

Proof. Suppose that (i) holds. For any nonnegative $\alpha_1, \ldots, \alpha_m$, with at least one $\alpha_i > 0$, we get

$$\sum_{i=1}^{m} \alpha_i f_i(x^0) < 0 \tag{4.72}$$

and (ii) cannot hold.

Suppose now that (i) does not hold. Let

$$Y = \{y : y \in R^m, \text{ there is an } x \in C \text{ such that } f_i(x) < y_i, \; i = 1, \ldots, m\}. \tag{4.73}$$

That is, every $y \in Y$ must have a positive component. Since the f_i are proper convex functions, the set Y is nonempty. Moreover, it is a convex set, for let $y^1 \in Y$, $y^2 \in Y$ and $x^1 \in C$, $x^2 \in C$ such that

$$f_i(x^1) < y_i^1 \qquad f_i(x^2) < y_i^2, \qquad i = 1, \ldots, m. \tag{4.74}$$

Then for every vector of weights q we have $(q_1x^1 + q_2x^2) \in C$ and

$$f_i(q_1x^1 + q_2x^2) \leq q_1f_i(x^1) + q_2f_i(x^2) < q_1y_i^1 + q_2y_i^2, \qquad i = 1, \ldots, m. \tag{4.75}$$

Thus $(q_1y^1 + q_2y^2) \in Y$. Let $N \subset R^m$ be the nonpositive orthant (a convex set); that is,

$$N = \{y : y \in R^m, y \leq 0\}. \tag{4.76}$$

The sets Y and N are two nonempty disjoint convex sets, and by Theorem 4.7 there exists a hyperplane that separates them. That is, there exists a nonzero vector $\alpha \in R^m$ and a real number b such that for every $y \in Y$

$$\sum_{i=1}^{m} \alpha_i y_i \geq b \tag{4.77}$$

and for every $y \in N$

$$\sum_{i=1}^{m} \alpha_i y_i \leq b. \tag{4.78}$$

If $\alpha_k < 0$, then for the vector $y = (0, \ldots, 0, y_k, 0, \ldots, 0) \in N$, with y_k sufficiently negative, (4.78) would be violated; that is, $\alpha_i \geq 0$ for all $i = 1, \ldots, m$ and also $b \geq 0$. For each $x \in C$ let y be defined by

$$y_i = f_i(x) + \epsilon, \qquad i = 1, \ldots, m \tag{4.79}$$

where ϵ is an arbitrary positive number. Then $y \in Y$ and

$$\sum_{i=1}^{m} \alpha_i f_i(x) + \sum_{i=1}^{m} \alpha_i \epsilon \geq b \geq 0. \tag{4.80}$$

It follows that for every $x \in C$

$$\sum_{i=1}^{m} \alpha_i f_i(x) \geq 0 \tag{4.81}$$

and (ii) holds. ■

Rockafellar [13] shows by an example that the condition $C \subset \bigcap_i \text{ED}(f_i)$, or some weaker version of it, is necessary for the preceding theorem. Let $f_1(x) = -(x)^{1/2}$ if $x \geq 0$ and $f_1(x) = +\infty$ for $x < 0$. Let $f_2(x) = x$ and $C = R$. There exists no $x \in C$ such that $f_1(x) < 0$ and $f_2(x) < 0$, and there exist no nonnegative numbers α_1 and α_2, not both zero, such that $\alpha_1 f_1(x) + \alpha_2 f_2(x) \geq 0$ for every $x \in C$. Thus neither (i) nor (ii) holds. The reason is that ED $(f_1) \cap$ ED (f_2) is the nonnegative orthant and C is not contained in it.

Numerous further generalizations of Theorem 4.19 are possible. One of them can be viewed as a generalized version of the Farkas lemma [3] and will be useful later in this chapter.

Theorem 4.20

Let $f_0, f_1, \ldots, f_m$, be proper convex functions and let C be a nonempty convex set such that $C \subset \bigcap_{i=0}^{m} \text{ED}\,(f_i)$. If the system of inequalities

$$f_0(x) < 0 \tag{4.82}$$

$$f_i(x) \leq 0, \qquad i = 1, \ldots, m \tag{4.83}$$

has no solution $x \in C$, while there exists an $x^0 \in C$ such that

$$f_i(x^0) < 0, \qquad i = 1, \ldots, m \tag{4.84}$$

then either

$$f_0(x) \geq 0 \tag{4.85}$$

for every $x \in C$ or there exist nonnegative numbers $\alpha_1, \ldots, \alpha_m$, not all zero, such that

$$f_0(x) + \sum_{i=1}^{m} \alpha_i f_i(x) \geq 0 \tag{4.86}$$

for every $x \in C$.

Proof. If (4.82) and (4.83) have no solution in C, then also (4.82) and $f_i(x) < 0$, $i = 1, \ldots, m$, have no solution in C. By Theorem 4.19, there exist nonnegative numbers $\lambda_0, \lambda_1, \ldots, \lambda_m$, not all zero, such that

$$\lambda_0 f_0(x) + \sum_{i=1}^{m} \lambda_i f_i(x) \geq 0 \tag{4.87}$$

for every $x \in C$. Without loss of generality, we can assume that $\lambda_0 + \lambda_1 + \cdots + \lambda_m = 1$. Suppose that $\lambda_0 = 1$. Then (4.85) holds. Suppose now that $0 < \lambda_0 < 1$. Divide (4.87) by λ_0 and let $\alpha_i = \lambda_i/\lambda_0$ for $i = 1, \ldots, m$. Then (4.86) holds.

Finally, suppose that $\lambda_0 = 0$. Then

$$\sum_{i=1}^{m} \lambda_i f_i(x) \geq 0 \tag{4.88}$$

for every $x \in C$. Since $x^0 \in C$ satisfies (4.84), then (4.88) can hold only if $\lambda_1 = 0, \ldots, \lambda_m = 0$. This is a contradiction, for the λ_i are not all zero. ■

4.3 DIFFERENTIAL PROPERTIES OF CONVEX FUNCTIONS

We begin this section by recalling a few notions from the differential calculus of real functions and applying them to convex functions. Let us define a direction as any vector $y \in R^n$. Given a real function f defined on a subset S of R^n and a point x^0 in the interior of S, the derivative of f at x^0 in the direction of y or the **directional derivative** of f at x^0 in the direction of y is defined as

$$Df(x^0; y) = \lim_{t \to 0} \frac{f(x^0 + ty) - f(x^0)}{t} \tag{4.89}$$

if the limit exists.

A straightforward extension of this definition is as follows: Let f be any function defined on R^n with values in the extended reals. Let x^0 be a point where $f(x)$ has a finite value and let y be a direction. The **right-sided derivative** of f at x^0 in the direction of y is defined as

$$D^+f(x^0; y) = \lim_{t \to 0^+} \frac{f(x^0 + ty) - f(x^0)}{t} \tag{4.90}$$

if the limit, which can be $\pm\infty$, exists. Here $t \longrightarrow 0^+$ means that t approaches 0 through positive numbers. Similarly, the **left-sided derivative** of f at x^0 in the direction of y is defined as

$$D^-f(x^0; y) = \lim_{t \to 0^-} \frac{f(x^0 + ty) - f(x^0)}{t}. \tag{4.91}$$

For $y = 0$, both $D^+f(x^0; 0)$ and $D^-f(x^0; 0)$ are defined to be zero.

The reader can easily verify that

$$-D^+f(x^0; -y) = D^-f(x^0; y). \tag{4.92}$$

If for some x^0 and y

$$D^+f(x^0; y) = D^-f(x^0; y), \tag{4.93}$$

then we have a directional derivative in the sense of (4.89). Letting $y = (1, 0, \ldots, 0) \in R^n$, the directional derivative of f at x^0 in the direction of y is just $\partial f(x^0)/\partial x_1$, the partial derivative of f with respect to x_1, and we can similarly get the partial derivatives of f with respect to $x_2, \ldots, x_n$. If a function is **differentiable** [1] at a point x^0, then the directional derivatives of f at x^0 in all directions y are finite and are given by

$$Df(x^0; y) = y^T \nabla f(x^0), \tag{4.94}$$

where, as before, $\nabla f(x)$ is the gradient of f, evaluated at x. Before stating the first result on directional derivatives of convex functions, the reader might wish to recall that a function f on R^n is said to be positively homogeneous (of degree one) if for every $x \in R^n$ and positive numbers t

$$f(tx) = tf(x). \tag{4.95}$$

We have

Theorem 4.21

Let f be a convex function and let $x \in R^n$ be a point such that $f(x)$ is finite. Then the right- and left-sided derivatives of f at x exist in every direction y. Moreover, D^+f and D^-f are positively homogeneous convex functions of y and

$$D^+f(x; y) \geq D^-f(x; y). \tag{4.96}$$

Proof. Let $x^0 \in R^n$ be any point such that $f(x^0)$ is finite. Define $h(y) = f(x^0 + y) - f(x^0)$. The function h is convex, since $P(h)$ is obtained by translating the convex set $P(f)$ and $h(0) = 0$. Let $t_2 > t_1 > 0$ and $q_1 = t_1/t_2$, $q_2 = (t_2 - t_1)/t_2$. Then $q_1 > 0$, $q_2 > 0$, and $q_1 + q_2 = 1$. Hence

$$h(q_1 t_2 y + q_2 0) = h(t_1 y) \leq q_1 h(t_2 y) + q_2 h(0) \tag{4.97}$$

and

$$\frac{h(t_1 y)}{t_1} \leq \frac{h(t_2 y)}{t_2}. \tag{4.98}$$

The last inequality indicates that $h(ty)/t$ is a nondecreasing function of $t > 0$, and, consequently, the difference quotient appearing in (4.90) is a nondecreasing function of $t > 0$. Thus for a convex function f we can write for every y

$$D^+f(x^0; y) = \inf_{t > 0} \frac{f(x^0 + ty) - f(x^0)}{t} \tag{4.99}$$

and $D^+f(x^0; y)$ exists (although it may not be finite). Since $D^+f(x^0; y)$ exists for every y, it follows from (4.92) that $D^-f(x^0; y)$ also exists for every y. Let λ be a positive number. Then

$$D^+f(x^0; \lambda y) = \lim_{t\lambda \to 0^+} \frac{\lambda[f(x^0 + t\lambda y) - f(x^0)]}{t\lambda} \tag{4.100}$$

$$= \lambda D^+f(x^0; y). \tag{4.101}$$

Hence D^+f is positively homogeneous. The same is true for D^-f. From the homogeneity property it follows that $D^+f(x^0, 0) = D^-f(x^0, 0) = 0$.

Next we show that D^+f is a convex function of y. Let $(y^1, \alpha^1) \in P(D^+f)$, $(y^2, \alpha^2) \in P(D^+f)$. Then

$$D^+f(x^0; y^1) \leq \alpha^1 < +\infty \tag{4.102}$$

$$D^+f(x^0; y^2) \leq \alpha^2 < +\infty. \tag{4.103}$$

By the convexity of f, we have for every q and $t > 0$

$$f(x^0 + t(q_1y^1 + q_2y^2)) = f(\tfrac{1}{2}(x^0 + 2tq_1y^1) + \tfrac{1}{2}(x^0 + 2tq_2y^2)) \tag{4.104}$$

or

$$f(x^0 + t(q_1y^1 + q_2y^2)) \leq \tfrac{1}{2}f(x^0 + 2tq_1y^1) + \tfrac{1}{2}f(x^0 + 2tq_2y^2). \tag{4.105}$$

Note that, by (4.102) and (4.103), the values of f at the right-hand side of (4.105) are not $+\infty$ for sufficiently small t; therefore the undefined expression $\infty + (-\infty)$ is avoided. Hence subtracting $f(x^0)$ from both sides of (4.105), dividing by t, and taking limits as $t \to 0^+$, we obtain

$$D^+f(x^0; q_1y^1 + q_2y^2) \leq D^+f(x^0; q_1y^1) + D^+f(x^0; q_2y^2). \tag{4.106}$$

Multiplying (4.102) and (4.103) by q_1 and q_2, respectively, using the homogeneity of D^+f, and adding the resulting expressions, we finally have

$$D^+f(x^0; q_1y^1 + q_2y^2) \leq q_1\alpha^1 + q_2\alpha^2. \tag{4.107}$$

That is, $(q_1y^1 + q_2y^2, q_1\alpha^1 + q_2\alpha^2) \in P(D^+f)$ as required. By the homogeneity and convexity of D^+f, it follows for every y that

$$D^+f(x^0; y) + D^+f(x^0; -y) \geq D^+f(x^0; 0) = 0 \tag{4.108}$$

and so

$$D^+f(x^0; y) \geq D^-f(x^0; y). \tag{4.109}$$

■

The next concept to be introduced is the subgradient, which is related to the ordinary gradient in the case of differentiable convex functions and to the directional derivatives in the more general case. A **subgradient** of a convex function f at a point $x \in R^n$ is a vector $\xi \in R^n$ satisfying

$$f(y) \geq f(x) + \xi^T(y - x) \tag{4.110}$$

for every $y \in R^n$. For a convex function f it is possible that, at some point x, (a) no vector satisfying (4.110) exists, or (b) there is a unique ξ satisfying (4.110), or (c) there exist more than one such ξ. Following Rockafellar [13], who has thoroughly studied the differential properties of convex functions, we denote by $\partial f(x)$ the set of all subgradients of a convex function f at x. Some basic properties of subgradients can be summarized as follows: The set $\partial f(x)$, also called the **subdifferential** of f, is a closed convex set. It contains a single vector $\xi \in R^n$ if and only if the convex function f is differentiable (in the ordinary sense) at x and then $\xi = \nabla f(x)$. That is,

$$\xi_j = \frac{\partial f(x)}{\partial x_j}, \qquad j = 1, \ldots, n. \tag{4.111}$$

Subgradients can be characterized by the directional derivatives, as will be shown in the next theorem.

Theorem 4.22

A vector $\xi \in R^n$ is a subgradient of a convex function f at a point x where $f(x)$ is finite if and only if

$$D^+f(x; z) \geq \xi^T z \tag{4.112}$$

for every direction z.

Proof. If ξ is a subgradient of f at x, then it satisfies (4.110). Let $y = x + tz$, where t is a positive number. Then

$$f(x + tz) \geq f(x) + t\xi^T z \tag{4.113}$$

for every $z \in R^n$ and $t > 0$.

Dividing both sides of (4.113) by t and rearranging, we get

$$\frac{f(x + tz) - f(x)}{t} \geq \xi^T z \tag{4.114}$$

for every $z \in R^n$ and $t > 0$. The inequality in (4.112) then holds by noting that $D^+f(x; z)$ is the infimum of the difference quotient in (4.114). Conversely, if (4.112) holds for every $z \in R^n$, then (4.113) holds by the same argument as above, and, consequently, (4.110) holds. This, in turn, implies that ξ is a subgradient of f at x. ∎

From the last two theorems we have the following

Corollary 4.23

Let f be a convex function on R^n and suppose that $f(x)$ is finite. Then

$$f(y) \geq f(x) + D^+f(x; y - x) \tag{4.115}$$

for every $y \in R^n$. In particular, if f is differentiable at x, then

$$f(y) \geq f(x) + (y - x)^T \nabla f(x). \tag{4.116}$$

The proof of this corollary is left for the reader.

Corollary 4.24

Let f be a convex function on R^n and suppose that $f(x)$ and $f(y)$ are finite. Then

$$D^+f(y; y - x) \geq D^+f(x; y - x) \tag{4.117}$$

and

$$D^-f(y; y - x) \geq D^-f(x; y - x). \tag{4.118}$$

In particular, if f is differentiable at x and y, then

$$(y - x)^T[\nabla f(y) - \nabla f(x)] \geq 0. \tag{4.119}$$

Proof. From Corollary 4.23 we have

$$f(y) \geq f(x) + D^+f(x; y - x) \tag{4.120}$$

$$f(x) \geq f(y) + D^+f(y; x - y). \tag{4.121}$$

Thus

$$-D^+f(y; x - y) \geq D^+f(x; y - x) \tag{4.122}$$

and by (4.92) and Theorem 4.21,

$$D^+f(y; y - x) \geq D^-f(y; y - x) = -D^+f(y; x - y). \tag{4.123}$$

Hence

$$D^+f(y; y - x) \geq D^+f(x; y - x). \tag{4.124}$$

The proof of the similar result for the left-sided derivatives is identical. ■

Theorem 4.15 showed that the convexity of a function f on R^n is equivalent to the convexity of its restriction to any line segment in R^n. Therefore,

in some cases, it is sufficient to study the behavior of convex functions on the real line R. Such a study is also desirable from another point of view, for often the results are considerably simpler than in the multidimensional case. For example, in the one-dimensional case where f is defined on R, all the right- and left-sided derivatives of f at a point x can be computed from $D^+f(x; 1)$ and $D^-f(x; 1)$, respectively, because of the homogeneity of the derivatives. We shall now show that these derivatives are monotone nondecreasing functions of x.

Theorem 4.25

Let f be a convex function on R and let x^1 and x^2, $x^2 > x^1$ be two points such that $f(x^1)$ and $f(x^2)$ are both finite. Then

$$D^+f(x^2; 1) \geq D^-f(x^2; 1) \geq D^+f(x^1; 1) \geq D^-f(x^1; 1). \tag{4.125}$$

Proof. The first and last inequalities follow from Theorem 4.21. The remaining inequality follows from the relation

$$D^-f(x^2; 1) \geq \frac{f(x^1) - f(x^2)}{x^1 - x^2} = \frac{f(x^2) - f(x^1)}{x^2 - x^1} \geq D^+f(x^1; 1) \tag{4.126}$$

■

Example 4.3.1

To illustrate the notions and results obtained so far, consider the convex function f defined on R by

$$f(x) = \begin{cases} +\infty & x < -1 \\ 2 & x = -1 \\ (x)^2 & -1 < x \leq 0 \\ x & 0 \leq x \leq 1 \\ +\infty & 1 < x. \end{cases} \tag{4.127}$$

Using the definitions of the right- and left-sided derivatives and noting that they can be computed from $D^+f(x; 1)$ and $D^-f(x; 1)$, respectively, we obtain

$$D^+f(x; 1) = \begin{cases} \text{undefined} & x < -1 \\ -\infty & x = -1 \\ 2x & -1 < x < 0 \\ 1 & 0 \leq x < 1 \\ +\infty & x = 1 \\ \text{undefined} & 1 < x \end{cases} \tag{4.128}$$

and

$$D^-f(x;1) = \begin{cases} \text{undefined} & x < -1 \\ -\infty & x = -1 \\ 2x & -1 < x \leq 0 \\ 1 & 0 < x \leq 1 \\ \text{undefined} & 1 < x. \end{cases} \tag{4.129}$$

We can see that $D^+f(x;1) = D^-f(x;1)$ for $-1 \leq x < 0$ and $0 < x < 1$, and $D^+f(x;1) > D^-f(x;1)$ for $x = 0$ and $x = 1$. By Theorem 4.22, the number ξ is a subgradient of f if and only if

$$D^+f(x;z) \geq \xi z \tag{4.130}$$

for every $z \in R$. Since one-sided derivatives are positively homogeneous, we get

$$D^+f(x;z) = \begin{cases} zD^+f(x;1) & z > 0 \\ 0 & z = 0 \\ -zD^+f(x;-1) & z < 0. \end{cases} \tag{4.131}$$

From the last two relations we conclude that $\xi \in \partial f(x)$ if and only if

$$D^+f(x;1) \geq \xi \geq D^-f(x;1). \tag{4.132}$$

Consequently,

$$\partial f(x) = \begin{cases} \emptyset & x \leq -1 \\ 2x & -1 < x < 0 \\ \{\xi : 0 \leq \xi \leq 1\} & x = 0 \\ 1 & 0 < x < 1 \\ \{\xi : \xi \geq 1\} & x = 1 \\ \emptyset & 1 < x. \quad \blacksquare \end{cases} \tag{4.133}$$

In the next few results we restrict our attention to real-valued differentiable convex functions. Most of the following results can be extended to convex functions defined in the wider sense at the beginning of Section 4.2 by the general concepts of one-sided (right and left) derivatives and subgradients. Proofs of such more general results, however, often become rather complicated because of the behavior of the functions involved on the boundary or outside their effective domain. For this reason, we have chosen to present such results in the rest of this section with some loss of generality. They can, however, be extended by the interested reader by using a little caution at "slippery" points.

Theorem 4.26

Let f be a real-valued differentiable function on an open interval $D \subset R$. Then f', the first derivative of f, is a nondecreasing function on D if and only if f is convex on D.

Proof. If f is convex, the result follows from Theorem 4.25. Let $x^1 \in D$, $x^2 \in D$ such that $x^2 > x^1$ and let $x^3 = q_1x^1 + q_2x^2$. By the Mean Value Theorem [1],

$$f(x^2) = f(x^3) + q_1(x^2 - x^1)f'(\tilde{x}), \qquad x^2 \geq \tilde{x} \geq x^3 \tag{4.134}$$

$$f(x^3) = f(x^1) + q_2(x^2 - x^1)f'(\tilde{\tilde{x}}), \qquad x^3 \geq \tilde{\tilde{x}} \geq x^1. \tag{4.135}$$

If f' is nondecreasing on R, we have

$$f(x^3) \leq f(x^1) + q_2(x^2 - x^1)f'(\tilde{x}). \tag{4.136}$$

Multiplying (4.136) and (4.134) by q_1 and $-q_2$, respectively, and adding up, we get

$$q_1f(x^3) - q_2f(x^2) \leq q_1f(x^1) - q_2f(x^3) \tag{4.137}$$

or

$$f(x^3) \leq q_1f(x^1) + q_2f(x^2). \tag{4.138}$$

That is, f is a convex function. ■

Most differential results established so far were necessary conditions for convex functions. The next theorem shows that, in the special case of differentiable convex functions, these conditions are also quite easily established as sufficient ones.

Theorem 4.27

Let f be a real-valued differentiable function on R^n. If

$$f(x^2) \geq f(x^1) + (x^2 - x^1)^T\nabla f(x^1) \tag{4.139}$$

for every two points $x^1 \in R^n$, $x^2 \in R^n$, then f is convex on R^n.

Proof. Let x^1 and x^2 be any two points in R^n and let $x^3 = q_1x^1 + q_2x^2$. Then

$$f(x^1) \geq f(x^3) + q_2(x^1 - x^2)^T\nabla f(x^3) \tag{4.140}$$

$$f(x^2) \geq f(x^3) + q_1(x^2 - x^1)^T\nabla f(x^3). \tag{4.141}$$

Multiplying (4.141) and (4.140) by q_1 and q_2, respectively, and adding up, we get

$$q_1 f(x^1) + q_2 f(x^2) \geq f(x^3) \tag{4.142}$$

and f is convex. ■

Corollary 4.23 and Theorem 4.27 have a simple geometric meaning. A differentiable function f is convex on R^n if and only if the first two terms of the Taylor expansion of f at a point x^0—that is, the linear function

$$f(x^0) + (x - x^0)^T \nabla f(x^0) \tag{4.143}$$

has values less than or equal to $f(x)$ for any $x \in R^n$.

The next theorem deals with the continuity of the first partial derivatives of convex functions. The proof of this theorem would require additional background material and is omitted. The interested reader is referred to Fenchel [8], Rockafellar [13], or Stoer and Witzgall [15].

Theorem 4.28

Let f be a real-valued convex function on an open convex set $C \subset R^n$. If f is differentiable on C, then f has continuous first partial derivatives on C.

We conclude this section with some results involving twice differentiable convex functions.

Theorem 4.29

Let f be a real-valued twice differentiable function on an open interval $D \subset R$. Then f is convex on D if and only if f'', the second derivative of f, evaluated at every $x \in D$ is nonnegative.

Proof. By Theorem 4.26, f is convex on D if and only if f' is nondecreasing; that is, $f''(x) \geq 0$ for every $x \in D$. ■

Now we extend the last result to the multidimensional case.

Theorem 4.30

Let f be a real-valued function on an open convex set $C \subset R^n$ with continuous second partial derivatives. Then f is convex on C if and only if the Hessian of f evaluated at every $x \in C$ is positive semidefinite. That is, for each $x \in C$, we have

$$y^T \nabla^2 f(x) y \geq 0 \tag{4.144}$$

for every $y \in R^n$.

Proof. By Theorem 4.15, the function f is convex on C if and only if for every $x^1 \in C$, $x^2 \in C$

$$\phi(\lambda) = f(\lambda x^1 + (1 - \lambda)x^2) \tag{4.145}$$

is convex for $1 \geq \lambda \geq 0$. By Theorem 4.29, this condition is equivalent to the condition that the function

$$\phi''(\lambda) = \sum_{j=1}^{n} \sum_{k=1}^{n} (x_j^1 - x_j^2)(x_k^1 - x_k^2) \frac{\partial^2 f(\lambda x^1 + (1 - \lambda)x^2)}{\partial x_j \partial x_k} \tag{4.146}$$

is nonnegative for every $x^1 \in C$, $x^2 \in C$ and $0 \leq \lambda \leq 1$. Letting $y = x^1 - x^2$, and since $(\lambda x^1 + (1 - \lambda)x^2) \in C$, we can write (4.146) as

$$\sum_{j=1}^{n} \sum_{k=1}^{n} y_j y_k \frac{\partial^2 f(x)}{\partial x_j \partial x_k} \geq 0 \tag{4.147}$$

for every $y \in R^n$ and $x \in C$. ∎

A word of caution is in order here. Theorem 4.30 cannot be fully sharpened in the case of strictly convex functions by replacing the words "positive semidefinite" in the statement of the theorem by "positive definite." In fact, the reader can easily find examples of strictly convex functions whose Hessians are not positive definite.

The situation is, however, somewhat better in the converse case; that is, under the hypotheses of the theorem a positive definite Hessian matrix does imply strict convexity. For a thorough treatment of these points, the reader is referred to [4].

4.4 EXTREMA OF CONVEX FUNCTIONS

As indicated earlier, convex functions and their generalizations play a central role in nonlinear programming. The main importance of convex functions lies in some basic properties that are summarized below.

Theorem 4.31

Let f be a proper convex function on R^n. Then every local minimum of f is a global minimum of f on R^n.

Proof. If x^* is a local minimum, then

$$f(x) \geq f(x^*) \tag{4.148}$$

for all x in a sufficiently small neighborhood $N_\delta(x^*)$. Let z be any point in R^n. Then $((1-\lambda)x^* + \lambda z) \in N_\delta(x^*)$ for sufficiently small $0 < \lambda < 1$, and

$$f((1-\lambda)x^* + \lambda z) \geq f(x^*). \tag{4.149}$$

Since f is a proper convex function,

$$(1-\lambda)f(x^*) + \lambda f(z) \geq f((1-\lambda)x^* + \lambda z). \tag{4.150}$$

Adding the last two inequalities and dividing the result by λ, we obtain

$$f(z) \geq f(x^*). \tag{4.151}$$

That is, x^* is a global minimum. ■

Suppose that in problem (P), defined in Section 3.1, the functions $g_1, \ldots, g_m$ are all concave functions and the $h_1, \ldots, h_p$ are all linear functions. Then the feasible set X is a convex set. If the objective function f to be minimized is a proper convex function we can define a new objective function $\hat{f}$, given by

$$\hat{f}(x) = \begin{cases} f(x) & \text{if } x \in X \\ +\infty & \text{if } x \notin X, \end{cases} \tag{4.152}$$

and $\hat{f}$ will be a proper convex function on R^n, coinciding with f on X. From the previous theorem we conclude that every local minimum of $\hat{f}$ is also a global minimum, or if X is nonempty, every local minimum of f at some point $x \in X$ is also a global minimum of f on all X. Formally, we have

Theorem 4.32

Let f be a proper convex function on R^n and let $X \subset R^n$ be a convex set. Then every local minimum of f at $x \in X$ is a global minimum of f over all X.

Note that generally the minimal value of a convex function can be attained at more than one point. We shall now show that the set of minimizing points of a proper convex function is a convex set. First we have the following result.

Lemma 4.33

Let f be a convex function on R^n and let α be a real number. Then the level sets of f, given by

$$S(f, \alpha) = \{x : x \in R^n, f(x) \leq \alpha\}, \tag{4.153}$$

are convex sets for any α.

Proof. Suppose that $x^1 \in S(f, \alpha)$ and $x^2 \in S(f, \alpha)$. It follows that

$$f(q_1x^1 + q_2x^2) \leq q_1f(x^1) + q_2f(x^2) \tag{4.154}$$

and

$$\leq q_1\alpha + q_2\alpha = \alpha. \tag{4.155}$$

Hence $S(f, \alpha)$ is convex. ■

Theorem 4.34

Let f be a convex function on R^n. The set of points at which f attains its minimum is convex.

Proof. Let α^* be the value of f at the minimizing points. Then the set $\{x : x \in R^n, f(x) \leq \alpha^*\}$ is precisely the set of points at which f attains its minimum; and by the preceding lemma, this is a convex set. ■

Another result, quite important in some applications, is worth mentioning here.

Corollary 4.35

Let f be a strictly convex function defined on a convex set $X \subset R^n$. If f attains its minimum on X, it is attained at a unique point of X.

Proof. Suppose that the minimum is attained at two distinct points $x^1 \in X$, $x^2 \in X$ and let $f(x^1) = f(x^2) = \alpha$. It follows from Theorem 4.34 that for every q we have $f(q_1x^1 + q_2x^2) = \alpha$, contradicting that f is strictly convex. ■

In many applications, when the minimum of a differentiable function is sought, one looks for the stationary points of the function—that is, points at which the gradient vanishes. This situation can be justified in the case of convex functions by the following

Theorem 4.36

Let f be a convex function. Then $0 \in \partial f(x^*)$ if and only if f attains its minimum at x^*.

Proof. By the definition of subgradients, $0 \in \partial f(x^*)$ if and only if

$$f(y) \geq f(x^*) \tag{4.156}$$

for every $y \in R^n$; that is, x^* is a minimum of f. ■

Corollary 4.37

Let f be a differentiable convex function on R^n. Then

$$\nabla f(x^*) = 0 \tag{4.157}$$

if and only if f attains its minimum at x^*.

Proof. $\partial f(x) = \{\nabla f(x)\}$ if and only if f is differentiable at x. Then the corollary is a direct consequence of the preceding theorem. ■

The last corollary will generally remain valid if we replace R^n by some open convex subset $X \subset R^n$ such that $x^* \in X$. It also indicates that in seeking an unconstrained minimum of a convex function, no second-order conditions need be checked at points where the gradient vanishes. This conclusion, of course, also follows from Theorem 4.30.

4.5 OPTIMALITY CONDITIONS FOR CONVEX PROGRAMS

Here we consider convex programs, a special case of the general nonlinear programming problem (P), introduced in Chapter 3. The optimality conditions derived there become simpler for convex programs. Consider, then, the following nonlinear program, called a **convex program**:

(CP)
$$\min f(x) \tag{4.158}$$

subject to the constraints

$$g_i(x) \geq 0, \qquad i = 1, \ldots, m \tag{4.159}$$

$$h_j(x) = 0, \qquad j = 1, \ldots, p \tag{4.160}$$

where f is a proper convex function on R^n, the g_i are proper concave functions, and the h_j are linear (affine) functions of the form

$$h_j(x) = \sum_{k=1}^{n} a_{jk}x_k - b_j. \tag{4.161}$$

Such a program is called convex because the objective function is a convex function, and the set of all $x \in R^n$ satisfying the inequalities in (4.159) and the equations in (4.160) is a convex set (see Exercise 4.Q). Note that generally the set of $x \in R^n$ satisfying an equation $h(x) = 0$, where h is a nonlinear convex or concave function, is not a convex set.

We shall now show that with an appropriate assumption of differentiability, the Kuhn-Tucker necessary conditions for optimality, as stated in Theorem 3.8, are also sufficient when applied to a convex program.

Theorem 4.38

Suppose that the functions f and $g_1, \ldots, g_m$ are real-valued, differentiable, convex and concave functions on R^n, respectively, and let $h_1, \ldots, h_p$ be linear. If there exist vectors x^*, λ^*, μ^*, with x^* satisfying (4.159) *and* (4.160), *and*

$$\nabla f(x^*) - \sum_{i=1}^{m} \lambda_i^* \nabla g_i(x^*) - \sum_{j=1}^{p} \mu_j^* \nabla h_j(x^*) = 0 \tag{4.162}$$

$$\lambda_i^* g_i(x^*) = 0, \qquad i = 1, \ldots, m \tag{4.163}$$

$$\lambda^* \geq 0, \tag{4.164}$$

then x^* is a global optimum of (CP).

Proof. Let x be any point satisfying (4.159) and (4.160). Then

$$f(x) \geq f(x) - \sum_{i=1}^{m} \lambda_i^* g_i(x) - \sum_{j=1}^{p} \mu_j^* h_j(x). \tag{4.165}$$

Applying Corollary 4.23 to f, g_i, and h_j and using (4.165), we obtain

$$\begin{aligned} f(x) \geq f(x^*) &+ (x - x^*)^T \nabla f(x^*) - \sum_{i=1}^{m} \lambda_i^* g_i(x^*) - \sum_{i=1}^{m} \lambda_i^* (x - x^*)^T \nabla g_i(x^*) \\ &- \sum_{j=1}^{p} \mu_j^* h_j(x^*) - \sum_{j=1}^{p} \mu_j^* (x - x^*)^T \nabla h_j(x^*). \end{aligned} \tag{4.166}$$

Rearranging yields

$$\begin{aligned} f(x) \geq f(x^*) &- \sum_{i=1}^{m} \lambda_i^* g_i(x^*) - \sum_{j=1}^{p} \mu_j^* h_j(x^*) \\ &+ (x - x^*)^T \left[\nabla f(x^*) - \sum_{i=1}^{m} \lambda_i^* \nabla g_i(x^*) - \sum_{j=1}^{p} \mu_j^* \nabla h_j(x^*) \right] \end{aligned} \tag{4.167}$$

and by (4.160), (4.162), and (4.163),

$$f(x) \geq f(x^*). \tag{4.168}$$

■

It is clear from this result that under the same hypotheses the Fritz John conditions, as stated in Theorem 3.4, are also sufficient, provided that $\lambda_0^* > 0$.

The regularity condition (3.71) assumed in the necessary conditions for optimality in the general case can be replaced in convex programs by a simpler and easier computable, although stronger condition, due to Slater [12, 14]. A general mathematical program (P), as defined in Chapter 3, is said to be **strongly consistent** if there exists a point $x^0 \in R^n$ satisfying

$$g_i(x^0) > 0, \qquad i = 1, \ldots, m \tag{4.169}$$

$$h_j(x^0) = 0, \qquad j = 1, \ldots, p. \tag{4.170}$$

Such a program is also said to satisfy Slater's condition. An additional condition that we impose on the constraints of (CP) is that the vectors of coefficients $a^j = (a_{j1}, \ldots, a_{jn})^T$ in the linear functions $h_j(x) = (a^j)^T x - b_j$ be linearly independent. This is not a serious limitation, for if there is a linear dependence between the vectors of coefficients, we can always use some elementary algebraic operations on these vectors, without changing the feasible set of the problem, so that one or more equations will become $0^T x = 0$, satisfied by every $x \in R^n$, where 0 is an n vector, each component of which is 0. In this way, it is possible to eliminate all the linearly dependent constraints.

We can now prove in a rather straightforward way the Kuhn-Tucker [11] necessary conditions for optimality in a convex program (CP).

Theorem 4.39

Suppose that the functions f and $g_1, \ldots, g_m$ are real-valued, differentiable, convex and concave functions on R^n, respectively, and let

$$h_j(x) = \sum_{k=1}^{n} a_{jk} x_k - b_j, \qquad j = 1, \ldots, p \tag{4.171}$$

be such that the vectors $a^j = (a_{j1}, \ldots, a_{jn})$, $j = 1, \ldots, p$, are linearly independent. Suppose that problem (CP) *is strongly consistent. If x^* is a solution of* (CP), *then there exist vectors $\lambda^* = (\lambda_1^*, \ldots, \lambda_m^*)$ and $\mu^* = (\mu_1^*, \ldots, \mu_p^*)$ such that*

$$\nabla f(x^*) - \sum_{i=1}^{m} \lambda_i^* \nabla g_i(x^*) - \sum_{j=1}^{p} \mu_j^* \nabla h_j(x^*) = 0 \tag{4.172}$$

$$\lambda_i^* g_i(x^*) = 0, \qquad i = 1, \ldots, m \tag{4.173}$$

$$\lambda^* \geq 0. \tag{4.174}$$

Proof. Under the hypotheses on f, g_i, and h_j it follows from Theorem 4.28 that the hypotheses of Theorem 3.4 also hold. Hence there exist vectors

$\alpha^* = (\alpha_0^*, \alpha_1^*, \ldots, \alpha_m^*)$ and $\beta^* = (\beta_1^*, \ldots, \beta_p^*)$ such that

$$\alpha_0^* \nabla f(x^*) - \sum_{i=1}^{m} \alpha_i^* \nabla g_i(x^*) - \sum_{j=1}^{p} \beta_j^* \nabla h_j(x^*) = 0 \tag{4.175}$$

$$\alpha_i^* g_i(x^*) = 0, \qquad i = 1, \ldots, m \tag{4.176}$$

$$(\alpha^*, \beta^*) \neq 0, \qquad \alpha^* \geq 0. \tag{4.177}$$

If $\alpha_0^* > 0$, define $\lambda_i^* = \alpha_i^*/\alpha_0^*$, $\mu_j^* = \beta_j^*/\alpha_0^*$ and the theorem is proved. We shall now show that assuming $\alpha_0^* = 0$ leads to contradictions. Suppose that $\alpha_0^* = 0$ and $(\alpha_1^*, \ldots, \alpha_m^*) \neq 0$. Let x^0 satisfy (4.169) and (4.170). Then by the concavity of the g_i,

$$\sum_{i=1}^{m} \alpha_i^* g_i(x^0) \leq \sum_{i=1}^{m} \alpha_i^* g_i(x^*) + \sum_{k=1}^{n} (x_k^0 - x_k^*)\left[\sum_{i=1}^{m} \alpha_i^* \frac{\partial g_i(x^*)}{\partial x_k}\right] \tag{4.178}$$

and by (4.175) and (4.176),

$$\sum_{i=1}^{m} \alpha_i^* g_i(x^0) \leq \sum_{k=1}^{n} (x_k^0 - x_k^*)\left[-\sum_{j=1}^{p} \beta_j^* \frac{\partial h_j(x^*)}{\partial x_k}\right]. \tag{4.179}$$

Hence by the concavity of the (linear) h_j,

$$\sum_{i=1}^{m} \alpha_i^* g_i(x^0) \leq \sum_{j=1}^{p} \beta_j^*[h_j(x^*) - h_j(x^0)] = 0, \tag{4.180}$$

contradicting (4.169), since if $g_i(x^0) > 0$ for $i = 1, \ldots, m$ and $(\alpha_1^*, \ldots, \alpha_m^*) \neq 0$, then

$$\sum_{i=1}^{m} \alpha_i^* g_i(x^0) > 0. \tag{4.181}$$

Suppose now that $\alpha^* = 0$—that is, $\beta^* \neq 0$. Then from (4.175) we have

$$\sum_{j=1}^{p} \beta_j^* \nabla h_j(x^*) = 0 \tag{4.182}$$

or

$$\sum_{j=1}^{p} \beta_j^* a_{jk} = 0, \qquad k = 1, \ldots, n \tag{4.183}$$

implying that the vectors a_j are linearly dependent, a contradiction. ■

We conclude this chapter with a converse of Theorem 3.11 for the case of convex programs. We shall prove that every solution of a convex program is a saddlepoint of the corresponding Lagrangian. A slightly different version of Theorem 4.20 is needed first, however.

Theorem 4.40

Let $g_1, \ldots, g_m$ be proper concave functions and let $h_1, \ldots, h_p$ be linear functions. Let $C \subset R^n$ be a nonempty convex set such that

$$C \subset \bigcap_{i=1}^{m} \text{ED}\,(g_i). \tag{4.184}$$

If there exists no $x \in C$ satisfying

$$g_i(x) > 0, \qquad i = 1, \ldots, m \tag{4.185}$$

$$h_j(x) = 0, \qquad j = 1, \ldots, p \tag{4.186}$$

then there exist numbers $\alpha_1, \ldots, \alpha_m, \beta_1, \ldots, \beta_p$, not all zero, such that $\alpha_i \geq 0$ for $i = 1, \ldots, m$ *and for every $x \in C$*

$$\sum_{i=1}^{m} \alpha_i\, g_i(x) + \sum_{j=1}^{p} \beta_j h_j(x) \leq 0. \tag{4.187}$$

The proof of this theorem is similar to that of Theorem 4.20 and will be left for the reader. Now we have the following generalized Kuhn-Tucker saddlepoint theorem.

Theorem 4.41

Let the functions $f, g_1, \ldots, g_m, h_1, \ldots, h_p$ satisfy the hypotheses of Theorem 4.39, possibly without the differentiability assumption. Suppose that (CP) *is strongly consistent. If x^* is a solution of* (CP), *then there exist vectors $\lambda^* = (\lambda_1^*, \ldots, \lambda_m^*)$ and $\mu^* = (\mu_1^*, \ldots, \mu_p^*)$ such that (x^*, λ^*, μ^*) is a solution of problem* (S), *as defined in Section 3.3—that is, $\lambda^* \geq 0$ and*

$$L(x^*, \lambda, \mu) \leq L(x^*, \lambda^*, \mu^*) \leq L(x, \lambda^*, \mu^*). \tag{4.188}$$

Furthermore,

$$\lambda_i^* g_i(x^*) = 0, \qquad i = 1, \ldots, m. \tag{4.189}$$

Proof. The system

$$f(x^*) - f(x) > 0 \tag{4.190}$$

$$g_i(x) \geq 0, \qquad i = 1, \ldots, m; \qquad h_j(x) = 0, \qquad j = 1, \ldots, p \tag{4.191}$$

has no solution $x \in R^n$. It follows from Theorem 4.40 that there exist numbers $\alpha_0 \geq 0, \alpha_1 \geq 0, \ldots, \alpha_m \geq 0, \beta_1, \ldots, \beta_p$, not all zero, such that

$$\alpha_0(f(x^*) - f(x)) + \sum_{i=1}^{m} \alpha_i g_i(x) + \sum_{j=1}^{p} \beta_j h_j(x) \leq 0 \tag{4.192}$$

for every x, or

$$\alpha_0 f(x^*) \le \alpha_0 f(x) - \sum_{i=1}^{m} \alpha_i g_i(x) - \sum_{j=1}^{p} \beta_j h_j(x). \tag{4.193}$$

Suppose that $\alpha_0 > 0$. Then divide (4.193) by α_0 and let $\lambda_i^* = \alpha_i/\alpha_0$ for $i = 1, \ldots, m$ and $\mu_j^* = \beta_j/\alpha_0$ for $j = 1, \ldots, p$ to obtain

$$f(x^*) \le L(x, \lambda^*, \mu^*) \tag{4.194}$$

for every $x \in R^n$. Since $g_i(x^*) \ge 0$ and $h_j(x^*) = 0$, we get

$$f(x^*) - \sum_{i=1}^{m} \lambda_i g_i(x^*) - \sum_{j=1}^{p} \mu_j h_j(x^*) = L(x^*, \lambda, \mu) \le f(x^*) \tag{4.195}$$

for every $\lambda \in R^m$, $\lambda \ge 0$, and $\mu \in R^p$. Substituting x^*, λ^*, μ^* everywhere in (4.194) and (4.195), we conclude that $f(x^*) = L(x^*, \lambda^*, \mu^*)$. This result, in turn, implies that $\lambda_i^* g_i(x^*) = 0$ for $i = 1, \ldots, m$. The proof of the theorem is thus complete for the case of $\alpha_0 > 0$.

Suppose now that $\alpha_0 = 0$. Then

$$\sum_{i=1}^{m} \alpha_i g_i(x) + \sum_{j=1}^{p} \beta_j h_j(x) \le 0 \tag{4.196}$$

for every $x \in R^n$. Since (CP) is strongly consistent, we must have $\alpha_1 = \alpha_2 = \cdots = \alpha_m = 0$. Then $\beta \ne 0$ and

$$\sum_{k=1}^{n} \left(\sum_{j=1}^{p} \beta_j a_{jk} \right) x_k \le \sum_{j=1}^{p} \beta_j b_j \tag{4.197}$$

for every $x \in R^n$. If $\sum_{j=1}^{p} \beta_j a_{jk_0} \ne 0$ for some k_0, $1 \le k_0 \le p$, we can choose x_{k_0} with $|x_{k_0}|$ sufficiently large and $x_k = 0$, $k \ne k_0$, such that (4.197) is violated. Hence

$$\sum_{j=1}^{p} \beta_j a_{jk} = 0, \qquad k = 1, \ldots, n. \tag{4.198}$$

Since $\beta \ne 0$, the last equation contradicts the assumption that the $a^j = (a_{j1}, \ldots, a_{jn})$ are linearly independent. Hence α_0 cannot vanish. ■

EXERCISES

4.A. Show by an example that the union of convex sets is not necessarily convex. Give examples of convex sets whose complements are also convex.

4.B. Prove Theorem 4.3.

4.C. Let $C \subset R^n$ be a convex set. Show that

$$\bar{X} = \{x : x \in R^n, x = A\rho, \rho \in C\}, \tag{4.199}$$

where A is a given $n \times p$ real matrix, is a convex set in R^n.

4.D. Given a linear (affine) function on R^n

$$h(x) = a^T x + b, \tag{4.200}$$

show that it is both convex and concave.

4.E. Given a continuous real-valued function f on a convex set $C \subset R^n$ such that

$$f(\tfrac{1}{2}x^1 + \tfrac{1}{2}x^2) \leq \tfrac{1}{2}f(x^1) + \tfrac{1}{2}f(x^2) \tag{4.201}$$

for every $x^1 \in C$, $x^2 \in C$, show that f is a convex function.

4.F. Many well-known inequalities can be proven by convexity. Use Theorem 4.9 to derive the famous arithmetic-geometric mean inequality

$$\sum_{i=1}^{n} \alpha_i a_i \geq \prod_{i=1}^{n} (a_i)^{\alpha_i}, \tag{4.202}$$

where the a_i are given positive numbers and the α_i are arbitrary nonnegative weights satisfying

$$\sum_{i=1}^{n} \alpha_i = 1. \tag{4.203}$$

4.G. Prove Theorem 4.10.

4.H. Complete the proof of Theorem 4.15.

4.I. Show that a real-valued, positively homogeneous function on R^n is convex if and only if

$$f(x + y) \leq f(x) + f(y) \tag{4.204}$$

for every $x \in R^n$, $y \in R^n$.

4.J. Let f be a real-valued convex function on R^n. Show that if f has a global maximum at some $x \in R^n$, then $f(x)$ is constant for all $x \in R^n$.

4.K. A point x^0 belonging to a convex set $C \subset R^n$ is called an **extreme point** of C if there are no points $x^1 \in C$, $x^2 \in C$, $x^1 \neq x^2$, such that x^0 can be written as $x^0 = \lambda x^1 + (1 - \lambda)x^2$ with $0 < \lambda < 1$. Prove that a real-valued convex function on R^n attains its maximum on a compact convex set $K \subset R^n$ at some extreme point of K. You can assume the following result: Every compact convex set in R^n is the convex hull of its extreme points [15].

4.L. (a) Show that a proper convex function f on R^n is closed if and only if the set $\{x : x \in R^n, f(x) \leq \alpha\}$ is a closed set for every real α.
(b) Show that cl f is a proper convex function if and only if f is proper convex.

4.M. Let f be a convex function, finite at $x^0 \in R^n$, such that the right- and left-sided derivatives exist at x^0. Show that

$$-D^+ f(x^0, -y) = D^- f(x^0, y) \tag{4.205}$$

for every y.

4.N. Show that if f is a convex function, then $\partial f(x)$ is a closed convex set. Also show that if $\partial f(x^0)$ is nonempty, then $\text{cl}\, f(x^0) = f(x^0)$. Let $f(x) = ||x||$. Show that f is convex and compute $\partial f(x)$ for every $x \in R^n$.

4.O. Prove Corollary 4.23.

4.P. Show that the real-valued function

$$f(x) = \sum_{j=1}^{n} x_j \ln x_j - \sum_{j=1}^{n} x_j \ln \left(\sum_{j=1}^{n} x_j \right) \tag{4.206}$$

is convex on the set

$$C(\alpha) = \{x : x \in R^n, x_1 + \cdots + x_n = \alpha, x > 0, \alpha > 0\}. \tag{4.207}$$

Is it also strictly convex?

4.Q. Show that the feasible set of (CP), that is the set of $x \in R^n$ satisfying (4.159) and (4.160) is a closed convex set.

4.R. Consider the nonlinear program

$$\text{(NLP)} \qquad \min f(x) \tag{4.208}$$

subject to

$$Ax = b \tag{4.209}$$

$$x \geq 0 \tag{4.210}$$

where f is a real-valued continuously differentiable convex function on R^n, A is a given real $m \times n$ matrix with $m < n$, and b is a given n vector. One way of solving this program is by searching for an $x^* \in R^n$ that satisfies the preceding constraints and such that an optimal solution to the linear program

$$\text{(LP)} \qquad \min \bar{f}(y) = y^T \nabla f(x^*) \tag{4.211}$$

subject to

$$Ay = b \tag{4.212}$$

$$y \geq 0 \tag{4.213}$$

is $y = x^*$. Show that if $y = x^*$ solves (LP), then x^* solves (NLP).

REFERENCES

1. BARTLE, R. G., *The Elements of Real Analysis*, 2nd ed., John Wiley & Sons, New York, 1976.
2. BERGE, C., *Topological Spaces*, Oliver & Boyd Ltd., Edinburgh, 1963.
3. BERGE, C., and A. GHOUILA-HOURI, *Programming, Games and Transportation Networks*, Methuen & Co. Ltd., London, 1965.
4. BERNSTEIN, B., and R. A. TOUPIN, "Some Properties of the Hessian Matrix of a Strictly Convex Function," *J. für die Reine und Angew. Math.*, **210**, 67–72 (1962).
5. EGGLESTON, H. G., *Convexity*, Cambridge University Press, Cambridge, England, 1958.
6. FAN, K., I. GLICKSBERG, and A. J. HOFFMAN, "Systems of Inequalities Involving Convex Functions," *Proc. Am. Math. Soc.*, **8**, 617–622 (1957).
7. FARKAS, J., "Über die Theorie der Einfachen Ungleichungen," *J. für die Reine und Angew. Math.*, **124**, 1–27 (1902).
8. FENCHEL, W., *Convex Cones, Sets and Functions*, Mimeographed lecture notes, Princeton University, Princeton, N.J. 1951.
9. HÖLDER, O., "Über einen Mittelwertsatz," *Göttinger Nachrichten*, pp. 38–47 (1889).
10. JENSEN, J. L. W. V., "Sur les Fonctions Convexes et les Inégalités Entre les Valeurs Moyennes," *Acta Math.*, **30**, 175–193 (1906).
11. KUHN, H. W., and A. W. TUCKER, "Nonlinear Programming," in *Proceedings of the Second Berkeley Symposium on Mathematical Statistics and Probability*, J. Neyman (Ed.), University of California Press, Berkeley, Calif., 1951.
12. MANGASARIAN, O. L., *Nonlinear Programming*, McGraw-Hill Book Co., New York, 1969.
13. ROCKAFELLAR, R. T., *Convex Analysis*, Princeton University Press, Princeton, N.J., 1970.
14. SLATER, M., "Lagrange Multipliers Revisited: A Contribution to Nonlinear Programming," Cowles Commission Discussion Paper, Math. 403, November 1950.
15. STOER, J., and C. WITZGALL, *Convexity and Optimization in Finite Dimensions I*, Springer-Verlag, Berlin, 1970.
16. VALENTINE, F. A., *Convex Sets*, McGraw-Hill Book Co., New York, 1964.

5 DUALITY IN NONLINEAR CONVEX PROGRAMMING

In discussing duality for nonlinear programming problems, it is necessary to start with linear programming, where the first and most complete duality results in mathematical programming were obtained. According to Dantzig [5], the notion of duality was first introduced into linear programming by Von Neumann in 1947 and was subsequently formulated in a precise form by Gale, Kuhn, and Tucker [16]. The idea of duality is to associate with each linear program, called the **primal**, another linear program, called the **dual program**. Dual linear programs have some interesting properties that, in addition to being elegant from a theoretical point of view, are also significant for computational purposes and economic interpretations. For a comprehensive study of duality in linear programming, the reader is referred to Dantzig [5] and Kuhn and Tucker [23].

The most important properties of primal-dual linear programs are as follows:

1. If the primal program is a minimization of a linear function over a set of linear constraints, then the dual is a maximization of another linear function over another set of linear constraints. Given a (primal) linear program, its dual can easily be formulated from the data of the primal alone. Primal variables do not appear in the dual program and vice versa. Relations between primal and dual programs are "involutions"; that is, the dual of the dual is again the primal.

2. For all primal and dual feasible solutions, the primal objective function (to be minimized) has a value greater than or equal to the value of the dual objective function (to be maximized). If the primal has an optimal solution, then the dual also has an optimal solution and the respective objective functions have the same value at optimum. By knowing the optimal

solution of either the primal or the dual, it is possible to obtain an optimal solution of the other program without solving it. It is possible, therefore, to solve a linear program via its dual if the dual program is computationally more attractive.

3. Linear programming plays an important role in economic theory. An optimal dual solution can be considered the vector of Lagrange multipliers or "shadow prices" of the primal program and vice versa. Many economic problems can be stated as linear programs. The dual programs in these cases turn out to represent related economic problems. A discussion of the relations between linear programming and economics can be found in [9].

The elegant duality theorems in linear programming, as well as the landmark paper of Kuhn and Tucker [22], have encouraged many researchers to extend the notion of duality to nonlinear programming problems. Early results in this direction were not encouraging. Although it has been shown that one can formulate a corresponding dual program for certain nonlinear programs, there was little similarity to results in linear programming. It quickly became clear that in order to obtain significant results, only **convex** primal programs could be considered. Unfortunately, the dual of such a program, as formulated in the earlier works [10, 21, 24, 36], was usually a nonconvex program, and there was no "involution" between primal and dual programs as in the linear case, where the dual of the dual is the primal.

During the late 1960s a more complete duality theory for convex programs emerged, the mean features of which are similar to those of the primal-dual linear programs listed above. This theory can be derived along different lines. The one chosen here is based on the concept of conjugate functions, introduced by Fenchel [13] and developed to its full strength by Rockafellar in a series of works, some of which are listed in the references at the end of this chapter. Most of Rockafellar's results in convex duality are summarized in [32]. This theory is quite recent and should have a great potential, but its real impact on nonlinear optimization is not yet fully understood. This chapter presents the essentials of duality theory for convex programs.

According to the modern approach, nonlinear convex programs will be formulated together with certain kinds of perturbations in the data of the problem. Such perturbations usually represent small changes in the parameters of the program and are important from both a theoretical and a practical point of view. It is quite important to know, for example, that for a given programming problem in which optimal solutions exist and the minimum of the objective function is finite, a small change in a parameter could cause an infinitely steep decrease in the value of the objective function. This approach will allow us, therefore, to study questions on the sensitivity of optimal solutions to small perturbations in the parameters of the program, in addi-

tion to the more standard questions on existence and characterization of feasible and optimal solutions.

Little can be said about duality in general nonconvex programming, for this subject is at about the same stage as convex duality was in the early 1950s. A few works on the subject exist, such as [27, 28], but results are not very satisfactory. Some special cases will be mentioned in Chapters 6 and 7. Additional references are mentioned at the end of Section 6.2.

5.1 CONJUGATE FUNCTIONS

In the previous chapter we defined the closure of a convex function f on R^n as the pointwise supremum of all linear (affine) functions h with values $h(x) \leq f(x)$. Suppose that $h(x) = \xi^T x - b$. We can ask the following question. Given a vector of coefficients $\xi \in R^n$, for what values of b will $h(x)$ be less than or equal to $f(x)$ for every $x \in R^n$? Clearly, b must satisfy

$$b \geq \xi^T x - f(x) \tag{5.1}$$

for every $x \in R^n$. Hence

$$b \geq \sup_x \{\xi^T x - f(x)\}, \tag{5.2}$$

where, unless explicitly mentioned, the supremum is understood to be taken on all R^n. Generally, the greatest lower bound of b satisfying (5.2) is a function of ξ, called the **conjugate function** of f and denoted f^*. Thus

$$f^*(\xi) = \sup_x \{\xi^T x - f(x)\}. \tag{5.3}$$

The geometric interpretation of conjugate functions can be seen in Figure 5.1. The intercept with the f axis of the highest linear function having a vector

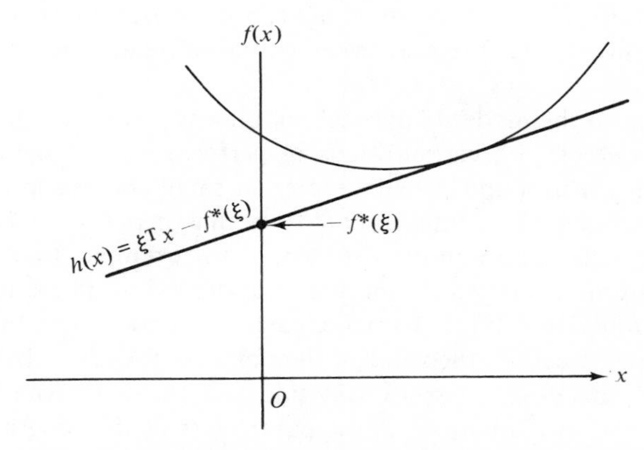

Fig. 5.1 Geometric interpretation of conjugate functions.

of coefficients ξ and lying below the function f is equal to $-f^*(\xi)$. Changing ξ will accordingly raise or lower the value of $-f^*(\xi)$ and shift the point x where the linear function h is tangent to f (if there is such a point). Although conjugate functions may look like a pure mathematical concept, they have sound economic interpretations. Suppose, for example, that the cost of manufacturing quantities $x_1, x_2, \ldots, x_n$ of n goods is given by $f(x)$. The goods can be sold at prices $\xi_1, \xi_2, \ldots, \xi_n$, respectively. The manufacturer's problem is to choose values of $x_1, \ldots, x_n$ such that his profit from selling the goods is maximized. Clearly, his profit is given by the difference between his revenues $\sum_j \xi_j x_j$ and the production cost $f(x)$. The highest possible value of his profit, as a function of the prices, is then given by $f^*(\xi)$, as can be seen from (5.3).

Let us examine some properties of conjugate functions. By (4.36), the epigraph of f^* is

$$P(f^*) = \{(\xi, \alpha) : \xi \in R^n, \alpha \in R, \sup_x \{\xi^T x - f(x)\} \leq \alpha\} \tag{5.4}$$

or

$$P(f^*) = \{(\xi, \alpha) : \xi \in R^n, \alpha \in R, \xi^T x - \alpha \leq f(x), \text{ for every } x \in R^n\}. \tag{5.5}$$

Comparing the last expression with (4.63), we can conclude that $P(f^*)$ is equal to the support set $L(f)$. The reader can easily verify that $P(f^*)$ is a convex set and, consequently, that f^* is a convex function. Let us now compute f^{**}, the conjugate function of f^*. By (5.3),

$$f^{**}(x) = \sup_\xi \{\xi^T x - f^*(\xi)\} \tag{5.6}$$

$$= \sup_{\{(\xi, \alpha^*) : f^*(\xi) \leq \alpha^*\}} \{\xi^T x - \alpha^*\} \tag{5.7}$$

$$= \sup_{\{(\xi, \alpha^*) : \xi^T y - \alpha^* \leq f(y) \text{ for every } y \in R^n\}} \{\xi^T x - \alpha^*\} \tag{5.8}$$

$$= \sup_{(\xi, \alpha^*) \in L(f)} \{\xi^T x - \alpha^*\} \tag{5.9}$$

$$= \operatorname{cl} f(x). \tag{5.10}$$

We can summarize these results in the following

Theorem 5.1

Let f be a convex function on R^n. The conjugate function f^* is also a convex function and $P(f^*) = L(f)$. Moreover, $f^{**} = \operatorname{cl} f$.

It is easy to show that $L(f^*) = P(f^{**}) = P(\operatorname{cl} f)$ and $\operatorname{cl} f^* = (\operatorname{cl} f)^*$. Since $\operatorname{cl} f(x) \leq f(x)$ for every $x \in R^n$, we have

$$\xi^T x - \operatorname{cl} f(x) \geq \xi^T x - f(x) \tag{5.11}$$

and $(\text{cl } f)^* = \text{cl } f^* \geq f^*$. But we also have that $\text{cl } f^* \leq f^*$. Hence $\text{cl } f^* = f^*$ and f^* is actually a closed convex function.

The only closed improper convex functions are either $f(x) = +\infty$ for all $x \in R^n$ or $f(x) = -\infty$ for all x. In the first case, $f^*(\xi) = -\infty$ for all $\xi \in R^n$, whereas in the second case $f^*(\xi) = +\infty$ for all ξ. We have thus established

Theorem 5.2

The conjugate function f^* is a closed convex function. It is proper if and only if f is proper.

The following elementary properties of conjugate functions can easily be demonstrated by the reader. Let f be a convex function on R^n, α a real number, and z a vector in R^n.

(i) If $\phi(x) = f(x) + \alpha$, then $\phi^*(\xi) = f^*(\xi) - \alpha$. (5.12)

(ii) If $\phi(x) = f(x + z)$, then $\phi^*(\xi) = f^*(\xi) - \xi^T z$. (5.13)

(iii) If $\phi(x) = f(\alpha x)$, $\alpha \neq 0$, then $\phi^*(\xi) = f^*\left(\frac{\xi}{\alpha}\right)\cdot$ (5.14)

(iv) If $\phi(x) = \alpha f(x)$, $\alpha > 0$, then $\phi^*(\xi) = \alpha f^*\left(\frac{\xi}{\alpha}\right)$ (5.15)

Example 5.1.1

Consider a few simple pairs of conjugate functions.

(a) Let $\phi(x) = a^T x - b$, $x \in R^n$. Compute first the conjugate function of $f(x) = a^T x$:

$$f^*(\xi) = \sup_x \{\xi^T x - a^T x\} = \begin{cases} 0 & \text{if } \xi = a \\ +\infty & \text{if } \xi \neq a. \end{cases} \tag{5.16}$$

Hence, by (5.12), $\phi^*(\xi) = b$ if $\xi = a$ and $\phi^*(\xi) = +\infty$ otherwise.

(b) Similar to (a) above, we have the following pair. Let $\phi(x) = 0$ if $x \in C \subset R^n$ and $\phi(x) = +\infty$ otherwise, where C is a given convex set. This convex function is called the **indicator function** of the set C. Its conjugate is

$$\phi^*(\xi) = \sup_{x \in C} \xi^T x \tag{5.17}$$

and ϕ^* is called the **support function** of C.

Note that the indicator function of the empty set is identically $+\infty$ and its conjugate, the support function, is identically $-\infty$. Also note that the

support function is positively homogeneous. It is proper if and only if C is a nonempty set. The reader will be asked to show that, conversely, given a positively homogeneous proper convex function, its conjugate is always an indicator function.

(c) Let $\phi(x) = \frac{1}{2}x^T x$, $x \in R^n$, with $f(x) = x^T x$. Then

$$f^*(\xi) = \sup_x \{\xi^T x - x^T x\}. \tag{5.18}$$

We can see by simple differentiation that the supremum is achieved for $x = \frac{1}{2}\xi$. Thus $f^*(\xi) = \frac{1}{4}\xi^T\xi$ and, by (5.15), $\phi^*(\xi) = \frac{1}{2}\xi^T\xi$.

(d) Let $x \in R$ and let $\phi(x) = -\log x$ if $x > 0$ and $\phi(x) = +\infty$ otherwise. Then $\phi^*(\xi) = \sup_{x>0} \{\xi x + \log x\}$. For $\xi \geq 0$ we have $\phi^*(\xi) = +\infty$, since $\xi x + \log x$ can be made arbitrarily large by choosing $x \longrightarrow +\infty$. For $\xi < 0$ we can find the supremum by differentiation and obtain $\phi^*(\xi) = -1 - \log(-\xi)$. Thus

$$\phi^*(\xi) = \begin{cases} -1 - \log(-\xi) & \text{if } \xi < 0 \\ +\infty & \text{if } \xi \geq 0. \end{cases} \tag{5.19}$$

The reader is urged to find the conjugates of the conjugate functions in the preceding four cases. ▌

Conjugate functions can also be regarded as an extension of the classical Legendre transformation [4, 17]. Suppose that f is a closed convex function with effective domain ED $(f) \subset R^n$ and that it is differentiable on C, the nonempty interior of ED (f). Further suppose that ∂f, the subdifferential of f is a one-to-one mapping. This is equivalent to the condition that f is strictly convex and become infinitely steep near boundary points of its effective domain [30].

Under the foregoing assumptions on f, at every point $x \in C$ the subdifferential ∂f contains a single vector $\xi = \nabla f(x)$. If ∇f is one to one, then it has an inverse function ∇f^{-1} whose domain is D, the range of ∇f—that is, the image of C under the gradient mapping. The function ∇f^{-1} is also one to one [2]. Define

$$H(\xi) = \xi^T \nabla f^{-1}(\xi) - f(\nabla f^{-1}(\xi)). \tag{5.20}$$

Then H is also a differentiable convex function on D and $\nabla H = \nabla f^{-1}$. The transformation from the pair $(C, f(x))$ to the pair $(D, H(\xi))$ is called the **Legendre transformation.**

Let $\xi \in D \subset R^n$. Then by (4.110),

$$\xi^T x - f(x) \geq \xi^T y - f(y) \tag{5.21}$$

for every $y \in R^n$. Thus

$$\xi^T x - f(x) = \sup_y \{\xi^T y - f(y)\} = f^*(\xi). \tag{5.22}$$

Since $x = \nabla f^{-1}(\xi)$, we conclude that $H(\xi)$ coincides with $f^*(\xi)$ on D. Under the same assumptions on ∇f and in the special case where $C = R^n$, f is a strictly convex function and

$$\lim_{\lambda \to \infty} \frac{f(\lambda x)}{\lambda} = +\infty \tag{5.23}$$

for every $x \neq 0$. Then $D = R^n$ and $H(\xi) = f^*(\xi)$ for every $\xi \in R^n$ [30].

Conjugate functions are, therefore, an extension of the Legendre transformation for convex functions that have values in the extended reals and are not necessarily continuous or differentiable.

There is also a close relationship between subgradients and conjugate functions, as can be seen in the following results.

Theorem 5.3

Let f be a convex function on R^n. Then $\xi \in \partial f(x)$ if and only if $f^*(\xi) = \xi^T x - f(x)$.

Proof. By (4.110), $\xi \in \partial f(x)$ if and only if (5.21), and hence (5.22), holds. ■

Theorem 5.4

Let f be a convex function on R^n. If $\xi \in \partial f(x)$, then $x \in \partial f^*(\xi)$. If f is also closed at x and $x \in \partial f^*(\xi)$, then $\xi \in \partial f(x)$.

Proof. By the previous theorem, $\xi \in \partial f(x)$ if and only if $f^*(\xi) = \xi^T x - f(x)$ and $x \in \partial f^*(\xi)$ if and only if $f^{**}(x) = \xi^T x - f^*(\xi)$. The reader was asked to show in Exercise 4.N. that if $\partial f(\hat{x})$ is nonempty at some $\hat{x}$, then $\text{cl } f(\hat{x}) = f(\hat{x})$. Hence $\xi \in \partial f(x)$ implies

$$\text{cl } f(x) = f^{**}(x) = f(x) = \xi^T x - f^*(\xi) \tag{5.24}$$

and $x \in \partial f^*(\xi)$. Conversely, let f be closed and suppose that $x \in \partial f^*(\xi)$. Then (5.24) holds—that is,

$$f^*(\xi) = \xi^T x - f(x) \tag{5.25}$$

and $\xi \in \partial f(x)$. ■

Let us briefly discuss the conjugates of **concave** functions. Generally speaking, every result in conjugate function theory for convex functions can

be similarly developed for concave functions, remembering that a function g is concave if and only if $f = -g$ is convex. In particular, the conjugate function g^* of a concave function g is defined as

$$g^*(\xi) = \inf_x \{\xi^T x - g(x)\}. \tag{5.26}$$

Here g^* is a closed concave function and its conjugate is, in turn, cl g, with the closure operation defined in analogy to the convex case. This g^* is, however, not the same as $-f^*$, where $f = -g$.

In fact,

$$f^*(\xi) = \sup_x \{\xi^T x - f(x)\} \tag{5.27}$$

$$= \sup_x \{-[-\xi^T x - g(x)]\} \tag{5.28}$$

$$= -\inf_x \{-\xi^T x - g(x)\} = -g^*(-\xi). \tag{5.29}$$

Finally, we note that conjugate functions can be defined as above for functions that are neither convex nor concave. Suppose that f is an arbitrary function defined on R^n with values in the extended reals. Then its **convex conjugate** f^* is defined by (5.3) and it is a closed convex function. It is the conjugate of the closure of the greatest convex function f_c satisfying $f_c \leq f$. In other words, f^* is the conjugate of the closure of the convex function that is the lower envelope of the convex hull of $P(f)$ (see Theorem 4.14). For such an arbitrary function, we can define a concave conjugate in an obvious analogy.

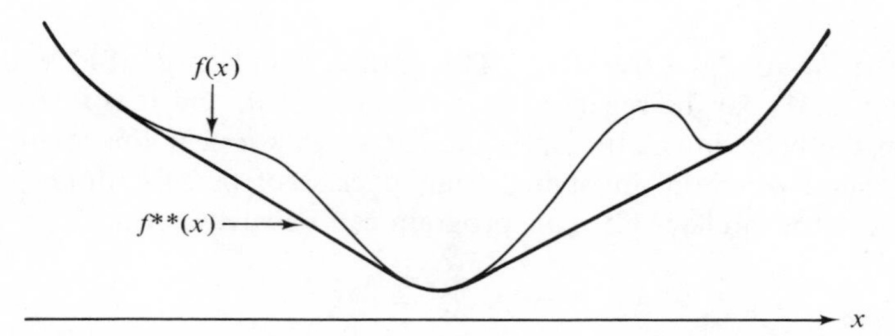

Fig. 5.2 The conjugate of the convex conjugate of a nonconvex function.

Just as conjugates of convex and concave functions play an important role in duality theorems of nonlinear convex programming, as will be demonstrated in the next section, some extensions of duality relations to special types of nonconvex programming problems have also been obtained by conjugates of certain nonconvex functions (to be discussed in the next two chapters).

5.2 DUAL CONVEX PROGRAMS

As noted earlier, we shall develop duality relations for nonlinear convex programming problems in a somewhat broad sense that will give us insight into the sensitivity of the optima to certain perturbations in the parameters of the programs. We deal in this section with the following version of (CP), introduced in the previous chapter:

$$\text{(SP)} \qquad \min_x f(x) \tag{5.30}$$

subject to

$$g_i(x) \geq 0, \qquad i = 1, \ldots, m \tag{5.31}$$

where f is a proper convex function on R^n and the g_i are proper concave functions on R^n. We call program (SP) the **standard primal** convex programming problem. This program is called the "canonical" primal problem by Geoffrion [18] when f and g_i are real valued and the "ordinary" convex program by Rockafellar [32]. It differs from the (CP) formulation of Chapter 4 in assuming either that no linear equality constraints are present or that they appear as two inequality constraints. The perturbations mentioned earlier can be introduced, for example, at the right-hand side of the constraints. That is, the inequalities in (5.31) can be replaced by the more general constraints

$$g_i(x) \geq u_i, \qquad i = 1, \ldots, m \tag{5.32}$$

where the u_i are real parameters. The optimal solution of (SP) with (5.32) replacing (5.31) can be regarded as a function of u, and it is interesting to examine the behavior of the optimum for vectors u near the origin. A still more general program, involving a larger class of perturbations, was also studied by Rockafellar [32]. This program can be written as

$$\text{(QP)} \qquad \min f(x - y^0) \tag{5.33}$$

subject to

$$g_i(x - y^i) \geq u_i, \qquad i = 1, \ldots, m. \tag{5.34}$$

Here the $y^i \in R^n$, $i = 0, 1, \ldots, m$, are the perturbations by which the functions f and g_i are translated, and the u_i are, as before, the perturbations of the constraint right-hand sides. Program (SP) is obviously a special case of (QP) with $y^i = 0$ and $u_i = 0$.

Denoting by C the effective domain of f, we then define a new convex function ϕ on R^n by the relation

$$\phi(x) = \begin{cases} f(x) & \text{if } x \in C \text{ and } g_i(x) \geq 0,\ i = 1, \ldots, m \\ +\infty & \text{otherwise} \end{cases} \tag{5.35}$$

and (SP) can be equivalently written as the (unconstrained) minimization problem of ϕ. If we denote some vector of perturbations by $w \in R^k$, we can generalize (5.35) to

$$\phi(x, w) = \begin{cases} f(x, w) & \text{if } (x, w) \in \text{ED}(f) \subset R^{n+k} \text{ and } g_i(x, w) \geq 0, \\ & \qquad i = 1, \ldots, m \\ +\infty & \text{otherwise} \end{cases} \tag{5.36}$$

and by a proper choice of the form in which w is incorporated in the functions involved, (SP) can become the problem of minimizing $\phi(x, 0)$ over all $x \in R^n$. The formulation given in (5.36) motivates the forthcoming general analysis, which is based on the works of Geoffrion [18], Hamala [19] and Rockafellar [31, 32].

Let ϕ be a proper convex function of two vectors $x \in R^n$ and $w \in R^k$. The conjugate function of ϕ is defined as

$$\phi^*(\xi, \lambda) = \sup_{x, w} \{\xi^T x + \lambda^T w - \phi(x, w)\}. \tag{5.37}$$

Hence for every $x \in R^n$, $\xi \in R^n$, $w \in R^k$, and $\lambda \in R^k$, we have

$$\phi(x, w) + \phi^*(\xi, \lambda) \geq \xi^T x + \lambda^T w. \tag{5.38}$$

In particular, we can set $w = 0$, $\xi = 0$, and for every $x \in R^n$ and $\lambda \in R^k$, we obtain

$$\phi(x, 0) + \phi^*(0, \lambda) \geq 0. \tag{5.39}$$

As a result of the last relation, we define the following pair of convex programs:

(P_ϕ) $\qquad \min_x \phi(x, 0) \qquad$ (primal program) $\qquad$ (5.40)

(D_{ϕ^*}) $\qquad \max_\lambda -\phi^*(0, \lambda) \qquad$ (dual program). $\qquad$ (5.41)

It should be pointed out that in these two programs we are looking for those x and λ that give a minimal value of $\phi(x, 0)$ and a maximal value of $-\phi^*(0, \lambda)$, respectively; but in many cases no such x and λ exist, although

$\phi(x, 0)$ may have a finite infimum and $-\phi^*(0, \lambda)$ a finite supremum. We shall call the infimum of $\phi(x, 0)$ and the supremum of $-\phi^*(0, \lambda)$ the **optimal values** of (P_ϕ) and (D_{ϕ^*}), respectively. The following result is a straightforward consequence of (5.39).

Theorem 5.5 (Weak Duality Theorem)

$$\inf_x \phi(x, 0) \geq \sup_\lambda - \phi^*(0, \lambda). \tag{5.42}$$

Corollary 5.6

If

$$\phi(x^*, 0) = -\phi^*(0, \lambda^*), \tag{5.43}$$

then x^* and λ^* are optimal solutions of (P_ϕ) and (D_{ϕ^*}), respectively.

Program (P_ϕ) is said to be **consistent** if there exists an $\hat{x}$ such that $\phi(\hat{x}, 0) < +\infty$ or, equivalently, if $(\hat{x}, 0) \in \text{ED}(\phi)$. Similarly, (D_{ϕ^*}) is consistent if there exists a $\hat{\lambda}$ such that $\phi^*(0, \hat{\lambda}) < +\infty$. If for every $x \in R^n$, $(x, 0) \notin \text{ED}(\phi)$, then (P_ϕ) is said to be **inconsistent**. If $(0, \lambda) \notin \text{ED}(\phi^*)$ for every $\lambda \in R^k$, then (D_{ϕ^*}) is inconsistent. Using this terminology, we obtain from Theorem 5.5 the following result.

Corollary 5.7

If

$$\inf_x \phi(x, 0) = -\infty, \tag{5.44}$$

then (D_{ϕ^*}) is inconsistent. If

$$\sup_\lambda - \phi^*(0, \lambda) = +\infty, \tag{5.45}$$

then (P_ϕ) is inconsistent.

Convexity plays a central role in the forthcoming results, but it is interesting to note here that relations (5.37) to (5.43) also hold for functions ϕ that are not necessarily convex, provided that the undefined sum $+\infty + (-\infty)$ is avoided. This condition is ensured, for example, if $\phi(x, 0)$ has finite values on R^n. Any feasible solution λ of a dual program provides a value of $-\phi^*(0, \lambda)$ that can serve as a lower bound on the optimal value of (P_ϕ). It would be desirable, however, to have a stronger result than (5.42), one stating that the optimal values of (P_ϕ) and (D_{ϕ^*}) are actually equal. Here is one of the main differences between convex and nonconvex programs: for a convex ϕ, under suitable regularity assumptions, (5.42) does hold as an equality, whereas for ϕ nonconvex one can get a **duality gap**, which is the difference

between the optimal values of (P_ϕ) and (D_{ϕ^*}) when (5.42) holds as a strict inequality.

Suppose now that ϕ is a closed proper convex function on R^{n+k}. Then $\phi^{**} = \phi$, and by reformulating (D_{ϕ^*}) as a primal program

$$(\mathrm{P}_{\phi^*}) \qquad \min_{\lambda} \phi^*(0, \lambda), \tag{5.46}$$

we get a dual program

$$(\mathrm{D}_\phi) \qquad \max_{x} - \phi^{**}(x, 0) = \max_{x} - \phi(x, 0), \tag{5.47}$$

which is equivalent to the original primal program (5.40) in the sense that the vector x is optimal for (P_ϕ) if and only if it is optimal for (D_ϕ). Thus the duality relations for closed proper convex functions are symmetric: the dual of (D_{ϕ^*}) is (P_ϕ).

Associated with ϕ and ϕ^*, we define the primal and dual **perturbation functions** Φ and Ψ as follows:

$$\Phi(w) = \inf_{x} \phi(x, w) \tag{5.48}$$

$$\Psi(\xi) = \inf_{\lambda} \phi^*(\xi, \lambda). \tag{5.49}$$

Here is an important property of perturbation functions:

Theorem 5.8

Let ϕ be a proper convex function on R^{n+k}. Then the perturbation function Φ is convex on R^k.

Proof. We must show that $P(\Phi) \subset R^{k+1}$ is a convex set. Since ϕ is proper, $P(\phi)$ is nonempty and $P(\Phi)$ is also nonempty. Let $(w^1, \alpha^1) \in P(\Phi)$, $(w^2, \alpha^2) \in P(\Phi)$. It follows that for every $\epsilon > 0$ we can find vectors x^1, x^2 such that

$$\phi(x^1, w^1) < \inf_{x} \phi(x, w^1) + \epsilon = \Phi(w^1) + \epsilon \leq \alpha^1 + \epsilon \tag{5.50}$$

$$\phi(x^2, w^2) < \inf_{x} \phi(x, w^2) + \epsilon = \Phi(w^2) + \epsilon \leq \alpha^2 + \epsilon. \tag{5.51}$$

Since $P(\phi)$ is a convex set, we have

$$\phi(q_1 x^1 + q_2 x^2, q_1 w^1 + q_2 w^2) < q_1 \alpha^1 + q_2 \alpha^2 + \epsilon \tag{5.52}$$

for every $q_1 \geq 0, q_2 \geq 0, q_1 + q_2 = 1$, and $\epsilon > 0$. Hence

$$\Phi(q_1 w^1 + q_2 w^2) = \inf_{x} \phi(x, q_1 w^1 + q_2 w^2) \leq \phi(q_1 x^1 + q_2 x^2, q_1 w^1 + q_2 w^2). \tag{5.53}$$

Combining (5.52) and (5.53), we conclude that $\Phi(q_1 w^1 + q_2 w^2) \leq q_1 \alpha^1 + q_2 \alpha^2$ and $P(\Phi)$ is a convex set. ■

It is easy to show that the effective domain of Φ is given by

$$\text{ED}(\Phi) = \{w : w \in R^k, \phi(x, w) < +\infty \text{ for some } x\}. \tag{5.54}$$

Note also that $\Psi(\xi)$ is convex by arguments similar to the preceding ones.

Subgradients and directional derivatives of convex functions, defined in Chapter 4, play an important role in duality relations. We begin with a criterion for the existence of subgradients, which will be useful in the sequel.

Lemma 5.9

Let f be a convex function and let $x \in R^n$ be a point where $f(x)$ is finite. Then f has a subgradient at x; that is, $\partial f(x) \neq \emptyset$ if and only if

$$D^+ f(x\,; y) > -\infty \tag{5.55}$$

for every $y \in R^n$.

Proof. We shall actually prove the contrapositive—that is, $\partial f(x) = \emptyset$ if and only if

$$D^+ f(x\,; y) = D^- f(x\,; y) = -\infty \tag{5.56}$$

for some $y \in R^n$.

Let $\hat{x}$ be a point such that $f(\hat{x})$ is finite. By Theorem 4.21, the function $d(y) = D^+ f(\hat{x}\,; y)$ is a positively homogeneous convex function. Its conjugate d^* is a closed convex indicator function (see Exercise 5.C). Thus there exists a convex set E such that

$$d^*(\eta) = \begin{cases} 0 & \text{if } \eta \in E \\ +\infty & \text{if } \eta \notin E. \end{cases} \tag{5.57}$$

For every point $\eta \in E$ we have

$$d^*(\eta) = \sup_y \{\eta^T y - d(y)\} = 0. \tag{5.58}$$

Hence the set E is given by

$$E = \{\eta : D^+ f(\hat{x}\,; y) \geq \eta^T y \text{ for every } y\}. \tag{5.59}$$

By Theorem 4.22, we conclude that the set E is the set of all the subgradients of f at $\hat{x}$; that is, $E = \partial f(\hat{x})$. The conjugate of d^* is the support function of $\partial f(\hat{x})$, and it is $d^{**} = \text{cl}\, d = \text{cl}\, D^+ f(x\,; y)$. The indicator func-

tion of the empty set is identically $+\infty$. Its conjugate, the support function, is identically $-\infty$. Thus $\partial f(\hat{x}) = \emptyset$ if and only if cl $D^+f(x; y) = -\infty$ for every y. That is, $D^+f(\hat{x}; y)$ must have the value $-\infty$ for some y. ■

Consider the convex perturbation function Φ. Primal program (P_ϕ) is said to be **stable** if either (a) $\Phi(0)$ is finite and there is no direction y such that $D^+\Phi(0; y) = -\infty$ or (b) $\Phi(0) = -\infty$. In other words, (P_ϕ) is stable either if $\inf_x \phi(x, 0)$ is finite and the perturbation function $\Phi(w)$ does not decrease infinitely steeply for any w in the neighborhood of 0 or if the optimal value of (P_ϕ) is unbounded from below. As a result of Lemma 5.9, we can give an alternative characterization of stability: Program (P_ϕ) is stable if and only if $\partial\Phi(0)$ is a nonempty set. Note that if $\Phi(0) = -\infty$, then $\partial\Phi(0) = R^k$. Stability in nonlinear programming was also studied by Evans and Gould [12] without convexity assumptions and by Dantzig, Folkman, and Shapiro [7] for certain classes of convex programs.

Example 5.2.1

Let us consider some examples of stable and unstable programs.

(a) Let

$$\phi_1(x, w) = \begin{cases} x & \text{if } x \geq w \\ +\infty & \text{if } x < w \end{cases} \tag{5.60}$$

where $x \in R$, $w \in R$. It is easy to show that

$$\Phi_1(w) = \inf_x \phi_1(x, w) = w \tag{5.61}$$

for every $w \in R$ and $\Phi_1(0) = 0$. Hence Φ is finite at 0. Indeed, $\partial\Phi_1(0)$ consists of a single point

$$\partial\Phi_1(0) = \nabla\Phi_1(0) = 1 \tag{5.62}$$

and (P_{ϕ_1}) is stable.

(b) Let

$$\phi_2(x, w) = \begin{cases} -x & \text{if } x \geq w \\ +\infty & \text{if } x < w. \end{cases} \tag{5.63}$$

Now we get

$$\Phi_2(w) = \inf_x \phi_2(x, w) = -\infty \tag{5.64}$$

for every $w \in R$. Hence $\Phi_1(0) = -\infty$ and (P_{ϕ_2}) is again stable.

(c) Next take

$$\phi_3(x, w) = \begin{cases} x & \text{if } (x)^2 \leq w \\ +\infty & \text{if } (x)^2 > w. \end{cases} \tag{5.65}$$

Then

$$\Phi_3(w) = \begin{cases} -\sqrt{w} & \text{if } w \geq 0 \\ +\infty & \text{if } w < 0. \end{cases} \tag{5.66}$$

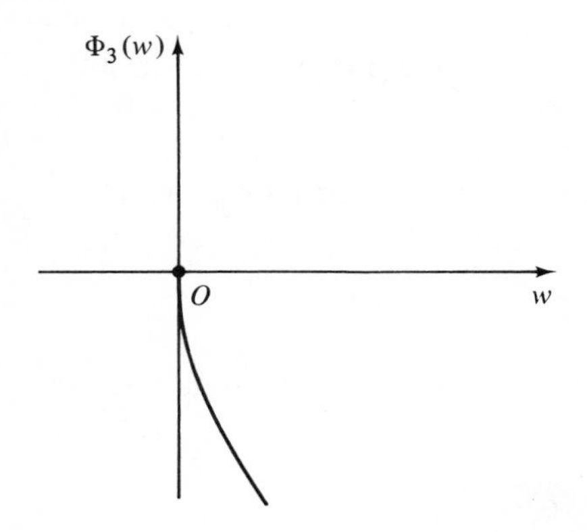

Fig. 5.3 Perturbation function for an unstable program.

Here Φ_3 is a closed proper convex function, as can be seen in Figure 5.3. At $w = 0$, we get $\Phi_3(0) = 0$, but

$$D^+\Phi_3(0; 1) = \lim_{w\to 0^+} \frac{\Phi_3(w) - \Phi_3(0)}{w} = \lim_{w\to 0^+} \frac{-\sqrt{w}}{w} = -\infty. \tag{5.67}$$

Thus (P_{ϕ_3}) is unstable. ∎

Stability is a condition on (P_ϕ) that we shall use in the next theorem in order to establish equality for the optimal values of (P_ϕ) and (D_{ϕ^*}), the existence of an optimal solution for the dual program (D_{ϕ^*}), and a characterization of such a solution in terms of subgradients of Φ. We need first the following

Lemma 5.10

Let $\phi(x, w)$ be a proper convex function on R^{n+k} and define $\Phi(w)$ as in (5.48). *Then*

$$\phi^*(0, \lambda) = \Phi^*(\lambda). \tag{5.68}$$

Proof. We have

$$\phi^*(0, \lambda) = \sup_{x,w} (0^T x + \lambda^T w - \phi(x, w)\} \tag{5.69}$$

$$= \sup_{w} \{\lambda^T w - \inf_{x} \phi(x, w)\} \tag{5.70}$$

$$= \sup_{w} \{\lambda^T w - \Phi(w)\} = \Phi^*(\lambda). \tag{5.71}$$

■

We now state and prove a strong duality theorem based on stability.

Theorem 5.11 (Strong Duality Theorem)

Let ϕ be a proper convex function on R^{n+k} and let $\Phi(0)$ be finite. Then program (D_{ϕ^*}) has an optimal solution λ^*, and

$$\Phi(0) = \inf_{x} \phi(x, 0) = \max_{\lambda} - \phi^*(0, \lambda) = -\phi^*(0, \lambda^*) \tag{5.72}$$

if and only if (P_ϕ) is stable. Moreover, λ^* is a subgradient of $\Phi(0)$ if and only if (5.72) holds.

Proof. Program (P_ϕ) is stable if and only if $\partial\Phi(0)$ is nonempty. By the definition of subgradients, $\lambda^* \in \partial\Phi(0)$ if and only if

$$\Phi(w) \geq \Phi(0) + \lambda^{*T} w \tag{5.73}$$

for every $w \in R^k$. Hence

$$\Phi(0) \geq \inf_{w} \{\Phi(w) - \lambda^{*T} w\} = -\Phi^*(\lambda^*) \geq \Phi(0). \tag{5.74}$$

From Lemma 5.10 it follows that

$$-\Phi^*(\lambda^*) = -\phi^*(0, \lambda^*) = \Phi(0) = \inf_{x} \phi(x, 0); \tag{5.75}$$

and by Theorem 5.5 we conclude that

$$-\phi^*(0, \lambda^*) = \max_{\lambda} - \phi^*(0, \lambda) = \inf_{x} \phi(x, 0). \tag{5.76}$$

■

In complete analogy with the definition of stability for (P_ϕ), we can also define stability for (D_{ϕ^*}): Program (D_{ϕ^*}) is stable if and only if $\partial\Psi(0) \neq \emptyset$. If we assume that $\phi(x, w)$ is a closed proper convex function, we can obtain dual results analogous to Lemma 5.10 and Theorem 5.11.

Lemma 5.12

Let $\phi(x, w)$ be a closed proper convex function on R^{n+k}. Then

$$\phi(x, 0) = \Psi^*(x). \tag{5.77}$$

Proof. Since ϕ is closed and convex,

$$\phi(x, 0) = \phi^{**}(x, 0) = \sup_{\xi, \lambda} \{\xi^T x + \lambda^T 0 - \phi^*(\xi, \lambda)\} \tag{5.78}$$

$$= \sup_{\xi} \{\xi^T x - \inf_{\lambda} \phi^*(\xi, \lambda)\} \tag{5.79}$$

$$= \sup_{\xi} \{\xi^T x - \Psi(\xi)\} = \Psi^*(x). \tag{5.80}$$

■

Thus we have the following dual result.

Theorem 5.13

Let ϕ be a closed proper convex function and let $\Psi(0)$ be finite. Then program (P_ϕ) has an optimal solution x^* and

$$\phi(x^*, 0) = \min_x \phi(x, 0) = \sup_{\lambda} - \phi^*(0, \lambda) \tag{5.81}$$

if and only if (D_{ϕ^*}) is stable. Moreover, x^* is a subgradient of $\Psi(0)$ if and only if (5.81) holds.

Proof. Analogous to the proof of Theorem 5.11. ■

The next two corollaries deal with the existence of optimal solutions of (P_ϕ) and (D_{ϕ^*}) in case ϕ is a closed proper convex function.

Corollary 5.14

Let ϕ be a closed proper convex function on R^{n+k} and let $\Phi(0)$ be finite. If (P_ϕ) and (D_{ϕ^*}) are both stable, then both have optimal solutions and

$$+\infty > \min_x \phi(x, 0) = \max - \phi^*(0, \lambda) > -\infty. \tag{5.82}$$

Proof. By Theorems 5.11 and 5.13, we have that if both (P_ϕ) and (D_{ϕ^*}) are stable, then

$$\min_x \phi(x, 0) = \max_{\lambda} - \phi^*(0, \lambda). \tag{5.83}$$

Since ϕ is proper, the minimum value of $\phi(x, 0)$ attained cannot be $-\infty$, and, similarly, the maximum value of $-\phi^*(0, \lambda)$ cannot be $+\infty$. ■

Corollary 5.15

Let ϕ be a closed proper convex function. Then (P_ϕ) is stable and has an optimal solution x^* if and only if (D_{ϕ^*}) is stable and has an optimal solution λ^*.

Proof. By Theorems 5.11 and 5.13, program (P_ϕ) is stable and has an optimal solution x^* if and only if

$$\phi(x^*, 0) = \min_x \phi(x, 0) = \max_\lambda - \phi^*(0, \lambda) = -\phi^*(0, \lambda^*). \tag{5.84}$$

But this is equivalent to the condition that program (D_{ϕ^*}) is stable and has a solution λ^*. ■

Example 5.2.2

Consider again the functions that appeared in the previous example.

(a) First let

$$\phi_1(x, w) = \begin{cases} x & \text{if } x \geq w \\ +\infty & \text{if } x < w. \end{cases} \tag{5.85}$$

Then

$$\phi_1^*(\xi, \lambda) = \begin{cases} 0 & \text{if } \xi + \lambda = 1, \lambda \geq 0 \\ +\infty & \text{otherwise.} \end{cases} \tag{5.86}$$

Hence the stable program (P_{ϕ_1}) can be written as

$$\min_x \phi_1(x, 0) = \min_{x \geq 0} x \tag{5.87}$$

and $(D_{\phi_1^*})$ is given by

$$\max_\lambda - \phi_1^*(0, \lambda) = \max_{\lambda = 1} 0. \tag{5.88}$$

Clearly, $\min_{x \geq 0} x = 0$, and the optimal solutions of (P_{ϕ_1}) and $(D_{\phi_1^*})$ are $x^* = 0, \lambda^* = 1$, respectively. We have already seen that $\phi_1(w) = w$ for every $w \in R$, $\Phi_1(0) = 0$, and $\partial\Phi_1(0) = 1$.

Similarly,

$$\Psi_1(\xi) = \begin{cases} 0 & \text{if } \xi \leq 1 \\ +\infty & \text{if } \xi > 1. \end{cases} \tag{5.89}$$

It follows that $\Psi_1(0) = 0$ and $\partial\Psi_1(0) = 0$. Hence $(D_{\phi_1^*})$ is also stable. In addition, we have $x^* \in \partial\Psi(0)$ and $\lambda^* \in \partial\Phi(0)$.

(b) Next, take

$$\phi_2(x, w) = \begin{cases} -x & \text{if } x \geq w \\ +\infty & \text{if } x < w. \end{cases} \tag{5.90}$$

Then

$$\phi_2^*(\xi, \lambda) = \begin{cases} 0 & \text{if} \quad \xi + \lambda = -1, \lambda \geq 0 \\ +\infty & \text{otherwise.} \end{cases} \tag{5.91}$$

The optimal value of (P_{ϕ_2}) approaches $-\infty$; that is,

$$\inf_x \phi_2(x, 0) = \inf_{x \geq 0} - x = -\infty. \tag{5.92}$$

Turning now to the dual problem $(D_{\phi_2^*})$, we can see that $(0, \lambda) \notin ED(\phi_2^*)$ for every real λ (since $\lambda = -1$, $\lambda \geq 0$ has no solution).

(c) Let

$$\phi_3(x, w) = \begin{cases} x & \text{if } (x)^2 \leq w \\ +\infty & \text{if } (x)^2 > w. \end{cases} \tag{5.93}$$

We saw in Example 5.2.1 that program (P_{ϕ_3}) is unstable. Computing the conjugate function of ϕ_3, it can be shown that

$$\phi_3^*(\xi, \lambda) = \begin{cases} 0 & \text{if} \quad \xi = 1, \lambda = 0 \\ -\dfrac{(\xi - 1)^2}{4\lambda} & \text{if} \quad \lambda < 0 \\ +\infty & \text{otherwise.} \end{cases} \tag{5.94}$$

The optimal solution of (P_{ϕ_3}) is $x^* = 0$ and

$$\min_x \phi_3(x, 0) = 0. \tag{5.95}$$

Turning to the dual program $(D_{\phi_3^*})$, we obtain

$$\sup_{\lambda < 0} - \phi_3^*(0, \lambda) = \sup_{\lambda < 0} \frac{1}{4\lambda} = 0, \tag{5.96}$$

but there is no real λ^* for which the supremum is attained. ∎

We have seen that stability is a very important property for duality of convex programs. Actually, we shall see that it plays an equally important role in establishing the existence of Lagrange (Kuhn-Tucker) multipliers for convex programs. Thus it is important to study conditions that imply stability. In the previous chapter we defined strongly consistent programs that we redefine now for our more general convex program (P_ϕ) as given in (5.40).

Program (P_ϕ) is said to be **strongly consistent** if there exists a positive number ϵ such that for every w satisfying $\|w\| < \epsilon$, an x^0 can be found such

that $\phi(x^0, w) < +\infty$. The reader can show that for a convex program of type (SP), given by (5.30) and (5.31), strong consistency, as defined in the previous chapter, means that there exists an x^0 such that

$$g_i(x^0) > 0, \qquad i = 1, \ldots, m \tag{5.97}$$

and this condition, in turn, implies that the corresponding (P_ϕ), derived from a program formulation in which the perturbation vector w is on the right-hand sides of the constraints, is strongly consistent in the sense just stated. The reader will be asked to prove the following

Theorem 5.16

Program (P_ϕ) *is strongly consistent if and only if* 0 *is in the interior of* ED (Φ). *If* (P_ϕ) *is strongly consistent, then* (P_ϕ) *is stable.*

The converse of the latter statement is not true, for there are stable programs that are consistent but not strongly consistent.

We mentioned briefly the concept of duality gap—that is, the case where the optimal value of a primal program is strictly greater than the optimal value of the corresponding dual program. In cases where stability is lacking we can construct such examples, as illustrated below.

Example 5.2.3 [29]

Consider the problem of minimizing $\phi(x, w)$, where $x \in R^2$, $w \in R^2$, and

$$\phi(x, w) = \begin{cases} \max\,[-1, -(-w_1 x_2)^{1/2}] & \text{if } \; -w_1 = x_1 \geq 0 \\ & \text{and } \; x_2 \geq 0 \\ +\infty & \text{otherwise.} \end{cases} \tag{5.98}$$

Clearly, $\phi(x, 0)$ is either 0 or $+\infty$, and the optimal value of (P_ϕ) is

$$\inf_x \phi(x, 0) = \min_x \phi(x, 0) = \phi(x^*, 0) = 0, \tag{5.99}$$

where, for example, $x^* = (0, 0)$. The reader can verify that

$$\Phi(w) = \begin{cases} -1 & \text{if } w_1 < 0 \\ 0 & \text{if } w_1 = 0 \\ +\infty & \text{if } w_1 > 0 \end{cases} \tag{5.100}$$

and $\Phi(0) = 0$. Program (P_ϕ) is unstable, since $\partial\Phi(0)$ is empty, or, equivalently, since $\Phi(w)$ decreases infinitely steeply in every neighborhood of the origin

(in the direction of negative w_1). The reader can also verify that

$$\phi^*(0, \lambda) = \begin{cases} 1 & \text{if } \lambda_1 \geq 0, \lambda_2 = 0 \\ +\infty & \text{otherwise} \end{cases} \tag{5.101}$$

and the optimal value of (D_{ϕ^*}) is given by

$$\sup_\lambda - \phi^*(0, \lambda) = \max_\lambda - \phi^*(0, \lambda) = -\phi^*(0, \lambda^*) = -1, \tag{5.102}$$

where, for example, $\lambda^* = (0, 0)$ and we obtain a duality gap, since

$$\min_x \phi(x, 0) = 0 > -1 = \max_\lambda - \phi^*(0, \lambda). \tag{5.103}$$

∎

In Example 5.2.2(c) we had an unstable primal program without duality gap, whereas Example 5.2.3 demonstrated an unstable primal program with duality gap. The main difference between the two primal programs lies in the corresponding perturbation functions: the primal perturbation function of the program pair without duality gap is a closed convex function, whereas the primal perturbation function of Example 5.2.3 is not closed at $w = 0$. Note, however, that the primal objective functions ϕ in both examples are closed convex functions. It turns out that the closedness of the primal and dual perturbation functions at the origin is a necessary and sufficient condition for equal optimal values of the primal and dual programs.

Before proving this result, we need the following

Lemma 5.17

Let $\phi(x, w)$ be a closed proper convex function on R^{n+k}. Then

$$\Phi(0) = -\text{cl}\, \Psi(0) \tag{5.104}$$

$$\Psi(0) = -\text{cl}\, \Phi(0). \tag{5.105}$$

Proof. We shall prove (5.104) and the reader will be asked to do the rest. We have

$$-\text{cl}\, \Psi(0) = -\Psi^{**}(0) = -\sup_x \{0^T x - \Psi^*(x)\} = \inf_x \Psi^*(x). \tag{5.106}$$

From Lemma 5.12 we obtain

$$\inf_x \Psi^*(x) = \inf_x \phi(x, 0) = \Phi(0). \tag{5.107}$$

∎

Now we can state and prove

Theorem 5.18

Let $\phi(x, w)$ be a closed proper convex function on R^{n+k}. Then the following statements are equivalent.

(i)

$$\Phi(0) = \inf_x \phi(x, 0) = \sup_\lambda - \phi^*(0, \lambda) = -\Psi(0) \tag{5.108}$$

(ii) *The primal perturbation function Φ is closed at $w = 0$.*

(iii) *The dual perturbation function Ψ is closed at $\xi = 0$.*

Proof. From Lemma 5.17 and the definition of closedness, we have that $\Phi(0) = -\Psi(0)$ if and only if $\Phi(0) = \text{cl}\,\Phi(0)$. Similarly, $\Phi(0) = -\Psi(0)$ if and only if $\Psi(0) = \text{cl}\,\Psi(0)$. ■

Program (P_ϕ) is called **normal** by Rockafellar [31] if part (ii) of Theorem 5.18 holds. Similarly, (D_{ϕ^*}) is called normal if part (iii) holds. Using these definitions, we obtain the following corollary of the last theorem.

Corollary 5.19

Let $\phi(x, w)$ be a closed proper convex function on R^{n+k}. Program (P_ϕ) is normal and its optimal value is finite if and only if (D_{ϕ^*}) is normal and its optimal value is finite. In that case, the optimal values are equal. Conversely, if $\inf_x \phi(x, 0) = \sup_\lambda - \phi^*(0, \lambda)$, then both (P_ϕ) and (D_{ϕ^*}) are normal.

5.3 OPTIMALITY CONDITIONS AND LAGRANGE MULTIPLIERS

In this section we attempt to connect the duality relations for convex programs derived in the previous section with optimality conditions presented in Chapters 3 and 4. Our first problem is to find an appropriate definition of Lagrange multipliers and Lagrangian functions for a convex program of type (P_ϕ). As before, consider a standard convex program

$$\text{(SP)} \qquad \min f(x) \tag{5.109}$$

subject to

$$g_i(x) \geq 0, \qquad i = 1, \ldots, m \tag{5.110}$$

and one possible type of perturbed problem

$$\text{(PSP)} \qquad \min f(x) \tag{5.111}$$

subject to

$$g_i(x) \geq w_i, \qquad i = 1, \ldots, m. \tag{5.112}$$

In both problems f and g_i are, respectively, proper convex and concave functions on R^n. Define

$$\phi_1(x, w) = \begin{cases} f(x) & \text{if } x \in \text{ED}(f) \text{ and } g_i(x) \geq w_i, \ i = 1, \ldots, m \\ +\infty & \text{otherwise} \end{cases} \tag{5.113}$$

and define the perturbation function Φ_1 by

$$\Phi_1(w) = \inf_x \phi_1(x, w). \tag{5.114}$$

Suppose now that f is a cost function and, instead of minimizing f in the unperturbed problem (SP), we can solve the more general problem (PSP) and be paid an amount $\sum_i \lambda_i w_i$, where $\lambda_i \geq 0$ is the price paid per unit perturbation in constraint i. We can then ask: Under what conditions will the solution of the unperturbed problem (SP) remain the best choice? The answer is that it will remain the best choice if

$$\Phi_1(w) - \sum_{i=1}^{m} \lambda_i w_i \geq \Phi_1(0) \tag{5.115}$$

for every $w \in R^m$. This inequality is satisfied if and only if

$$f(x) - \sum_{i=1}^{m} \lambda_i w_i \geq \Phi_1(0) \tag{5.116}$$

for all $x \in R^n$ and $w \in R^m$ such that $g_i(x) \geq w_i$, $i = 1, \ldots, m$. If we assume that $\Phi_1(0)$ is finite, then (5.116) is equivalent to the condition that

$$f(x) - \sum_{i=1}^{m} \lambda_i g_i(x) \geq \Phi_1(0) \tag{5.117}$$

for all $x \in R^n$.

The left-hand side of (5.117) is the Lagrangian associated with (SP), as defined in earlier chapters, and the λ_i are the Lagrange multipliers. The nonnegative Lagrange multipliers λ_i^* for which the infimum with respect to x of the left-hand side of (5.117) coincides with $\Phi_1(0)$, the infimum of (SP), are also called "equilibrium prices" [31]. These interpretations of (5.117) will be shown to be in agreement with the saddlepoint relations developed in previous chapters.

Motivated by the preceding discussion, we extend the definition of Lagrange multipliers to the more general case. Let (P_ϕ) be a program as

defined in (5.40). Suppose that $\Phi(0)$ is finite. Then $\lambda^* \in R^k$ is said to be a vector of Lagrange multipliers for (P_ϕ) if

$$\Phi(w) - \lambda^{*T}w \geq \Phi(0) \tag{5.118}$$

for every $w \in R^k$. Similarly, if $\Psi(0)$ is finite, we can define $x^* \in R^n$ to be the Lagrange multipliers for (D_{ϕ^*}) if

$$\Psi(\xi) - \xi^T x^* \geq \Psi(0) \tag{5.119}$$

for every $\xi \in R^n$. Comparing the last two inequalities with (4.110), we conclude that (5.118) and (5.119) are equivalent to the conditions $\lambda^* \in \partial\Phi(0)$ and $x^* \in \partial\Psi(0)$, respectively.

Similarly, we define the Lagrangian that corresponds to (P_ϕ) as

$$L(x, \lambda) = \inf_w \{\phi(x, w) - \lambda^T w\}. \tag{5.120}$$

This function is also called the "Kuhn-Tucker function" corresponding to (P_ϕ); see [19, 31]. In the rest of this section we give a number of results that connect the program pair (P_ϕ) and (D_{ϕ^*}) with Lagrange multipliers and saddlepoints of the Lagrangian. Some results will be formulated in different terms, and inevitably there will be a certain overlap in some theorems.

The first result deals with the existence of Lagrange multipliers in convex programs.

Theorem 5.20

Assume that (P_ϕ) *has an optimal solution. Then there exists a vector of Lagrange multipliers* λ^* *if and only if* (P_ϕ) *is stable. The vector* λ^* *is a vector of Lagrange multipliers for* (P_ϕ) *if and only if* $\lambda^* \in \partial\Phi(0)$.

Proof. If (P_ϕ) has an optimal solution, then $\Phi(0)$ is finite. It follows that $\partial\Phi(0)$ is nonempty if and only if (P_ϕ) is stable. Furthermore, from the definition of Lagrange multipliers it follows that λ^* is a vector of Lagrange multipliers if and only if λ^* is a subgradient of $\Phi(0)$. ■

This theorem has an interesting implication. Since it can be considered a general result for convex programs, we conclude that, for such programs, stability is a **necessary and sufficient** qualification for the existence of Lagrange multipliers or, in other words, for satisfying the Kuhn-Tucker conditions. It follows, therefore, that all the different kinds of constraint qualifications must imply stability in the case of convex programs. The dual formulation of the last theorem is

Theorem 5.21

Assume that (D_{ϕ^*}) *has an optimal solution. Then there exists a vector of Lagrange multipliers* x^* *if and only if* (D_{ϕ^*}) *is stable. The vector* x^* *is a vector of Lagrange multipliers for* (D_{ϕ^*}) *if and only if* $x^* \in \partial\Psi(0)$.

Turning now to a characterization of Lagrange multipliers and optimal solutions, let ϕ be a proper convex function on R^{n+k}.

Theorem 5.22

The vector λ^* *is a vector of Lagrange multipliers for* (P_ϕ) *if and only if* λ^* *is optimal for* (D_{ϕ^*}) *and* $\Phi(0) = -\phi^*(0, \lambda^*)$. *Dually, let* ϕ *be closed. Then* x^* *is a vector of Lagrange multipliers for* (D_{ϕ^*}) *if and only if* x^* *is optimal for* (P_ϕ) *and* $\phi(x^*, 0) = -\Psi(0)$.

Proof. By Theorem 5.20, if λ^* is a vector of Lagrange multipliers for (P_ϕ), then $\lambda^* \in \partial\Phi(0)$. By Theorem 5.4, we then have $0 \in \partial\Phi^*(\lambda^*)$. From the definition of subgradients it follows that Φ^* attains its minimum at λ^* [see (4.110)], and, by Lemma 5.10, this is equivalent to λ^* being an optimal solution for (D_{ϕ^*}). From Exercise 4.N., if $\partial\Phi(0) \neq \emptyset$, then $\Phi(0) = \text{cl } \Phi(0)$. Now

$$\text{cl } \Phi(0) = \Phi^{**}(0) = \sup_\lambda \{0^T\lambda - \Phi^*(\lambda)\}. \tag{5.121}$$

From Lemma 5.10 we obtain

$$\sup_\lambda \{0^T\lambda - \Phi^*(\lambda)\} = \sup_\lambda - \phi^*(0, \lambda) = -\phi^*(0, \lambda^*). \tag{5.122}$$

Suppose that λ^* is optimal for (D_{ϕ^*}) and $\Phi(0) = -\phi^*(0, \lambda^*)$. By Lemma 5.10, this means that Φ^* attains its minimum at λ^*. Hence $0 \in \partial\Phi^*(\lambda^*)$. By (5.121) and (5.122), $\Phi(0) = \text{cl } \Phi(0)$; and, by Theorem 5.4, $\lambda^* \in \partial\Phi(0)$. Thus λ^* is a vector of Lagrange multipliers for (P_ϕ). The proof of the dual part is similar. ∎

Consider the Lagrangian functions as defined above. We shall prove saddlepoint results for this function that are similar to those in Chapter 3 and 4. First we look at a convexity-concavity property of L.

Lemma 5.23

The Lagrangian function corresponding to (P_ϕ) *is convex in* $x \in R^n$ *for any fixed* λ *and concave in* $\lambda \in R^k$ *for any fixed* x.

Proof. The proof of the convexity of L in x for any fixed λ is the same as the proof of Theorem 5.8.

Suppose that $(x, \lambda^1, \alpha^1) \in Q(L)$, $(x, \lambda^2, \alpha^2) \in Q(L)$, where $Q(L)$ is the hypograph of L. Then

$$\alpha^1 \le L(x, \lambda^1) = \inf_w \{\phi(x, w) - \lambda^{1T} w\} \tag{5.123}$$

$$\alpha^2 \le L(x, \lambda^2) = \inf_w \{\phi(x, w) - \lambda^{2T} w\}. \tag{5.124}$$

Hence

$$q_1\alpha^1 + q_2\alpha^2 \le \inf_w \{q_1\phi(x, w) - q_1\lambda^{1T} w\} + \inf_w \{q_2\phi(x, w) - q_2\lambda^{2T} w\} \tag{5.125}$$

and

$$q_1\alpha^1 + q_2\alpha^2 \le \inf_w \{\phi(x, w) - (q_1\lambda^1 + q_2\lambda^2)^T w\} = L(x, q_1\lambda^1 + q_2\lambda^2). \tag{5.126}$$

That is, $(x, q_1\lambda^1 + q_2\lambda^2, q_1\alpha^1 + q_2\alpha^2) \in Q(L)$ and L is concave in λ for any fixed x. ■

The Lagrangian corresponding to (P_ϕ) was defined in (5.120) as a function of ϕ. It is possible to turn this relation around, as we shall see below. The proof of the next lemma uses **partial conjugacy**. Suppose that f is a convex function on R^{n+k}. Then the partial conjugate $f_p^*(x, \eta)$ of $f(x, y)$ for each fixed $x \in R^n$ is defined as

$$f_p^*(x, \eta) = \sup_{y \in R^k} \{\eta^T y - f(x, y)\} \tag{5.127}$$

and the partial conjugate of $f_p^*(x, \eta)$ for each fixed x is, in turn, $\text{cl}\, f(x, y)$, where the closure operation refers to the second argument.

Lemma 5.24

Suppose that ϕ is a closed proper convex function on R^{n+k}. Then

$$\phi(x, w) = \sup_\lambda \{L(x, \lambda) + \lambda^T w\} \tag{5.128}$$

$$-\phi^*(\xi, \lambda) = \inf_x \{L(x, \lambda) - \xi^T x\}. \tag{5.129}$$

Proof. We have

$$\sup_\lambda \{L(x, \lambda) + \lambda^T w\} = \sup_\lambda \{\inf_u [\phi(x, u) - \lambda^T u] + \lambda^T w\} \tag{5.130}$$

$$= \sup_\lambda \{\lambda^T w - \sup_u [\lambda^T u - \phi(x, u)]\} \tag{5.131}$$

$$= \sup_\lambda \{\lambda^T w - \phi_p^*(x, \lambda)\} = \phi_p^{**}(x, w) = \phi(x, w). \tag{5.132}$$

Similarly,

$$\inf_x \{L(x, \lambda) - \xi^T x\} = \inf_x \inf_u \{\phi(x, u) - \xi^T x - \lambda^T u\} \tag{5.133}$$

$$= -\phi^*(\xi, \lambda). \tag{5.134}$$

■

As an immediate consequence of this lemma, we get

$$\phi(x, 0) = \sup_\lambda L(x, \lambda) \tag{5.135}$$

and

$$-\phi^*(0, \lambda) = \inf_x L(x, \lambda). \tag{5.136}$$

The first relation between saddlepoints of the Lagrangian and optimal solutions of the corresponding convex programs is given in

Theorem 5.25

Let ϕ be a closed proper convex function on R^{n+k} and let L be the Lagrangian corresponding to (P_ϕ). If (x^*, λ^*) is a saddlepoint of L, then x^* is an optimal solution of (P_ϕ), λ^* is an optimal solution of (D_{ϕ^*}), and

$$\phi(x^*, 0) = L(x^*, \lambda^*) = -\phi^*(0, \lambda^*) \tag{5.137}$$

Proof. If

$$L(x^*, \lambda) \leq L(x^*, \lambda^*) \leq L(x, \lambda^*) \tag{5.138}$$

for every $x \in R^n$ and $\lambda \in R^k$, then, by (5.135) and (5.136),

$$L(x^*, \lambda^*) = \sup_\lambda L(x^*, \lambda) = \phi(x^*, 0) \tag{5.139}$$

and

$$L(x^*, \lambda^*) = \inf_x L(x, \lambda^*) = -\phi^*(0, \lambda^*). \tag{5.140}$$

From Corollary 5.6 it follows that x^* and λ^* are optimal solutions of (P_ϕ) and (D_{ϕ^*}), respectively. ■

The converse of this result is

Theorem 5.26

Let ϕ be a closed proper convex function. If (P_ϕ) is stable and x^* is optimal for (P_ϕ), then there exists a λ^* such that (x^*, λ^*) is a saddlepoint of the Lagrangian L. Dually, if (D_{ϕ^*}) is stable and λ^* is optimal for (D_{ϕ^*}), then there exists an x^* such that (x^*, λ^*) is a saddlepoint of the Lagrangian L.

Proof. If the assumptions on (P_ϕ) hold, then, by Theorem 5.11, there exists a λ^* optimal for (D_{ϕ^*}), and $\phi(x^*, 0) = -\phi^*(0, \lambda^*)$. By (5.135) and (5.136),

$$\sup_\lambda L(x^*, \lambda) = \inf_x L(x, \lambda^*). \tag{5.141}$$

Clearly,

$$\sup_\lambda L(x^*, \lambda) \geq L(x^*, \lambda^*) \geq \inf_x L(x, \lambda^*). \tag{5.142}$$

Hence for every $x \in R^n$ and $\lambda \in R^k$ we have

$$L(x^*, \lambda) \leq \sup_\lambda L(x^*, \lambda) = L(x^*, \lambda^*) = \inf_x L(x, \lambda^*) \leq L(x, \lambda^*). \tag{5.143}$$

The proof of the dual part is similar. ■

The following theorem summarizes the main results of this section. Its proof is based on previously stated results and is left for the reader.

Theorem 5.27 (Equivalence Theorem)

Let ϕ be a closed proper convex function on R^{n+k}. The following five statements are equivalent.

(i) *The vector (x^*, λ^*) is a saddlepoint of the Lagrangian L corresponding to* (P_ϕ).

(ii) *The vector x^* is an optimal solution of* (P_ϕ) *and λ^* is a vector of Lagrange multipliers for* (P_ϕ).

(iii) *The vector λ^* is an optimal solution of* (D_{ϕ^*}) *and x^* is a vector of Lagrange multipliers for* (D_{φ^*}).

(iv) *The vectors λ^* and x^* are vectors of Lagrange multipliers for* (P_ϕ), *and* (D_{ϕ^*}), *respectively, and $\phi(x^*, 0) = -\phi^*(0, \lambda^*)$.*

(v) *Programs* (P_ϕ) *and* (D_{ϕ^*}) *are normal, and x^*, λ^* are optimal solutions of* (P_ϕ) *and* (D_{ϕ^*}), *respectively.*

5.4 DUALITY AND OPTIMALITY FOR STANDARD CONVEX PROGRAMS

The results of the previous sections will now be illustrated by applying them to standard convex programs as defined earlier in this chapter. For convenience, we restate the standard primal program as follows:

$$\text{(SP)} \qquad \min f(x) \tag{5.144}$$

subject to

$$g_i(x) \geq 0. \qquad i = 1, \ldots, m \tag{5.145}$$

where f and g_i are proper convex and concave functions on R^n, respectively. Note that problem (SP) could have been slightly modified by adding an "implicit" constraint, such as restricting x to be in a given convex set $X \subset R^n$, as well as satisfying (5.145). Such a modification would also allow us to transfer one or more constraints from (5.145) into the set X and would result in a different dual program with a smaller number of dual variables. Generally this operation is not advantageous, for the maximand of the dual program, as we shall see below, is less amenable, in this case, to being expressed explicitly. We shall assume that, unless specifically mentioned, X is equal to R^n.

Let us return to (SP) and introduce the perturbations on the right-hand sides of (5.145). The perturbed problem is given by

$$\min f(x) \tag{5.146}$$

subject to

$$g_i(x) \geq w_i, \qquad i = 1, \ldots, m \tag{5.147}$$

and the perturbation function is

$$\Phi(w) = \inf_x \{f(x) : g_i(x) \geq w_i, i = 1, \ldots, m\}. \tag{5.148}$$

In order to get the dual program corresponding to (SP), we write

$$\phi(x, w) = \begin{cases} f(x) & \text{if } g_i(x) \geq w_i, i = 1, \ldots, m \\ +\infty & \text{otherwise.} \end{cases} \tag{5.149}$$

Assume in the sequel that ϕ is a closed convex function and compute its conjugate

$$\phi^*(\xi, \lambda) = \sup_{x,w} \{\xi^T x + \lambda^T w - \phi(x, w)\} \tag{5.150}$$

$$= \sup_{\substack{g_i(x) \geq w_i \\ i=1,\ldots,m}} \{\xi^T x + \lambda^T w - f(x)\}. \tag{5.151}$$

Introducing nonnegative "slack variables" s_i, we can convert constraints (5.147) into

$$w_i = g_i(x) - s_i, \qquad i = 1, \ldots, m \tag{5.152}$$

$$s_i \geq 0 \qquad i = 1, \ldots, m \tag{5.153}$$

and obtain

$$\phi^*(\xi, \lambda) = \sup_x \left\{\xi^T x + \sum_{i=1}^m \lambda_i g_i(x) - f(x)\right\} + \sup_{s \geq 0} \{-\lambda^T s\}. \tag{5.154}$$

Hence

$$\phi^*(\xi, \lambda) = \begin{cases} \sup_x \left\{\xi^T x + \sum_{i=1}^m \lambda_i g_i(x) - f(x)\right\} & \text{if } \lambda \geq 0 \\ +\infty & \text{otherwise.} \end{cases} \tag{5.155}$$

The dual problem of (SP), consisting of maximizing $-\phi^*(0, \lambda)$ is given by

$$\text{(DSP)} \qquad \max_{\lambda \geq 0} \left\{\inf_x \left[f(x) - \sum_{i=1}^m \lambda_i g_i(x)\right]\right\}. \tag{5.156}$$

The dual perturbation function Ψ is

$$\Psi(\xi) = \inf_{\lambda \geq 0} \sup_x \left\{\xi^T x + \sum_{i=1}^m \lambda_i g_i(x) - f(x)\right\}. \tag{5.157}$$

Let us compute the Lagrangian associated with (SP). By (5.120) and (5.150),

$$L(x, \lambda) = \inf_w \{f(x) - \lambda^T w : g_i(x) \geq w_i, i = 1, \ldots, m\}; \tag{5.158}$$

and by introducing the slack variables s_i as before, we get

$$L(x, \lambda) = \begin{cases} f(x) - \sum_{i=1}^m \lambda_i g_i(x) & \text{if } \lambda \geq 0 \\ -\infty & \text{otherwise.} \end{cases} \tag{5.159}$$

For those x and λ that yield finite values of $L(x, \lambda)$, the preceding formula corresponds to the Lagrangian associated with problem (SP), as defined in Chapters 2, 3, and 4.

Kuhn-Tucker type optimality conditions for (SP) can now be readily derived from the general results of this chapter. By our use of the notion of stability, these conditions become stronger than similar results in previous chapters.

Theorem 5.28

Assume that (SP) *has an optimal solution x^*. Then there exists a vector $\lambda^* \in R^m$ such that*

$$\lambda^* \geq 0 \tag{5.160}$$

$$\lambda_i^* g_i(x^*) = 0, \qquad i = 1, \ldots, m \tag{5.161}$$

and the Lagrangian L has a saddlepoint at (x^*, λ^*). *That is,*

$$L(x^*, \lambda) \leq L(x^*, \lambda^*) \leq L(x, \lambda^*) \tag{5.162}$$

for all $\lambda \geq 0$ and $x \in R^n$ if and only if (SP) *is stable.*

Proof. Suppose that (SP) is stable. By Theorem 5.20, there exists a vector of Lagrange multipliers λ^* such that $\lambda^* \in \partial\Phi(0)$; that is,

$$\Phi(w) - \lambda^{*T}w \geq \Phi(0) \tag{5.163}$$

for all $w \in R^m$. Choose $w^1 = (-1, 0, 0, \ldots, 0)^T$. Then

$$\lambda_1^* \geq \Phi(0) - \Phi(w^1). \tag{5.164}$$

The right-hand side of this inequality is nonnegative, for lowering the first component of w from 0 to -1 cannot increase the value of Φ. Hence $\lambda_1^* \geq 0$. Letting $w^2 = (0, -1, 0, \ldots, 0)^T$, we get $\lambda_2^* \geq 0$, and, similarly, $\lambda_i^* \geq 0$ for $i = 1, \ldots, m$. Since x^* is feasible for (SP), we have

$$\lambda_i^* g_i(x^*) \geq 0, \qquad i = 1, \ldots, m. \tag{5.165}$$

Substituting $\tilde{w}^1 = (g_1(x^*), 0, \ldots, 0)^T$ yields

$$\Phi(\tilde{w}^1) - \Phi(0) \geq \lambda_1^* g_1(x^*). \tag{5.166}$$

The left-hand side of this inequality vanishes, for if $g_1(x^*) = 0$, then the preceding difference is identically zero; and if $g_1(x^*) > 0$, then letting w_1 increase from 0 to $g_1(x^*)$ will not alter the optimal solution of (SP), and so Φ will not change its value from $\Phi(0)$. Thus $0 \geq \lambda_1^* g_1(x^*)$, and we similarly get $0 \geq \lambda_i^* g_i(x^*)$ for $i = 1, \ldots, m$, and (5.161) must hold. By Theorem 5.26, the vector (x^*, λ^*) is a saddlepoint of L. Conversely, let λ^* be a vector of Lagrange multipliers. If (x^*, λ^*) is a saddlepoint of L for all $\lambda \geq 0$ and $x \in R^n$, then

$$f(x) - \sum_{i=1}^{m} \lambda_i^* g_i(x) \geq f(x^*) - \sum_{i=1}^{m} \lambda_i^* g_i(x^*) = f(x^*), \tag{5.167}$$

where the equality follows from (5.161). Since $\lambda^* \geq 0$, we obtain

$$f(x) - \sum_{i=1}^{m} \lambda_i^* w_i \geq f(x^*) \tag{5.168}$$

for all x and w satisfying $g_i(x) \geq w_i$, $i = 1, \ldots, m$.

Since $f(x^*) = \Phi(0)$, and by taking the infimum of the left-hand side of (5.168) with respect to x, we obtain

$$\Phi(w) - \sum_{i=1}^{m} \lambda_i^* w_i \geq \Phi(0) \tag{5.169}$$

for all $w \in$ ED (Φ). If $w \notin$ ED (Φ), we have $\Phi(w) = +\infty$, and from the definition of subgradients it follows that $\lambda^* \in \partial\Phi(0)$. Thus (SP) is stable. ■

This theorem is the analog of Theorem 4.41, in which we assumed strong consistency, which implies stability. We cannot prove the converse of Theorem 4.41, since stability does not imply strong consistency. Note, however, that, for closed f and g_i, it is possible to state and prove a dual result similar to the last theorem for the (DSP) problem.

We turn now to the optimality conditions for the case where the functions f and g_i are differentiable. The next theorem constitutes a set of Kuhn-Tucker type optimality conditions for differentiable (SP) programs.

Theorem 5.29

Suppose that f and $g_1, \ldots, g_m$ are real-valued differentiable convex and concave functions on R^n, respectively. Assume that (SP) *has an optimal solution x^*. Then there exists a vector $\lambda^* \in R^m$ such that*

$$\nabla f(x^*) - \sum_{i=1}^{m} \lambda_i^* \nabla g_i(x^*) = 0 \tag{5.170}$$

$$\lambda_i^* g_i(x^*) = 0, \qquad i = 1, \ldots, m \tag{5.171}$$

$$\lambda^* \geq 0 \tag{5.172}$$

if and only if (SP) *is stable.*

Proof. Suppose that (SP) is stable. We need to show that (5.170) holds, since (5.171) and (5.172) hold by the preceding theorem. By Theorem 5.11, there is an optimal solution λ^* for (DSP), and the optimal values of (SP) and (DSP) are equal.
Hence

$$f(x^*) = \inf_x \left\{ f(x) - \sum_{i=1}^{m} \lambda_i^* g_i(x) \right\} \tag{5.173}$$

or

$$f(x^*) \leq f(x) - \sum_{i=1}^{m} \lambda_i^* g_i(x) \tag{5.174}$$

for all x. Substituting $x = x^*$, we can see that the function appearing on the right-hand side of (5.174) attains its unconstrained minimum at x^*. Hence its gradient vanishes there and (5.170) holds. The proof of the converse part parallels the proof of the previous theorem and is left for the reader. ■

Dual programs of (SP) have been investigated by many researchers, and several formulations were proposed, each of which can possibly be derived from the results of this chapter. It is clear that for every primal convex program we can formulate many different dual programs, one for each type of perturbation introduced.

For (SP), the type of perturbation used in this section results in a dual program (DSP) whose objective function [see (5.156)] is given in terms of an additional optimization problem—that is, that of

$$\inf_x \left\{ f(x) - \sum_{i=1}^{m} \lambda_i g_i(x) \right\} \cdot \tag{5.175}$$

There are, however, special cases of (SP), or other types of perturbations, in which this additional optimization problem can be explicitly solved. This point will be illustrated in the following example and also in Chapter 7.

Example 5.4.1

Let us derive the dual of a linear program by the results of this chapter. The primal linear program is given by

(LP)
$$\min_x c^T x \tag{5.176}$$

subject to

$$Ax \geq b \tag{5.177}$$

$$x \geq 0, \tag{5.178}$$

where A is an $m \times n$ real matrix and b and c are m and n vectors, respectively. Introducing the perturbations on the right-hand sides of the constraints, we get

$$\phi(x, w, v) = \begin{cases} c^T x & \text{if } Ax - b \geq w,\ x \geq v \\ +\infty & \text{otherwise} \end{cases} \tag{5.179}$$

and

$$\phi^*(\xi, \lambda, \nu) = \sup_{\substack{x, w, v \\ Ax - b \geq w \\ x \geq v}} \{\xi^T x + \lambda^T w + \nu^T v - c^T x\}. \tag{5.180}$$

This expression reduces to

$$\phi^*(\xi, \lambda, \nu) = \begin{cases} \sup_x \{(\xi - c + A^T \lambda + \nu)^T x - \lambda^T b\} & \text{if } \lambda \geq 0,\ \nu \geq 0 \\ +\infty & \text{otherwise.} \end{cases} \tag{5.181}$$

Hence

$$\phi^*(\xi, \lambda, \nu) = \begin{cases} -\lambda^T b & \text{if } \lambda \geq 0,\ \nu \geq 0,\ A^T \lambda + \nu = c - \xi \\ +\infty & \text{otherwise.} \end{cases} \tag{5.182}$$

The dual linear program becomes

(DLP)
$$\max \lambda^T b \tag{5.183}$$

subject to

$$A^T \lambda \leq c \tag{5.184}$$

$$\lambda \geq 0, \tag{5.185}$$

where the vector of dual variables ν has been eliminated from the problem, for it is simply a vector of slack variables.

The (LP) problem is clearly a special case of (SP) above, and the (DLP) could also have been obtained from the results of this section. To do so, let $f(x) = c^T x$, $g(x) = Ax - b$, and $X = \{x : x \geq 0\}$. Then for the dual program (using the (SP) problem modified by the addition of the implicit constraint discussed earlier), we have

$$\max_{\lambda \geq 0} \{ \inf_{x \in X} [c^T x - \lambda^T (Ax - b)] \}. \tag{5.186}$$

By rearranging (5.186), we obtain

$$\max_{\lambda \geq 0} \{ \inf_{x \geq 0} [(c - A^T \lambda)^T x + \lambda^T b] \} \tag{5.187}$$

and the maximand takes on the value $\lambda^T b$ if $c \geq A^T \lambda$ and $-\infty$ otherwise.

Optimality conditions and duality theorems for linear programs, can, of course, be also derived from the general results of this chapter. ▮

We conclude this section by comparing dual programs as derived in this chapter with an earlier work in duality, obtained without using conjugate functions. A typical example is the following pair of dual problems introduced by Wolfe [36]:

(WP)
$$\min_x f(x) \tag{5.188}$$

subject to

$$g_i(x) \geq 0, \qquad i = 1, \ldots, m \tag{5.189}$$

and

(WD)
$$\max_{x, \lambda} f(x) - \sum_{i=1}^{m} \lambda_i g_i(x) \tag{5.190}$$

subject to

$$\nabla f(x) = \sum_{i=1}^{m} \lambda_i \nabla g_i(x), \qquad \lambda \geq 0, \tag{5.191}$$

where f and g_i are real-valued, differentiable, convex and concave functions on R^n, respectively. We can immediately note that Wolfe's primal problem (WP) has precisely our (SP) formulation. Let us relate his dual problem (WD) to (DSP). For convenience, we restate the dual of (SP) as

$$\text{(DSP)} \qquad \max_{\lambda \geq 0} \left\{ \inf_x \left[f(x) - \sum_{i=1}^{m} \lambda_i g_i(x) \right] \right\}. \tag{5.192}$$

First we note that given any $\hat{\lambda} \geq 0$, the vector $\hat{x}$ is feasible for Wolfe's dual if and only if $\hat{x}$ is a global minimum of $f(x) - \sum_i \hat{\lambda}_i g_i(x)$. This condition follows from the convexity of f, the concavity of the g_i, and the gradient conditions of (5.191). Thus (5.190) and (5.191) can be equivalently rewritten as the problem

$$\text{(WSD)} \qquad \max_{\lambda \geq 0} \left\{ \min_x \left[f(x) - \sum_{i=1}^{m} \lambda_i g_i(x) \right] \right\}, \tag{5.193}$$

where λ is further restricted to those values for which the preceding minimum is attained for some x.

We can see that if λ is feasible for (WSD), then it is also feasible for (DSP) and the maximands are equal. The optimal value of (DSP) may, however, be higher than in (WSD) if the optimal λ^* is not feasible for (WSD). However, under the conditions studied by Wolfe, this situation cannot occur. We first establish the following lemma, which follows directly from previous results.

Lemma 5.30

Suppose that (WP) *is stable and x^* is an optimal solution. Then x^* is a global minimum of $f(x) - \sum_i \lambda_i^* g_i(x)$, where λ^* is optimal for* (DSP).

Proof. Suppose that x^* is optimal for (WP). By Theorem 5.11, (DSP) has an optimal solution λ^*, and we have

$$f(x^*) = \inf_x \left\{ f(x) - \sum_{i=1}^{m} \lambda_i^* g_i(x) \right\}. \tag{5.194}$$

From Theorems 5.22 and 5.28 it follows that $\lambda_i^* g_i(x^*) = 0$ for $i = 1, \ldots, m$. Hence

$$f(x^*) - \sum_{i=1}^{m} \lambda_i^* g_i(x^*) = \inf_x \left\{ f(x) - \sum_{i=1}^{m} \lambda_i^* g_i(x) \right\} \tag{5.195}$$

and we can replace infimum by minimum. ■

From this lemma we conclude that if (WP) is stable and has an optimal solution, then the optimal solutions of (WSD) and (DSP) are identical.

Wolfe states his strong duality theorem under the condition that the Kuhn-Tucker, or some other constraint qualification [22, 25], holds for (WP), which, in turn, implies that (WP) is stable [18]:

Theorem 5.31 (Wolfe)

If x^* *solves* (WP), *then there exists a* λ^* *such that* (x^*, λ^*) *solves* (WD) *and the optimal values are equal.*

Problem (WD) as stated above is not very attractive computationally except in special cases. Although (WP) is a convex program, (WD) is generally nonconvex. It contains more variables then the primal, its objective function is not concave in (x, λ), and the feasible set is not necessarily convex. A similar discussion of Wolfe's duality relations can be also found in Geoffrion [18] and Whinston [34].

If the constraints (5.189) in (WP) are all linear, we obtain the pair of dual problems studied by Dennis [8] and Dorn [10]. The former work is interesting, since it uses the Legendre transformation, which we have seen to be closely related to conjugate functions. Wolfe's treatment of duality in nonlinear convex programming is a representative example of works in this area prior to Rockafellar's results. All these earlier works rely heavily on the Kuhn-Tucker theory [22] of nonlinear programming with or without differentiability, which itself is an extension of the Lagrange multipliers theory of classical optimization. Among them, we can mention the works of Cottle [3], Dantzig, Eisenberg, and Cottle [6], Eisenberg [11], Hanson [20], Huard [21], Mangasarian [24], Mangasarian and Ponstein [26], and Stoer [33]. Interestingly, in contrast to these earlier results, we first established duality relations and from them obtained optimality conditions of the Kuhn-Tucker type.

Economic interpretations of duality in linear programming have been extended to nonlinear convex programs. Interested readers might wish to read the works of Balinski and Baumol [1], Gale [14, 15], and Williams [35] on this subject. Williams' work is also based on conjugate functions, and it presents economic meanings to many results derived in this chapter.

EXERCISES

5.A. Let f be any function with real or infinite values on R^n. Show that its conjugate function f^* is convex.

5.B. Given two proper convex functions f_1 and f_2 on R^n, define their **convolution function** as

$$(f_1 \square f_2)(x) = \inf_y \{f_1(x - y) + f_2(y)\}. \tag{5.196}$$

Show that $f_1 \square f_2$ is convex and prove that its conjugate is

$$(f_1 \square f_2)^*(\xi) = f_1^*(\xi) + f_2^*(\xi). \tag{5.197}$$

Find an economic interpretation of the convolution function and its conjugate.

5.C. Given a positively homogeneous convex function, show that its conjugate is an indicator function. Find an indicator function whose conjugate is also an indicator function.

5.D. Show that the function f, given by

$$f(x) = \begin{cases} \left(\sum_{j=1}^{n} (x_j)^p\right)^{1/p} & \text{if } x \geq 0 \\ +\infty & \text{otherwise,} \end{cases} \tag{5.198}$$

where p is a nonzero real constant, is either convex or concave, depending on the value of p, and find the conjugate function of f. Find the limiting cases of f and f^* as $p \longrightarrow \pm\infty$.

5.E. A function f on R^n is called **separable** if

$$f(x) = f_1(x_1) + f_2(x_2) + \cdots + f_n(x_n). \tag{5.199}$$

Show that its conjugate function is also separable.

5.F. Let A be an $n \times n$ real nonsingular matrix, b and c vectors in R^n, and α a real number. Define $\hat{f}$ by

$$\hat{f}(x) = f(Ax - b) + c^T x + \alpha, \tag{5.200}$$

where f is a convex function. Find $\hat{f}^*$.

5.G. Prove that if the perturbation function Φ of a convex program (P_ϕ) is proper and $0 \in \text{ri}\,[\text{ED}\,(\Phi)]$, then program (P_ϕ) is stable.

5.H. Prove Theorem 5.16.

5.I. Discuss the stability of program $(D_{\phi_3^*})$ in Example 5.2.2.

5.J. Discuss the properties of (P_ϕ) and its dual for

$$\phi(x, w) = \begin{cases} e^{-\sqrt{xw}} & \text{if } x \in R, x \geq 0, w \in R, w \geq 0 \\ +\infty & \text{otherwise.} \end{cases} \tag{5.201}$$

5.K. Why do we need ϕ to be closed for the second part of Theorem 5.22? Complete its proof.

5.L. Prove Theorem 5.27.

5.M. Complete the proof of Theorem 5.29.

5.N. Given a primal convex program

$$\min_{x} \{f(x) - g(x)\}, \tag{5.202}$$

where f and g are proper convex and concave functions on R^n, respectively, find the dual program corresponding to the perturbation that replaces $f(x)$ by $f(x + w)$.

5.O. Given a primal convex program

$$\min_{x \in K} f(x), \tag{5.203}$$

where f is a closed proper convex function and $K \subset R^n$ is a nonempty closed convex cone, show that a dual program can be written as

$$\max_{\lambda \in K^*} \; - f^*(\lambda), \tag{5.204}$$

where K^* is the positively normal cone to K (see Chapter 3); that is,

$$K^* = \{\lambda : \lambda \in R^n, \lambda^T x \geq 0 \text{ for all } x \in K\}. \tag{5.205}$$

REFERENCES

1. BALINSKI, M. L., and W. J. BAUMOL, "The Dual in Nonlinear Programming and Its Economic Interpretation," *Review of Economic Studies*, **35**, 237–256 (1968).
2. BARTLE, R. G., *The Elements of Real Analysis*, 2nd ed., John Wiley & Sons, New York, 1976.
3. COTTLE, R. W., "Symmetric Dual Quadratic Programs," *Quart. Appl. Math.*, **21**, 237–243 (1963).
4. COURANT, R., and D. HILBERT, *Methods of Mathematical Physics*, Interscience Publishers, New York, Vol. I (1953), Vol. II (1962).
5. DANTZIG, G. B., *Linear Programming and Extensions*, Princeton University Press, Princeton, N.J., 1963.
6. DANTZIG, G. B., E. EISENBERG, and R. W. COTTLE, "Symmetric Dual Nonlinear Programs," *Pacific J. of Math.*, **15**, 809–812 (1965).
7. DANTZIG, G. B., J. FOLKMAN, and N. Z. SHAPIRO, "On the Continuity of the Minimum Set of a Continuous Function," *J. Math. Anal. & Appl.*, **17**, 519–548 (1967).
8. DENNIS, J. B., *Mathematical Programming and Electrical Networks*, John Wiley & Sons, New York, 1959.
9. DORFMAN, R., P. SAMUELSON, and R. SOLOW, *Linear Programming and Economic Analysis*, McGraw-Hill Book Co., New York, 1958.

10. DORN, W. S., "A Duality Theorem for Convex Programs," *IBM J. Res. & Dev.*, **4**, 407–413 (1960).

11. EISENBERG, E., "Duality in Homogeneous Programming," *Proc. Am. Math. Soc.*, **12**, 783–787 (1961).

12. EVANS, J. P., and F. J. GOULD, "Stability in Nonlinear Programming," *Operations Research*, **18**, 107–118 (1970).

13. FENCHEL, W., "On Conjugate Convex Functions," *Can. J. Math.*, **1**, 73–77 (1949).

14. GALE, D., "A Geometric Duality Theorem with Economic Applications," *Review of Economic Studies*, **34**, 19–24 (1967).

15. GALE, D., "Non-Linear Duality and Qualitative Properties of Optimal Growth," in *Integer and Nonlinear Programming*, J. Abadie (Ed.), North-Holland Publishing Co., Amsterdam, 1970.

16. GALE, D., H. W. KUHN, and A. W. TUCKER, "Linear Programming and the Theory of Games," in *Activity Analysis of Production and Allocation*, T. C. Koopmans (Ed.), John Wiley & Sons, New York, 1951.

17. GELFAND, I. M., and S. V. FOMIN, *Calculus of Variations*, Prentice-Hall, Englewood Cliffs, N.J., 1963.

18. GEOFFRION, A. M., "Duality in Nonlinear Programming: A Simplified Applications-Oriented Development," *SIAM Review*, **13**, 1–37 (1971).

19. HAMALA, M., "Geometric Programming in Terms of Conjugate Functions," Center for Operations Research and Econometrics, Université Catholique de Louvain Discussion Paper No. 6811, Louvain, Belgium, June 1968.

20. HANSON, M. A., "A Duality Theorem in Non-Linear Programming with Non-Linear Constraints," *Australian J. Statistics*, **3**, 64–72 (1961).

21. HUARD, P., "Dual Programs," *IBM J. Res. & Dev.*, **6**, 137–139 (1962).

22. KUHN, H. W., and A. W. TUCKER, "Nonlinear Programming," in *Proceedings of the Second Berkeley Symposium on Mathematical Statistics and Probability*, J. Neyman (Ed.), University of California Press, Berkeley, Calif., 1951.

23. KUHN, H. W., and A. W. TUCKER (Eds.), *Linear Inequalities and Related Systems*, Princeton University Press, Princeton, N.J., 1956.

24. MANGASARIAN, O. L., "Duality in Nonlinear Programming," *Quart. Appl. Math.*, **20**, 300–302 (1962).

25. MANGASARIAN, O. L., *Nonlinear Programming*, McGraw-Hill Book Co., New York, 1969.

26. MANGASARIAN, O. L., and J. PONSTEIN, "Minmax and Duality in Nonlinear Programming," *J. Math. Anal. & Appl.*, **11**, 504–518 (1965).

27. RISSANEN, J., "On Duality without Convexity," *J. Math. Anal. & Appl.*, **18**, 269–275 (1967).

28. RISSANEN, J., "Reproof of a Theorem on Duality without Convexity," *J. Math. Anal. & Appl.*, **29**, 429–431 (1970).

29. ROCKAFELLAR, R. T., "Duality and Stability in Extremum Problems Involving Convex Functions," *Pacific J. Math.*, **21**, 167–187 (1967).

30. ROCKAFELLAR, R. T., "Conjugates and Legendre Transforms of Convex Functions," *Can. J. Math.*, **19**, 200–205 (1967).

31. ROCKAFELLAR, R. T., "Duality in Nonlinear Programming," in *Mathematics of the Decision Sciences*, Part 1, G. B. Dantzig and A. F. Veinott (Eds.), American Mathematical Society, Providence, R. I., 1968.

32. ROCKAFELLAR, R. T., *Convex Analysis*, Princeton University Press, Princeton, N.J., 1970.

33. STOER, J., "Duality in Nonlinear Programming and the Minimax Theorem," *Numerische Mathematik*, **5**, 371–379 (1963).

34. WHINSTON, A., "Some Applications of the Conjugate Function Theory to Duality," in *Nonlinear Programming*, J. Abadie (Ed.), North-Holland Publishing Co., Amsterdam, 1967.

35. WILLIAMS, A. C., "Nonlinear Activity Analysis and Duality," in *Proceedings of the Princeton Symposium on Mathematical Programming*, H. W. Kuhn (Ed.), Princeton University Press, Princeton, N.J., 1970.

36. WOLFE, P., "A Duality Theorem for Non-Linear Programming," *Quart. Appl. Math.*, **19**, 239–244 (1961).

6 GENERALIZED CONVEXITY

In the preceding two chapters we discussed some of the most important aspects of convexity from a mathematical programming point of view. For the results presented there, convexity was a sufficient condition, but, as it turns out, it is by no means a necessary one. In this chapter we extend the notions of convexity so that most of the previously obtained results can be readily generalized.

The subjects discussed here are still evolving, and, indeed, research in this direction constitutes one of the current trends in mathematical programming. Fortunately, the tools introduced in the study of convex sets and functions are usually sufficient to generalize results of the last two chapters for certain types of nonconvex programs. Such generalizations should be quite straightforward, for we know from convex analysis what type of results can be expected. For this reason, we present only some basic ideas on the various generalizations of convexity, and interested readers can continue the work of deriving a more complete theory. In the next chapter we shall analyze some particular nonlinear programs, using the results obtained in this and in preceding chapters.

Besides the additional research needed for a generalized convex analysis, there is even a greater need for research in finding efficient numerical methods to solve nonconvex optimization problems. In the subject area of nonconvex algorithms, the impact of generalized convexity is not yet felt. We hope that the results presented in this chapter will stimulate future work in developing new numerical methods, as well as in extending convergence results of existing methods for generalized convex optimization problems.

6.1 QUASICONVEX AND PSEUDOCONVEX FUNCTIONS

Given a real-valued function f on $X \subset R^n$ and a number $\alpha \in R$, we define a **level set** $S(f, \alpha)$ of f as

$$S(f, \alpha) = \{x : x \in X, f(x) \leq \alpha\}. \tag{6.1}$$

It is easy to show that if X is a convex set and f is a convex function, then $S(f, \alpha)$ is a convex set for every $\alpha \in R$. Conversely, suppose now that, for some f, the set $S(f, \alpha)$ is known to be convex for every $\alpha \in R$. Does this fact imply that f is convex? The reader can readily disprove this assertion by the simple example of $X = \{x : x \in R, x \geq 0\}$ and $f(x) = (x)^{1/2}$, a concave function. We can, nevertheless, characterize functions whose level sets are convex by the first generalization of convexity to be presented here.

A real-valued function f, defined on a convex set $X \subset R^n$, is said to be **quasiconvex** if

$$f(q_1x^1 + q_2x^2) \leq \max\,[f(x^1), f(x^2)] \tag{6.2}$$

for every $x^1 \in X$, $x^2 \in X$, and weights q (recall the properties of weights from Chapter 4). Clearly, every real convex function is quasiconvex, but not conversely. A function f is said to be **quasiconcave** if $-f$ is quasiconvex. The first result for quasiconvex functions is

Theorem 6.1

Let f be a real-valued function defined on a convex set $X \subset R^n$. The level sets of f are convex for every $\alpha \in R$ if and only if f is a quasiconvex function.

Proof. Suppose that $S(f, \alpha)$ is a convex set for every $\alpha \in R$ and let $x^1 \in X$, $x^2 \in X$, $\bar{\alpha} = \max\,[f(x^1), f(x^2)]$. Then $x^1 \in S(f, \bar{\alpha})$, $x^2 \in S(f, \bar{\alpha})$, and since $S(f, \bar{\alpha})$ is convex, $(q_1x^1 + q_2x^2) \in S(f, \bar{\alpha})$ for arbitrary weights q also.

Hence

$$f(q_1x^1 + q_2x^2) \leq \bar{\alpha} = \max\,[f(x^1), f(x^2)]. \tag{6.3}$$

Conversely, let $S(f, \alpha)$ be any level set of f. Let $x^1 \in S(f, \alpha)$, $x^2 \in S(f, \alpha)$. Then

$$f(x^1) \leq \alpha \qquad f(x^2) \leq \alpha \tag{6.4}$$

and since f is quasiconvex, we have

$$f(q_1x^1 + q_2x^2) \leq \alpha \tag{6.5}$$

and $(q_1x^1 + q_2x^2) \in S(f, \alpha)$. ■

Functions having convex level sets have been studied by de Finetti [15], whose results were subsequently generalized and corrected by Fenchel [17].

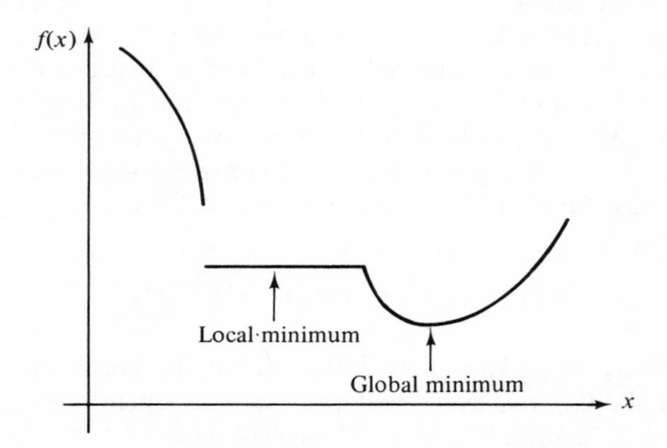

Fig. 6.1 A quasiconvex function.

Figure 6.1 illustrates some of the main differences between convex and quasiconvex functions. In contrast to convex functions, quasiconvex functions can have discontinuities in the interior of their domain, and not every local minimum is a global one. Local minima, however, that are not global cannot be strict minima, as we can see in the following result [24].

Theorem 6.2

Let f be a quasiconvex function on a convex set $X \subset R^n$. If x^* is a strict local minimum of f, then x^* is also a strict global minimum of f on X.

Proof. Suppose that x^* is a strict local minimum—that is, there is a $\delta > 0$ such that for every x in the set $X \cap N_\delta(x^*)$, where, as before,

$$N_\delta(x^*) = \{x : x \in R^n, ||x - x^*|| < \delta\}, \tag{6.6}$$

we have

$$f(x) > f(x^*). \tag{6.7}$$

Suppose now that x^* is not a strict global minimum of f over X—that is, there exists an $\bar{x} \in X$, $\bar{x} \neq x^*$, such that

$$f(\bar{x}) \leq f(x^*). \tag{6.8}$$

Then by quasiconvexity,

$$f(q_1\bar{x} + q_2x^*) \leq f(x^*) \tag{6.9}$$

for all weights q. But for sufficiently small q_1 it follows that $(q_1\bar{x} + q_2x^*) \in X \cap N_\delta(x^*)$, contradicting (6.7). ■

If a quasiconvex function f is differentiable, then further characterizations can be derived.

Theorem 6.3

Let f be differentiable on an open convex set $X \subset R^n$. Then f is quasiconvex if and only if for any $x^1 \in X$, $x^2 \in X$ such that

$$f(x^1) \leq f(x^2), \tag{6.10}$$

we have

$$(x^1 - x^2)^T\nabla f(x^2) \leq 0. \tag{6.11}$$

Proof. Suppose that f is quasiconvex. Let $x^1 \in X$, $x^2 \in X$. Then $f(x^1) \leq f(x^2)$ implies

$$f(\lambda x^1 + (1 - \lambda)x^2) \leq f(x^2) \tag{6.12}$$

for $1 \geq \lambda \geq 0$. Since f is differentiable,

$$Df(x^2; x^1 - x^2) = (x^1 - x^2)^T\nabla f(x^2) \tag{6.13}$$

and

$$Df(x^2; x^1 - x^2) = \lim_{\lambda \to 0^+} \frac{f(x^2 + \lambda(x^1 - x^2)) - f(x^2)}{\lambda}. \tag{6.14}$$

By (6.12) to (6.14), we obtain

$$Df(x^2; x^1 - x^2) = (x^1 - x^2)^T\nabla f(x^2) \leq 0. \tag{6.15}$$

Conversely, let $f(x^1) \leq f(x^2)$ and suppose that there exists an $x^3 = \lambda x^1 + (1 - \lambda)x^2$ such that

$$f(x^2) < f(x^3) \tag{6.16}$$

for some $1 > \lambda > 0$. Then by the hypothesis,

$$(x^2 - x^3)^T \nabla f(x^3) \leq 0 \tag{6.17}$$

$$(x^1 - x^3)^T \nabla f(x^3) \leq 0. \tag{6.18}$$

Writing x^3 as a convex combination of x^1 and x^2, substituting into (6.17) and (6.18), and rearranging, we get

$$(x^2 - x^1)^T \nabla f(x^3) \leq 0 \tag{6.19}$$

$$(x^1 - x^2)^T \nabla f(x^3) \leq 0. \tag{6.20}$$

Hence

$$(x^2 - x^1)^T \nabla f(x^3) = 0. \tag{6.21}$$

Define

$$U = \{x : f(x) \leq f(x^2), x = \mu x^2 + (1 - \mu)x^3, 1 \geq \mu \geq 0\}. \tag{6.22}$$

Then we can find an $x^0 \in U$, nearest to x^3 and an $\tilde{x}$ between x^0 and x^3 such that, by the Mean Value Theorem [8],

$$f(x^3) = f(x^0) + (x^3 - x^0)^T \nabla f(\tilde{x}) \tag{6.23}$$

$$= f(x^0) + \mu^0 (x^3 - x^2)^T \nabla f(\tilde{x}) \tag{6.24}$$

for some $1 \geq \mu^0 > 0$, and

$$f(x^3) = f(x^0) + \lambda \mu^0 (x^1 - x^2)^T \nabla f(\tilde{x}). \tag{6.25}$$

Since $\tilde{x}$ is also some convex combination of x^1 and x^2 and $f(x^2) < f(\tilde{x})$, it follows from (6.16) through (6.21) that

$$(x^1 - x^2)^T \nabla f(\tilde{x}) = 0. \tag{6.26}$$

Hence $f(x^3) = f(x^0) \leq f(x^2)$, contradicting (6.16). ■

For the quasiconcave case, the senses of the inequalities in (6.10) and (6.11) are reversed.

In the case of twice differentiable functions, Arrow and Enthoven [2] derived necessary and sufficient conditions for quasiconvexity on the nonnegative orthant of R^n in terms of **bordered determinants** associated with the functions involved. These conditions were subsequently extended by Ferland [18] to functions on more general convex sets. The kth-order bordered deter-

minant $D_k(f, x)$ of a twice differentiable function f at a point $x \in R^n$ is defined as

$$D_k(f, x) = \det \begin{bmatrix} 0 & \frac{\partial f}{\partial x_1} & \cdots & \frac{\partial f}{\partial x_k} \\ \frac{\partial f}{\partial x_1} & \frac{\partial^2 f}{\partial x_1 \partial x_1} & \cdots & \frac{\partial^2 f}{\partial x_1 \partial x_k} \\ \cdot & \cdot & & \cdot \\ \cdot & \cdot & & \cdot \\ \cdot & \cdot & & \cdot \\ \frac{\partial f}{\partial x_k} & \frac{\partial^2 f}{\partial x_k \partial x_1} & \cdots & \frac{\partial^2 f}{\partial x_k \partial x_k} \end{bmatrix} \qquad k = 1, \ldots, n. \tag{6.27}$$

A set $X \subset R^n$ is called **solid** if it has a nonempty interior (the nonnegative orthant of R^n is clearly a solid set). A necessary condition for f to be quasiconvex on a solid convex set $X \subset R^n$ is that

$$D_k(f, x) \leq 0, \qquad k = 1, \ldots, n \tag{6.28}$$

for every $x \in X$.
Similarly, if f is quasiconcave on a solid convex set $X \subset R^n$, then

$$(-1)^k D_k(f, x) \geq 0, \qquad k = 1, \ldots, n \tag{6.29}$$

for every $x \in X$. Conversely, a real-valued function f, twice continuously differentiable on an open set containing the convex set $X \subset R^n$ such that

$$D_k(f, x) < 0, \qquad k = 1, \ldots, n \tag{6.30}$$

for every $x \in X$, is quasiconvex on X. If

$$(-1)^k D_k(f, x) < 0, \qquad k = 1, \ldots, n \tag{6.31}$$

for every $x \in X$, then f is quasiconcave on X.

Let us strengthen quasiconvexity. A real-valued function f defined on a convex set $X \subset R^n$ is said to be **strongly quasiconvex** if

$$f(q_1 x^1 + q_2 x^2) < \max\,[f(x^1), f(x^2)] \tag{6.32}$$

for any two points $x^1 \in X$, $x^2 \in X$, $x^1 \neq x^2$ and for any weights $q_1 > 0$, $q_2 > 0$, $q_1 + q_2 = 1$. For such functions we have the following result, which follows immediately from the foregoing definition.

Theorem 6.4

A strongly quasiconvex function f defined on a convex set $X \subset R^n$ attains its minimum on X at no more than one point.

The reader can easily verify that every strictly convex function is also strongly quasiconvex.

We can slightly weaken the concept of strong quasiconvexity by defining a real-valued function f on a convex set $X \subset R^n$ to be **strictly quasiconvex** if and only if (6.32) holds for any two points $x^1 \in X$, $x^2 \in X$, such that $f(x^1) \neq f(x^2)$, and for any $q_1 > 0$, $q_2 > 0$, $q_1 + q_2 = 1$ [20]. It is easy to show that convex and strongly quasiconvex functions are also strictly quasiconvex. Although strongly quasiconvex functions are also quasiconvex, such is not necessarily the case for strictly quasiconvex functions, for

$$f(x) = \begin{cases} 1 & \text{if } x = 0 \\ 0 & \text{if } x \neq 0 \end{cases} \tag{6.33}$$

is strictly quasiconvex on $X = R$, but it is not quasiconvex. It has been shown, however, that strict quasiconvexity implies quasiconvexity if the function is lower semicontinuous [23]. A nice property of strictly quasiconvex functions is given in

Theorem 6.5

Let f be a strictly quasiconvex function on a convex set $X \subset R^n$ and let $x^* \in X$ be a local minimum of f. Then x^* is also a global minimum of f over X.

The proof is very similar to that of Theorem 6.2 and will be omitted. Later in this chapter we shall characterize functions with the property stated in this theorem.

We saw in Chapter 3 that, for a general nonlinear program, the Kuhn-Tucker necessary conditions for optimality may not hold without some regularity conditions. In the case of a convex program, Slater's condition of strong consistency, as stated in Chapter 4, ensured that these necessary conditions do indeed hold at an optimum. Slater's condition, and hence Theorem 4.39, has been extended to more general programs by Arrow and Enthoven [2] and by Arrow, Hurwicz, and Uzawa [4]. Suppose that we have a nonlinear program (P), defined in Section 3.1, whose constraints are given by

$$g_i(x) \geq 0, \qquad i = 1, \ldots, m \tag{6.34}$$

$$h_j(x) = 0, \qquad j = 1, \ldots, p \tag{6.35}$$

where the g_i and h_j are differentiable quasiconcave and linear functions, respectively. If such a program is strongly consistent—that is, there exists an $x^0 \in R^n$ satisfying (6.34) and (6.35) such that the inequalities are strict, the coefficient vectors of h_j are linearly independent, and

$$\nabla g_i(x^*) \neq 0, \qquad i \in I(x^*) = \{i : g_i(x^*) = 0\} \tag{6.36}$$

then Theorem 4.39, originally derived for convex programs, also holds in the more general case.

We note here that the set of $x \in R^n$ satisfying (6.34) and (6.35) is a convex set. Thus if a convex function is minimized subject to these constraints, then every local minimum is global. This result will be extended in the sequel to other families of objective functions. Also note that general nonlinear programs having quasiconcave constraints may have points other than the global solution that satisfy the necessary optimality conditions. For an interesting comparison of the properties of convex and quasiconvex functions, the reader is referred to Greenberg and Pierskalla [19].

Let us turn now to another extension of convexity that was introduced by Mangasarian [25, 26].

A real-valued differentiable function f is said to be **pseudoconvex** on an open convex set $X \subset R^n$ if

$$(x^1 - x^2)^T \nabla f(x^2) \geq 0 \quad \text{implies} \quad f(x^1) \geq f(x^2) \tag{6.37}$$

for any two $x^1 \in X$, $x^2 \in X$. A function f is called **pseudoconcave** if $-f$ is pseudoconvex or, equivalently, both inequalities in (6.37) are reversed. If f is both pseudoconvex and pseudoconcave, it is sometimes called **pseudolinear**. Interestingly, certain properties of mathematical programs involving linear functions can be extended to programs that involve pseudolinear functions [29]. Differentiable convex functions are pseudoconvex, and pseudoconvex functions are, in turn, strictly quasiconvex. Thus every local minimum of a pseudoconvex function is also global. Although strictly quasiconvex functions may have stationary points where the gradient vanishes that are not global minima, this situation is impossible for pseudoconvex functions.

Theorem 6.6

Let f be a pseudoconvex function on an open convex set $X \subset R^n$ and suppose that $\nabla f(x^*) = 0$ for some $x^* \in X$. Then x^* is a global minimum of f over X.

Proof. Follows immediately from the definition of pseudoconvex functions. ■

Ferland [18] showed that, for twice continuously differentiable functions, the earlier stated sufficient condition for quasiconvexity in terms of determinants, is also sufficient for pseudoconvexity. He proved that if f is a twice differentiable quasiconvex function on a solid convex set $X \subset R^n$, then f is pseudoconvex at any point $x^0 \in X$, where $\nabla f(x^0) \neq 0$. That is, the condition $(x - x^0)^T \nabla f(x^0) \geq 0$ implies $f(x) \geq f(x^0)$ for every $x \in X$.

We can place a further restriction on a pseudoconvex function by requiring the implied inequality in (6.37) to be a strict inequality for $x^1 \neq x^2$. In this case, the function is called **strictly pseudoconvex**, and Theorem 6.6 can be strengthened to state that if $\nabla f(x^*) = 0$ then x^* is the unique global minimum of f over X. For a review of the relationship between the various families of generalized convex functions defined so far, the reader is referred to Ponstein [37], where additional families are also investigated.

Let us now show that the Kuhn-Tucker necessary conditions for optimality in a nonlinear program, as stated in Theorem 3.8, are also sufficient when the objective and constraint functions are certain generalized convex and concave functions. The following result is then a generalization of Theorem 4.38. Consider the nonlinear program

$$\text{(PQP)} \qquad \min f(x) \tag{6.38}$$

subject to the constraints

$$g_i(x) \geq 0, \qquad i = 1, \ldots, m \tag{6.39}$$

$$h_j(x) = 0, \qquad j = 1, \ldots, p. \tag{6.40}$$

As before, let

$$I(x^0) = \{i : g_i(x^0) = 0, i = 1, \ldots, m\}. \tag{6.41}$$

We then have [25] the following result, a somewhat restricted version of which can be also found in [2], where economic aspects of quasiconvexity are also discussed.

Theorem 6.7

Let $f, g_1, \ldots, g_m$ and $h_1, \ldots, h_p$ be real-valued differentiable functions on some open convex set $X \subset R^n$. Suppose that f is pseudoconvex, the g_i, $i \in I(x^*)$, are quasiconcave, and the h_j are both quasiconvex and quasiconcave. If there exist vectors $\lambda^* \in R^m$, $\mu^* \in R^p$ such that

$$\nabla f(x^*) - \sum_{i=1}^{m} \lambda_i^* \nabla g_i(x^*) - \sum_{j=1}^{p} \mu_j^* \nabla h_j(x^*) = 0 \tag{6.42}$$

$$\lambda_i^* g_i(x^*) = 0, \qquad i = 1, \ldots, m \tag{6.43}$$

$$g_i(x^*) \geq 0, \qquad i = 1, \ldots, m \tag{6.44}$$

$$h_j(x^*) = 0, \qquad j = 1, \ldots, p \tag{6.45}$$

$$\lambda^* \geq 0, \tag{6.46}$$

then x^* is a global optimum of (PQP).

Proof. Let $x \in X$ be any point satisfying (6.39) and (6.40). Then

$$g_i(x) \geq g_i(x^*), \qquad i \in I(x^*) \tag{6.47}$$

$$h_j(x) = h_j(x^*), \qquad j = 1, \ldots, p. \tag{6.48}$$

Hence

$$(x - x^*)^T \nabla g_i(x^*) \geq 0, \qquad i \in I(x^*) \tag{6.49}$$

$$(x - x^*)^T \nabla h_j(x^*) = 0, \qquad j = 1, \ldots, p. \tag{6.50}$$

Note that $\lambda_i^* = 0$ for $i \notin I(x^*)$. It follows that

$$(x - x^*)^T \left\{ \sum_{i=1}^{m} \lambda_i^* \nabla g_i(x^*) + \sum_{j=1}^{p} \mu_j^* \nabla h_j(x^*) \right\} \geq 0 \tag{6.51}$$

and by (6.42),

$$(x - x^*)^T \nabla f(x^*) \geq 0. \tag{6.52}$$

Since f is pseudoconvex, we have, by (6.37), $f(x) \geq f(x^*)$. ■

Minch [34] has extended the notions of differentiable convex, pseudoconvex, and quasiconvex functions by defining the **symmetric gradient** $f^s(x)$ of a real-valued function f. In the case of a function of one variable, it is given by

$$f^s(x) = \lim_{h \to 0} \frac{f(x + h) - f(x - h)}{2h}, \tag{6.53}$$

provided that this limit exists. For example, the function $f(x) = |x|$ is not differentiable at $x = 0$, but it has a symmetric gradient there and $f^s(0) = 0$. In replacing ordinary gradients of convex functions and their extensions discussed so far by symmetric gradients, Minch has obtained results similar to the differentiable case.

Parenthetically, we can mention that convex functions also play an important role in the theory of games in so-called minimax theorems, the first of which was stated by von Neumann in 1928. These theorems were subsequently generalized by also proving them for quasiconvex functions; see Berge [12] and Nikaidô [35].

In spite of the apparent usefulness of the results stated above, the difficult problem of establishing the quasiconvexity or pseudoconvexity of a function still remains. The difficulty lies in the definitions of these families of functions, since they generally involve checking infinitely many inequalities. Clearly, the same difficulty also arises in recognizing a convex function. Under suitable conditions, however, it is possible to identify the family

to which a **composite** function belongs. We now study this aspect in a more detailed form. The forthcoming results are mainly due to Mangasarian [27]. Additional, more general results were obtained by Schaible [43, 44].

A real-valued function ϕ defined on a set $T \subset R^m \times R^k$ is said to be **increasing-decreasing** (incr-decr) on T if and only if for every $(y^1, z^1) \in T$ and $(y^2, z^2) \in T$

$$y^2 \geq y^1 \quad \text{and} \quad z^2 \leq z^1 \quad \text{imply} \quad \phi(y^2, z^2) \geq \phi(y^1, z^1). \tag{6.54}$$

Similarly, ϕ is said to be **y-increasing** (y-incr) on T if and only if for every $(y^1, z) \in T$, $(y^2, z) \in T$

$$y^2 \geq y^1 \quad \text{implies} \quad \phi(y^2, z) \geq \phi(y^1, z) \tag{6.55}$$

and ϕ is **y-decreasing** (y-decr) on T if and only if for every $(y^1, z) \in T$ and $(y^2, z) \in T$

$$y^2 \leq y^1 \quad \text{implies} \quad \phi(y^2, z) \geq \phi(y^1, z). \tag{6.56}$$

The next result follows immediately from these definitions.

Lemma 6.8

Let ϕ be a real-valued differentiable function on an open convex set $T \subset R^m \times R^k$. Then ϕ is incr-decr on T if and only if

$$\nabla_y \phi(y, z) \geq 0 \qquad \nabla_z \phi(y, z) \leq 0 \tag{6.57}$$

for every $(y, z) \in T$.

We then have

Theorem 6.9

Let $X \subset R^n$ be a convex set, let $f(x) = (f_1(x), \ldots, f_m(x))$ and $g(x) = (g_1(x), \ldots, g_k(x))$ be defined on X, and let ϕ be a real-valued function on $R^m \times R^k$. Define

$$\Phi(x) = \phi(f(x), g(x)) \tag{6.58}$$

and let any one of the following assumptions hold:

(i) *f is convex, g is concave, ϕ is incr-decr*
(ii) *f is linear, g is linear*
(iii) *f is convex, g is linear, ϕ is y-incr*
(iv) *f is concave, g is linear, ϕ is y-decr.*

Then

(a) *If ϕ is convex, then Φ is convex.*

(b) *If X is open, f and g are differentiable on X, and ϕ is pseudoconvex, then Φ is pseudoconvex.*

(c) *If ϕ is quasiconvex, then Φ is quasiconvex.*

Proof. Part (a) can be proved by some easy modifications of Theorem 4.12. We shall now prove part (b) under assumption (i). Let $x^1 \in X, x^2 \in X$. Then

$$(x^2 - x^1)^T\nabla\Phi(x^1) = (x^2 - x^1)^T[\nabla f(x^1)\nabla_f\phi(f(x^1), g(x^1)) + \nabla g(x^1)\nabla_g\phi(f(x^1), g(x^1))]. \tag{6.59}$$

By (4.116) and Lemma 6.8, it follows that

$$(x^2 - x^1)^T\nabla\Phi(x^1) \leq [f(x^2) - f(x^1)]^T\nabla_f\phi(f(x^1), g(x^1)) + [g(x^2) - g(x^1)]^T\nabla_g\phi(f(x^1), g(x^1)). \tag{6.60}$$

Hence if the left-hand side of (6.60) is nonnegative, so is the right-hand side; and by the pseudoconvexity of ϕ, we obtain

$$\phi(f(x^1), g(x^1)) \leq \phi(f(x^2), g(x^2)) \tag{6.61}$$

or

$$\Phi(x^1) \leq \Phi(x^2) \tag{6.62}$$

and Φ is pseudoconvex. The reader can easily complete the proof of part (b) under conditions (ii) to (iv).

Let us now prove part (c) under assumption (i).

Let $x^1 \in X$, $x^2 \in X$ and let q_1, q_2 be weights. If $\Phi(x^1) \leq \Phi(x^2)$, then

$$\phi(f(x^1), g(x^1)) \leq \phi(f(x^2), g(x^2)). \tag{6.63}$$

Since ϕ is quasiconvex, we have

$$\phi(q_1 f(x^1) + q_2 f(x^2), q_1 g(x^1) + q_2 g(x^2)) \leq \phi(f(x^2), g(x^2)) \tag{6.64}$$

and, by the hypotheses, it follows that

$$\phi(f(q_1 x^1 + q_2 x^2), g(q_1 x^1 + q_2 x^2)) \leq \phi(f(x^2), g(x^2)) \tag{6.65}$$

or

$$\Phi(q_1 x^1 + q_2 x^2) \leq \Phi(x^2) \tag{6.66}$$

and Φ is quasiconvex. The rest of the proof is similar. ∎

Theorem 6.9 can be applied to a large class of functions. Consider first **nonlinear fractional functions**. Let X be a convex set in R^n, let h_1 and h_2 be

real-valued functions on X, and let

$$\Phi(x) = \frac{h_1(x)}{h_2(x)}. \tag{6.67}$$

Suppose that any one of the following assumptions holds on X:

1. h_1 is nonnegative and convex; h_2 is positive and concave.
2. h_1 is nonpositive and concave; h_2 is negative and convex.
3. h_1 is nonpositive and convex; h_2 is positive and convex.
4. h_1 is nonnegative and concave; h_2 is negative and concave.
5. h_1 is linear; h_2 is nonzero and linear.
6. h_1 is nonpositive and linear; h_2 is nonzero and convex.
7. h_1 is nonnegative and linear; h_2 is nonzero and concave.
8. h_1 is convex; h_2 is positive and linear.
9. h_1 is concave; h_2 is negative and linear.

Then Φ is quasiconvex on X. In addition, if X is open and h_1 and h_2 are differentiable on X, then Φ is also pseudoconvex.

Next, consider **reciprocal functions**. Assume now that

$$\Phi(x) = \frac{1}{h(x)}. \tag{6.68}$$

If h is positive and concave on X, then Φ is convex on X. If h is negative and convex on X, then Φ is concave on X.

Theorem 6.9 can be also applied to **bi-nonlinear functions**. Let $X \subset R^n$ be an open convex set. Let h_1 and h_2 be differentiable on X and define

$$\Phi(x) = h_1(x)h_2(x). \tag{6.69}$$

Then Φ is pseudoconvex under assumptions (1) or (2) below:

1. h_1 is nonpositive and convex; h_2 is positive and concave.
2. h_1 is negative and convex; h_2 is nonnegative and concave.

Extensions of these results can be found in [43, 44].

Convexity of a real-valued **quadratic** function on R^n can be verified by checking a finite number of conditions on its Hessian matrix. There are, however, quadratic functions that are nonconvex but only pseudoconvex or quasiconvex. For example, the function of two variables

$$g(x) = -x_1x_2 \tag{6.70}$$

is quasiconvex for $x_1 \geq 0$, $x_2 \geq 0$, but it is not convex. Finite tests, similar to the convex case, have been derived by Cottle and Ferland [13, 14], Ferland [18], and Martos [32] for checking the pseudo-or quasiconvexity of quadratic

functions. The following discussion is based on these references, in which proofs of the results can be found.

We begin with a few definitions of certain classes of matrices. A real symmetric matrix Q and its corresponding quadratic form $x^T Qx$ are called **positive subdefinite** [32] if

$$x^T Qx < 0 \quad \text{implies} \quad Qx \geq 0 \quad \text{or} \quad Qx \leq 0 \tag{6.71}$$

and **strictly positive subdefinite** if

$$x^T Qx < 0 \quad \text{implies} \quad Qx > 0 \quad \text{or} \quad Qx < 0. \tag{6.72}$$

Recall that Q and $x^T Qx$ are called **positive semidefinite** if $x^T Qx \geq 0$ for every $x \in R^n$, or, in a form similar to the last two definitions,

$$x^T Qx \leq 0 \quad \text{implies} \quad Qx = 0. \tag{6.73}$$

Positive semidefiniteness is a necessary and sufficient condition for the convexity of the quadratic form $x^T Qx$. A comparison of the preceding three classes shows that positive semidefinite quadratic forms are (trivially) strictly positive subdefinite forms, which are, in turn, positive subdefinite. A real symmetric matrix is said to be (strictly) **merely** positive subdefinite if it is (strictly) positive subdefinite but not positive semidefinite. Similarly, we call a function merely quasiconvex (pseudoconvex) if it is quasiconvex (pseudoconvex) but not convex. We have then the following basic result.

Theorem 6.10

The quadratic form $x^T Qx$ is quasiconvex on the nonnegative orthant if and only if it is positive subdefinite. It is pseudoconvex on the nonnegative orthant, not including the origin, if and only if it is strictly positive subdefinite.

The proof of this theorem can be found in [32].

A simple way to test whether a merely quasiconvex quadratic form is also pseudoconvex is by the following result of Martos.

Theorem 6.11

A merely quasiconvex quadratic form $x^T Qx$ on the nonnegative orthant is merely pseudoconvex on the nonnegative orthant, not including the origin, if and only if the matrix Q does not contain a row of zeros.

Let us turn now to characterizing subdefinite matrices and quadratic forms.

Theorem 6.12

The quadratic form $x^T Qx$ is merely positive subdefinite if and only if
(i) *every element of the matrix Q is nonpositive and there is at least one negative element in Q.*
(ii) *the determinants of all the principal minors of Q are nonpositive.*

The proof of this theorem can be found in Cottle and Ferland [13]. Finally, we give a sufficient condition for mere pseudoconvexity.

Theorem 6.13

Suppose that every element of the real symmetric matrix Q is nonpositive and there is at least one negative element in Q. If the determinants of all the leading principal minors of Q are negative, then $x^T Qx$ is strictly merely positive subdefinite.

Theorem 6.12 provides a finite test for quasiconvexity of quadratic forms on the nonnegative orthant via the determinants of the principal minors. This test complements, in a certain sense, the similar test for positive semidefiniteness. Another interesting property of merely positive subdefinite quadratic forms $x^T Qx$ is that Q has exactly one (simple) negative eigenvalue, and the corresponding eigenvector is either nonnegative or nonpositive, but it does not vanish. We mention here briefly the conditions for mere quasiconcavity:

1. Every element of the matrix Q is nonnegative, and there is at least one positive element in Q.
2. Denote by D_{ik} the determinant of the ith principal minor of order k in Q. Then $(-1)^k D_{ik}$ is nonpositive for every i.

Quadratic functions generally consist of a sum of a quadratic form and a linear function. If the quadratic form is convex, the above sum will also be convex. Unfortunately, the sum of a pseudoconvex or quasiconvex function and a linear function is not necessarily pseudoconvex or quasiconvex. The previous results for quadratic forms have been extended, however, to quadratic functions [13, 14, 18], and the results are summarized below.

Theorem 6.14

The quadratic function

$$f(x) = b^T x + \tfrac{1}{2} x^T Qx, \qquad x \in R^n \tag{6.74}$$

is convex on R^n if and only if it is quasiconvex. It is merely quasiconvex on the nonnegative orthant if and only if the matrix

$$\bar{Q} = \begin{bmatrix} Q & b \\ b^T & 0 \end{bmatrix} \tag{6.75}$$

is merely positive subdefinite. If f is merely quasiconvex on the nonnegative orthant and $b \neq 0$, then f is also pseudoconvex on the nonnegative orthant.

Example 6.1.1 [18]

Consider the quadratic form

$$f_1(x) = -(x_1 + x_2)^2. \tag{6.76}$$

If we wish to write

$$f_1(x) = x^T Q x, \tag{6.77}$$

we obtain

$$Q = \begin{bmatrix} -1 & -1 \\ -1 & -1 \end{bmatrix}. \tag{6.78}$$

The eigenvalues of Q are $\lambda^1 = -2$ and $\lambda^2 = 0$, and the eigenvector corresponding to λ^1 is $v^1 = (1, 1)^T$. This quadratic form is nonconvex, since Q is not positive semidefinite (it is actually negative semidefinite). However, by Theorem 6.12, the matrix Q is merely positive subdefinite. Hence, by Theorem 6.10, the function f_1 is merely quasiconvex on the nonnegative orthant. Moreover, by Theorem 6.11, it is also merely pseudoconvex. Consider now the quadratic function

$$f_2(x) = -x_1 - x_2 - \tfrac{1}{2}(x_1 + x_2)^2. \tag{6.79}$$

From (6.75) we then have

$$\bar{Q} = \begin{bmatrix} -1 & -1 & -1 \\ -1 & -1 & -1 \\ -1 & -1 & 0 \end{bmatrix}. \tag{6.80}$$

The reader can easily verify that $\bar{Q}$ is merely positive subdefinite. Hence, by Theorem 6.15, this quadratic function is merely quasiconvex and merely pseudoconvex on the nonnegative orthant. ∎

It should be clear that when convexity is sufficient but not necessary, some analytical or computational advantages may result from deriving such generalizations as pseudoconvexity or quasiconvexity. The study of these generalizations is especially helpful if the functions involved in a mathematical program are nonconvex but "merely" belong to one of the families

mentioned in the preceding discussions. Such functions can sometimes be transformed into convex functions by a nonlinear transformation. This question will be studied in the next section.

We close this section by mentioning the works of Beckenbach [9] and Beckenbach and Bing [10], where convex functions are generalized by using the notion of "sub-Φ functions." The main results of these works can also be found in Berge [12].

6.2 ARCWISE-CONNECTED SETS AND CONVEX-TRANSFORMABLE FUNCTIONS

Let us now generalize convexity from another point of view. A subset C of R^n was defined in Chapter 4 to be a convex set if for any two points in C the line segment joining them is also in C. This definition can be generalized if, instead of the line segment, we require that for every two points there is a continuous curve joining them that is entirely contained in the set. This idea is well known in analysis and can be stated as follows:

A set $S \subset R^n$ is said to be **arcwise connected** if for every pair of points $x^1 \in S$, $x^2 \in S$ there is a continuous vector-valued function p_{x^1,x^2}, called a **connecting function** or **arc**, defined on the unit interval $[0, 1] \subset R$ and with values in S such that

$$p_{x^1,x^2}(0) = x^1 \qquad p_{x^1,x^2}(1) = x^2. \tag{6.81}$$

Note that p_{x^1,x^2} generally depends on the two points x^1, x^2 and the set S, and for a pair of points x^1, x^2 in an arcwise-connected set there may be more than a single connecting function. For $x^1 = x^2 \in S$ we can choose

$$p_{x^1,x^2}(\theta) = x^1, \qquad 0 \leq \theta \leq 1. \tag{6.82}$$

Example 6.2.1

It is clear that convex sets are also arcwise connected and that the function

$$p_{x^1x^2}(\theta) = (1-\theta)x^1 + \theta x^2, \qquad 0 \leq \theta \leq 1 \tag{6.83}$$

is a connecting function for any two points in a convex set. Consider now the set of points $S_1 \subset R^2$ lying outside a circle centered at the origin and having radius r. That is,

$$S_1 = \{(x_1, x_2) : (x_1)^2 + (x_2)^2 \geq (r)^2\}. \tag{6.84}$$

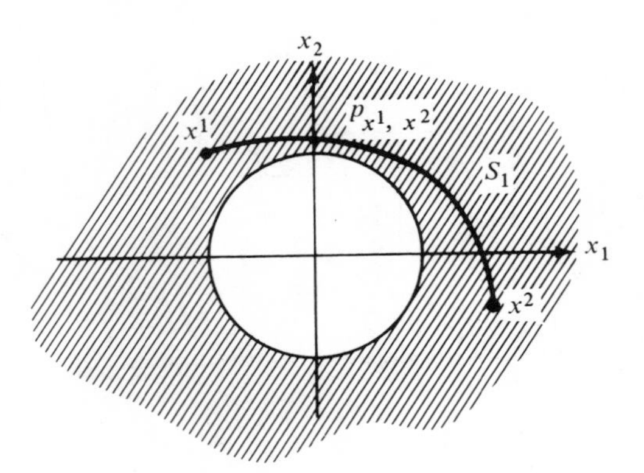

Fig. 6.2 Complement of circle is arcwise connected.

Then for any two points $x^1 \in S_1$, $x^2 \in S_1$, the function, expressed in polar coordinates

$$p_{x^1,x^2}(\theta) = \begin{bmatrix} [(1-\theta)r^1 + \theta r^2] \cos((1-\theta)\alpha^1 + \theta\alpha^2) \\ [(1-\theta)r^1 + \theta r^2] \sin((1-\theta)\alpha^1 + \theta\alpha^2) \end{bmatrix}, \qquad 0 \leq \theta \leq 1 \tag{6.85}$$

where

$$x_1^i = r^i \cos \alpha^i \qquad x_2^i = r^i \sin \alpha^i, \qquad i = 1, 2 \tag{6.86}$$

is a connecting function and S_1 is an arcwise-connected set. See also Figure 6.2. ∎

Arcwise-connected sets are closely related to connected sets. In particular, every arcwise-connected set is connected, and every open connected set is also arcwise connected. We remind the reader that a set $D \subset R^n$ is said to be disconnected if there exist two open sets A, B such that $A \cap D$ and $B \cap D$ are disjoint, nonempty sets whose union is D. A subset of R^n that is not disconnected is said to be **connected**. For further details, the reader is referred to Apostol [1], Bartle [8], Berge [12], or Singer and Thorpe [45]. Some examples of arcwise-connected sets are shown in Figure 6.3.

The epigraph and effective domain of a convex function are convex sets; in addition, the domains of the generalized convex functions discussed so far are convex. Now we begin to study some new generalizations of convex functions that are related to arcwise-connected sets.

It was shown in Theorem 4.15 that a function is convex if and only if its restriction to each line segment in its domain is also convex. This statement, of course, does not mean that restrictions of convex functions to other arcs in a convex set are also convex. Suppose that we consider families of

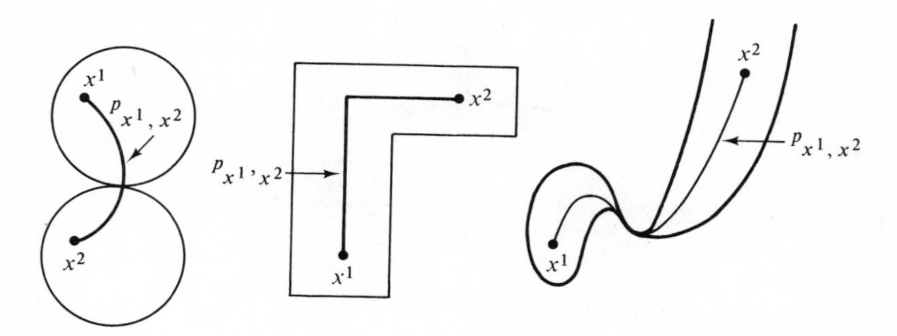

Fig. 6.3 Arcwise-connected sets.

real functions defined on arcwise-connected sets such that the restrictions of the functions to certain arcs in the domains of the functions are convex (ordinary convex functions constitute one such family with line segments taken as arcs). Let S be the (arcwise-connected) domain of such a function, and for every pair of points $x^1 \in S$, $x^2 \in S$, let H_{x^1,x^2} be the arc with the foregoing property. We can then write

$$f(H_{x^1,x^2}(\theta)) \leq (1-\theta)f(x^1) + \theta f(x^2), \qquad 0 \leq \theta \leq 1. \tag{6.87}$$

This idea can be further extended by considering functions that may be nonconvex, but it is possible to transform them into convex functions by some continuous strictly monotone function. In fact, for every pair of real numbers α^1 and α^2, define $\Phi_{\alpha^1\,\alpha^2}$ to be a strictly monotone function on the interval $[0, 1] \subset R$ with values in R such that

$$\Phi_{\alpha^1,\alpha^2}(0) = \alpha^1 \qquad \Phi_{\alpha^1,\alpha^2}(1) = \alpha^2. \tag{6.88}$$

A real-valued function f, defined on an arcwise-connected set S, is then said to be **arcwise convex** if for every pair of points $x^1 \in S$, $x^2 \in S$, and $0 \leq \theta \leq 1$, we have

$$f(H_{x^1,x^2}(\theta)) \leq \Phi_{f(x^1),f(x^2)}(\theta). \tag{6.89}$$

The classical inequality (4.42), defining real-valued convex functions, is a special case of (6.89), obtained by letting

$$H_{x^1,x^2}(\theta) = (1-\theta)x^1 + \theta x^2 \tag{6.90}$$

and

$$\Phi_{f(x^1),f(x^2)}(\theta) = (1-\theta)f(x^1) + \theta f(x^2). \tag{6.91}$$

Additional material on arcwise-convex functions and related subjects can be found in Avriel and Zang [7] and Zang [46].

In the remainder of this section we consider an important subclass of arcwise-convex functions. The particular H and Φ used in the definition of convex functions are actually the weighted arithmetic means of a pair of points x^1, x^2 and their corresponding function values $f(x^1)$, $f(x^2)$, respectively, as can be seen in (6.90) and (6.91). This fact suggests considering the class of arcwise-convex functions for which H_{x^1,x^2} and $\Phi_{f(x^1),f(x^2)}$ are more general **mean value functions** [21]. We now explore this avenue of generalized convexity in more detail.

As in the case of convex functions, we can assume that f is defined on R^n with values in the extended reals; that is, $f(x)$ is either a real number or it is $\pm\infty$. Notions defined earlier, such as closed, proper, and effective domain, will also be used here. Let h be a continuous one-to-one and onto function defined on a subset of R^n, including the effective domain of f and with values in R^n. Similarly, let ϕ be continuous one-to-one and onto, hence a strictly monotone function defined on a subset of the extended reals, including the finite range of f and with values in R. Note that both h and ϕ have one-to-one inverse functions h^{-1} and ϕ^{-1}, respectively, that satisfy $h^{-1}h(x) = hh^{-1}(x) = x$ and $\phi^{-1}\phi(\alpha) = \phi\phi^{-1}(\alpha) = \alpha$. Assume now that H_{x^1,x^2} is an **h-mean value function** given by

$$H_{x^1,x^2}(\theta) = h^{-1}[(1-\theta)h(x^1) + \theta h(x^2)], \qquad 0 \leq \theta \leq 1. \tag{6.92}$$

Similarly, $\Phi_{f(x^1),f(x^2)}$ is assumed to be a **ϕ-mean value** function

$$\Phi_{f(x^1),f(x^2)}(\theta) = \phi^{-1}[(1-\theta)\phi f(x^1) + \theta\phi f(x^2)], \qquad 0 \leq \theta \leq 1. \tag{6.93}$$

Accordingly, a proper arcwise-convex function f on R^n is said to be **(h, ϕ)-convex** if for every $x^1 \in R^n$, $x^2 \in R^n$, and $0 \leq \theta \leq 1$ we have

$$f[h^{-1}((1-\theta)h(x^1) + \theta h(x^2))] \leq \phi^{-1}[(1-\theta)\phi f(x^1) + \theta\phi f(x^2)]. \tag{6.94}$$

We shall now see a few examples of (h, ϕ)-convex functions.

Example 6.2.2

(a) First we take $h(x) = x$, $x \in R^n$, and $\phi(\alpha) = e^{r\alpha}$, where r is a real nonzero parameter. Then substituting into (6.94), we obtain

$$f((1-\theta)x^1 + \theta x^2) \leq \log[(1-\theta)e^{rf(x^1)} + \theta e^{rf(x^2)}]^{1/r}. \tag{6.95}$$

Functions satisfying (6.95) are also called **r-convex** and were introduced by Avriel [5] and Martos [30].

(b) Next, take Rosenbrock's curved valley function on R^2, given by

$$f(x) = 100[x_2 - (x_1)^2]^2 + (1 - x_1)^2. \tag{6.96}$$

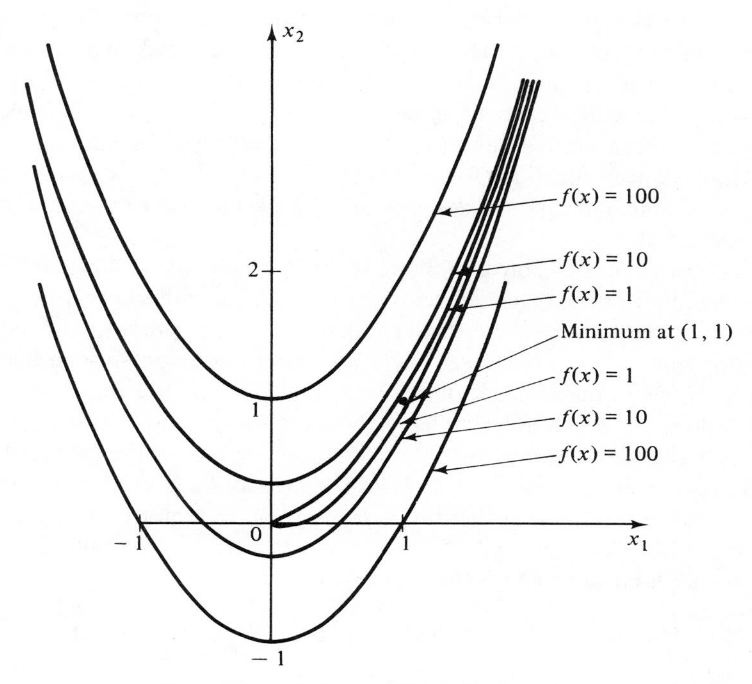

Fig. 6.4 Rosenbrock's curved valley function.

This is a continuously differentiable nonconvex function having a unique minimum at $x^* = (1, 1)$. However, it does not belong to any family of generalized convex functions discussed in the preceding section, such as pseudo- or quasiconvex functions. In fact, its level sets are nonconvex "banana-shaped" sets, as can be seen in Figure 6.4. Letting

$$h(x) = \begin{pmatrix} 10[x_2 - (x_1)^2] \\ 1 - x_1 \end{pmatrix} \tag{6.97}$$

$$h^{-1}(t) = \begin{pmatrix} 1 - t_2 \\ \frac{1}{10}t_1 + (1 - t_2)^2 \end{pmatrix} \tag{6.98}$$

and $\phi(\alpha) = \alpha$, we obtain for every $x^1 \in R^2$, $x^2 \in R^2$

$$H_{x^1,x^2}(\theta) = \begin{pmatrix} (1-\theta)x_1^1 + \theta x_1^2 \\ (1-\theta)[x_2^1 - (x_1^1)^2] + \theta[x_2^2 - (x_1^2)^2] + [(1-\theta)x_1^1 + \theta x_1^2]^2 \end{pmatrix} \tag{6.99}$$

and the reader can verify that for $0 \leq \theta \leq 1$,

$$f(H_{x^1, x^2}(\theta)) \leq (1 - \theta)f(x^1) + \theta f(x^2). \tag{6.100}$$

In fact, from (6.96) and (6.98) we get

$$\phi f h^{-1}(t) = (t_1)^2 + (t_2)^2, \tag{6.101}$$

a convex quadratic function! We shall see below that this result holds for every (h, ϕ)-convex function; that is, they can be transformed into convex functions. In part (a) we had an example with h being the identity function, and in part (b) we had an example of an (h, ϕ)-convex function with ϕ taken as the identity function. In the next example neither h nor ϕ will be the identity functions.

(c) Let f be a positive-valued function on the positive orthant of R^n. Define f_1 by

$$f_1(x) = \begin{cases} f(x) & x > 0 \\ +\infty & \text{otherwise.} \end{cases} \tag{6.102}$$

Taking $h(x) = (\log x_1, \ldots, \log x_n)$ for $x > 0$ and $\phi(\alpha) = \log \alpha$ for $\alpha > 0$, we can define f_1 to be (log, log)-convex if for every two points x^1, x^2 in the effective domain of f_1 and $0 \leq \theta \leq 1$ we have, by (6.94) and after rearrangement,

$$f[(x_1^1)^{1-\theta}(x_1^2)^\theta, \ldots, (x_n^1)^{1-\theta}(x_n^2)^\theta] \leq (f(x^1))^{1-\theta}(f(x^2))^\theta. \tag{6.103}$$

Posynomials [16], which appear in a special branch of nonlinear programming called geometric programming, are defined by

$$g(x) = \sum_{i=1}^{m} c_i \prod_{j=1}^{n} x_j^{a_{ij}}, \tag{6.104}$$

where $c_i > 0$ and $a_{ij} \in R$ are constants. Such functions are (log, log)-convex. ∎

The most important feature of (h, ϕ)-convex functions is that they are convex transformable. In other words, they can be transformed into convex functions, as we shall see in the next theorem. For the rest of this section, assume that ϕ is a strictly increasing function. Our basic result is

Theorem 6.15

The function f is proper (h, ϕ)-convex on R^n if and only if $\hat{f}$, given by

$$\hat{f}(y) = \phi f h^{-1}(y), \tag{6.105}$$

is proper convex on R^n.

Proof. The function f is proper (h, ϕ)-convex on R^n if and only if

$$f[h^{-1}((1-\theta)h(x^1) + \theta h(x^2))] \leq \phi^{-1}[(1-\theta)\phi f(x^1) + \theta\phi f(x^2)] \quad (6.106)$$

for every $x^1 \in R^n$, $x^2 \in R^n$, and $0 \leq \theta \leq 1$. Since ϕ is strictly increasing, we obtain

$$\phi f[h^{-1}((1-\theta)h(x^1) + \theta h(x^2))] \leq (1-\theta)\phi f(x^1) + \theta\phi f(x^2). \quad (6.107)$$

Letting $h(x) = y$ and substituting into the last inequality yields

$$\phi f h^{-1}((1-\theta)y^1 + \theta y^2) \leq (1-\theta)\phi f h^{-1}(y^1) + \theta\phi f h^{-1}(y^2). \quad (6.108)$$

That is, $\phi f h^{-1}$ is proper convex. ■

Many properties of (h, ϕ)-convex functions can be conveniently derived by introducing certain generalized operations of addition and multiplication, due to Ben-Tal [11]. Let z be an n vector-valued continuous function, defined on a subset Z of R^n and possessing an inverse function z^{-1}. Define the **z-vector addition** of $x \in Z$ and $y \in Z$ as

$$x \oplus y = z^{-1}[z(x) + z(y)] \quad (6.109)$$

and the **z-scalar multiplication** of $x \in Z$ and $\lambda \in R$ as

$$\lambda \odot x = z^{-1}[\lambda z(x)]. \quad (6.110)$$

Similarly, let ω be a real-valued continuous function, defined on $\Omega \subset R$ and possessing an inverse function ω^{-1}. Then the **ω addition** of two numbers, $\alpha \in \Omega$ and $\beta \in \Omega$, is given by

$$\alpha\, [+]\, \beta = \omega^{-1}[\omega(\alpha) + \omega(\beta)] \quad (6.111)$$

and the **ω-scalar multiplication** of $\alpha \in \Omega$ and $\lambda \in R$ as

$$\lambda\, [\cdot]\, \alpha = \omega^{-1}[\lambda\omega(\alpha)]. \quad (6.112)$$

Finally, the **(z, ω)-inner product** of vectors $x \in Z$, $y \in Z$ is defined as

$$(x^T y)_{z,\omega} = \omega^{-1}[(z(x))^T z(y)], \quad (6.113)$$

provided that the right-hand side is well defined.

A complete treatment of these operations from an algebraic point of view would require considerably more extensive background material. Interested readers are referred to [11].

Example 6.2.3

Let

$$z(x) = (\log x_1, \dots, \log x_n) \tag{6.114}$$

and

$$Z = \{x : x \in R^n, x > 0\}. \tag{6.115}$$

Then

$$x \oplus y = [\exp(\log x_1 + \log y_1), \dots, \exp(\log x_n + \log y_n)]^T \tag{6.116}$$

$$= (x_1 y_1, \dots, x_n y_n)^T \tag{6.117}$$

and for $\lambda \in R$

$$\lambda \odot x = [\exp(\lambda \log x_1), \dots, \exp(\lambda \log x_n)]^T \tag{6.118}$$

$$= ((x_1)^\lambda, \dots, (x_n)^\lambda)^T. \tag{6.119}$$

∎

It is easy to verify that mean value functions can be written in this form as well—for example,

$$H_{x^1, x^2}(\theta) = [(1 - \theta) \odot x^1] \oplus (\theta \odot x^2) \tag{6.120}$$

and

$$\Phi_{\alpha^1, \alpha^2}(\theta) = [(1 - \theta)\,[\cdot]\,\alpha^1]\,[+]\,(\theta\,[\cdot]\,\alpha^2). \tag{6.121}$$

It follows then that the defining inequality (6.94) for proper (h, ϕ)-convex functions can be written as

$$f[((1 - \theta) \odot x^1) \oplus (\theta \odot x^2)] \le ((1 - \theta)\,[\cdot]\,f(x^1))\,[+]\,(\theta\,[\cdot]\,f(x^2)). \tag{6.122}$$

Thus (h, ϕ)-convex functions are "convex" under the foregoing generalized algebraic operations.

Using these operations, let us see how to generalize results presented in the preceding chapters for convex functions to (h, ϕ)-convex functions. For example, Theorem 4.10 states that if f and g are convex functions and λ is a nonnegative number, then $f + g$ and λf are also convex.

The corresponding results for functions f and g that are (h, ϕ)-convex and for $\lambda \geq 0$ are that $f\,[+]\,g$ and $\lambda\,[\cdot]\,f$ are also (h, ϕ)-convex. We have seen that f is convex if and only if $-f$ is concave. Similarly, f is (h, ϕ)-convex if and only if $(-1)\,[\cdot]\,f$ is (h, ϕ)-concave. In Chapter 4 we proved that a

differentiable function on a convex set $C \subset R^n$ is convex if and only if

$$f(x^2) \geq f(x^1) + (x^2 - x^1)^T \nabla f(x^1) \tag{6.123}$$

for any two points $x^1 \in C$, $x^2 \in C$. Similarly, a differentiable function is (h, ϕ)-convex on R^n if and only if

$$f(x^2) \geq f(x^1) \, [+] \, [(x^2 \ominus x^1)^T \nabla^* f(x^1)]_{h,\phi}, \tag{6.124}$$

where

$$x^2 \ominus x^1 = x^2 \oplus (-1) \odot x^1 = h^{-1}[h(x^2) - h(x^1)] \tag{6.125}$$

and

$$\nabla^* f(x^1) = h^{-1}[\nabla \phi f h^{-1}(t)|_{t=h(x^1)}]. \tag{6.126}$$

The inequality of (6.124) can be translated back into the more familiar form of ordinary algebraic operations, and then we receive

$$\phi f(x^2) \geq \phi f(x^1) + \phi'(f(x^1)) \sum_{i=1}^{n} \sum_{j=1}^{n} \frac{\partial f(x^1)}{\partial x_j} \frac{\partial h_j^{-1}(h(x^1))}{\partial x_i} (h_i(x^2) - h_i(x^1)). \tag{6.127}$$

Properties of a subclass of (h, ϕ)-convex functions, where $h(x) = x$, were derived by Avriel and Zang [6] without using generalized algebraic operations. Some aspects of nonlinear programming with such functions are also discussed there.

It is also possible to define conjugate functions of (h, ϕ)-convex functions and derive duality relations for nonlinear programs involving this type of generalized convexity. We shall only mention this topic very briefly. The conjugate function of an (h, ϕ)-convex function f on R^n is defined as

$$f^*(\xi) = \sup_x \{(\xi^T x)_{h,\phi} \, [-] \, f(x)\} \tag{6.128}$$

or

$$f^*(\xi) = \sup_x \{\phi^{-1}[h(\xi)^T h(x) - \phi f(x)]\} \tag{6.129}$$

provided that the right-hand side is well defined. Note that the range of an (h, ϕ)-convex function may include $\pm\infty$. Monotonicity of functions on $R \cup \pm\infty$ must, therefore, be appropriately interpreted.

Example 6.2.4

Let $x \in R$ and

$$f(x) = -xe^{-x}. \tag{6.130}$$

Letting $h(t) = t$ and

$$\phi(\alpha) = \begin{cases} -\log(-\alpha) & \alpha < 0 \\ +\infty & \alpha \geq 0, \end{cases} \tag{6.131}$$

we can verify that f is (h, ϕ)-convex. Indeed,

$$\hat{f}(x) = \phi f h^{-1}(x) = \phi f(x) = \begin{cases} x - \log x & x > 0 \\ +\infty & x \leq 0 \end{cases} \tag{6.132}$$

is convex on R. These functions are illustrated in Figure 6.5.

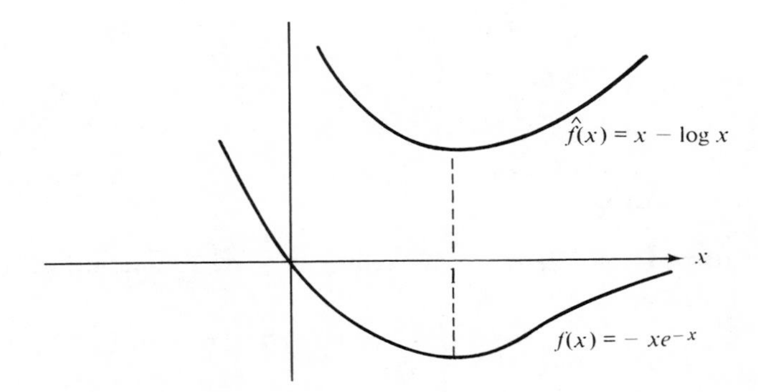

Fig. 6.5 The (h, ϕ)-convex function $f(x) = -xe^{-x}$ and its convex transformation.

Now

$$\phi^{-1}(\alpha) = -e^{-\alpha} \tag{6.133}$$

and

$$f^*(\xi) = \sup_x \{-\exp(-\xi x + \phi f(x))\} \tag{6.134}$$

$$= \sup_{x>0} \{-\exp(-\xi x + x - \log x)\}. \tag{6.135}$$

Hence

$$f^*(\xi) = \begin{cases} e\xi - e & \xi < 1 \\ 0 & \xi \geq 1. \end{cases} \tag{6.136}$$

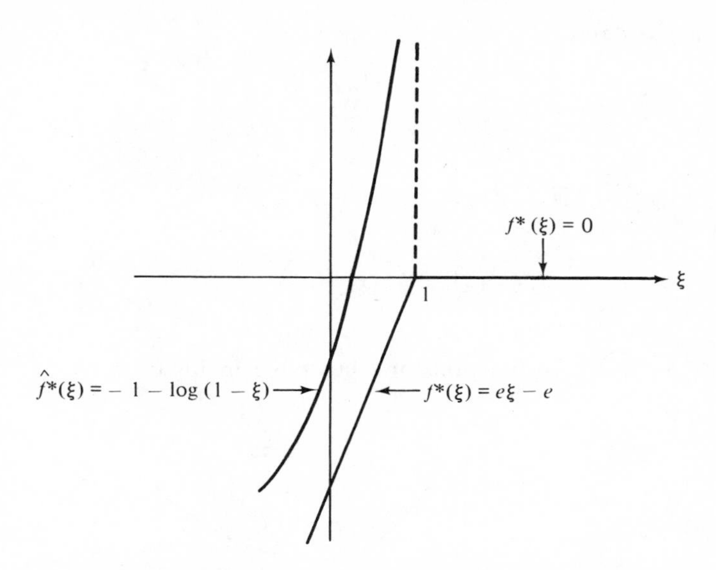

Fig. 6.6 The (h, ϕ)-convex conjugate function f^* and its convex transformation.

It is interesting to note that f^* is actually a concave function, but it is also (h, ϕ)-convex, since

$$\hat{f}^*(\xi) = \phi f^*(\xi) = \begin{cases} -1 - \log(1 - \xi) & \xi < 1 \\ +\infty & \xi \geq 1 \end{cases} \tag{6.137}$$

is convex (see Figure 6.6). The reader can verify that f^{**}, the conjugate of f^*, is $\text{cl} f$ just as in the convex case. ▮

It is evident that meaningful applications of particular (h, ϕ)-convex functions in nonlinear programming analysis can be demonstrated only if the verification of this type of generalized convexity does not require a prohibitive amount of work. Consequently, the question arises as to how one can find functions h and ϕ, possessing the required properties, such that the (h, ϕ)-convexity of a function f could be established. At present, only partial answers to this question are available. In certain special cases, we can find the required functions with reasonably little effort. We have, for example, the following result, due to Avriel and the proof of which can be found in [5].

Theorem 6.16

Let f be a twice continuously differentiable, real quasiconvex function defined on an open convex set $C \subset R^n$. If there exists a real number r^* satisfying

$$r^* = \sup_{\substack{x \in C \\ \|z\|=1}} \frac{-z^T \nabla^2 f(x) z}{[z^T \nabla f(x)]^2} \tag{6.138}$$

whenever $z^T \nabla f(x) \neq 0$, *then* f *is* r^**-convex; that is, it is* (h, ϕ)*-convex with* $h(t) = t$ *and* $\phi(\alpha) = e^{r^*\alpha}$.

Example 6.2.5

Let $f(x) = \log x$. This is a well-known concave function, defined on $C = \{x : x \in R, x > 0\}$. We have

$$\nabla f(x) = \frac{1}{x}, \qquad \nabla^2 f(x) = -\frac{1}{(x)^2} \tag{6.139}$$

and by (6.138),

$$r^* = \sup_{\substack{x > 0 \\ \|z\|=1}} \frac{[1/(x)^2](z)^2}{(1/x)^2(z)^2} = 1. \tag{6.140}$$

and $\log x$ is 1-convex—that is, (h, ϕ)-convex with $h(t) = t$ and $\phi(\alpha) = e^{\alpha}$.

Take $f(x) = (x)^3$ on $C \subset R$. This is a monotone increasing function having an inflection point at $x = 0$. Here we obtain

$$\sup_{x \neq 0} \frac{-6x(z)^2}{9(x)^4(z)^2} = \sup_{x \neq 0} \frac{-2}{3(x)^3} = +\infty \tag{6.141}$$

as $x \longrightarrow 0^-$. Hence there is no real r^* for which $(x)^3$ would be r^*-convex. ∎

Given a function f, there may be many different h and ϕ for which f could be (h, ϕ)-convex. We shall deal with this property in the case of r-convex functions in the Exercises. Analysis of certain nonlinear programs involving the type of generalized convex functions introduced in this section will be discussed in the next chapter. A discussion of certain types of convex-transformable functions that use a different approach can be found in Fenchel [17].

Many of the results on convex functions presented in Chapters 4 and 5 can be extended to arcwise-convex functions. Similarly, results on quasi-convex and other generalizations of convex functions obtained in the first section of this chapter can be further extended to "arcwise"-quasiconvex, "arcwise"-pseudoconvex, and other related types of functions. The reader can amuse himself by doing so.

Further extensions of convex programming analysis to general non-convex programs, especially in the areas of duality and Lagrangian theory, have been obtained. Some aspects of these results will be briefly mentioned in Chapter 12. For detailed discussions, the reader is referred to Arrow,

Gould, and Howe [3], Mangasarian [28], Rockafellar [39, 40, 41], and Roode [42].

6.3 LOCAL AND GLOBAL MINIMA

In this section we present characterizations of functions whose local minima are global. The principal tool in establishing the following results is the notion of point-to-set mappings. Let X and Y be two sets in R^p and R^q, respectively. If with each element $x \in X$ we associate a subset $F(x) \subset Y$, we call this correspondence a **point-to-set mapping** of points in X into subsets of Y. Point-to-set mappings are used in various branches of mathematical programming theory, which we shall not study here. The interested reader is referred to Hogan [22] and Zangwill [49]. For topological properties of point-to-set mappings, see Berge [12].

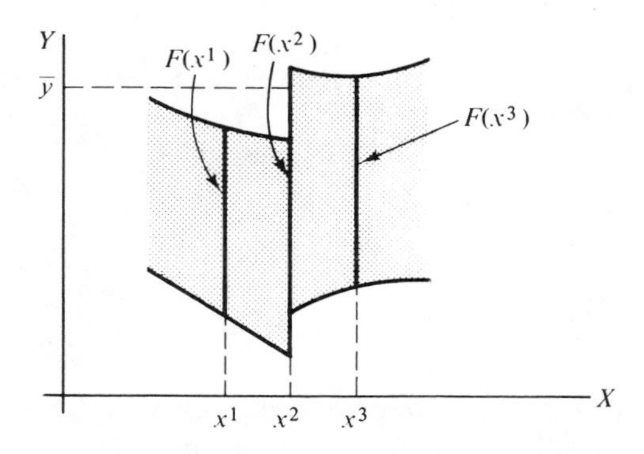

Fig. 6.7 A point-to-set mapping.

In Figure 6.7 we have an example of a point-to-set mapping where $X \subset R$ and $F(x)$ are subsets of R.

The following important concept will be needed in the sequel [12]. The point-to-set mapping F is said to be **lower semicontinuous** (lsc) at a point $\bar{x} \in X$ if for every $y \in F(\bar{x})$ and every sequence $\{x^i\} \subset X$ converging to $\bar{x}$ there exists a natural number K and a sequence $\{y^i\}$, converging to y, such that

$$y^i \in F(x^i), \qquad i = K, K+1, \ldots. \tag{6.142}$$

If F is lsc at every $x \in X$, it is said to be lsc on X. Referring to Figure 6.7, we can see that the point-to-set mapping illustrated there is not lower semicontinuous, since it is not lsc at x^2. If we take $\bar{y} \in F(x^2)$ and a sequence of increasing x^i from x^1 and converging to x^2, there will be no $y^i \in F(x^i)$ that would converge to $\bar{y}$. On the other hand, F can easily be seen to be lsc at x^1 or x^3.

A special type of point-to-set mapping associated with real-valued functions is considered next.

Let f be a real function on a subset C of R^n and let α be a real number. Consider the level sets of f

$$S(f, \alpha) = \{x : x \in C, f(x) \leq \alpha\} \tag{6.143}$$

and the set

$$G_f = \{\alpha : \alpha \in R, S(f, \alpha) \neq \emptyset\}. \tag{6.144}$$

If follows that $S(f, \alpha)$ can be considered a point-to-set mapping of points in G_f into subsets of R^n.

Example 6.3.1

Suppose that $C = R$ and

$$f(x) = (x)^2. \tag{6.145}$$

Then G_f is the nonnegative orthant of R.

Take, for example, $\alpha = 4$. Thus $\alpha \in G_f$ and

$$S(f, 4) = \{x : x \in R, |x| \leq 2\}. \tag{6.146}$$

∎

The preceding definition of lower semicontinuity can be specialized to level-set mappings as follows:

The point-to-set mapping $S(f, \alpha)$ is said to be lower semicontinuous (lsc) at a point $\bar{\alpha} \in G_f$ if for every $x \in S(f, \bar{\alpha})$ and any sequence $\{\alpha^i\} \subset G_f$, converging to $\bar{\alpha}$, there exists a natural number K and a sequence $\{x^i\}$, converging to x, such that

$$x^i \in S(f, \alpha^i), \qquad i = K, K+1, \ldots. \tag{6.147}$$

If $S(f, \alpha)$ is lsc at every $\alpha \in G_f$, it is said to be lsc on G_f.

We can also define lower semicontinuity in terms of open neighborhoods. The point-to-set mapping $S(f, \alpha)$ is said to be lower semicontinuous (lsc) at a point $\bar{\alpha} \in G_f$ if for every open set $A \subset R^n$, such that

$$A \cap S(f, \bar{\alpha}) \neq \emptyset, \tag{6.148}$$

there is an open neighborhood $N_\delta(\bar{\alpha})$ such that for every $\tilde{\alpha} \in N_\delta(\bar{\alpha}) \cap G_f$ we have

$$A \cap S(f, \tilde{\alpha}) \neq \emptyset. \tag{6.149}$$

We shall use both definitions of lower semicontinuity in forthcoming theorems.

We can now state and prove necessary and sufficient conditions, due to Zang and Avriel [47] and Zang, Choo and Avriel [48], for every local minimum of a function to be global. We start with the following sufficient conditions.

Theorem 6.17

Let f be a real function on $C \subset R^n$ and let $\bar{\alpha} = f(\bar{x})$, $\bar{x} \in C$. Suppose that $S(f, \alpha)$ is lsc at $\bar{\alpha}$. If $\bar{x}$ is a local minimum of f, then it is also a global minimum of f on C.

Proof. Suppose that the hypotheses hold and $\bar{x}$ is not a global minimum of f on C. Hence there exists a point $\tilde{x} \in C$ such that

$$f(\tilde{x}) < f(\bar{x}). \tag{6.150}$$

Define the sequence $\{\alpha^i\}$ by

$$\alpha^i = \left[\frac{1}{i} f(\tilde{x}) + \left(1 - \frac{1}{i}\right) f(\bar{x})\right], \qquad i = 1, 2, \ldots. \tag{6.151}$$

Clearly,

$$\lim_{i \to \infty} \{\alpha^i\} = f(\bar{x}) = \bar{\alpha} \tag{6.152}$$

and $\bar{x} \in S(f, \bar{\alpha})$. From (6.150) and (6.151) it follows that

$$f(\tilde{x}) \leq \alpha^i < f(\bar{x}), \qquad i = 1, 2, \ldots \tag{6.153}$$

and $\tilde{x} \in S(f, \alpha^i)$, $i = 1, 2, \ldots$. Hence $\{\alpha^i\} \subset G_f$.

Since $S(f, \alpha)$ is assumed to be lsc at $\bar{\alpha}$, there exists a natural number K and a sequence $\{x^i\}$, converging to $\bar{x}$, such that $x^i \in S(f, \alpha^i)$ for $i = K$, $K + 1, \ldots$. Hence

$$f(x^i) \leq \alpha^i, \qquad i = K, K + 1, \ldots \tag{6.154}$$

and by (6.150),

$$f(x^i) < f(\bar{x}), \qquad i = K, K + 1, \ldots. \tag{6.155}$$

Since $\{x^i\} \longrightarrow \bar{x}$, for a sufficiently small $\delta > 0$ there exists a natural number K_δ such that $x^i \in C \cap N_\delta(\bar{x})$, $i = K_\delta, K_\delta + 1, \ldots$ which, by the hypotheses, will also satisfy

$$f(x^i) \geq f(\bar{x}), \qquad i = K_\delta, K_\delta + 1, \ldots \tag{6.156}$$

contradicting (6.155). ∎

Corollary 6.18

Let f be a real function on $C \subset R^n$. If $S(f, \alpha)$ is lsc on G_f, then every local minimum of f is also a global minimum of f on C.

We now prove a converse result to Theorem 6.17.

Theorem 6.19

Let f be a real function on $C \subset R^n$ and let $\bar{\alpha} \in G_f$. Then $S(f, \alpha)$ is lsc at $\bar{\alpha}$ if any of the following assumptions holds,

(i) *Every $x \in C$ satisfying $f(x) = \bar{\alpha}$ is a global minimum of f on C.*

(ii) *None of the points $x \in C$ satisfying $f(x) = \bar{\alpha}$ is a local minimum of f.*

Proof. Suppose that the hypotheses hold and $S(f, \alpha)$ is not lsc at $\bar{\alpha}$. Then there exists an open set $A \subset R^n$ such that

$$A \cap S(f, \bar{\alpha}) \neq \emptyset \tag{6.157}$$

and for every $\delta > 0$ there exists an $\alpha(\delta) \in N_\delta(\bar{\alpha}) \cap G_f$ such that

$$A \cap S(f, \alpha(\delta)) = \emptyset. \tag{6.158}$$

Therefore we can find a sequence $\{\alpha^i\} \subset G_f$ converging to $\bar{\alpha}$ such that

$$A \cap S(f, \alpha^i) = \emptyset, \qquad i = 1, 2, \ldots. \tag{6.159}$$

It follows that $\alpha^i < \bar{\alpha}$ for all i; otherwise if $\alpha^k \geq \bar{\alpha}$ for some k, then $S(f, \bar{\alpha}) \subset S(f, \alpha^k)$, and, by (6.157), we get a contradiction to (6.159). For every $\bar{x} \in A \cap S(f, \bar{\alpha})$ it follows that $\bar{x} \notin S(f, \alpha^i)$ for every i; and since $\{\alpha^i\} \longrightarrow \bar{\alpha}$, we have $f(\bar{x}) = \bar{\alpha}$. Moreover, for every $x \in A$,

$$f(x) \geq \bar{\alpha} = f(\bar{x}). \tag{6.160}$$

Since A is an open set, it follows that $\bar{x}$ must be a local minimum of f, and, clearly, $\bar{x}$ is not a global minimum of f on C. This result contradicts the hypotheses. ∎

Corollary 6.20

Let f be a real function on $C \subset R^n$. If every local minimum of f is a global minimum of f on C, then $S(f, \alpha)$ is lsc on G_f.

Finally, as an immediate result of Corollaries 6.18 and 6.20, we obtain

Corollary 6.21

Let f be a real function on $C \subset R^n$. Every local minimum of f is a global minimum of f on C if and only if $S(f, \alpha)$ is lsc on G_f.

Example 6.3.2

Let us illustrate the preceding results. The function appearing in Figure 6.8 has a nonglobal local minimum at $\hat{x}$ and a global minimum at x^*. Also note that $f(\hat{x}) = f(\bar{x}) = 0$. Let $\bar{\alpha} = 0$ and take the sequence $\{\alpha^i\}$, whose elements are

$$\alpha^i = \frac{1}{i} f(x^*), \qquad i = 1, 2, \ldots. \tag{6.161}$$

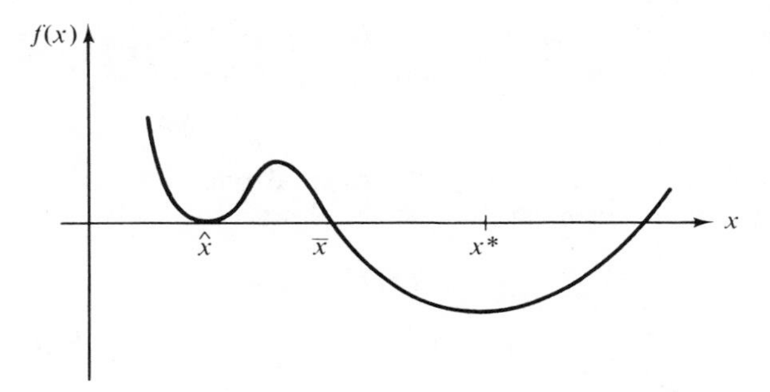

Fig. 6.8 Function having a nonglobal local minimum.

Clearly, $\{\alpha^i\} \subset G_f$ and $\{\alpha^i\} \longrightarrow \bar{\alpha}$. The point $\hat{x}$ is contained in $S(f, 0)$, but in every sequence $\{x^i\}$, such that $x^i \in S(f, \alpha^i)$ for sufficiently large i, we must have $x^i > \bar{x}$, and there is no such sequence that would converge to $\hat{x}$. Hence $S(f, \alpha)$ is not lsc at $\bar{\alpha}$. ∎

In some cases, it is not too difficult to verify the lower semicontinuity of the level-set mapping of certain functions, thus ensuring the desirable property of having only global minima (if they exist). Such a verification is given below for the family of functions discussed in the preceding section.

Example 6.3.3

Let us show that if f is a real (h, ϕ)-convex function defined on an arcwise-connected set $C \subset R^n$, where C contains all the arcs given by (6.92), then $S(f, \alpha)$ is lower semicontinuous on G_f. Take $\bar{\alpha} \in G_f$ and take any sequence $\{\alpha^i\} \subset G_f$, converging to $\bar{\alpha}$. It is sufficient to consider points $\bar{x} \in S(f, \bar{\alpha})$ that satisfy $f(\bar{x}) = \bar{\alpha}$, for if $f(\bar{x}) < \bar{\alpha}$, then, by taking $x^i = \bar{x}$ for all i, the sequence $\{x^i\}$ satisfies (6.147). If $\bar{x}$ is a global minimum of f on C, then we must have

$$\alpha^i \geq \bar{\alpha}, \qquad i = 1, 2, \ldots \tag{6.162}$$

and we can again choose $x^i = \bar{x}$ for every i. If $\bar{x}$ is not a global minimum of

of f on C, then there exists a point $\tilde{x} \in C$ such that

$$f(\tilde{x}) < f(\bar{x}) = \bar{\alpha}. \tag{6.163}$$

Define a sequence $\{\hat{x}^k\}$ by

$$\hat{x}^k = h^{-1}\left[\frac{1}{k}h(\tilde{x}) + \left(1 - \frac{1}{k}\right)h(\bar{x})\right], \qquad k = 1, 2, \ldots. \tag{6.164}$$

Clearly, $\{\hat{x}^k\}$ converges to $\bar{x}$ and $\{\hat{x}^k\} \subset C$. We may assume that $\| \hat{x}^k - \bar{x} \| < 1/k$ for every k.

From the (h, ϕ)-convexity of f it follows for $k = 1, 2, \ldots,$ that

$$\begin{aligned} f(\hat{x}^k) &= f\left[\left(\frac{1}{k} \odot \tilde{x}\right) \oplus \left(\left(1 - \frac{1}{k}\right) \odot \bar{x}\right)\right] \\ &\leq \left[\frac{1}{k} [\cdot] f(\tilde{x})\right] [+] \left[\left(1 - \frac{1}{k}\right) [\cdot] f(\bar{x})\right]. \end{aligned} \tag{6.165}$$

From (6.163) and (6.165) we obtain

$$f(\hat{x}^k) < f(\bar{x}) = \bar{\alpha}. \tag{6.166}$$

To complete the proof, we construct a sequence $\{x^i\}$ with the required properties for lower semicontinuity of $S(f, \alpha)$. From (6.166) and the convergence of $\{\alpha^i\}$ to $\bar{\alpha}$, if follows that there exists a K_1 such that

$$f(\hat{x}^1) \leq \alpha^i, \qquad i = K_1, K_1 + 1, \ldots. \tag{6.167}$$

For every α^i, $i = K_1 + 1, \ldots$ of the sequence $\{\alpha^i\}$, let x^i be defined in the following way. Choose $\bar{k}(i)$ such that

$$\bar{k}(i) = \sup\{k : f(\hat{x}^k) \leq \alpha^i\} \tag{6.168}$$

if a finite supremum exists; otherwise let $\bar{k}(K_1) = 1$ and

$$\bar{k}(i) = \bar{k}(i - 1) + 1. \tag{6.169}$$

Moreover, let

$$x^i = \begin{cases} \hat{x}^1 & i = 1, \ldots, K_1 \\ \hat{x}^{\bar{k}(i)} & i = K_1 + 1, \ldots. \end{cases} \tag{6.170}$$

Let us now show that $\{x^i\}$ converges to $\bar{x}$. That is, for any positive λ there exists a natural number $K(\lambda)$ such that

$$x^i \in N_\lambda(\bar{x}), \qquad i = K(\lambda), K(\lambda) + 1, \ldots. \tag{6.171}$$

Define

$$\delta(k) = \frac{1}{k}, \qquad k = 1, 2, \ldots \tag{6.172}$$

and

$$\hat{k}(\lambda) = \begin{cases} \max\left\{k : \delta(k-1) = \dfrac{1}{k-1} > \lambda\right\} & \text{if } \lambda < 1 \\ 1 & \text{if } \lambda \geq 1. \end{cases} \tag{6.173}$$

Then $N_{\delta(\hat{k}(\lambda))}(\bar{x}) \subset N_\lambda(\bar{x})$,

$$f(\hat{x}^{\hat{k}(\lambda)}) < \bar{\alpha} \tag{6.174}$$

and we can find a $\bar{K}(\lambda) \geq K_1$ and $\bar{K}(\lambda) \geq \hat{k}(\lambda)$ such that

$$f(\hat{x}^{\hat{k}(\lambda)}) \leq \alpha^i, \qquad i = \bar{K}(\lambda), \bar{K}(\lambda) + 1, \ldots. \tag{6.175}$$

Consequently, one of the following situations may occur.

1. There exists an $\bar{i} \geq \bar{K}(\lambda)$ for which $\bar{k}(\bar{i})$ is obtained by (6.168). Then for $i \geq \bar{i}$ we get from (6.168), (6.169), and (6.175) that $\bar{k}(i) \geq \hat{k}(\lambda)$ and

$$\hat{x}^{\bar{k}(i)} \in N_{\delta(\bar{k}(i))}(\bar{x}) \subset N_{\delta(\hat{k}(\lambda))}(\bar{x}) \subset N_\lambda(\bar{x}). \tag{6.176}$$

Hence

$$x^i \in N_\lambda(\bar{x}), \qquad i = \bar{i}, \bar{i} + 1, \ldots. \tag{6.177}$$

2. For all $i \geq \bar{K}(\lambda)$, the elements of the sequence $\{x^i\}$ are chosen by (6.169) and (6.170). Clearly, for $i \geq \bar{K}(\lambda)$, the x^i are consecutively taken from the sequence $\{\hat{x}^k\}$ that converges to $\bar{x}$.

In both situations, therefore, $\{x^i\}$ converges to $\bar{x}$ and

$$x^i \in S(f, \alpha^i), \qquad i = K_1, K_1 + 1, \ldots. \tag{6.178}$$

∎

EXERCISES

6.A. Show that an alternative definition of quasiconvex functions can be given by requiring the set

$$S^s(f, \alpha) = \{x : x \in X \subset R^n, f(x) < \alpha\} \tag{6.179}$$

to be convex for every $\alpha \in R$. (*Hint:* Use Theorem 4.2.) Compare with Theorem 6.1.

6.B. A real function f, defined on $X \subset R^n$, is called **quasimonotonic** [31] if f is both quasiconvex and quasiconcave. Show that if f is quasimonotonic, then

$$\bar{S}(f, \alpha) = \{x : x \in X, f(x) = \alpha\} \tag{6.180}$$

is a convex set for every $\alpha \in R$. Conversely, show that if $\bar{S}(f, \alpha)$ is convex for every $\alpha \in R$ and if f is continuous, then f is quasimonotonic. Illustrate these results on the **linear fractional function** g, defined on a convex set $C \subset R^n$ and given by

$$g(x) = \frac{a^T x + b}{c^T x + d}, \tag{6.181}$$

where the denominator never vanishes.

6.C. It is well known that a linear function attains its extrema on a convex subset $X \subset R^n$ at an extreme point of X (if the extrema exist). Extend this result to **pseudolinear** functions—that is, functions that are both pseudoconvex and pseudoconcave. Are linear fractional functions, defined in the preceding exercise, pseudolinear?

6.D. Show that if the quadratic function

$$f(x) = b^T x + \tfrac{1}{2} x^T Q x, \qquad x \in R^n \tag{6.182}$$

is merely quasiconvex, then $b \leq 0$.

6.E. (a) Establish the positive (negative) definiteness or subdefiniteness of the following matrices:

$$\text{(i)} \begin{bmatrix} 1 & 1 & 2 \\ 1 & 1 & 2 \\ 2 & 2 & 0 \end{bmatrix} \quad \text{(ii)} \begin{bmatrix} 0 & -1 & -1 \\ -1 & 0 & -1 \\ -1 & -1 & 0 \end{bmatrix} \quad \text{(iii)} \begin{bmatrix} 1 & 1 & 1 \\ 1 & 0 & 0 \\ 1 & 0 & 0 \end{bmatrix} \tag{6.183}$$

(b) Let

$$Q = \begin{bmatrix} -1 & -1 \\ -1 & 0 \end{bmatrix} \qquad b = (0 \quad -2)^T. \tag{6.184}$$

Is $f(x) = b^T x + \frac{1}{2} x^T Q x$ quasiconvex on the nonnegative orthant?

6.F. (a) Let $f(x) = -(x)^2 - x$ on $X = \{x : x \in R, 0 \leq x \leq 1\}$. Show that f is strictly pseudoconvex on X. Is it strongly quasiconvex?

(b) Now let $f(x) = -(x)^2$. Show that f is not pseudoconvex on the same X. Is it quasiconvex?

6.G. (a) Ortega and Rheinboldt [36] define pseudoconvex functions as follows: A real function f on a convex set $X \subset R^n$ is called pseudoconvex if for any $x^1 \in X$, $x^2 \in X$ such that

$$f(x^1) > f(x^2) \tag{6.185}$$

there exist positive numbers α and τ, $\tau \leq 1$, both depending, in general,

on x^1 and x^2 and satisfying

$$f((1-\theta)x^1+\theta x^2)\leq f(x^1)-\theta\alpha \tag{6.186}$$

for every $0\leq\theta\leq\tau$. Show that for differentiable functions the preceding definition is equivalent to the one given in this chapter.

(b) If (6.186) holds whenever $x^1\neq x^2$ and $f(x^1)\geq f(x^2)$, then Ortega and Rheinboldt call f strictly pseudoconvex. On the other hand, Ponstein [37] defines differentiable functions to be strictly pseudoconvex if $x^1\neq x^2$ and $f(x^1)\geq f(x^2)$ imply

$$(x^2-x^1)^T\nabla f(x^1)<0. \tag{6.187}$$

Show that these two definitions are equivalent for differentiable functions. Show that the function

$$f(x)=\begin{cases} x & x<0\\ (x)^2+1 & x\geq 0\end{cases} \tag{6.188}$$

is strictly pseudoconvex in the sense of Ortega-Rheinboldt.

6.H. Show that a continuous real-valued function f is strictly quasiconvex on R^n if and only if all its level sets are convex and either $\bar{S}(f,\alpha)$, the set of $x\in R^n$ such that $f(x)=\alpha$, is contained in the set of boundary points of $S(f,\alpha)$ or $\bar{S}(f,\alpha)=S(f,\alpha)$.

6.I. Prove that if f is a continuous real quasiconvex function on R^n whose local minima are global, then f is strictly quasiconvex. You may use in your proof the results of the preceding exercise.

6.J. Derive alternative definitions of differentiable (h,ϕ)-convex functions, generalizing (4.119) and (4.144).

6.K. Let r and s be real numbers, $s>r$, and let $\xi_1,\ldots,\xi_m$ be positive numbers. It is well known [21] that

$$\left\{\sum_{i=1}^m q_i(\xi_i)^r\right\}^{1/r}\leq\left\{\sum_{i=1}^m q_i(\xi_i)^s\right\}^{1/s} \tag{6.189}$$

Show that if f is an r-convex function, then it is also s-convex for every $s>r$. Using the relation

$$\lim_{r\to+\infty}\left\{\sum_{i=1}^m q_i(\xi_i)^r\right\}^{1/r}=\max(\xi_1,\ldots,\xi_m), \tag{6.190}$$

show that every r-convex function is also quasiconvex.

6.L. Prove that a twice continuously differentiable real-valued function f on R^n is r-convex if and only if the matrix Q, given by

$$Q(x)=\nabla^2 f(x)+r\nabla f(x)[\nabla f(x)]^T, \tag{6.191}$$

is positive semidefinite for every $x\in R^n$.

6.M. Given the function

$$f(x) = [x_2 - (x_1)^3 + (4 - x_1)^2]^{1/3} \tag{6.192}$$

defined on R^2, find h and ϕ for which f is (h, ϕ)-convex.

6.N. Show that if f is a strictly quasiconvex function on a convex subset of R^n, then $S(f, \alpha)$ is a lower semicontinuous point-to-set mapping on G_f.

REFERENCES

1. Apostol, T. M., *Mathematical Analysis*, 2nd ed., Addison-Wesley Publishing Co., Reading, Mass., 1974.
2. Arrow, K. J., and A. C. Enthoven, "Quasi-Concave Programming," *Econometrica*, **29**, 779–800 (1961).
3. Arrow, K. J., F. J. Gould, and S. M. Howe, "A General Saddle Point Result for Constrained Optimization," *Math. Prog.*, **5**, 225–234 (1973).
4. Arrow, K. J., L. Hurwicz, and H. Uzawa, "Constraint Qualifications in Maximization Problems," *Naval Res. Log. Quart.*, **8**, 175–191 (1961).
5. Avriel, M., "*r*-Convex Functions," *Math. Prog.*, **2**, 309–323 (1972).
6. Avriel, M., and I. Zang, "Generalized Convex Functions with Applications to Nonlinear Programming," Chapter 2 in *Mathematical Programs for Activity Analysis*, P. Van Moeseke (Ed.), North-Holland Publishing Co., Amsterdam, 1974.
7. Avriel, M., and I. Zang, "Generalized Convex Sets and Functions in Nonlinear Programming," in preparation.
8. Bartle, R. G., *The Elements of Real Analysis*, 2nd ed. John Wiley & Sons, New York, 1976.
9. Beckenbach, E. F., "Generalized Convex Functions," *Bull. Am. Math. Soc.*, **43**, 363–371 (1937).
10. Beckenbach, E. F., and R. H. Bing, "On Generalized Convex Functions," *Bull. Am. Math. Soc.*, **51**, 220–230 (1945).
11. Ben-Tal, A., "On Generalized Means and Generalized Convexity," *J. Optimization Theory & Appl.*, to appear.
12. Berge, C., *Topological Spaces*, Oliver & Boyd Ltd., Edinburgh, 1963.
13. Cottle, R. W., and J. A. Ferland, "Matrix-Theoretic Criteria for Quasi-Convexity and Pseudo-Convexity of Quadratic Functions," *Linear Alg. Appl.*, **5**, 123–136 (1972).
14. Cottle, R. W., and J. A. Ferland, "On Pseudo-Convex Functions of Nonnegative Variables," *Math. Prog.*, **1**, 95–101 (1971).
15. de Finetti, B., "Sulle Stratificazioni Convesse," *Ann. Math. Pura Appl.*, **30**, 173–183 (1949).

16. DUFFIN, R. J., E. L. PETERSON, and C. ZENER, *Geometric Programming—Theory and Applications*, John Wiley & Sons, New York, 1967.

17. FENCHEL, W., *Convex Cones, Sets and Functions*, Mimeographed lecture notes, Princeton University, Princeton, N.J., 1951.

18. FERLAND, J. A., "Quasi-Convex and Pseudo-Convex Functions on Solid Convex Sets," Ph.D. dissertation, Stanford University, Stanford Calif., 1971.

19. GREENBERG, H. J., and W. P. PIERSKALLA, "A Review of Quasi-Convex Functions," *Operations Research*, **19**, 1553–1570 (1971).

20. HANSON, M. A., "Bounds for Functionally Convex Optimal Control Problems," *J. Math. Anal. & Appl.*, **8**, 84–89 (1964).

21. HARDY, G. H., J. E. LITTLEWOOD, and G. PÓLYA, *Inequalities*, 2nd ed., Cambridge University Press, Cambridge, England, 1952.

22. HOGAN, W. W., "Point-to-Set Maps in Mathematical Programming," *SIAM Review*, **15**, 591–603 (1973).

23. KARAMARDIAN, S., "Duality in Mathematical Programming," *J. Math. Anal. & Appl.*, **20**, 344–358 (1967).

24. LUENBERGER, D. G., "Quasi-Convex Programming," *SIAM J. Appl. Math.*, **16**, 1090–1095 (1968).

25. MANGASARIAN, O. L., "Pseudo-Convex Functions," *J. SIAM Control, Ser. A*, **3**, 281–290 (1965).

26. MANGASARIAN, O. L., *Nonlinear Programming*, McGraw-Hill Book Co., New York, 1969.

27. MANGASARIAN, O. L., "Convexity, Pseudo-Convexity and Quasi-Convexity of Composite Functions," *Cahiers du Centre d'Etude de Rech. Oper.*, **12**, 114–122 (1970)

28. MANGASARIAN, O. L., "Unconstrained Lagrangians in Nonlinear Programming," *SIAM J. Control*, **13**, 772–791 (1975).

29. MARTOS, B., "Hyperbolic Programming," *Naval Res. Log. Quart.*, **11**, 135–155 (1964).

30. MARTOS, B., "Nem-Lineáris Programmozási Módszerek Hatóköre (The Power of Nonlinear Programming Methods)," MTA Közgazdaságtudományi Intézetének Közleményei No. 20, Budapest, 1966 (in Hungarian).

31. MARTOS, B., "Quasi-Convexity and Quasi-Monotonicity in Nonlinear Programming," *Studia Sci. Math. Hung.*, **2**, 265–273 (1967).

32. MARTOS, B., "Subdefinite Matrices and Quadratic Forms," *SIAM J. Appl. Math.*, **17**, 1215–1223 (1969).

33. MEYER, R., "The Validity of a Family of Optimization Methods," *SIAM J. Control*, **8**, 41–54 (1970).

34. MINCH, R. A., "Applications of Symmetric Derivatives in Mathematical Programming," *Math. Prog.*, **1**, 307–320 (1971).

35. Nikaidô, H., "On von Neumann's Minimax Theorem," *Pacific J. Math.*, **4**, 65–72 (1954).

36. Ortega, J. M., and W. C. Rheinboldt, *Iterative Solution of Nonlinear Equations in Several Variables*, Academic Press, New York, 1970.

37. Ponstein, J., "Seven Kinds of Convexity," *SIAM Review*, **9**, 115–119 (1967).

38. Rockafellar, R. T., *Convex Analysis*, Princeton University Press, Princeton, N.J., 1970.

39. Rockafellar, R. T., "Penalty Methods and Augmented Lagrangians in Nonlinear Programming," in *Proceedings of the 5th IFIP Conference on Optimization Techniques*, R. Conti and A. Ruberti (Eds.), Springer-Verlag, Berlin, 1973.

40. Rockafellar, R. T., "Augmented Lagrange Multiplier Functions and Duality in Nonconvex Programming," *SIAM J. Control*, **12**, 268–285 (1974).

41. Rockafellar, R. T., "A Dual Approach to Solving Nonlinear Programming Problems by Unconstrained Optimization," *Math. Prog.*, **5**, 354–373 (1973).

42. Roode, J. D., "Generalized Lagrangian Functions in Mathematical Programming," Doctoral dissertation, University of Leiden, The Netherlands, 1968.

43. Schaible, S., "Beiträge zur quasikonvexen Programmierung," Doctoral dissertation, University of Cologne, Germany, 1971.

44. Schaible, S., "Quasiconvex Optimization in General Real Linear Spaces," *Zeitschrift für Operations Research*, **16**, 205–213 (1972).

45. Singer, I. M., and J. A. Thorpe, *Lecture Notes on Elementary Topology and Geometry*, Scott, Foresman and Co., Glenview, Ill., 1967.

46. Zang, I., "Generalized Convex Programming," D.Sc. dissertation, Technion, Israel Institute of Technology, Haifa, 1974.

47. Zang, I., and M. Avriel, "On Functions whose Local Minima are Global," *J. Optimization Theory & Appl.*, **16**, 183–190 (1975).

48. Zang, I., E. U. Choo, and M. Avriel, "A Note on Functions whose Local Minima are Global," *J. Optimization Theory & Appl.*, **18**, 555–559 (1976).

49. Zangwill, W. I., *Nonlinear Programming: A Unified Approach*, Prentice-Hall, Englewood Cliffs, N.J., 1969.

7 ANALYSIS OF SELECTED NONLINEAR PROGRAMMING PROBLEMS

In concluding Part I, we have chosen three representative nonlinear programs in order to illustrate the methods and results of the previous chapters. These programs have been extensively studied and analyzed by many scholars using different techniques. We shall try to present our analysis so that it will generally follow, at least verbally, results published in the literature, but the reader can, in most cases, prove the results presented here without difficulty by consulting earlier chapters.

In the first section we deal with quadratic programming, which is perhaps the kind of nonlinear program closest to linear programming from an analytical and computational point of view. In the second section a certain type of stochastic programming problem is presented. One of the reasons we have included it—and, in fact, all three types of problems—here is to call the reader's attention to the existence of these program formulations. Stochastic programming represents an important class of nonlinear optimization problems, one on which numerous special monographs and textbooks have been written. In the last section we describe some aspects of geometric programming and, in particular, derive a dual program for it. Geometric programming and its extensions, because of their apparently large potential for real-life applications, have been the subject of many studies in recent years.

Unfortunately, it is beyond the scope of this book to discuss more extensively these three types of problems as well as several others that should be included. We can only advise interested readers to look further in the more specialized mathematical programming literature.

7.1 QUADRATIC PROGRAMMING

Consider the nonlinear program

(PQP) $$\min f(x) = a + c^T x + \tfrac{1}{2} x^T Q x \tag{7.1}$$

subject to

$$A^T x \geq b, \tag{7.2}$$

where $a \in R$, $c \in R^n$, $b \in R^m$ are given vectors, Q is an $n \times n$ symmetric real matrix, and A is an $n \times m$ real matrix. Program (PQP) is called a **quadratic program**. For simplicity, assume that $m < n$ and that A has rank m. Clearly, the objective function and constraint functions are twice continuously differentiable. Moreover, the gradients of the constraint functions

$$g_i(x) = (a^i)^T x - b_i, \qquad i = 1, \ldots, m \tag{7.3}$$

where a^i is the ith column of A, are linearly independent. Hence the first- and second-order constraint qualifications for necessary optimality conditions hold at every feasible point. The Lagrangian associated with (PQP) is given by

$$L(x, \lambda) = a + c^T x + \tfrac{1}{2} x^T Q x - \sum_{i=1}^{m} \lambda_i [(a^i)^T x - b_i]. \tag{7.4}$$

Let us now state the necessary optimality conditions that follow from Theorems 3.8 and 3.10.

Theorem 7.1

Let x^* be a solution of problem (PQP). *Then there exists a vector $\lambda^* \in R^m$ such that*

$$c + Qx^* - A\lambda^* = 0 \tag{7.5}$$

$$\lambda_i^*[(a^i)^T x^* - b_i] = 0, \qquad i = 1, \ldots, m \tag{7.6}$$

$$\lambda^* \geq 0 \tag{7.7}$$

and for every $z \neq 0$ satisfying

$$z^T a^i = 0, \qquad i \in I(x^*) \tag{7.8}$$

where

$$I(x^*) = \{i : (a^i)^T x^* = b_i, i = 1, \ldots, m\}, \tag{7.9}$$

we have

$$z^T Qz \geq 0. \tag{7.10}$$

Turning to sufficient conditions, we first define

$$\hat{I}(x^*) = \{i : i \in I(x^*), \lambda_i^* > 0\}. \tag{7.11}$$

Then from Theorem 3.11 we have

Theorem 7.2

Let x^* be feasible for problem (PQP). *If there exists a vector λ^* satisfying* (7.5) *to* (7.7) *and for every $z \neq 0$, such that*

$$z^T a^i = 0, \qquad i \in \hat{I}(x^*) \tag{7.12}$$

$$z^T a^i \geq 0, \qquad i \in I(x^*), \quad i \notin \hat{I}(x^*) \tag{7.13}$$

it follows that

$$z^T Qz > 0, \tag{7.14}$$

then x^* is a strict local minimum of (PQP).

Since the objective function is twice differentiable and the constraints are linear, convexity properties of program (PQP) can be completely characterized by the matrix Q.

In Chapters 4 and 6 we stated conditions on Q under which f is convex, pseudoconvex, or quasiconvex. Let us consider the simplest case only. For the rest of this section, assume that Q is positive definite. It follows then that f is strictly convex, and all the second-order conditions on $z^T Qz$ appearing in the preceding two theorems are satisfied. Thus the first-order conditions (7.5) to (7.7) are necessary and sufficient.

We can also state some duality results. To begin, we derive a dual program corresponding to (PQP). Following the derivations in Chapter 5, we transform (PQP) into an "unconstrained" program with perturbations:

$$\min \phi(x, w) = \begin{cases} f(x) & \text{if } A^T x - b \geq w \\ +\infty & \text{otherwise,} \end{cases} \tag{7.15}$$

where $x \in R^n$, $w \in R^m$. The conjugate function of ϕ is given by

$$\phi^*(\xi, \lambda) = \sup_{x,\, w} \{\xi^T x + \lambda^T w - \phi(x, w)\} \tag{7.16}$$

$$= \sup_{A^T x - b \geq w} \{\xi^T x + \lambda^T w - a - c^T x - \tfrac{1}{2} x^T Q x\} \tag{7.17}$$

and

$$\phi^*(\xi, \lambda) = \begin{cases} \sup_x \{(\xi + A\lambda - c)^T x - \frac{1}{2}x^T Q x - b^T\lambda - a\} & \text{if } \lambda \geq 0 \\ +\infty & \text{otherwise.} \end{cases} \tag{7.18}$$

The supremum of the strictly concave function in (7.18) can be found by differentiation, and we obtain

$$\phi^*(\xi, \lambda) = \begin{cases} \frac{1}{2}(\xi + A\lambda - c)^T Q^{-1}(\xi + A\lambda - c) - b^T\lambda - a & \text{if } \lambda \geq 0 \\ +\infty & \text{otherwise.} \end{cases} \tag{7.19}$$

The dual program of (PQP) becomes

(DQP) $$\max v(\lambda, \eta) = b^T\lambda - \tfrac{1}{2}\eta^T Q^{-1}\eta + a \tag{7.20}$$

subject to

$$A\lambda - \eta = c \tag{7.21}$$

$$\lambda \geq 0. \tag{7.22}$$

Thus (DQP) is also a quadratic program. Note that the vector variable η was merely introduced for notational convenience.

Let us examine some properties of this pair of dual quadratic programs. The primal perturbation function is given by

$$\Phi(w) = \inf_x \{f(x) : A^T x - b \geq w\}, \tag{7.23}$$

where the infimum over the empty set is interpreted as $+\infty$ and

$$\Phi(0) = \inf_x \{f(x) : A^T x - b \geq 0\}. \tag{7.24}$$

Hence if there is a feasible x for (PQP), then $\Phi(0) < +\infty$ and (PQP) is usually stable. A similar analysis can be carried out for (DQP). The feasible set of (DQP) is never empty, since for any $\lambda \geq 0$ we can always find an η satisfying (7.21). It may happen, however, that the optimal value of (DQP) is unbounded. To see this, we take an $n \times m$ matrix A with $n < m$, such as $A = [1 \quad -1]$, a vector $b = (1, 1)^T$, and $Q = 1$, $c = a = 0$. Then by maximizing the dual objective function

$$\max -\phi^*(0, \lambda) = \max_{\lambda \geq 0} \{b^T\lambda - \tfrac{1}{2}(A\lambda - c)^T Q^{-1}(A\lambda - c) + a\} \tag{7.25}$$

and choosing $\lambda_1 = \lambda_2$, both approaching $+\infty$, we get $-\phi^*(0, \lambda) \to +\infty$.

Clearly, (PQP) has no feasible solution and $\Phi(0) = +\infty$; that is, (PQP) is not stable. From the results of Section 5.2 we obtain

Theorem 7.3

For any feasible solution x *of* (PQP) *and any feasible solution* (λ, η) *of* (DQP), *we have*

$$f(x) \geq v(\lambda, \eta). \tag{7.26}$$

Program (PQP) *is stable and has an optimal solution* x^* *if and only if* (DQP) *is stable and has an optimal solution* (λ^*, η^*). *Moreover, if both programs have optimal solutions, then*

$$f(x^*) = v(\lambda^*, \eta^*). \tag{7.27}$$

It should be mentioned that Cottle [7], Dennis [12], and Dorn [13, 14, 15] were among the first to state duality theorems for quadratic programs without using conjugate function theory as we have done here. The main importance of these early results is that they had a stimulating effect on the development of more recent and more general results in duality theory.

Optimality conditions are closely related to the Lagrangian, as we saw in Section 5.3. For our programs we have

Theorem 7.4

If (x^*, λ^*) *is a saddlepoint of the Lagrangian* L *given by* (7.4) *for every* x *and* $\lambda \geq 0$, *then* x^* *is an optimal solution of* (PQP) *and* λ^* *is an optimal solution of* (DQP). *Furthermore,*

$$f(x^*) = L(x^*, \lambda^*) = v(\lambda^*, \eta^*), \tag{7.28}$$

where η^* *satisfies* (7.21). *Conversely, if* (PQP) *is stable and has an optimal solution* x^*, *then there exists a* λ^* *such that* (x^*, λ^*) *is a saddlepoint of* L. *Dually, if* (DQP) *is stable and* λ^* *is an optimal solution, then there exists an* x^* *such that* (x^*, λ^*) *is a saddlepoint of* L.

Quadratic programming problems often arise in practice, either as representations of some economic model or as subproblems in the course of solving more general nonlinear programs. Many numerical solution methods have been proposed for their solution, one of which is discussed in Chapter 13. Interested readers may also consult the books of Boot [6] and Dantzig [11] or the papers of Beale [5], Cottle [8], Cottle and Mylander [10], Keller [21], and the references cited there.

7.2 STOCHASTIC LINEAR PROGRAMMING WITH SEPARABLE RECOURSE FUNCTIONS

The linear programming problem in its traditional form as an activity analysis model is considered next. Thus we deal with commodities numbered $1, \ldots, m$ and with activities numbered $1, \ldots, n$. Let a_{ij} be the amount of commodity i used up (for $a_{ij} > 0$) or produced (for $a_{ij} < 0$) by activity j, and let p_j be the unit profit for the jth activity. Then if each activity is operated at level $x_j \geq 0$, the quantities y_i, defined by

$$y_i = \sum_{j=1}^{n} a_{ij}x_j, \tag{7.29}$$

will be the total amount of commodity i used up ($y_i > 0$) or produced ($y_i < 0$). An ordinary linear program, of an activity analysis model, is then obtained by specifying the amounts of supplies and demands, represented by $b = (b_1, \ldots, b_m)^T$, and writing

$$\max f(x) = p^T x \tag{7.30}$$

subject to

$$Ax = b \tag{7.31}$$

$$x \geq 0. \tag{7.32}$$

In stochastic linear programming the vector b is regarded as a random variable. Consequently, new problems of formulation, as well as of analysis, arise. Here we shall deal with the recourse-type formulation, details of which can be found in Walkup and Wets [27], Wets [28], and Williams [29]. Our analysis will primarily follow the approach of Avriel and Williams [3, 4], which is based on conjugate function theory.

We assume that a decision must be made—that is, choose x first. This step leads to some y and profit $p^T x$. Next, the actual value of b is revealed, and some corrective action is taken to resolve the difference between y and the revealed b. It is also assumed that we are given a **separable recourse function** g, depending on the vector $b - y$, such that

$$g(b - y) = \sum_{i=1}^{m} g_i(b_i - y_i). \tag{7.33}$$

This recourse function specifies additional cost if $g > 0$ or revenue if $g < 0$. Our **stochastic linear program with separable recourse** consists then of choosing x in the first place so as to maximize the overall expected profit. Formally,

we have

(PRP) $$\max_{x \geq 0} p^T x - \sum_{i=1}^{m} E[g_i(t_i - y_i)], \tag{7.34}$$

where E is the expectation operator for the random variable t_i, and y_i is given by (7.29). It is assumed that $E[g_i(t_i - y_i)]$ is finite for every y_i. Our analysis of problem (PRP) will consist of existence, characterization, and duality theorems. Assuming that the g_i are proper convex functions of y_i, we maximize a proper concave function over the nonnegative orthant. Since program (PRP) can become a convex program by an obvious change of sign, we can use results from Chapters 4 and 5 in our analysis. Before doing so, however, we derive several more general results.

Consider the concave nonlinear program

(CP) $$\max \psi(x) = u(x) - z(Ax), \tag{7.35}$$

where $x \in R^n$, u is a closed proper concave function on R^n, A is an $m \times n$ real matrix, and z is a closed proper convex function on R^m. To compute a dual of (7.35), we first introduce a perturbation vector $w \in R^m$, thereby obtaining

$$\psi(x, w) = u(x) - z(Ax - w). \tag{7.36}$$

The next step is to compute the concave conjugate function ψ^*:

$$\psi^*(\xi, \lambda) = \inf_{x, w} \{\xi^T x + \lambda^T w - \psi(x, w)\} \tag{7.37}$$

$$= \inf_{x, w} \{\xi^T x + \lambda^T w - u(x) + z(Ax - w)\}. \tag{7.38}$$

Substituting

$$w = Ax - y, \tag{7.39}$$

we obtain

$$\psi^*(\xi, \lambda) = \inf_{x, y} \{\xi^T x - \lambda^T y + \lambda^T Ax - u(x) + z(y)\} \tag{7.40}$$

$$= \inf_{x} \{(\xi + A^T\lambda)^T x - u(x)\} - \sup_{y} \{\lambda^T y - z(y)\} \tag{7.41}$$

and

$$\psi^*(\xi, \lambda) = u^*(\xi + A^T\lambda) - z^*(\lambda). \tag{7.42}$$

Thus the dual program of (7.35) is given by

(CD) $$\min - \psi^*(0, \lambda) = z^*(\lambda) - u^*(A^T\lambda). \tag{7.43}$$

In association with these programs, we can define another problem (which is not an optimization problem):

(CC) Find an $x \in R^n$ and a $\lambda \in R^m$ such that

$$A^T\lambda \in \partial u(x) \tag{7.44}$$

and

$$Ax \in \partial z^*(\lambda), \tag{7.45}$$

where $\partial u(x)$ and $\partial z^*(\lambda)$ are the subdifferentials of u and z^* at x and λ, respectively.

The three problems are related to each other by the following theorems [23].

Theorem 7.5

In order for x^* and λ^* to be vectors such that

$$u(x^*) - z(Ax^*) = \max\{u(x) - z(Ax)\} \tag{7.46}$$

$$= \min\{z^*(\lambda) - u^*(A^T\lambda)\} = z^*(\lambda^*) - u^*(A^T\lambda^*), \tag{7.47}$$

it is necessary and sufficient that x^* and λ^* satisfy

$$A^T\lambda^* \in \partial u(x^*) \qquad Ax^* \in \partial z^*(\lambda^*). \tag{7.48}$$

The proof of this theorem follows from arguments similar to those in Chapter 5.

For the preceding three problems, we have an existence theorem.

Theorem 7.6

Suppose that there exist vectors x and λ in the effective domains of ψ and ψ^*, respectively. If (CP) *and* (CD) *are both stable then* (CP), (CD), *and* (CC) *have solutions.*

Next, a characterization theorem is given.

Theorem 7.7

Suppose that (CP) *is normal. A vector pair (x^*, λ^*) solves* (CC) *if and only if x^* is optimal in* (CP) *and λ^* is optimal in* (CD).

Finally, we can state a strong duality theorem.

Theorem 7.8

Program (CP) *is stable and has an optimal solution if and only if program* (CD) *is stable and has an optimal solution. In this case, the maximum value of ψ in* (CP) *equals the minimum value of $-\psi^*$ in* (CD).

Duality theorems in the special case where $A = I$, the identity matrix, were first studied by Fenchel [17] and then extended and sharpened by Rockafellar [23].

For **separable** functions u and z

$$u(x) = \sum_{j=1}^{n} u_j(x_j) \tag{7.49}$$

$$z(y) = \sum_{i=1}^{m} z_i(y_i) \tag{7.50}$$

the conjugate functions are also separable (see Exercise 5.E). Hence we have the pair of separable dual programs

(PSP)
$$\max \sum_{j=1}^{n} u_j(x_j) - \sum_{i=1}^{m} z_i\,(Ax)_i \tag{7.51}$$

and

(DSP)
$$\min \sum_{i=1}^{m} z_i^*(\lambda_i) - \sum_{j=1}^{n} u_j^*(A^T\lambda)_j. \tag{7.52}$$

Turning now to the stochastic program with separable recourse, we note that it is a separable program as given by (7.51). Letting

$$u_j(x_j) = \begin{cases} p_j x_j & \text{if } x_j \geq 0, \\ -\infty & \text{if } x_j < 0, \end{cases} \qquad j = 1, \ldots, n \tag{7.53}$$

$$z_i(y_i) = E[g_i(t_i - y_i)], \qquad i = 1, \ldots, m \tag{7.54}$$

we obtain

$$u_j^*(A^T\lambda)_j = \begin{cases} 0 & \text{if } (A^T\lambda)_j \geq p_j \\ -\infty & \text{if } (A^T\lambda)_j < p_j \end{cases} \tag{7.55}$$

$$z_i^*(\lambda_i) = \sup_{y_i} \{\lambda_i y_i - E[g_i(t_i - y_i)]\}. \tag{7.56}$$

The dual recourse program then becomes

(DRP)
$$\min_{\lambda} \sum_{i=1}^{m} \sup_{y_i} \{\lambda_i y_i - E[g_i(t_i - y_i)]\} \tag{7.57}$$

subject to

$$A^T\lambda \geq p. \tag{7.58}$$

Note that, by assumption, ED $(E[g_i]) = R$, the whole real line.

Let us look a little more closely at the dual objective function (7.57). The supremum with respect to y_i can be explicitly computed if and only if

$$\lambda_i \in \partial E[g_i(t_i - y_i)]. \tag{7.59}$$

Since the domain of the functions involved is R, this condition is equivalent to the condition

$$D^+E[g_i(t_i - y_i; 1)] \geq \lambda_i \geq D^-E[g_i(t_i - y_i; 1)]. \tag{7.60}$$

It can be shown that

$$D^+E[g_i(t_i - y_i; 1)] = -E[D^-g_i(t_i - y_i; 1)] \tag{7.61}$$

and

$$D^-E[g_i(t_i - y_i; 1)] = -E[D^+g_i(t_i - y_i; 1)]. \tag{7.62}$$

Hence (7.60) can be written as

$$-E[D^-g_i(t_i - y_i; 1)] \geq \lambda_i \geq -E[D^+g_i(t_i - y_i; 1)]. \tag{7.63}$$

Since the g_i are proper convex functions, the $-g_i$ are proper concave. Therefore $-D^+g_i(t_i - y_i; 1)$ and $-D^-g_i(t_i - y_i; 1)$ are nonincreasing with $t_i - y_i$; that is, nondecreasing with y_i. The limits of the right and left derivatives as $y_i \longrightarrow +\infty$ (or $-\infty$) are equal and may have infinite values. Assume that

$$\lim_{y_i \to +\infty} - D^+g_i(t_i - y_i; 1) = \lim_{y_i \to +\infty} - D^-g_i(t_i - y_i; 1) = \gamma_i, \qquad i = 1, \ldots, m \tag{7.64}$$

$$\lim_{y_i \to -\infty} - D^+g_i(t_i - y_i; 1) = \lim_{y_i \to -\infty} - D^-g_i(t_i - y_i; 1) = \delta_i, \qquad i = 1, \ldots, m. \tag{7.65}$$

Therefore for any y_i and t_i

$$\gamma_i \geq -D^-g_i(t_i - y_i; 1) \geq -D^+g_i(t_i - y_i; 1) \geq \delta_i, \quad i = 1, \ldots, m \tag{7.66}$$

and for any y_i

$$\gamma_i \geq -E[D^-g_i(t_i - y_i; 1)] \geq -E[D^+g_i(t_i - y_i; 1)] \geq \delta_i, \qquad i = 1, \ldots, m. \tag{7.67}$$

The bounds in (7.63) then imply

$$\gamma_i \geq \lambda_i \geq \delta_i \tag{7.68}$$

and for $i = 1, \ldots, m$

$$\lambda_i = \begin{cases} \delta_i & \text{only if } \delta_i = -E[D^+ g_i(t_i - y_i; 1)] \text{ for some } y_i \\ \gamma_i & \text{only if } \gamma_i = -E[D^- g_i(t_i - y_i; 1)] \text{ for some } y_i \end{cases} \tag{7.69}$$

We are now in a position to state the basic existence theorem for stochastic linear programs with separable recourse.

Theorem 7.9

Program (PRP) *always has a feasible solution. If there exists a* λ *satisfying* (7.58), (7.68) *and* (7.69) *and* (DRP) *is stable then* (PRP) *has an optimal solution.*

We also have a characterization theorem for (PRP).

Theorem 7.10

Suppose that (PRP) *is stable. The nonnegative vector* x^* *is optimal for* (PRP) *if and only if for some vector* λ^* *satisfying*

$$A^T \lambda^* \geq p \tag{7.70}$$

$$\gamma \geq \lambda^* \geq \delta \tag{7.71}$$

and

$$\lambda_i^* = \begin{cases} \delta_i & \textit{only if } \delta_i = -E[D^+ g_i(t_i - y_i; 1)] \textit{ for some } y_i \\ \gamma_i & \textit{only if } \gamma_i = -E[D^- g_i(t_i - y_i; 1)] \textit{ for some } y_i \end{cases} \tag{7.72}$$

we have

$$-E[D^- g_i(t_i - (Ax^*)_i; 1)] \geq \lambda_i^* \geq -E[D^+ g_i(t_i - (Ax^*)_i; 1)], \quad i = 1, \ldots, m \tag{7.73}$$

and

$$x_j^* > 0 \tag{7.74}$$

only if

$$\sum_{i=1}^{m} a_{ij} \lambda_i^* = p_j. \tag{7.75}$$

The reader will be asked to show that (7.70) to (7.75) are the conditions (expressed for stochastic programming with recourse) of problem (CC) as given in (7.44) and (7.45).

Finally, we state a strong duality theorem.

Theorem 7.11

Program (PRP) *has an optimal solution* x^* *if and only if program* (DRP) *has an optimal solution* λ^*, *and, in this case,*

$$p^T x^* - \sum_{i=1}^{m} E[g_i(t_i - (Ax^*)_i)] = \sum_{i=1}^{m} \sup_{y_i} \{\lambda_i^* y_i - E[g_i(t_i - y_i)]\}. \tag{7.76}$$

Let us specify the recourse functions g_i more precisely. Many different assumptions can be made on these functions, but we shall mention only the simplest form of recourse—namely, the **linear recourse** case in which

$$-g_i(b_i - y_i) = \begin{cases} \gamma_i(b_i - y_i) & \text{if } y_i \geq b_i \\ \delta_i(b_i - y_i) & \text{if } y_i \leq b_i \end{cases} \tag{7.77}$$

for $i = 1, \ldots, m$, where the vectors $\gamma = (\gamma_1, \ldots, \gamma_m)^T$, $\delta = (\delta_1, \ldots, \delta_m)^T$ with $\gamma_i \geq \delta_i$ are given. Williams [29] showed that the stochastic linear program with linear recourse is given by

$$\text{(PLRP)} \quad \max_{x \geq 0} p^T x - \delta^T Ax - \sum_{i=1}^{m} \left\{(\gamma_i - \delta_i)\left[\int_{-\infty}^{(Ax)_i} F_i(t)\, dt\right] + \delta_i E(b_i)\right\}, \tag{7.78}$$

where the F_i are the marginal distributions of

$$F(t) = \text{probability } \{b \leq t\}. \tag{7.79}$$

The reader will be asked to derive the dual program corresponding to (PLRP) as

$$\text{(DLRP)} \quad \min \sum_{i=1}^{m} \left\{(\lambda_i - \delta_i)\hat{y}_i - (\gamma_i - \delta_i)\left[\int_{-\infty}^{\hat{y}_i} F_i(t)\, dt\right] + \delta_i E(b_i)\right\} \tag{7.80}$$

subject to

$$\gamma \geq \lambda \geq \delta \tag{7.81}$$

$$A^T \lambda \geq p \tag{7.82}$$

where the $\hat{y}_i$ satisfy (see the meanings of F_i^+ and F_i^- below)

$$[\gamma_i F_i^+(\hat{y}_i) + \delta_i(1 - F_i^+(\hat{y}_i))] \geq \lambda_i \geq [\gamma_i F_i^-(\hat{y}_i) + \delta_i(1 - F_i^-(\hat{y}_i))]. \tag{7.83}$$

Since the F_i can be discontinuous, it is possible that, for a given number τ, the equation

$$F_i(y_i) = \tau \tag{7.84}$$

has no solution even when $0 < \tau < 1$. For this reason, we introduced the notation

$$F^+(\hat{y}) = \inf_{y \to \hat{y}^+} F(y) \tag{7.85}$$

$$F^-(\hat{y}) = \sup_{y \to \hat{y}^-} F(y). \tag{7.86}$$

Then

$$F_i^+(\hat{y}_i) \geq \tau \geq F_i^-(\hat{y}_i) \tag{7.87}$$

always has a solution $\hat{y}_i$ for $0 \leq \tau \leq 1$.

Duality theorems for stochastic programs with separable recourse are useful in the numerical solution of these problems, for they provide certain bounds on the optimal values of approximating linear programs and indicate the deviation of the approximating solutions from the true optimum. For details, the reader is referred to Williams [30], where the linear recourse case is discussed. The solution of such problems by a different approach was proposed by Ziemba [33]. Readers interested in the general area of stochastic programming may find expository material in the books of Sengupta [25], Vajda [26], and in the numerous surveys published in the literature.

7.3 GEOMETRIC PROGRAMMING

One of the most widely studied branches of nonlinear programming, from an analytical and computational point of view, is geometric programming and its extensions. The great interest in this class of nonlinear programs lies in two aspects. On the one hand, almost every analytical result and computational algorithm derived for nonlinear programs can be elegantly illustrated by geometric programming problems. On the other, many practical optimization problems from various disciplines can be formulated as geometric programs or as extensions of them. We present here some of the analytical aspects of geometric programming that demonstrate theoretical results derived in previous chapters. Readers interested in this class of nonlinear programs may find a great number of articles dealing with geometric programming in the literature. For an introduction to the subject, the books of Duffin, Peterson, and Zener [16] and Zener [32] are recommended. For a more extended view, see Avriel, Rijckaert, and Wilde [1], where later results are also presented.

We begin the study of geometric programming by defining a **posynomial** g as a real function consisting of a finite sum of terms

$$g(x) = \sum_i c_i \prod_{j=1}^n (x_j)^{a_{ij}}, \tag{7.88}$$

where $c_i > 0$ and $a_{ij} \in R$ are given constants and $x > 0$. Posynomials are generally neither convex nor concave. They are, however, (log, log)-convex, as we shall now see.

Letting $h(t) = \log t$, $h^{-1}(y) = e^y$, and $\phi(t) = \log t$, we obtain

$$\hat{g}(y) = \phi g h^{-1}(y) = \log \sum_i c_i \exp \sum_{j=1}^n a_{ij} y_j. \tag{7.89}$$

Now, by Theorem 6.15, the real function g is (h, ϕ)-convex if and only if $\phi g h^{-1}$ is convex. We shall prove the convexity of the function in (7.89) by using the well-known Hölder inequality [19, 20]: Let $\alpha_i > 0$, $\beta_i > 0$ be positive numbers and let $p > 0$, $r > 0$ such that $(p)^{-1} + (r)^{-1} = 1$. Then

$$\sum_i \alpha_i \beta_i \leq \Big(\sum_i (\alpha_i)^p\Big)^{1/p} \Big(\sum_i (\beta_i)^r\Big)^{1/r}. \tag{7.90}$$

Lemma 7.12

Let g be a posynomial defined on the positive orthant of R^n. Then $\hat{g}(y)$, as given by (7.89), is a convex function on R^n.

Proof. Let $y^1 \in R^n$, $y^2 \in R^n$ be any two points. Also let

$$\alpha_i = \Big\{c_i \exp \sum_{j=1}^n a_{ij} y_j^1\Big\}^{q_1} \qquad \beta_i = \Big\{c_i \exp \sum_{j=1}^n a_{ij} y_j^2\Big\}^{q_2} \tag{7.91}$$

and

$$(p)^{-1} = q_1 \qquad (r)^{-1} = q_2. \tag{7.92}$$

Substituting into (7.90), we obtain

$$\sum_i \Big\{c_i \exp \sum_{j=1}^n a_{ij} y_j^1\Big\}^{q_1} \Big\{c_i \exp \sum_{j=1}^n a_{ij} y_j^2\Big\}^{q_2} \leq \Big(\sum_i \Big\{c_i \exp \sum_{j=1}^n a_{ij} y_j^1\Big\}\Big)^{q_1} \Big(\sum_i \Big\{c_i \exp \sum_{j=1}^n a_{ij} y_j^2\Big\}\Big)^{q_2}. \tag{7.93}$$

Taking logarithms of both sides yields

$$\log \sum_i c_i \exp \sum_{j=1}^n a_{ij}(q_1 y_j^1 + q_2 y_j^2) \leq q_1 \log \sum_i c_i \exp \sum_{j=1}^n a_{ij} y_j^1 + q_2 \log \sum_i c_i \exp \sum_{j=1}^n a_{ij} y_j^2 \tag{7.94}$$

and $\hat{g}$ is convex. ■

We can now define geometric programs. A nonlinear program given by

(PGP) $$\min g_0(x) \tag{7.95}$$

subject to

$$g_k(x) \leq 1, \qquad k = 1, \ldots, m \tag{7.96}$$

$$x > 0, \tag{7.97}$$

where $g_0, g_1, \ldots, g_m$ are posynomials, is called a **geometric program**. Accordingly, let

$$g_k(x) = \sum_{i \in I_k} c_i \prod_{j=1}^{n} (x_j)^{a_{ij}} \tag{7.98}$$

where

$$I_k = \{r_k, \ldots, s_k\}, \qquad k = 0, 1, \ldots, m \tag{7.99}$$

$$r_0 = 1 \tag{7.100}$$

$$r_k = s_{k-1} + 1, \qquad k = 1, \ldots, m \tag{7.101}$$

$$s_m = M. \tag{7.102}$$

Hence (PGP) consists of an objective function and m posynomial constraints in n positive variables. The kth posynomial is a sum of $s_k - r_k + 1$ terms, called **monomials**. The total number of monomials in the program is M.

Since x is restricted to be positive, we can make the change of variables

$$x_j = e^{y_j}, \qquad j = 1, \ldots, n \tag{7.103}$$

and by taking logarithms of $g_0, g_1, \ldots, g_m$ (a monotone transformation that does not alter the location of extrema), we obtain a convex program that is amenable to the analysis presented in Chapters 4 and 5 [19]. Although this approach is used by most authors when analyzing geometric programs, we shall pursue here the direct (and equivalent) approach of considering (PGP) as a (log, log)-convex program, thereby illustrating some of the results of Chapter 6 as applied to duality.

Before presenting our results, we must outline the relevant extensions of convex duality to (h, ϕ)-convex programs.

The important fact to remember is that by using the generalized algebraic operations defined in the preceding chapter and some easy modifications, all convex duality results also apply for the more general case.

A standard (h, ϕ)-convex program is given by [31]

(SHP) $$\min_x f(x) \tag{7.104}$$

subject to

$$g_i(x) \geq 0_\phi, \qquad i = 1, \ldots, m \tag{7.105}$$

where f and $g_1, \ldots, g_m$ are proper (h, ϕ)-convex and (h, ϕ)-concave functions on R^n, respectively, and

$$0_\phi = \phi^{-1}(0). \tag{7.106}$$

An (h, ϕ)-convex function f on R^n is called proper if $\mathrm{ED}(f) \neq \emptyset$ and for every $x \in R^n$ we have $f(x) > -\infty_\phi$, where

$$\mathrm{ED}(f) = \{x : x \in R^n, f(x) < +\infty_\phi\} \tag{7.107}$$

and

$$+\infty_\phi = \lim_{y \to +\infty} \phi^{-1}(y) \qquad -\infty_\phi = \lim_{y \to -\infty} \phi^{-1}(y). \tag{7.108}$$

The limits in (7.108) can, of course, be infinite. For example, if $\phi(t) = \log t$, then

$$+\infty_\phi = \lim_{y \to +\infty} e^y = +\infty \qquad -\infty_\phi = \lim_{y \to -\infty} e^y = 0. \tag{7.109}$$

Introduce now perturbations in (SHP) in the form of a vector $w = (w_1, \ldots, w_l)^T$ such that $f(x, w)$ and $g_i(x, w)$ are $((h, v), \phi)$-convex and $((h, v), \phi)$-concave, respectively. That is, $v_1(w_1), \ldots, v_l(w_l)$ are mean value functions having inverses $v_1^{-1}(u_1), \ldots, v_l^{-1}(u_l)$ such that $\phi f(h^{-1}(y), v^{-1}(u))$ and $\phi g_i(h^{-1}(y), v^{-1}(u))$ are, respectively, ordinary convex and concave functions. Similar to (5.36), we define

$$\theta(x, w) = \begin{cases} f(x, w) & \text{if } (x, w) \in \mathrm{ED}(f), g_i(x, w) \geq 0_\phi, i = 1, \ldots, m \\ +\infty_\phi & \text{otherwise.} \end{cases} \tag{7.110}$$

It follows then that, by a proper choice of the way w is incorporated in the functions involved, (SHP) can be equivalent to minimizing $\theta(x, 0_v)$ with respect to x.

We saw earlier that conjugate functions are one of the major tools for deriving dual programs in the convex case. The notion of conjugate functions was further extended in Chapter 6 for (h, ϕ)-convex functions. Recalling the definition given in (6.128) as

$$f^*(\xi) = \sup_x \{(\xi^T x)_{h,\phi} [-] f(x)\} \tag{7.111}$$

if f is (h, ϕ)-convex, we can easily show that f^* is also (h, ϕ)-convex and $f^{**} = \mathrm{cl}\, f$, where the closure operation refers to closed epigraphs. It turns out that we can define an infinite variety of conjugate functions for a given f. Let a generalized inner product be

$$(x^T y)_{h^1, h^2, \phi} = \phi^{-1}[(h^1(x))^T h^2(x)], \tag{7.112}$$

where h^1 and h^2 are not necessarily the same mean value functions. If $h^1 = h^2 = h$, we obtain

$$(x^T y)_{h^1, h^2, \phi} = (x^T y)_{h, \phi} \tag{7.113}$$

where the right-hand side was defined in (6.113). Extending (6.128) and (6.129), we can write conjugate functions of an (h, ϕ)-convex f as

$$f^*(\xi) = \sup_x \{(\xi^T x)_{\bar{h}, h, \phi} [-] f(x)\} \tag{7.114}$$

or

$$f^*(\xi) = \sup_x \{\phi^{-1}[\bar{h}(\xi)^T h(x) - \phi f(x)]\}, \tag{7.115}$$

where, again, h and $\bar{h}$ are not necessarily the same mean value functions.

It can also be shown that f^*, defined by (7.114), is $(\bar{h}, \phi)$-convex and

$$f^{**}(x) = \sup_\xi \{(x^T \xi)_{h, \bar{h}, \phi} [-] f^*(\xi)\} = \text{cl} f. \tag{7.116}$$

The primal-dual program pair corresponding to (SHP) is then

$$(\text{P}_\theta) \qquad \min_x \theta(x, 0_v) \tag{7.117}$$

and

$$(\text{D}_{\theta^*}) \qquad \max_\lambda [-] \theta^*(0_{\bar{h}}, \lambda) = \eta(\lambda), \tag{7.118}$$

where θ^* is the conjugate function of θ, defined by

$$\theta^*(\xi, \lambda) = \sup_{x, w} \{((\xi^T x)_{\bar{h}, h, \phi} [+] (\lambda^T w)_{\bar{v}, v, \phi}) [-] \theta(x, w)\}, \tag{7.119}$$

and

$$[-] \theta^* = \phi^{-1}[-\phi(\theta^*)]. \tag{7.120}$$

Note that the function η in (7.118) is a $(\bar{v}, \phi)$-concave function of λ.

Before returning to geometric programming, where we shall use the preceding general dual pair, we illustrate a case that is a little simpler by a small example.

Example 7.3.1

Consider the problem

$$\min f(x) = (x)^{1/3} \tag{7.121}$$

subject to

$$g(x) = (x)^{1/6} \geq 1, \tag{7.122}$$

where $x \in R$. Note that f is actually a concave function and the feasible set is $\{x : x \in R, x \geq 1\}$. The optimal solution is clearly $x^* = 1$ and $f(x^*) = 1$. It is easy to show, however, that by letting $h(t) = t$ and $\phi(t) = (t)^3 - 1$ for all $t \in R$, we obtain a standard (h, ϕ)-convex program:

$$\min\ (x)^{1/3} \tag{7.123}$$

subject to

$$(x)^{1/6} \geq 0_\phi. \tag{7.124}$$

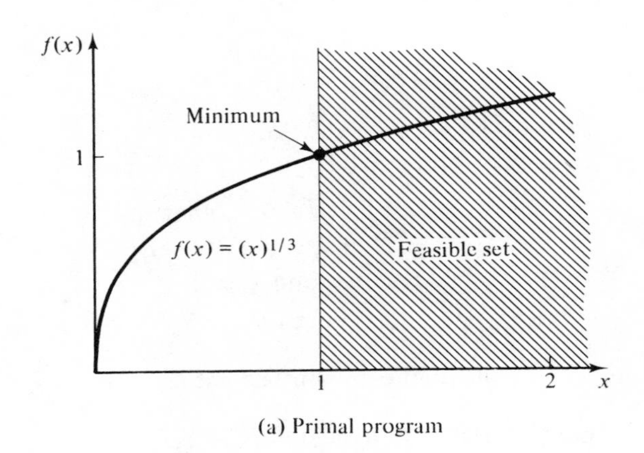

(a) Primal program

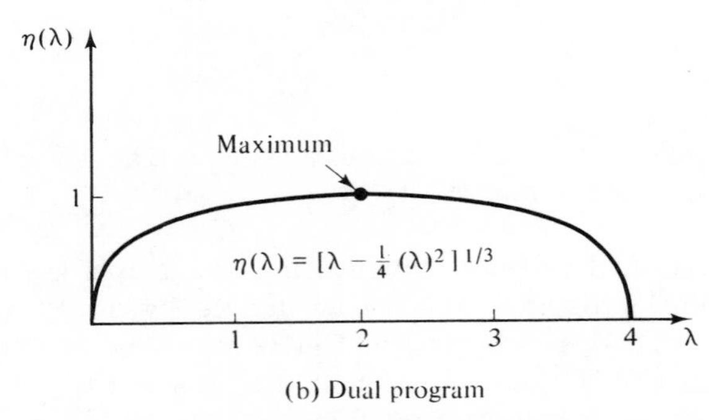

(b) Dual program

Fig. 7.1 A pair of (h, ϕ)-convex programs.

Note that $\phi^{-1}(y) = (y + 1)^{1/3}$ and $0_\phi = 1$, $+\infty_\phi = +\infty$. Introducing into (7.122) a perturbation variable w with $v(t) = t$, we get

$$\theta(x, w) = \begin{cases} (x)^{1/3} & \text{if } (x^{1/2} - w)^{1/3} \geq 0_\phi \\ +\infty_\phi & \text{otherwise.} \end{cases} \tag{7.125}$$

Next we compute θ^* by taking $\bar{h}(t) = h(t) = t$.

$$\theta^*(\xi, \lambda) = \sup_{x,w} \{((\xi x)_{h,\phi} [+] (\lambda w)_{v,\phi}) [-] \theta(x, w)\} \tag{7.126}$$

$$= \sup_{x,w} \{\phi^{-1}[\xi x + \lambda w - \phi\theta(x, w)]\} \tag{7.127}$$

$$= \sup_{((x)^{1/2}-w)^{1/3} \geq 0_\phi} \{\xi x + \lambda w - x + 2\}^{1/3}. \tag{7.128}$$

Introducing a slack variable $\mu \geq 0_\phi$ gives

$$(x^{1/2} - w)^{1/3} [-] \mu = 0_\phi \tag{7.129}$$

and

$$w = (x)^{1/2} - (\mu)^3. \tag{7.130}$$

Hence

$$\theta^*(\xi, \lambda) = \begin{cases} \sup_x \{\xi x + \lambda(x)^{1/2} - \lambda - x + 2\}^{1/3} & \text{if } (\lambda + 1)^{1/3} \geq 0_\phi \\ +\infty_\phi & \text{otherwise} \end{cases} \tag{7.131}$$

$$= \begin{cases} \left[\dfrac{(\lambda)^2}{4(1-\xi)} - \lambda + 2\right]^{1/3} & \text{if } (\lambda + 1)^{1/3} \geq 0_\phi \text{ and } \xi \neq 1 \\ +\infty_\phi & \text{otherwise.} \end{cases} \tag{7.132}$$

Moreover, letting $\xi = 0$, the dual program becomes

$$\max [-] \theta^*(0_h, \lambda) = \eta(\lambda) = [\lambda - \tfrac{1}{4}(\lambda)^2]^{1/3} \tag{7.133}$$

subject to

$$(\lambda + 1)^{1/3} \geq 1, \tag{7.134}$$

whose optimal solution is at $\lambda^* = 2$ and $\eta(\lambda^*) = 1$. The primal and dual programs are illustrated in Figure 7.1. ▮

Note that the dual program obtained in the example is only one of many possible dual programs that can be derived by using various perturbations. Although this example problem is also a geometric program (if we also restrict x to be positive), we did not treat it as a (log, log)-convex program. Thus it can be seen that the family of dual programs associated with a certain primal can be further enlarged by identifying different h and ϕ functions for which the primal-dual pair can become (h, ϕ)-convex.

Let us obtain now a dual program for geometric programming, identical to the original one derived by Duffin, Peterson, and Zener. Instead of the arithmetic-geometric mean inequality [16] or convex programming analysis [19, 22, 23] we use (h, ϕ)-convex duality.

We begin by reformulating (PGP) as follows: Let y^k be the vector-valued function of $x \in R^n$ whose components are y_i, $i \in I_k$, satisfying

$$y_i = \prod_{j=1}^{n} (x_j)^{a_{ij}}, \qquad i \in I_k, \quad k = 0, \ldots, m \tag{7.135}$$

or in terms of (log, log)-convex operations

$$y_i = (a_{i1} \odot x_1) \oplus \cdots \oplus (a_{in} \odot x_n). \tag{7.136}$$

We can then write

$$g_k(y^k) = \sum_{i \in I_k} c_i y_i, \qquad k = 0, \ldots, m \tag{7.137}$$

and program (PGP) becomes

(SGP)
$$\min g_0(y^0) \tag{7.138}$$

subject to

$$[-]g_k(y^k) \geq 0_{\log}, \qquad k = 1, \ldots, m \tag{7.139}$$

where g_0 is (log, log)-convex, $[-]g_1, \ldots, [-]g_m$ are (log, log)-concave, the y_i satisfy (7.136), and

$$0_{\log} = (e)^0 = 1. \tag{7.140}$$

Introducing perturbation vectors $u = (u^0, \ldots, u^m)^T$ and $w = (w_1, \ldots, w_m)^T$, where u^k is a vector whose components are u_i, $i \in I_k$, we obtain a perturbed problem

$$\min g_0(y^0) = \sum_{i \in I_0} c_i y_i \tag{7.141}$$

subject to

$$[-]g_k(y^k, w_k) = [-]\Big(\sum_{i \in I_k} c_i y_i [+] w_k\Big) \geq 0_{\log}, \qquad k = 1, \ldots, m \tag{7.142}$$

and

$$y_i \ominus u_i = (a_{i1} \odot x_1) \oplus \cdots \oplus (a_{in} \odot x_n), \qquad i \in I_k, k = 0, \ldots, m \tag{7.143}$$

Note that g_0 is (log, log)-convex, where the first "log" corresponds to the vector y and the g_k are $[(h, v), \phi]$-concave with respect to $(y^k, w_k)^T$, respectively, with $h_i(y_i) = \log y_i$ and $v(w_k) = w_k$. The operation $y_i \ominus u_i$ means y_i/u_i here, since we chose $h_i(u_i) = \log u_i$.

Thus it follows that the primal program (SGP) can be rewritten as

(P_θ)
$$\min_{x, y} \theta(x, y, 0_{\log}, 0). \tag{7.144}$$

Here

$$\theta(x, y, u, w) = g_0(y^0) \quad \text{if} \begin{cases} [-]g_k(y^k, w_k) \geq 0_{\log}, & k = 1, \ldots, m \\ y \ominus u = A \odot x, \end{cases} \tag{7.145}$$

where the second condition is simply shorthand for (7.143). If (7.145) does not hold, then $\phi(x, y, u, w) = +\infty_{\log}$.

Before turning to the dual program, we compute a conjugate function that will be needed in the sequel. Let g be an (h, ϕ)-convex function of y with $h(y) = (\log y_1, \ldots, \log y_n)^T$ and $\phi(t) = \log t$. Suppose that g is given by

$$g(y) = \sum_i c_i y_i, \tag{7.146}$$

where $c_i > 0$. Let $\bar{h}(\xi) = \xi$ and define

$$g^*(\xi) = \sup_y \{(\xi^T y)_{I, \log, \log} [-] \, g(y)\}, \tag{7.147}$$

where we use the notation $\bar{h} = I$ if $\bar{h}(y) = y$, the identity function.

The reader will be asked to prove that

$$g^*(\xi) = \begin{cases} \prod_i \left(\frac{\xi_i}{c_i}\right)^{\xi_i} & \text{if } \sum_i \xi_i = 1, \ \xi \geq 0 \\ +\infty_{\log} & \text{otherwise} \end{cases} \tag{7.148}$$

and g^* is $(\bar{h}, \phi)$-convex with $\bar{h}$ and ϕ as above.

Let us compute now a conjugate function of θ. Define

$$\theta^*(\xi, \eta, \delta, \lambda) = \sup_{x, y, u\, w} \{(\xi^T x)_{I, \log, \log} [+] \, (\eta^T y)_{I, \log \log} \\ [+] \, (\delta^T u)_{I, \log, \log} [+] \, (\lambda^T w)_{v, \log} [-] \, \theta(x, y, u, w)\} \tag{7.149}$$

$$= \sup_{\substack{[-] g_k(y^k, w_k) \geq 0_{\log} \\ y \ominus u = A \odot x}} \Big\{(\xi^T x)_{I, \log, \log} [+] \, (\eta^T y)_{I, \log \log} \\ [+] \, (\delta^T u)_{I, \log \log} [+] \, (\lambda^T w)_{I, \log} [-] \sum_{i \in I_0} c_i y_i\Big\}. \tag{7.150}$$

Introducing slack variables $\mu_k \geq 0_{\log}$, we obtain

$$\theta^*(\xi, \eta, \delta, \lambda) = \sup_{\substack{\mu_k \geq 0_{\log} \\ y \ominus u = A \odot x}} \Big\{(\xi^T x)_{I, \log, \log} [+] \, (\eta^T y)_{I, \log, \log} \\ [+] \, (\delta^T u)_{I, \log, \log} [-] \sum_{k=1}^m \lambda_k [\cdot] \Big(\sum_{i \in I_k} c_i y_i\Big) \\ [-] \, (\lambda^T \mu)_{I, \log, \log} [-] \sum_{i \in I_0} c_i y_i\Big\} \tag{7.151}$$

$$= \begin{cases} \sup_{y \ominus u = A \odot x} \Big\{(\xi^T x)_{I, \log, \log} [+] \, (\eta^T y)_{I, \log, \log} [+] \, (\delta^T u)_{I, \log, \log} \\ [-] \sum_{k=1}^m \lambda_k [\cdot] \Big(\sum_{i \in I_k} c_i y_i\Big) [-] \sum_{i \in I_0} c_i y_i\Big\} \quad \text{if } \lambda \geq 0 \\ +\infty_{\log} \qquad \text{otherwise.} \end{cases} \tag{7.152}$$

Consider the case of $\lambda \geq 0$. We can write

$$\theta^*(\xi, \eta, \delta, \lambda) = \sup_{y \ominus u = A \odot x} \Big\{(\xi^T x)_{I,\log,\log} [-] (\delta^T y \ominus u)_{I,\log,\log} [+] ((\eta + \delta)^T y)_{I,\log,\log} [-] \sum_{k=1}^{m} \lambda_k [\cdot] \Big(\sum_{i \in I_k} c_i y_i\Big) [-] \sum_{i \in I_0} c_i y_i \Big\} \tag{7.153}$$

$$= \sup_x \{(\xi^T x)_{I,\log,\log} [-] (\delta^T A \odot x)_{I,\log,\log}\} [+] \sup_y \Big\{((\eta + \delta)^T y)_{I,\log,\log} [-] \sum_{k=1}^{m} \lambda_k [\cdot] \Big(\sum_{i \in I_k} c_i y_i\Big) [-] \sum_{i \in I_0} c_i y_i \Big\}. \tag{7.154}$$

$$= S_1 [+] S_2. \tag{7.155}$$

The first term in (7.155) is

$$S_1 = \sup_x \Big\{\exp\Big[\Big(\xi_1 - \sum_{i=1}^{M} a_{i1}\delta_i\Big) \log x_1 + \cdots + \Big(\xi_n - \sum_{i=1}^{M} a_{in}\delta_i\Big) \log x_n\Big]\Big\} \tag{7.156}$$

and clearly,

$$S_1 = \begin{cases} 0_{\log} & \text{if } \sum_{i=1}^{M} a_{ij}\delta_i = \xi_j, \quad j = 1, \ldots, n \\ +\infty_{\log} & \text{otherwise.} \end{cases} \tag{7.157}$$

For the second term we obtain, after rearranging,

$$S_2 = \sup_{y_i, i \in I_0} \Big\{((\eta^0 + \delta^0)^T y^0)_{I,\log,\log} [-] \sum_{i \in I_0} c_i y_i \Big\} [+] \sup_{y_i, i \in I_1} \Big\{((\eta^1 + \delta^1)^T y^1)_{I,\log,\log} [-] \Big(\lambda_1 [\cdot] \sum_{i \in I_1} c_i y_i\Big)\Big\} [+] \cdots [+] \sup_{y_i, i \in I_m} \Big\{((\eta^m + \delta^m)^T y^m)_{I,\log,\log} [-] \Big(\lambda_m [\cdot] \sum_{i \in I_m} c_i y_i\Big)\Big\}. \tag{7.158}$$

Then by (7.146) to (7.148) and some modifications, we have

$$S_2 = \begin{cases} \prod_{i \in I_0} \Big(\dfrac{\eta_i + \delta_i}{c_i}\Big)^{(\eta_i + \delta_i)} [+] \prod_{i \in I_1} \Big(\dfrac{\eta_i + \delta_i}{c_i \lambda_1}\Big)^{(\eta_i + \delta_i)} [+] \cdots \\ [+] \prod_{i \in I_m} \Big(\dfrac{\eta_i + \delta_i}{c_i \lambda_m}\Big)^{(\eta_i + \delta_i)} & \text{if } \sum_{i \in I_k} (\eta_i + \delta_i) = \lambda_k, k = 1, \ldots, m \\ +\infty_{\log} & \text{otherwise.} \end{cases} \tag{7.159}$$

Finally,

$$\theta^*(0, 0, \delta, \lambda) = \prod_{i \in I_0} \left(\frac{\delta_i}{c_i}\right)^{\delta_i} \prod_{k=1}^{m} \prod_{i \in I_k} \left(\frac{\delta_i}{c_i \lambda_k}\right)^{\delta_i} \tag{7.160}$$

if

$$\delta \geq 0 \tag{7.161}$$

$$\sum_{i \in I_0} \delta_i = 1 \tag{7.162}$$

$$\sum_{k=0}^{m} \sum_{i \in I_k} a_{ij} \delta_i = 0, \qquad j = 1, \ldots, n \tag{7.163}$$

$$\sum_{i \in I_k} \delta_i = \lambda_k, \qquad k = 1, \ldots, m \tag{7.164}$$

and $\theta^*(0, 0, \delta, \lambda) = +\infty$ otherwise. A dual program of (P_θ) is then given by

$$(D_{\theta^*}) \qquad \max_{\delta, \lambda} [-] \, \theta^*(0, 0, \delta, \lambda) = \max_{\delta, \lambda} \Gamma(\delta, \lambda) \tag{7.165}$$

or

$$(\text{DGP}) \qquad \max_{\delta, \lambda} \Gamma(\delta, \lambda) = \prod_{i \in I_0} \left(\frac{c_i}{\delta_i}\right)^{\delta_i} \prod_{k=1}^{m} \prod_{i \in I_k} \left(\frac{c_i}{\delta_i}\right)^{\delta_i} (\lambda_k)^{\lambda_k}, \tag{7.166}$$

subject to (7.161) through (7.164). This is precisely the dual program obtained by Duffin, Peterson, and Zener [16] and many others. Note that Γ is (h, ϕ)-concave with $h(t) = t$ and $\phi(t) = \log t$.

This dual program has many interesting features. Perhaps the most significant is that the constraints (7.161) to (7.164) are **linear**, although the primal problem has a nonlinear objective function and nonlinear constraints. From this point of view, geometric programming belongs to a class of nonlinear programs whose duals are linearly constrained. This important class of problems has been studied by Rockafellar [24]. The importance of such programs lies in the claim that linearly constrained nonlinear programs are usually easier to solve than nonlinearly constrained ones. In the case of geometric programming, however, this advantage is not always apparent if one is interested in obtaining an optimal primal solution by solving the dual program. It is not always convenient to compute optimal primal variables when the optimal dual solution is known, especially if some primal constraints are inactive at the sought optimum. Nevertheless, at least one interesting situation occurs in which it is clearly easier to solve the dual program. If the total number of monomials M appearing in a primal geometric program exceeds the number of primal variables n by exactly one, the dual constraints consist of a square system of linear equations in nonnegative variables, that can have a unique solution if the equations are linearly independent. Let us illustrate this situation by an example.

Example 7.3.2

Consider the geometric program

$$\min g_0(x) = \frac{(x_4)^2}{(x_1)^4(x_2)} + \frac{3(x_1)^2}{(x_2)^2} \tag{7.167}$$

subject to $x > 0$ and

$$g_1(x) = \frac{(x_2)(x_3)}{3} + \frac{3}{(x_1)^{1/2}(x_2)^{3/4}(x_3)} + \frac{9(x_2)^{1/2}}{2(x_3)(x_4)^{1/2}} \leq 1. \tag{7.168}$$

Computing the dual program as developed above, we obtain

$$\max \Gamma(\delta, \lambda) = \left(\frac{1}{\delta_1}\right)^{\delta_1}\left(\frac{3}{\delta_2}\right)^{\delta_2}\left(\frac{1}{3\delta_3}\right)^{\delta_3}\left(\frac{3}{\delta_4}\right)^{\delta_4}\left(\frac{9}{2\delta_5}\right)^{\delta_5}(\lambda_1)^{\lambda_1} \tag{7.169}$$

subject to

$$\begin{aligned} \delta_1 + \delta_2 \qquad\qquad\qquad\qquad &= 1 \\ -4\delta_1 + 2\delta_2 \qquad - \tfrac{1}{2}\delta_4 \qquad &= 0 \\ -\delta_1 - 2\delta_2 + \delta_3 - \tfrac{3}{4}\delta_4 + \tfrac{1}{2}\delta_5 &= 0 \\ \delta_3 - \delta_4 - \delta_5 &= 0 \\ 2\delta_1 \qquad\qquad\qquad - \tfrac{1}{2}\delta_5 &= 0 \end{aligned} \tag{7.170}$$

$$\delta_3 + \delta_4 + \delta_5 = \lambda_1 \tag{7.171}$$

$$\delta \geq 0. \tag{7.172}$$

The square system of linear equations (7.170) has a unique solution $\delta_1^* = \frac{1}{4}$, $\delta_2^* = \frac{3}{4}$, $\delta_3^* = 2$, $\delta_4^* = 1$, $\delta_5^* = 1$. Hence $\lambda_1^* = 4$. Substituting into (7.169), we obtain $\Gamma(\delta^*, \lambda^*) = 384$. ∎

It is clear that we obtained a unique feasible and hence optimal solution to our dual program in this example because in the primal program we had $M = 5$ and $n = 4$. If $M > n + 1$, there will usually be an infinite number of feasible solutions to the dual program, and its objective function must be maximized in order to find an optimum. In any case, the optimal primal solution must be computed from the optimal dual solution. The relationship between primal and dual solutions can be found by the duality theorems that follow. The results will be given without proofs, for they are special cases of the general convex duality theory developed in Chapter 5, if we use the fact that the functions appearing in the primal-dual pair are convex

transformable. Equivalently, we can generalize the results of Chapter 5 to nonlinear programs with (h, ϕ)-convex functions. First,

Theorem 7.13 (Weak Duality Theorem)

If x satisfies the constraints of (PGP) *and (δ, λ) satisfies the constraints of* (DGP), *then*

$$g_0(x) \geq \Gamma(\delta, \lambda). \tag{7.173}$$

Now we have

Theorem 7.14 (Strong Duality Theorem)

If (PGP) *is strongly consistent and x^* is an optimal solution of* (PGP), *then*

(i) *Program* (DGP) *is consistent and there exists a feasible vector (δ^*, λ^*) such that*

$$g_0(x^*) = \Gamma(\delta^*, \lambda^*). \tag{7.174}$$

(ii) *There exists a $\lambda^* \geq 0$ such that the vector (x^*, λ^*) is a saddlepoint of the Lagrangian*

$$L(x, \lambda) = g_0(x)\,[-]\sum_{k=1}^{m} \lambda_k\,[\cdot]\,g_k(x) \tag{7.175}$$

for every $x > 0$ and $\lambda \geq 0$.

(iii) *The optimal (δ^*, λ^*) satisfies*

$$\delta_i^* = \begin{cases} c_i \prod_{j=1}^{n} \dfrac{(x_j^*)^{a_{ij}}}{g_0(x^*)} & i \in I_0 \\ \lambda_k^* c_i \prod_{j=1}^{n} (x_j^*)^{a_{ij}} & i \in I_k, \quad k = 1, \ldots, m. \end{cases} \tag{7.176}$$

(iv) *If (δ^*, λ^*) is optimal for* (DGP), *then*

$$c_i \prod_{j=1}^{n} (x_j^*)^{a_{ij}} = \begin{cases} \delta_i^*\,\Gamma(\delta^*, \lambda^*) & i \in I_0 \\ \dfrac{\delta_i^*}{\lambda_k^*} & i \in I_k, \quad k = 1, \ldots, m \end{cases} \tag{7.177}$$

provided that $\lambda_k^* > 0$.

Note that the algebraic operations in the square brackets in (7.175) correspond to $\phi(t) = \log t$.

The reader who may want to prove this theorem can start his proof at Theorem 5.16 (or at its equivalent), which states that strong consistency implies stability. Once stability is established, the rest of the results follow

from Chapter 5 and by computing subgradients. The last two theorems were first proved by Duffin, Peterson, and Zener [16], where additional duality results can also be found.

EXERCISES

7.A. Given a real vector $q \in R^p$ and a real $p \times p$ matrix M, suppose that we are seeking vectors w and z that satisfy the conditions

$$w = q + Mz, \qquad w \geq 0, \quad z \geq 0 \tag{7.178}$$

$$z^T w = 0. \tag{7.179}$$

This problem is termed the **fundamental problem of complementarity** by Cottle and Dantzig [9]. Show that the problem of solving a convex quadratic program in nonnegative variables $x \geq 0$ as given by (7.1) and (7.2) is equivalent to solving a corresponding complementarity problem as presented above. *Hint:* Define

$$u = c + Qx - A\lambda \qquad v = A^T x - b. \tag{7.180}$$

7.B. Find the dual of program (DQP) given by (7.20) to (7.22).

7.C. Show that (7.70) to (7.75) are the conditions for problem (CC) of stochastic linear programming with separable recourse.

7.D. Prove that (DLRP), the dual program of (PLRP), is indeed given by (7.80) through (7.83).

7.E. Formulate existence, characterization, and duality theorems for stochastic programming with **linear** recourse functions as given by (7.77).

7.F. Let g be a posynomial. Show that $1/g$ is a (log, log)-concave function.

7.G. Prove (7.148) and show that g^* is $(\bar{h}, \phi)$-convex.

7.H. Verify relations (7.156) through (7.166).

7.I. Continue Example 7.3.2. by computing x^* from the optimal dual solution. Verify that (7.174) holds in this case.

7.J. Indicate which results of Chapters 5 and 6 are needed for the proof of Theorem 7.14.

7.K. Suppose that we have the following linear minimax program

$$\text{(PM)} \qquad \min_x \left\{ \max_{1 \leq i \leq p} \left(\sum_{j=1}^{n} c_{ij} x_j + d_i \right) \right\} \tag{7.181}$$

subject to

$$\sum_{j=1}^{n} a_{kj} x_j \geq b_k, \qquad k = 1, \ldots, m. \tag{7.182}$$

Show that a convex dual program corresponding to (PM) is

$$\text{(DM)} \qquad \max \sum_{i=1}^{p} d_i\eta_i + \sum_{k=1}^{m} b_k\lambda_k \tag{7.183}$$

subject to

$$\sum_{i=1}^{p} \eta_i = 1 \tag{7.184}$$

$$\sum_{k=1}^{m} a_{kj}\lambda_k = \sum_{i=1}^{p} c_{ij}\eta_i, \qquad j = 1, \ldots, n \tag{7.185}$$

$$\eta \geq 0, \qquad \lambda \geq 0. \tag{7.186}$$

7.L. Consider the **algebraic program** [22]

$$\min G_0(x) \tag{7.187}$$

subject to

$$G_k(x) \leq 1, \qquad k = 1, \ldots, m \tag{7.188}$$

where, for $k = 0, 1, \ldots, m$,

$$G_k(x) = \left(\sum_{i \in I_k} \left\{w_i \left[\sum_{l \in L_i} (P_l(x))^{\alpha_l r_k} + (R_l(x))^{\beta_l r_k}\right]^{1/r_k}\right\}^{p_k}\right)^{1/p_k}, \tag{7.189}$$

the P_l, R_l are posynomials in $x > 0$, and α_l, β_l, p_k, r_k, w_i are positive constants. Find a dual program.

7.M. If we let $r_k \longrightarrow +\infty$ in the previous exercise, we obtain

$$G_k(x) = \left(\sum_{i \in I_k} \left\{w_i \sum_{l \in L_i} \max\,[(P_l(x))^{\alpha_l}, (R_l(x))^{\beta_l}]\right\}^{p_k}\right)^{1/p_k}. \tag{7.190}$$

How is the dual modified in this case? Formulate and solve a numerical example of an algebraic program in which the functions G_k are given by (7.190) via its dual, as in Example 7.3.2. More general algebraic programs involving signomials (generalized posynomials) were treated by Avriel and Gurovich [2] and Gurovich [18].

REFERENCES

1. Avriel, M., M. J. Rijckaert, and D. J. Wilde, (Eds.), *Optimization and Design*, Prentice-Hall, Englewood Cliffs, N.J., 1973.

2. Avriel, M., and V. Gurovich, "A Condensation Algorithm for a Class of Algebraic Programs," *Operations Research*, to appear.

3. Avriel, M., and A. C. Williams, "Stochastic Linear Programming with

Separable Recourse Functions," Mobil R & D Corp. Central Research Div. Progress Memorandum, Princeton, N.J. March 1967.

4. Avriel, M., and A. C. Williams, "Stochastic Programming with Separable Recourse Functions," paper presented at the 33rd ORSA National Meeting, San Francisco, May 1968.

5. Beale, E. M. L., "Numerical Methods," in *Nonlinear Programming*, J. Abadie (Ed.), North-Holland Publishing Co., Amsterdam, 1967.

6. Boot, J. G. C., *Quadratic Programming*, North-Holland Publishing Co., Amsterdam, 1964.

7. Cottle, R. W., "Symmetric Dual Programs," *Quart. Appl. Math.*, **21**, 237–243 (1963).

8. Cottle, R. W., "The Principal Pivoting Method of Quadratic Programming," in *Mathematics of the Decision Sciences, Part I*, G. B. Dantzig and A. F. Veinott (Eds.), American Mathematical Society, Providence, R.I., 1968.

9. Cottle, R. W., and G. B. Dantzig, "Complementary Pivot Theory of Mathematical Programming," *Linear Alg. Appl.*, **1**, 103–125 (1968).

10. Cottle, R. W., and W. C. Mylander, "Ritter's Cutting Plane Method for Nonconvex Quadratic Programming," in *Integer and Nonlinear Programming*, J. Abadie (Ed.), North-Holland Publishing Co., Amsterdam, 1970.

11. Dantzig, G. B., *Linear Programming and Extensions*, Princeton University Press, Princeton, N.J., 1963.

12. Dennis, J. B., *Mathematical Programming and Electrical Networks*, Technology Press, Cambridge, Mass., 1959.

13. Dorn, W. S., "Duality in Quadratic Programming," *Quart. Appl. Math.*, **18**, 155–162 (1960).

14. Dorn, W. S., "A Duality Theorem for Convex Programs," *IBM J. Res. Dev.*, **4**, 407–413 (1960).

15. Dorn, W. S., "Self-dual Quadratic Programs," *J. SIAM*, **9**, 51–54 (1961).

16. Duffin, R. J., E. L. Peterson, and C. Zener, *Geometric Programming—Theory and Applications*, John Wiley & Sons, New York, 1967.

17. Fenchel, W., "On Conjugate Convex Functions," *Can. J. Math.*, **1**, 73–77 (1949).

18. Gurovich, V., "Extensions of Signomial Programming: Numerical Solutions," M.Sc. thesis, Technion, Israel Institute of Technology, Haifa, 1974.

19. Hamala, M., "Geometric Programming in Terms of Conjugate Functions," Center for Operations Research and Econometrics, Université Catholique de Louvain Discussion Paper No. 6811, Louvain, Belgium, June 1968.

20. Hardy, G. H., J. E. Littlewood, and G. Pólya, *Inequalities*, 2nd ed., University Press, Cambridge, England, 1952.

21. KELLER, E. L., "The General Quadratic Optimization Problem," *Math. Prog.*, **5**, 311–337 (1973).

22. PETERSON, E. L., "Geometric Programming and Some of Its Extensions," in *Optimization and Design*, M. Avriel, M. J. Rijckaert, and D. J. Wilde (Eds.), Prentice-Hall, Englewood Cliffs, N.J., 1973.

23. ROCKAFELLAR, R. T., *Convex Analysis*, Princeton University Press, Princeton, N.J., 1970.

24. ROCKAFELLAR, R. T., "Some Convex Programs Whose Duals are Linearly Constrained," in *Nonlinear Programming*, J. B. Rosen, O. L. Mangasarian, and K. Ritter (Eds.), Academic Press, New York, 1970.

25. SENGUPTA, J. K., *Stochastic Programming—Methods and Applications*, North-Holland Publishing Co., Amsterdam, 1972.

26. VAJDA, S., *Probabilistic Programming*, Academic Press, London, 1972.

27. WALKUP, D. W., and R. J. B. WETS, "Stochastic Programs with Recourse," *SIAM J. Appl. Math.*, **15**, 1299–1314 (1967).

28. WETS, R. J. B., "Stochastic Programs with Recourse: A Survey I," Boeing Scientific Research Laboratories Document D1-82-0882, Seattle, Wash. August 1969.

29. Williams, A. C., "On Stochastic Linear Programming," *SIAM J. Appl. Math.*, **13**, 927–940 (1965).

30. WILLIAMS, A. C., "Approximation Formulas for Stochastic Linear Programming," *SIAM J. Appl. Math.*, **14**, 668–677 (1966).

31. ZANG, I., "Generalized Convex Programming," D.Sc. dissertation, Technion, Israel Institute of Technology, Haifa, 1974.

32. ZENER, C., *Engineering Design by Geometric Programming*, John Wiley & Sons, New York, 1971.

33. ZIEMBA, W. T., "Computational Algorithms for Convex Stochastic Programs with Simple Recourse," *Operations Research*, **18**, 414–431 (1970).

PART II

METHODS

Practical interest in nonlinear programming centers on the efficient finding of numerical solutions to a given problem. In contrast to the unifying role of the simplex method in linear programming, no such unified approach to obtaining an optimal solution of a nonlinear program exists. Whereas the simplex method can efficiently solve a linear program in thousands of variables and with hundreds of constraints, the question of how to minimize an unconstrained nonlinear function in a few variables is still an important one that concerns many scholars. Unconstrained optimization, from an algorithmic point of view, will be the subject of the first four chapters in Part II.

We start with the simplest case of minimizing single-variable real functions. Next we present nonderivative methods—that is, algorithms that find extrema without using derivatives of the function whose extremum is sought. These methods are important, for, in most practical applications, derivatives are either unavailable or difficult to compute. Unfortunately, the existing methods proposed for such problems are not too efficient and much work remains to be done in this area. On the other hand, significant progress has occurred since the 1960s in developing efficient algorithms for unconstrained optimization with derivatives. The first steps in devising numerical methods that take advantage of some parallel features in a new generation of computers will also be reported here.

The study of unconstrained optimization methods is important, since penalty methods, one of the most successful classes of algorithms for solving constrained nonlinear programs, are based on transforming constrained problems into unconstrained ones. Other methods handle constraints directly by modifying movement from one point in the space of variables to another, so that the constraints are taken into account. The discussion of constrained optimization methods in the last three chapters concludes the second part of this book.

8 ONE-DIMENSIONAL OPTIMIZATION

This chapter focuses on the problem of finding a minimum of a single-variable function f on the real line R or some subset of R. There are several reasons for discussing one-dimensional optimizations separately. First, some of the theoretical and numerical aspects of unconstrained optimization in R^n can be conveniently illustrated in one dimension. Secondly, several interesting search methods for functions of one variable apparently cannot be fully generalized for the problem of finding extrema of functions on R^n. Finally, some of the iterative methods for n-dimensional problems include steps in which extrema are sought along certain directions in R^n, and these steps are essentially equivalent to one-dimensional optimizations.

Consider, therefore, the problem of minimizing a real function f of a single real variable x. We saw in Chapter 2 that if f has a local extremum at x^* and is differentiable there, then the first derivative of f must vanish at x^*. That is, local extrema are solutions of the equation

$$f'(x) = 0. \tag{8.1}$$

We can try to solve this generally nonlinear equation and obtain all its solutions, among which will be local minima if such minima exist. If the second derivative of f is also available, then by evaluating f'' at the solutions of (8.1), we can usually determine which ones correspond to minima.

We shall discuss three types of methods for finding a minimum of a real-valued function along a line—that is, in one dimension. We begin with Newton's method and some related techniques, as applied to solving (8.1). The second type of method approximates the function whose minimum is sought by a low-order (usually not higher than third-order) polynomial. The minimum of the polynomial is then found analytically, and a new polynomial

approximation can be obtained. Continuing in this way, we can obtain an approximation for the true location of the minimum. The third type of optimum-seeking methods for single-variable functions is the class of "direct" methods. Here values of the function at different points of its domain are calculated in some systematic way, and, on the basis of these function evaluations and certain appropriate assumptions, we can bracket the location of a minimum within some small region.

8.1 NEWTON'S METHOD

In most cases, no analytic solution of (8.1) can be obtained, so this equation must be solved by some iterative method that generates a sequence of points $\{x^k\}$ and derivative values $\{f'(x^k)\}$ such that the limit of the latter sequence converges to zero. The classic technique for solving a nonlinear equation is Newton's method, which, in our case, requires the knowledge of $f''(x)$, the value of the second derivative. The idea behind Newton's method is to use a guess x^k for the solution of (8.1), linearize f' around x^k, and solve for the point where the linear function vanishes. This point is the next guess x^{k+1}. Formally, let x^k be the current guess for solving $f'(x) = 0$. The root of the linear equation

$$f'(x^k) + f''(x^k)(x - x^k) = 0 \tag{8.2}$$

is given by

$$x^{k+1} = x^k - \frac{f'(x^k)}{f''(x^k)}. \tag{8.3}$$

This iterative method is also illustrated in Figure 8.1.

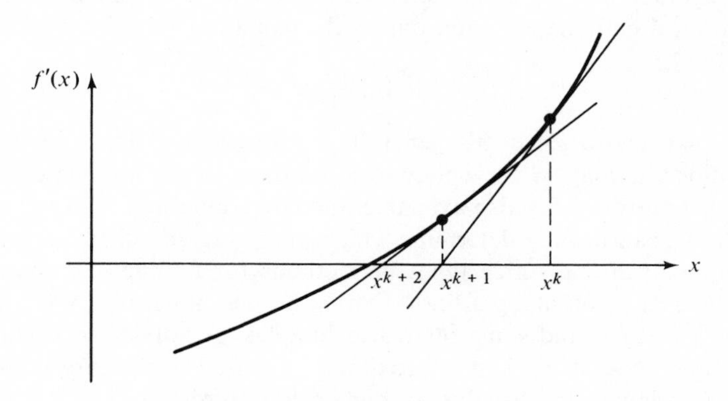

Fig. 8.1 Newton's method.

We now prove that, under suitable assumptions, the sequence of points $\{x^k\}$ generated by the preceding formula converges to a solution of (8.1). The subsequent discussion of Newton's method closely follows Goldstein's work [7], where the reader can find a more complete analysis.

Let $S = [a, b]$ be a closed interval on the real line and let ϕ be a continuous real-valued function on S. Let T be the range of ϕ. If $T \subset S$, then ϕ is called a mapping from S into itself, and a point $x \in S$, such that $\phi(x) = x$, is called a **fixed point** of ϕ. A map ϕ from S into itself is called a **contractor** if there exists a positive number $q < 1$ such that for every two points x^1 and x^2 in S

$$|\phi(x^1) - \phi(x^2)| \leq q|x^1 - x^2|. \tag{8.4}$$

Thus we have the following preliminary results.

Lemma 8.1

Let ϕ be a contractor on $S = [a, b] \subset R$ and let $x^0 \in S$, $x^{k+1} = \phi(x^k)$. Then there exists a unique fixed point $\bar{x}$ of ϕ, the sequence $\{x^k\}$ converges to $\bar{x}$, and

$$|x^{k+1} - \bar{x}| \leq (q)^{k+1}|x^0 - \bar{x}|, \qquad k = 0, 1, \ldots. \tag{8.5}$$

Proof. Every continuous function ϕ mapping S into itself has a fixed point [7]. Now

$$|x^1 - \bar{x}| = |\phi(x^0) - \phi(\bar{x})| \leq q|x^0 - \bar{x}|. \tag{8.6}$$

Thus (8.5) holds for $k = 0$. Suppose that (8.5) holds for $k = K \geq 1$. Then

$$|x^{K+1} - \bar{x}| \leq (q)^{K+1}|x^0 - \bar{x}| \tag{8.7}$$

and by (8.6) and (8.7),

$$|x^{K+2} - \bar{x}| = |\phi(x^{K+1}) - \phi(\bar{x})| \leq q|x^{K+1} - \bar{x}| \leq (q)^{K+2}|x^0 - \bar{x}|. \tag{8.8}$$

That is, (8.5) holds for any $k = 0, 1, \ldots$. Formula (8.5) implies that the sequence $\{x^k\}$ converges to $\bar{x}$, since $q < 1$. Finally, we need to show that $\bar{x}$ is unique. Assume that there are two distinct fixed points $\bar{x}^1$ and $\bar{x}^2$. Then

$$0 < |\bar{x}^1 - \bar{x}^2| = |\phi(\bar{x}^1) - \phi(\bar{x}^2)| \leq q|\bar{x}^1 - \bar{x}^2|, \tag{8.9}$$

which is a contradiction, since $q < 1$. ■

The next lemma deals with sufficient conditions on ϕ for being a contractor.

Lemma 8.2

Suppose that ϕ maps $S = [a, b] \subset R$ into itself and has a continuous derivative on S. If $|\phi'(x)| < 1$ for every $x \in S$, then ϕ is a contractor.

Proof. Let x^1 and x^2 be any two points in S. Then by the Mean Value Theorem [4],

$$\phi(x^1) = \phi(x^2) + \phi'(\tilde{x})(x^1 - x^2), \tag{8.10}$$

where $\tilde{x}$ is some point between x^1 and x^2. Hence,

$$|\phi(x^1) - \phi(x^2)| = |\phi'(\tilde{x})||x^1 - x^2|; \tag{8.11}$$

and by the hypotheses on ϕ', we can set q equal to the maximal value of $|\phi'(x)|$ on S and conclude that ϕ is a contractor. ■

With the aid of these two lemmas, we can prove a general result on the convergence of some root-finding methods.

Theorem 8.3

Let h and γ be continuously differentiable functions on $S = [a, b] \subset R$. Suppose that

$$h(a)h(b) < 0 \tag{8.12}$$

and for all $x \in S$

$$\gamma(x) > 0 \qquad h'(x) > 0 \tag{8.13}$$

$$0 \leq 1 - [\gamma(x)h(x)]' \leq q < 1. \tag{8.14}$$

Let

$$x^{k+1} = x^k - \gamma(x^k)h(x^k), \qquad k = 0, 1, \ldots \tag{8.15}$$

with $x^0 \in S$. Then the sequence $\{x^k\}$ converges to a solution $\bar{x}$ of $h(x) = 0$.

Proof. Define

$$\phi(x) = x - \gamma(x)h(x) \tag{8.16}$$

$$\phi'(x) = 1 - [\gamma(x)h(x)]'. \tag{8.17}$$

From (8.14) we have

$$0 \leq \phi'(x) \leq q < 1 \tag{8.18}$$

for all $x \in S$, and ϕ is monotone nondecreasing on S. By (8.12) and (8.13), it follows that h is monotone increasing on S and $h(a) < 0$, $h(b) > 0$. Hence $\phi(a) > a$ and $\phi(b) < b$. By the monotonicity of ϕ, we conclude that

$a < \phi(x) < b$ for all $x \in S$. Hence ϕ maps S into itself. Moreover, $|\phi'(x)| < 1$. Thus by Lemma 8.2, the function ϕ is a contractor on S. From (8.15) and (8.16) we have $x^{k+1} = \phi(x^k)$, $k = 0, 1, \ldots$. By Lemma 8.1, ϕ has a unique fixed point $\bar{x} \in S$. Finally, observe that $\bar{x}$ is a fixed point of ϕ if and only if $h(\bar{x}) = 0$. Thus $\{x^k\}$ converges to a solution of $h(x) = 0$. ■

Now we can state sufficient conditions on the convergence of Newton's method as described above.

Corollary 8.4

Let $h(x) = f'(x)$, $\gamma(x) = 1/f''(x)$, where f is a thrice continuously differentiable function on $S = [a, b]$. Assume that the hypotheses on h and γ stated in Theorem 8.3 hold. Then (8.15) *becomes the iteration formula of Newton's method and $\{x^k\}$ converges to a stationary point of f.*

Proof. Follows directly from Theorem 8.3. ■

Note that the preceding conditions on γ imply that f is strictly convex on S (a rather strong assumption). Although the conditions involve the third derivative of f, it is not used in the algorithm.

In addition to proving convergence of a certain algorithm, it is also important to know the **rate of convergence**. Suppose that we have a sequence $\{x^k\}$ with $x^k \in R^n$ that converges to a point $\bar{x}$ and $x^k \neq \bar{x}$ for all sufficiently large k. If there exists a number p and an $\alpha \neq 0$ such that

$$\lim_{k\to\infty} \frac{\|x^{k+1} - \bar{x}\|}{\|x^k - \bar{x}\|^p} = \alpha \tag{8.19}$$

then p is called the order of convergence of the sequence $\{x^k\}$ and $\|x^k - \bar{x}\|$ is called the error of the kth approximant. If $p = 1$, the rate of convergence of the sequence $\{x^k\}$ is said to be **linear.** If $p > 1$, then the rate of convergence is said to be **superlinear**. For the special case of $p = 2$, the convergence rate is said to be **quadratic**. A more general concept of linear convergence is as follows: There exists a natural number K, and numbers β and q such that $0 \leq q < 1$, satisfying

$$\|x^k - \bar{x}\| \leq \beta(q)^k, \qquad k = K, K+1, \ldots. \tag{8.20}$$

A weaker definition of superlinear convergence implied by (8.19) is

$$\lim_{k\to\infty} \frac{\|x^{k+1} - \bar{x}\|}{\|x^k - \bar{x}\|} = 0. \tag{8.21}$$

We have then

Theorem 8.5

Suppose that the hypotheses of Theorem 8.3 and Corollary 8.4 hold and the sequence $\{x^k\}$, $x^k \in R$, *generated by Newton's method, converges to a point* $\bar{x}$ *that satisfies* $h(x) = 0$. *Then the rate of convergence is quadratic.*

Proof. The point $\bar{x}$ solves $h(x) = 0$ if and only if $\bar{x}$ is a fixed point of

$$\phi(x^k) = x^k - \frac{h(x^k)}{h'(x^k)}. \tag{8.22}$$

By the Mean Value Theorem,

$$x^{k+1} - \bar{x} = \phi(x^k) - \phi(\bar{x}) = \phi'(\xi^k)(x^k - \bar{x}), \tag{8.23}$$

where ξ^k lies between x^k and $\bar{x}$. Then

$$|x^{k+1} - \bar{x}| = \frac{|h''(\xi^k)h(\xi^k)|}{[h'(\xi^k)]^2}|x^k - \bar{x}| \tag{8.24}$$

and

$$|h(\xi^k)| = |h(\xi^k) - h(\bar{x})| = |h'(\eta^k)||\xi^k - \bar{x}| \le |h'(\eta^k)||x^k - \bar{x}|, \tag{8.25}$$

where η^k lies between ξ^k and $\bar{x}$. Hence

$$|x^{k+1} - \bar{x}| \le \frac{|h''(\xi^k)h'(\eta^k)|}{[h'(\xi^k)]^2}|x^k - \bar{x}|^2. \tag{8.26}$$

Let

$$\beta = \sup_k \frac{|h''(\xi^k)h'(\eta^k)|}{[h'(\xi^k)]^2}. \tag{8.27}$$

Then

$$|x^{k+1} - \bar{x}| \le \beta |x^k - \bar{x}|^2. \tag{8.28}$$

and (8.19) holds. ■

A closely related root-finding method can be obtained by approximating the second derivative $f''(x)$ by the difference quotient

$$f''(x^k) \cong \frac{f'(x^k) - f'(x^{k-1})}{x^k - x^{k-1}}. \tag{8.29}$$

When this expression is substituted into (8.3), we get the iteration formula

of the **secant method**

$$x^{k+1} = x^k - \frac{f'(x^k)(x^k - x^{k-1})}{f'(x^k) - f'(x^{k-1})}. \tag{8.30}$$

It can be shown [10] that for the secant method

$$\lim_{k\to\infty} \frac{|x^{k+1} - \bar{x}|}{|x^k - \bar{x}|^\tau} = \left|\frac{2f''(\bar{x})}{f'''(\bar{x})}\right|^{1/\tau}, \tag{8.31}$$

where f''' is assumed to be nonzero and $\tau = (1 + \sqrt{5})/2 \cong 1.618$ is a solution of the equation $t^2 = t + 1$. Thus, for large values of k, the secant method is superlinear of asymptotic order 1.618. It is interesting to note here that the number τ will reappear later in this chapter in connection with another one-dimensional optimization method.

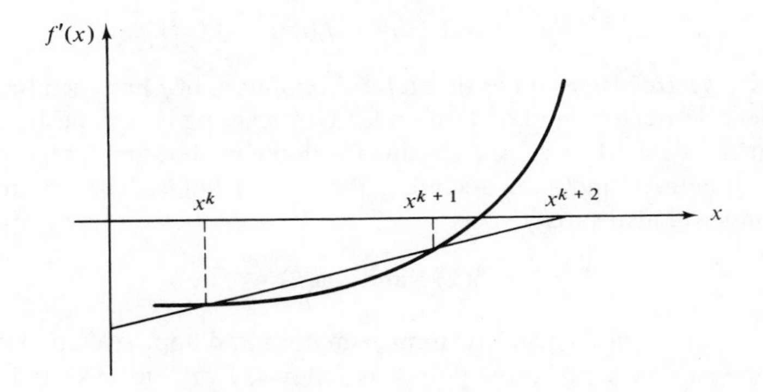

Fig. 8.2 The secant method.

Perhaps it is appropriate to mention that comparing two algorithms by their rates of convergence, as defined earlier, may be misleading. For example, the asymptotic convergence rate τ of the secant method is per function evaluation, since at each iteration the function whose root is sought is computed once. On the other hand, the quadratic rate of convergence of Newton's method is per two function evaluations, since at each iteration both the function and its first derivative must be evaluated.

8.2 POLYNOMIAL APPROXIMATION METHODS

The two most widely used polynomial approximation methods, the quadratic and the cubic, are discussed next. The basis for the **quadratic method** is to approximate f, the function whose minimum is sought, by a

quadratic function ϕ, given by

$$\phi(x) = a + bx + c(x)^2. \tag{8.32}$$

Suppose that we evaluate f at three points, x^1, x^2, x^3, such that $x^1 < x^2 < x^3$.

Letting $\phi(x^i) = f(x^i)$ for $i = 1, 2, 3$, we can solve for the coefficients a, b, and c. The minimum of the quadratic function ϕ (if it has a minimum) can be found analytically by setting $\phi'(x) = 0$, and, for a first approximation of a minimum of f, we obtain the point $\hat{x}$, given by

$$\hat{x} = -\frac{b}{2c}, \tag{8.33}$$

provided that $c > 0$. If $c < 0$, the quadratic function is actually a parabola with a maximum and so the point $\hat{x}$ obtained is unusable. A situation that will ensure that c is positive is

$$f(x^1) > f(x^2) \quad \text{and} \quad f(x^2) < f(x^3). \tag{8.34}$$

If it holds, we are also ensured that a local minimum of f has been bracketed somewhere between x^1 and x^3. Thus such a bracketing is very desirable, and in a moment we shall describe a routine for doing so, but first let us continue with the quadratic method, assuming that (8.34) holds. The minimum of ϕ so found will also satisfy

$$f(x^1) > \phi(\hat{x}) \quad \text{and} \quad \phi(\hat{x}) < f(x^3) \tag{8.35}$$

and we can find a new quadratic approximation and approximate minimum by choosing a new set of three points as follows: Evaluate $f(\hat{x})$ and choose as the new x^2 one of the four points at which f has been computed and which yielded the lowest value of f (it is either $\hat{x}$ or the "old" x^2). Let the new x^1 and x^3 be the two points adjacent to the new x^2 from the left and right, respectively, and repeat the iteration. This algorithm can be terminated if either the difference between the actual and approximate values of the function at the predicted minimum is less than some tolerance $\epsilon > 0$—that is, if

$$|f(\hat{x}) - \phi(\hat{x})| < \epsilon \tag{8.36}$$

or if estimates of the minimum point in two or more successive iterations are closer than some predetermined distance. There is, however, at least one situation in which the quadratic approximation method, as described, fails to converge to a minimum of f even though the inequalities of (8.34) hold. If $\hat{x} = x^2$, the algorithm will not evaluate new points, although x^2 may not be a local minimum of f. In such a degenerate case, some perturbations on $\hat{x}$ are needed in order for us to proceed with the computations.

The inequalities in (8.34) are sufficient (except in degenerate situations) for predicting a minimum of ϕ, but such a minimum can also be predicted if (8.34) does not hold. For example, if $f(x^1) > f(x^2) > f(x^3)$ and $f(x^2)$ lies below the line segment joining $f(x^1)$ and $f(x^3)$, the corresponding quadratic may still have a minimum at $\hat{x} > x^2$. However, this predicted minimum can be far enough to the right of the first three points that the approximation of f by ϕ may be very poor. This difficulty can be overcome by introducing a maximum step length, which can be used if such a predicted extrapolated minimum occurs. On the other hand, it is possible that the points x^1, x^2, x^3 are quite far from the minimum of f, and here an extrapolated minimum of ϕ can speed up the iterative procedure.

A bracketing method to attain three points such that (8.34) is satisfied will be described next. Choose any starting point x^0 and compute $f(x^0)$. Next, evaluate f at the point $x^0 + h^0$, where $h^0 > 0$ is some predetermined step length. If

$$f(x^0) > f(x^0 + h^0), \tag{8.37}$$

let $x^1 = x^0 + h^0$ and let $h^1 = \alpha h^0$ be the new step length, where $\alpha > 1$. Now compute $f(x^1 + h^1)$. If

$$f(x^1) > f(x^1 + h^1), \tag{8.38}$$

let $x^2 = x^1 + h^1$ and $h^2 = \alpha h^1$. Continue in this way until the value of f at $x^k + h^k$, the last point reached, is higher than $f(x^k)$.

If (8.37) does not hold, let $x^1 = x^0$ and $h^1 = -\beta h^0$, where $0 < \beta < 1$. Evaluate $f(x^1 + h^1)$. If

$$f(x^1 + h^1) > f(x^1), \tag{8.39}$$

we are finished, since a bracket $[x^0 - \beta h^0, x^0 + h^0]$ has been obtained. Otherwise let $x^2 = x^1 + h^1$, $h^2 = \alpha h^1$ and compute $f(x^2 + h^2)$, and so on, until a function increase occurs. In any case, the last three points reached will satisfy (8.34), and the quadratic method can be started. Values of α and β are preferably chosen so that $\alpha\beta < 1$. It should be pointed out that the bracketing procedure can itself be used as a method for locating a minimum.

In Figure 8.3 we demonstrate this situation by reversing the direction of search each time a function increase is obtained (see the small arrows next to the function values at the kth function evaluation). In this figure the values $\alpha = 2$, $\beta = \frac{1}{4}$ were used and 11 function evaluations were performed. For details of this method, the reader is referred to [12]. Note that both the quadratic and the bracketing methods can be modified and improved if the first derivative of the function f is known.

The next polynomial approximation method to be discussed is the **cubic method**, in which a given function f to be minimized is approximated by a

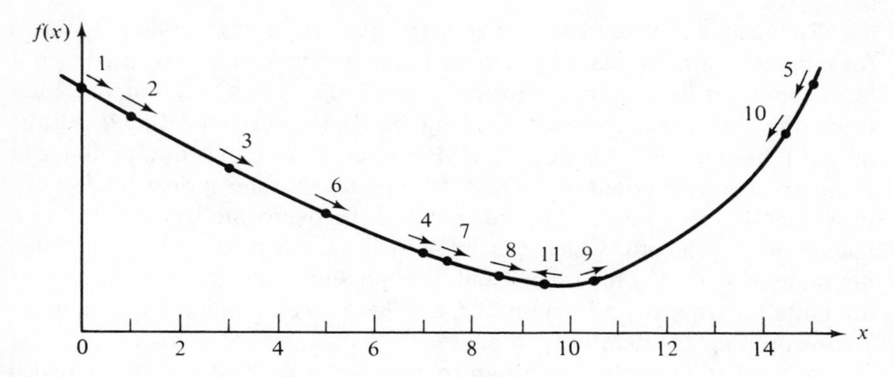

Fig. 8.3 Minimizing a function by the bracketing method.

third-order polynomial

$$\phi(x) = a + bx + c(x)^2 + d(x)^3. \tag{8.40}$$

The method described here was originally derived by Davidon [6]. The algorithm is based on the assumption that the first derivatives of f are available, but it can readily be modified for use without calculating derivatives. We start at an arbitrary point $x^1 \in R$ and compute $f(x^1)$, $f'(x^1)$. Assume that $f'(x^1)$ is negative. Then we find a point x^2, $x^2 > x^1$, by some iterative procedure such that either $f'(x^2)$ is nonnegative or $f(x^2) > f(x^1)$. The coefficients a, b, c, and d can now be computed by solving a system of four linear equations in four variables:

$$\begin{aligned} f(x^1) &= a + bx^1 + c(x^1)^2 + d(x^1)^3 \\ f'(x^1) &= b + 2cx^1 + 3d(x^1)^2 \\ f(x^2) &= a + bx^2 + c(x^2)^2 + d(x^2)^3 \\ f'(x^2) &= b + 2cx^2 + 3d(x^2)^2. \end{aligned} \tag{8.41}$$

Davidon found that the solution of these equations can be expedited by a simple change of variables. Define a new variable z and new functions g and ψ by

$$z = x - x^1 \tag{8.42}$$

and

$$g(z) = f(x^1 + z), \qquad \psi(z) = \phi(x^1 + z). \tag{8.43}$$

Then the first derivative of the cubic function ψ with respect to z is given by

$$\psi'(z) = g'(0) - \frac{2z}{\lambda}(g'(0) + \alpha) + \frac{(z)^2}{(\lambda)^2}(g'(0) + g'(\lambda) + 2\alpha), \quad (8.44)$$

where $\lambda = x^2 - x^1$ and

$$\alpha = \frac{3(g(0) - g(\lambda))}{\lambda} + g'(0) + g'(\lambda), \quad (8.45)$$

and the point $\hat{z}$, where $\psi'(z)$ vanishes and where a minimum is predicted, is given by

$$\hat{z} = \lambda(1 - \beta), \quad (8.46)$$

where

$$\beta = \frac{g'(\lambda) + [(\alpha)^2 - g'(0)g'(\lambda)]^{1/2} - \alpha}{g'(\lambda) - g'(0) + 2[(\alpha)^2 - g'(0)g'(\lambda)]^{1/2}} \quad (8.47)$$

If $|g'(\hat{z})| < \epsilon$, where ϵ is some predetermined tolerance, the procedure is terminated; otherwise the algorithm must be restarted by using a new set of two points selected by a procedure similar to that described for the quadratic approximation method. The cubic approximation method will usually have a faster convergence rate per iteration than the quadratic method. However, the required amount of computation at each iteration is greater in the cubic method.

Polynomial approximation methods seem to be very efficient for locating minima of reasonably "well-behaving" functions—that is, functions that can be reasonably closely approximated (in some sense) by quadratic or cubic functions. The efficiency of these algorithms greatly depends on the particular function whose minimum is sought. In the next methods to be described, a small interval of the real line that contains the minimum of a unimodal function (defined below) is found. Interestingly, the efficiency of these methods will be shown to be the same regardless of which particular one-dimensional function—belonging to the class of unimodal functions—is minimized.

8.3 DIRECT METHODS-FIBONACCI AND GOLDEN SECTION TECHNIQUES

We begin by defining unimodal functions, which are extremely important in establishing convergence and optimality of the methods presented below. A real-valued function f is said to be **unimodal** on a closed interval $L \subset R$ if there is an $x^* \in L$ such that x^* minimizes f on L and, for any two points

$x^1 \in L, x^2 \in L$ such that $x^1 < x^2$, we have

$$\begin{aligned} x^2 \leq x^* &\quad \text{implies that} \quad f(x^1) > f(x^2) \\ x^* \leq x^1 &\quad \text{implies that} \quad f(x^2) > f(x^1). \end{aligned} \tag{8.48}$$

Note that unimodal functions are not necessarily continuous or differentiable. Strictly convex functions and most of their generalizations are unimodal. Some examples of unimodal functions are shown in Figure 8.4.

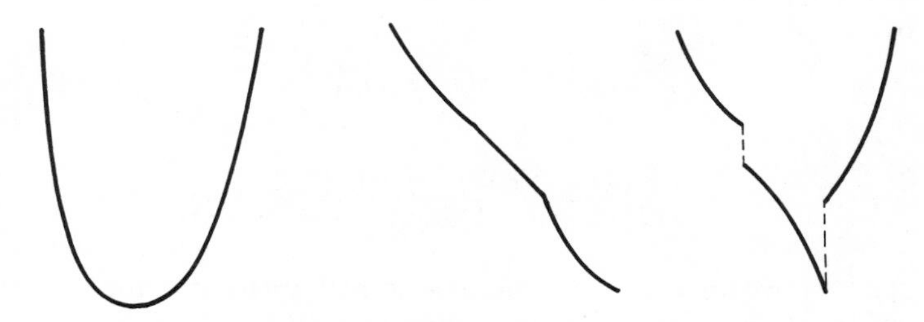

Fig. 8.4 Unimodal functions.

A very useful property of unimodal functions, one that will be fully exploited in the following methods, is the possibility of bracketing the location of the minimum by evaluating the function at two distinct points in L. Denote by l_1 and r_1 the left and right endpoints of the interval L, respectively; that is, $L = [l_1, r_1]$ or

$$L = \{x : x \in R, l_1 \leq x \leq r_1\}. \tag{8.49}$$

Suppose that we evaluate the function at two points x^1, x^2 in L such that $x^1 < x^2$ and find that $f(x^1) < f(x^2)$. It follows from the definition of unimodality (8.48) that $x^* \in [l_1, x^2]$. Similarly, if $f(x^1) > f(x^2)$, then we must have $x^* \in [x^1, r_1]$. If the function values at x^1 and x^2 happen to be equal, then $x^* \in [x^1, x^2]$, but, for simplicity, we shall regard x^* as belonging to one of the two larger intervals mentioned earlier. In any case, after the first two function evaluations, a portion of L to the right of x^2 or the left of x^1 can be eliminated from further search. If the remaining interval is $[l_1, x^2]$, then it will contain the point x^1, whereas if $[x^1, r_1]$ remains, it will contain x^2. Let us relabel the left and right endpoints of the remaining subinterval of L as $[l_2, r_2]$ and evaluate f at some point other than the one already present in it. Comparing the value of f at this point with the function value at the other point in $[l_2, r_2]$, we can further reduce this interval. We shall call every such interval reduction an iteration.

The method to be described next is called the **Fibonacci method**, for it uses the Fibonacci numbers F_k, defined by

$$F_0 = 0, \quad F_1 = 1, \quad F_k = F_{k-1} + F_{k-2}, \qquad k = 2, 3, \ldots. \tag{8.50}$$

The first few Fibonacci numbers are therefore, 0, 1, 1, 2, 3, 5, 8, 13, 21, 34, 55, 89, ...

The Fibonacci method is defined as follows: Let N be the total number of points at which the function f will be evaluated (note that, for N function evaluations, there are $N - 1$ interval reductions or iterations). At iteration k, the interval containing x^* is $[l_k, r_k]$. For $k = 1, 2, \ldots, N - 1$, the function values are compared at the two points

$$x_k^1 = l_k + \frac{F_{N-k}}{F_{N+2-k}}(r_k - l_k) \tag{8.51}$$

$$x_k^2 = l_k + \frac{F_{N+1-k}}{F_{N+2-k}}(r_k - l_k), \tag{8.52}$$

where, except for $k = 1$, one of the points x_k^1 or x_k^2 was already evaluated at a previous iteration.

Note that the points x_k^1 and x_k^2 are placed symmetrically in the interval $[l_k, r_k]$. From (8.51), (8.52), and the definition of the Fibonacci numbers, we have

$$x_k^2 - l_k = \frac{F_{N+1-k}}{F_{N+2-k}}(r_k - l_k) = \frac{F_{N+2-k} - F_{N-k}}{F_{N+2-k}}(r_k - l_k) \tag{8.53}$$

$$= r_k - l_k - \frac{F_{N-k}}{F_{N+2-k}}(r_k - l_k) = r_k - x_k^1. \tag{8.54}$$

(See also Figure 8.5.)

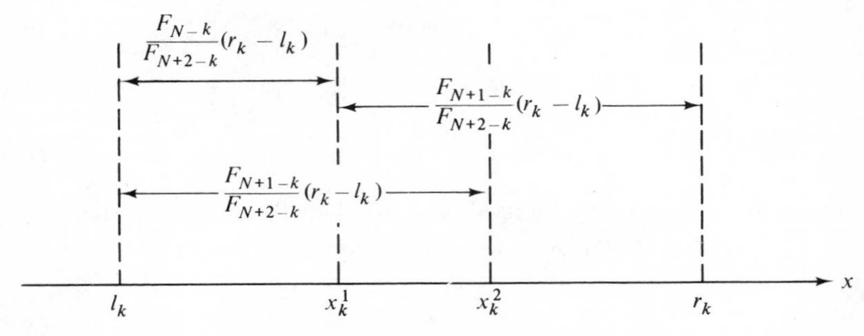

Fig. 8.5 Function evaluations at symmetric points by the Fibonacci method.

At the last iteration ($k = N - 1$), formulas (8.51) and (8.52) would place both x_{N-1}^1 and x_{N-1}^2 at

$$x_{N-1}^1 = x_{N-1}^2 = l_{N-1} + \tfrac{1}{2}(r_{N-1} - l_{N-1}), \tag{8.55}$$

and no further interval reduction is possible. Consequently, in order to reduce $[l_{N-1}, r_{N-1}]$ to about one-half its length by the last function evaluation, we place the last point, x_{N-1}^1 or x_{N-1}^2, at a distance ϵ from the existing point in the interval $[l_{N-1}, r_{N-1}]$, where ϵ is taken as the smallest practical distance between two adjacent points where the function is being evaluated so that the function values are distinguishable. More precisely, it means that if $x_k^2 - x_k^1 \geq \epsilon$, then $f(x_k^1) = f(x_k^2)$ only if x^* lies between x_k^1 and x_k^2.

After N function evaluations, the length of the interval containing x^* is given by

$$r_N - l_N = \frac{(r_1 - l_1)}{F_{N+1}} + \delta \tag{8.56}$$

where δ is either zero or ϵ. Thus we can bracket the minimum of any unimodal function within 1% of the starting interval by 11 function evaluations ($F_{12} = 144$) and within 0.1% by 16 function evaluations.

Example 8.3.1

Let us seek the minimum of $f(x) = (x - 3)^2$ by $N = 4$ function evaluations and suppose that $L = [0, 10]$.

At the end of the search, we should get, by the Fibonacci method, a final interval that contains the minimum of f at $x^* = 3$ and has a length

$$r_4 - l_4 = \frac{(10 - 0)}{5} + \epsilon = 2 + \epsilon. \tag{8.57}$$

The search will start as follows:

$$x_1^1 = 0 + \tfrac{2}{5}(10 - 0) = 4 \tag{8.58}$$

$$x_1^2 = 0 + \tfrac{3}{5}(10 - 0) = 6 \tag{8.59}$$

$$f(x_1^1) = 1 \qquad f(x_1^2) = 9. \tag{8.60}$$

We now reduce the starting interval, and, by the rules defined earlier, we get $l_2 = 0$, $r_2 = 6$. Proceeding with the second iteration, we obtain from (8.51) and (8.52)

$$x_2^1 = 0 + \tfrac{1}{3}(6 - 0) = 2 \tag{8.61}$$

$$x_2^2 = 0 + \tfrac{2}{3}(6 - 0) = 4. \tag{8.62}$$

Note that the point x_2^2 coincides in this case with x_1^1, where we have already evaluated f. Here

$$f(x_2^1) = 1 \qquad f(x_2^2) = 1. \tag{8.63}$$

Although (8.63) indicates that the minimum of f must lie between $x = 2$ and $x = 4$, we deliberately ignore this fortuitous situation (which seldom occurs in practice) and conclude that, say, $l_3 = 2, r_3 = 6$. Then

$$x_3^1 = 2 + \tfrac{1}{2}(6 - 2) = 4 \tag{8.64}$$

$$x_3^2 = 2 + \tfrac{1}{2}(6 - 2) = 4. \tag{8.65}$$

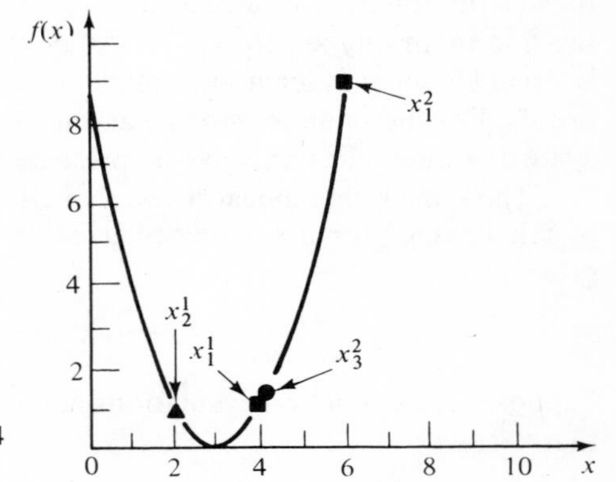

Fig. 8.6 Fibonacci search with $N = 4$ function evaluations.

Since the function value has already been evaluated at $x = 4$, we choose, say, $\epsilon = 0.01$ and obtain $f(4.01) = 1.0201$. Consequently, the final interval containing the minimum of f is [2, 4.01], as predicted. The search pattern is also illustrated in Figure 8.6. ∎

The Fibonacci search method was first derived by Kiefer [9], who also showed that the method is optimal in a certain sense. Among all nonrandomized search procedures with N function evaluations and which bracket the minimum of a unimodal function, the Fibonacci method minimizes the length of the maximum possible interval remaining after N function evaluations and containing the sought minimum. This statement can be reformulated by saying that the optimal search method maximizes the ratio of the lengths of the starting interval and the possible final intervals remaining after N function evaluations and containing the minimum. Following Avriel and Wilde [1], we present an optimality proof of the so-called symmetric Fibonacci method that differs slightly from Kiefer's method and that also ensures that, in the last iteration, the unimodal function is evaluated at two

points at a distance ϵ apart and placed symmetrically on opposite sides of the center of the interval $[l_{N-1}, r_{N-1}]$. This modification gives a slight improvement of order ϵ over Kiefer's method, and, for $\epsilon \longrightarrow 0$, the two methods become identical.

We define a search strategy $S(N, \epsilon)$, on a closed interval $[l_1, r_1]$ of the real line as a plan for evaluating the function at N distinct points $x^1, x^2, \ldots, x^N$ in this interval, where the location of x^{k+1} depends on the values of the unimodal function f at $x^1, \ldots, x^k$, and such that

$$|x^j - x^k| \geq \epsilon \tag{8.66}$$

for all $j \neq k$, where $\epsilon \leq \frac{1}{2}$. The plan terminates after successive reduction of the starting interval to a final interval $[l_N, r_N]$ of **unit length**, containing x^*, the minimum of f on $[l_1, r_1]$. We may note here that the choice of a final interval having unit length was made merely in order to simplify notation and proofs. For the same reason, we assume that $l_1 = 0$. By appropriate scaling, a starting interval can always be presented in this form.

The **symmetric Fibonacci search** method with N function evaluations is a search strategy for a starting interval $[0, r^F_{1,N}]$, where the right endpoint is given by

$$r^F_{1,N} = F_{N+1} - F_{N-1}\epsilon. \tag{8.67}$$

Suppose that k function evaluations have been performed and let x^{k*} be such that

$$f(x^{k*}) = \min \{f(x^1), \ldots, f(x^k)\}. \tag{8.68}$$

Define

$$l^F_k = \max [\{x^j : x^j < x^{k*}, j = 1, \ldots, k\} \cup \{0\}] \tag{8.69}$$

and

$$r^F_k = \min [\{x^j : x^j > x^{k*}, j = 1, \ldots, k\} \cup \{r^F_{1,N}\}]. \tag{8.70}$$

Since f is unimodal, it is clear that x^* must be in $[l^F_k, r^F_k]$. The function is evaluated at points that are given by the relations

$$x^1 = F_N - F_{N-2}\epsilon = r^F_{1,N-1} \tag{8.71}$$

and

$$x^{k+1} = l^F_k + r^F_k - x^{k*}, \qquad k = 1, \ldots, N-1. \tag{8.72}$$

Note that, by (8.67), the point x^1 is at $r^F_{1,N-1}$ for a Fibonacci search with $N-1$ function evaluations.

First we shall prove that the symmetric Fibonacci method is indeed a search strategy as defined above.

Lemma 8.6

For $\epsilon \leq \frac{1}{2}$, the symmetric Fibonacci search as outlined above is an $S(N, \epsilon)$ strategy on $[0, r_1^F]$ for any unimodal function.

Proof. We must show that, after N function evaluations, we obtain a final interval $[l_N^F, r_N^F]$ of unit length, and no adjacent points are placed closer than ϵ. The proof will be by induction on N. The lemma is trivially true for $N = 1$, since, by (8.67) through (8.71), we get $l_1^F = 0$, $r_{1,1}^F = 1$.

For $N = 2$, $r_{1,2}^F = 2 - \epsilon$ and $x^1 = 1$, implying that $x^2 = 0 + 2 - \epsilon - 1 = 1 - \epsilon$. Thus $|x^2 - x^1| = \epsilon$, and, for any unimodal function, either $x^{2*} = x^1$ or $x^{2*} = x^2$ and, in both cases, $r_2^F - l_2^F = 1$.

Now assume that the lemma is true for $N \geq 2$. Then for $N + 1$ function evaluations

$$r_{1,N+1}^F = F_{N+2} - F_N\epsilon \tag{8.73}$$

and

$$x^1 = F_{N+1} - F_{N-1}\epsilon = r_{1,N}^F \tag{8.74}$$

$$x^2 = F_{N+2} - F_N\epsilon - F_{N+1} + F_{N-1}\epsilon = F_N - F_{N-2}\epsilon = r_{1,N-1}^F. \tag{8.75}$$

Note that $x^1 - x^2 = r_{1,N}^F - r_{1,N-1}^F \geq 1 - \epsilon \geq \epsilon$. If $x^{2*} = x^2$, then $l_2^F = 0$ and $r_2^F = x^1 = r_{1,N}^F$. Thus $0 \leq x^* \leq r_{1,N}^F$ with one function evaluation at $r_{1,N-1}^F$, and, by the induction hypothesis, the lemma holds. If $x^{2*} = x^1$, then $l_2^F = r_{1,N-1}^F$ and $r_2^F = r_{1,N+1}^F$. By the simple change of variable

$$\tilde{x} = r_{1,N+1}^F - x, \tag{8.76}$$

we get $0 \leq \tilde{x}^* \leq (r_{1,N+1}^F - r_{1,N-1}^F) = r_{1,N}^F$ with one function evaluation at $r_{1,N-1}^F$, and again the lemma holds. ■

In the following theorem we shall show that, of all search strategies, the Fibonacci method offers the highest value of the right endpoint of the starting interval, which is equivalent to saying that the Fibonacci method yields the highest length ratio of starting-to-final intervals containing the minimum of any unimodal function. One exception, however, needs to be made. We mentioned earlier that if $f(x^{k+1}) = f(x^{k*})$, we can conclude that x^* must be between x^{k+1} and x^{k*}. Such a situation may require a modification of the search method that we shall not discuss here.

Theorem 8.7

Among all search strategies $S(N, \epsilon)$ on a starting interval $[0, r_{1,N}]$, the Fibonacci method $S_F(N, \epsilon)$ yields the highest value of the right endpoint, provided that $f(x^{k+1}) \neq f(x^{k*})$ for $k = 1, \ldots, N - 1$.

Proof. We again use induction on N. For $N = 1$, the proof is obvious because all $r_{1,1} = r^F_{1,1} = 1$. For $N = 2$, assume, without loss of generality, that $x^1 > x^2$ in any strategy $S(2, \epsilon)$. Then $x^1 \leq 1$, $r_{1,2} - x^1 \leq 1 - \epsilon$, and so $r_{1,2} \leq 2 - \epsilon = r^F_{1,2}$. Assume now that the theorem is true for $N \geq 2$. Then for an $S(N + 1, \epsilon)$ strategy, let $x^1 > x^2$ and $x^{2*} = x^2$. Hence $l_{2,N+1} = 0$, $r_{2,N+1} = x^1$, and we have an $S(N, \epsilon)$ strategy on $[0, x^1]$. By the induction hypothesis,

$$x^1 \leq r^F_{1,N}. \tag{8.77}$$

However,

$$x^2 \leq r^F_{1,N-1} \tag{8.78}$$

also be induction, for it may happen that $x^* < x^2$, and then x^* must be located by an $S(N - 1, \epsilon)$ strategy on $[0, x^2]$. If $x^{2*} = x^1$, then $l_{2,N+1} = x^2$, and $r_{2,N+1} = r_{1,N+1}$. By induction,

$$r_{1,N+1} - x^2 \leq r^F_{1,N}. \tag{8.79}$$

Addition of (8.78) and (8.79) yields

$$r_{1,N+1} \leq r^F_{1,N} + r^F_{1,N-1} = r^F_{1,N+1} \tag{8.80}$$

as asserted. Relations (8.77) to (8.79) hold as equalities for the Fibonacci search. Moreover, we observe that

$$r_{1,N+1} - x^1 \leq r^F_{1,N-1} \tag{8.81}$$

by an argument similar to that used for (8.78). It follows from (8.77) and (8.81) that this is the only way to achieve equality in (8.80). Thus the Fibonacci method is the only way to obtain the maximum value of $r_{1,N}$. ■

One of the disadvantages of the Fibonacci method is that the number of function evaluations N must be known in advance, prior to starting the search. This requirement is not necessary in a related technique, called the **golden section method**, which is a good approximation of the Fibonacci search. It can be shown that

$$\lim_{N\to\infty} \frac{F_{N-1}}{F_N} = \frac{1}{\tau} = \frac{\sqrt{5} - 1}{2} \cong 0.618. \tag{8.82}$$

It is possible to approximate the location of x^2_k, as given in (8.52), by

$$x^2_k \cong l_k + \frac{1}{\tau}(r_k - l_k). \tag{8.83}$$

The golden section method then places the points at which the function is to be evaluated at

$$x_{kG}^1 = l_k + \frac{\tau - 1}{\tau}(r_k - l_k) \tag{8.84}$$

$$x_{kG}^2 = l_k + \frac{1}{\tau}(r_k - l_k), \tag{8.85}$$

where the subscript G indicates that the points are placed by the golden section method and the rest of the notation was left unchanged.

From (8.84) and (8.85) it is clear that the golden section method also places the points symmetrically. Assuming $\epsilon \longrightarrow 0$, we can see from (8.56) that the Fibonacci method with N function evaluations reduces the interval containing the minimum of f by a factor of $1/F_{N+1}$. The reader will be asked to show that the golden section method reduces this interval by a factor of $1/(\tau)^{N-1}$. It can also be shown that

$$\lim_{N \to \infty} \frac{F_{N+1}}{(\tau)^{N-1}} = \frac{(\tau)^2}{\sqrt{5}} \cong 1.17. \tag{8.86}$$

Thus for a large N (and small ϵ) the golden section method yields a final interval that is some 17% longer than the Fibonacci method. Perhaps a better comparison of the two methods is illustrated by the fact that the golden section method using $N + 1$ function evaluations can always yield a final interval having the same length or a smaller one than the interval obtained by the Fibonacci search with N function evaluations.

The name "golden section" comes from a classical problem of dividing line segments or rectangles in a particular way. For a more complete discussion of the golden section method and its relation to the Fibonacci technique, the reader is referred to Wilde [15] and Wilde and Beightler [16].

8.4 OPTIMAL AND GOLDEN BLOCK SEARCH METHODS

The methods described in Section 8.3 are based on sequential search. In other words, the function to be minimized is evaluated at a certain point, and the choices of points for future evaluations depend on previous results. Another, much less efficient method (with respect to the reduction ratio of the starting-to-final interval containing the location of the minimum), in which all the function evaluations are performed **simultaneously**, has also been derived [15]. Such a search procedure can be useful in certain real-life

situations, where, for example, evaluating the function whose minimum is sought is a complicated and time-consuming task, and simultaneous measurements of the function values are possible.

The simultaneous method has been combined with the Fibonacci and golden section methods into **block search** techniques, in which a sequence of "blocks" of simultaneous function evaluations is performed. One possible way of having still larger and faster computers than those in operation today is to equip them with parallel processing units [13] so that arithmetic operations can be performed simultaneously. Block search methods were the first optimization techniques to be derived for this type of computer. The idea is to perform a block of simultaneous function evaluations in the parallel processing units, reduce the interval containing the minimum, and then perform a new block of evaluations in the remaining interval, and so forth. These methods were developed by Avriel and Wilde [2, 3] and are based on generalized Fibonacci numbers and golden sections. The optimality of these methods, in the same sense as in the case of the Fibonacci and golden section methods, can be proven similar to Lemma 8.6 and Theorem 8.7. For a proof along different lines, see Karp and Miranker [8]. We shall only outline the algorithms here; the optimality proofs can be found in the foregoing references.

The block search method consists of n sequential blocks of ξ simultaneous functions evaluations. The starting interval is $[0, r^B_{\xi,n}]$ (where the superscript B indicates "block search"), and after $N = \xi n$ function evaluations, it is reduced to unit length. Again, let $\epsilon \leq \frac{1}{2}$.

Depending on whether the number of simultaneous evaluations per block is $\xi = 2k - 1$ (odd) or $\xi = 2k$ (even), we have two different strategies as outlined below.

(a) Block Search Method for $\xi = 2k - 1$ (Odd)

Define the generalized Fibonacci sequence by the recurrence relation

$$A(k, h + 2) = k[A(k, h + 1) + A(k, h)] \tag{8.87}$$

with $A(k, 0) = 0$, $A(k, 1) = 1$. Note that the sequence is also defined for negative integers h. For $k = 1$, relation (8.87) becomes the well-known Fibonacci sequence. The right endpoint of the starting interval in the odd-block search is given by

$$r^B_{2k-1,n} = A(k, n + 1) - kA(k, n - 1)\epsilon. \tag{8.88}$$

Since the function to be minimized is assumed to be unimodal, we have, after the ith block of function evaluations, a reduced interval containing the point

x^{i*} at which the function has the lowest value among all the points of that block ($2k - 1$ points are evaluated, plus one point remaining from the previous block). Using earlier notation as much as possible, for $i = 1, 2, \ldots, n$, let $x_i^0 = l_{i-1}^B$ and let $x_i^1, x_i^2, \ldots, x_i^{2k}$ be the locations of the $2k - 1$ simultaneous function evaluations of the ith block and the previously performed evaluation at $x^{(i-1)*}$. Then

$$x_i^1 = l_{i-1}^B + A(k, n - i + 1) - kA(k, n - i - 1)\epsilon \tag{8.89}$$

$$x_i^{j+2} = x_i^j + \frac{1}{k}A(k, n - i + 2) - A(k, n - i)\epsilon, \qquad j = 0, \ldots, 2k - 2. \tag{8.90}$$

(b) Block Search Method for $\xi = 2k$ (Even)

The right endpoint of the starting interval is given by

$$r_{2k,n}^B = (k + 1)^n - \epsilon. \tag{8.91}$$

Again let $x_i^0 = l_{i-1}^B$. Then the $2k$ function evaluations of the ith block and the previously performed evaluation at $x^{(i-1)*}$ are placed as follows:

$$x_i^{j+2} = x_i^j + (k + 1)^{n-i}, \qquad j = 0, \ldots, 2k - 1 \tag{8.92}$$

$$x_1^1 = (k + 1)^{n-1} - \epsilon. \tag{8.93}$$

For $i = 2, \ldots, n$, either $x_i^1 = x^{(i-1)*} = l_{i-1}^B + \epsilon$ or $x_i^{2k+1} = x^{(i-1)*} = r_{i-1}^B - \epsilon$. Thus $x_i^2, x_i^4, \ldots, x_i^{2k}$ are found by using (8.92) recursively, starting with x_i^0, while $x_i^1, x_i^3, \ldots, x_i^{2k+1}$ are located by using (8.92) with $x^{(i-1)*}$ as the starting point.

Example 8.4.1

Suppose that we are to bracket the minimum of a unimodal function defined on the interval [0, 40] within a final interval of unit length when $\epsilon = 0.5$. For practical reasons, it is possible to evaluate the function in three blocks only, whereas the number of simultaneous function evaluations per block is unrestricted.

First we must determine ξ, the smallest number of simultaneous evaluations per block such that after three blocks a unit interval will result. Using (8.88) and (8.91) with $n = 3$, we obtain $r_{1,3}^B = 2.5$, $r_{2,3}^B = 7.5$, $r_{3,3}^B = 14$, $r_{4,3}^B = 26.5$, and $r_{5,3}^B = 40.5$. By performing $\xi = 5$ ($k = 3$) function evaluations per block, we are ensured of reducing the starting interval by a factor of 40. The function evaluations of the first block are placed by (8.89) and

(8.90) as follows:

$$
\begin{aligned}
x_1^1 &= 12 - (3)(1)(0.5) = 10.5 \\
x_1^2 &= \tfrac{1}{3}(45) - (3)(0.5) = 13.5 \\
x_1^3 &= 10.5 + \tfrac{1}{3}(45) - (3)(0.5) = 24.0 \\
x_1^4 &= 13.5 + \tfrac{1}{3}(45) - (3)(0.5) = 27.0 \\
x_1^5 &= 24 + \tfrac{1}{3}(45) - (3)(0.5) = 37.5.
\end{aligned}
\tag{8.94}
$$

Figure 8.7 shows the function values at the points of the first block. It can be seen that, after the first block, the remaining interval that contains the minimum is [13.5, 27.0].

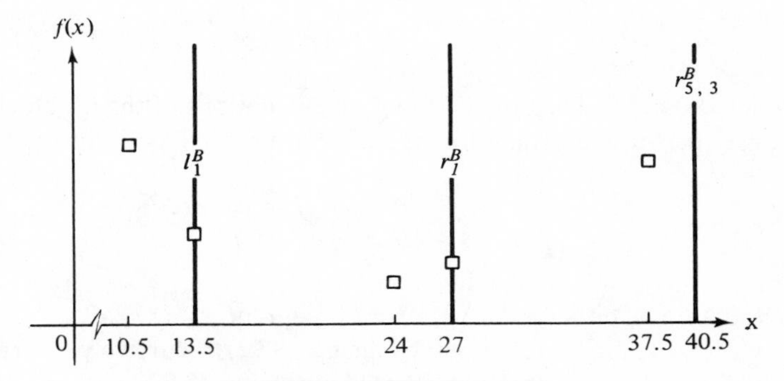

Fig. 8.7 First block of function evaluations.

The function is now evaluated at the following points of the second block (including the remaining point at 24.0).

$$
\begin{aligned}
x_2^1 &= 13.5 + 3 - (3)(0)(0.5) = 16.5 \\
x_2^2 &= 13.5 + \tfrac{1}{3}(12) - (1)(0.5) = 17.0 \\
x_2^3 &= 16.5 + \tfrac{1}{3}(12) - (1)(0.5) = 20.0 \\
x_2^4 &= 17.0 + \tfrac{1}{3}(12) - (1)(0.5) = 20.5 \\
x_2^5 &= 20.0 + \tfrac{1}{3}(12) - (1)(0.5) = 23.5 \\
x_2^6 &= 20.5 + \tfrac{1}{3}(12) - (1)(0.5) = 24.0.
\end{aligned}
\tag{8.95}
$$

The situation after the second block of evaluations is shown in Figure 8.8.

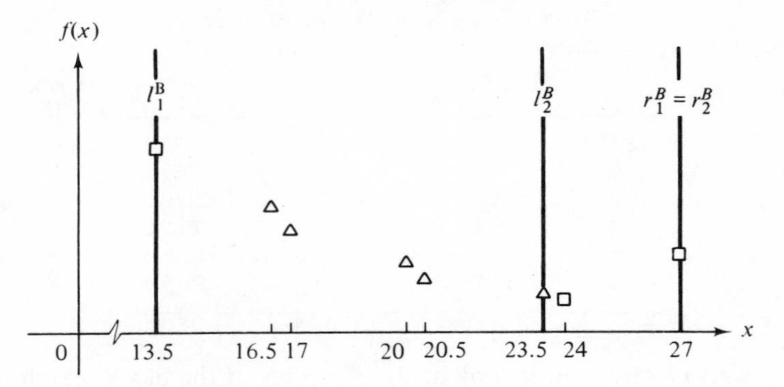

Fig. 8.8 Second block of function evaluations.

The remaining interval is [23.5, 27.0], and the remaining point is again at 24.0.

The reader can verify that, by (8.89) and (8.90), the function evaluations

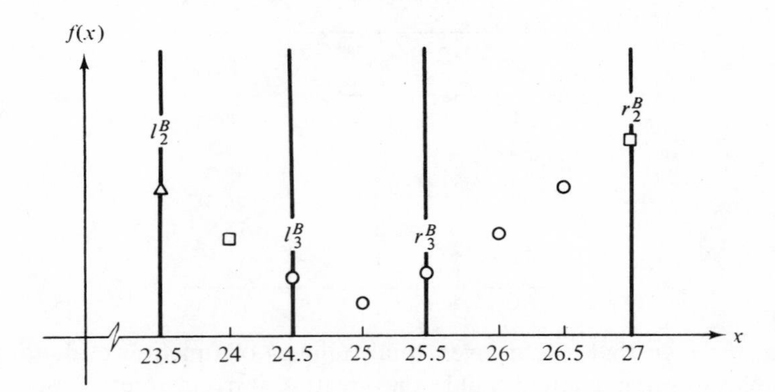

Fig. 8.9 Third block of function evaluations.

of the third block are placed at a distance $\frac{1}{2}$ units apart, as shown in Figure 8.9, and the final interval [24.5, 25.5] has unit length. ▮

We conclude the discussion of the optimal block search method by giving a few examples of its efficiency, based on the length of the starting interval, which can be reduced to a final interval of unit length. In Table 8.1 the length of the starting interval or, equivalently, the value of $r_{\xi,n}^B$ as a function of the number of simultaneous function evaluations per block is summarized for the case of 2, 5, and 8 blocks with $\epsilon = \frac{1}{2}$. This table shows that the length of the starting interval vastly increases with the number of blocks for a given ξ, and, for a given number of blocks, it also considerably increases with the number of simultaneous evaluations per block.

Table 8.1 LENGTH OF STARTING INTERVAL

ξ	$r^B_{\xi,2}$	$r^B_{\xi,5}$	$r^B_{\xi,8}$
1	1.5	6.5	27.5
2	3.5	31.5	255.5
3	5.0	104.0	2110.0
4	8.5	242.0	6560.5
5	10.5	580.5	31630.5
6	15.5	1023.5	65535.5

It is also interesting to look at the efficiency of the block search method for various distributions of a given total number of function evaluations. Table 8.2 compares the length of the starting intervals for 12 function evaluations arranged in various blocks. Here, again, $\epsilon = \frac{1}{2}$.

Table 8.2 LENGTH OF STARTING INTERVAL FOR 12 FUNCTION EVALUATIONS

ξ	n	$r^B_{\xi,n}$
1	12	188.5
2	6	63.5
3	4	38.0
4	3	26.5
6	2	15.5
12	1	6.5

It can be seen that, for a given total number of function evaluations, the Fibonacci search method yields the greatest starting interval that can be reduced to unit length, because it uses all the information available from the function evaluations, and no function evaluations are "wasted" by the simultaneous nature of the block search method.

A further extension of the optimal block search method is due to Beamer and Wilde [5], who allow the number of simultaneous function evaluations to vary from block to block. For further details, the reader is referred to [5].

The optimal block search method has similar shortcomings to the Fibonacci search method; that is, the number of blocks used in the search must be known in advance. Just as the golden section method is a good approximation to the Fibonacci technique, the **golden block search method,** which we describe next, is an even better approximation to the optimal block search.

It has been shown by Avriel and Wilde [3] that

$$\lim_{n\to+\infty} \frac{A(k, n+1)}{A(k, n)} = \frac{k + [(k)^2 + 4k]^{1/2}}{2} = \tau_k, \tag{8.96}$$

that is, for $k = 1, 2, \ldots$, the number τ_k is a solution of the equation

$$(t)^2 - kt - k = 0. \tag{8.97}$$

For $k = 1$, the value of τ_k is equal to τ, used in the golden section method. The value of τ_k for $k = 2$ is approximately 2.732, $\tau_3 = 3.791$, and so on.

Consider first an **odd** number of function evaluations per block. For $\epsilon \longrightarrow 0$ we get, for the optimal block search,

$$r^B_{2k-1,n} \cong A(k, n+1) \tag{8.98}$$

$$x^1_i \cong l^B_{i-1} + A(k, n-i+1) \tag{8.99}$$

and

$$x^{j+2}_i \cong x^j_i + \frac{1}{k}A(k, n-i+2), \qquad j = 0, \ldots, 2k-2. \tag{8.100}$$

For large n, we obtain from (8.96)

$$\frac{A(k, n)}{A(k, n-i)} \cong (\tau_k)^i. \tag{8.101}$$

Now suppose that the starting interval for a golden block search is equal to the starting interval of the optimal block search with n blocks of $2k - 1$ evaluations and $\epsilon = 0$. That is,

$$r^{\mathrm{GB}}_{2k-1} = r^B_{2k-1,n}, \tag{8.102}$$

where the superscript GB indicates "golden block." Then by (8.98) through (8.102), we evaluate the function according to the golden block search method as follows:

$$x^1_i = l^{\mathrm{GB}}_{i-1} + \left(\frac{1}{\tau_k}\right)^i r^{\mathrm{GB}}_{2k-1} \tag{8.103}$$

and

$$x^{j+2}_i = x^j_i + \frac{1}{k}(\tau_k)^{1-i} r^{\mathrm{GB}}_{2k-1}, \qquad j = 0, \ldots, 2k-2. \tag{8.104}$$

After n blocks of $2k - 1$ evaluations each, the length of the final interval containing the minimum of the unimodal function is given by $(1/k)(\tau_k)^{1-n} r^{\mathrm{GB}}_{2k-1}$.

The golden block search for an **even** number of function evaluations per block is obtained by rearranging the corresponding relations of the optimal even-block search. Then

$$x_1^1 = \frac{r_{2k}^{\mathrm{GB}} - k\epsilon}{k+1} \tag{8.105}$$

$$x_i^{j+2} = x_i^j + \frac{r_{2k}^{\mathrm{GB}} + \epsilon}{k+1}, \qquad j = 0, \ldots, 2k-1. \tag{8.106}$$

These expressions, involving only the starting interval $[0, r_{2k}^{\mathrm{GB}}]$ and the resolution ϵ, do not depend on the number of blocks. After n blocks of $2k$ simultaneous function evaluations, the length of the final interval is given by $(r_{2k}^{\mathrm{GB}} + \epsilon)/(k+1)^n$.

Example 8.4.2

Suppose that we are seeking the minimum of a unimodal function on the interval [0, 45]. It is easy to compute that for $\epsilon \longrightarrow 0$

$$r_{5,3}^{\mathrm{B}} \cong A(3, 4) = 45. \tag{8.107}$$

Thus three blocks of five simultaneous function evaluations placed by the optimal block search method can reduce the starting interval to a final interval of unit length. Let us perform a golden block search with five simultaneous evaluations on the same interval. Table 8.3 shows the locations of the evaluations in the first block, as given by (8.103) and (8.104). The numbers in parentheses indicate the locations of the function evaluations placed by the corresponding optimal block search.

Table 8.3 LOCATIONS OF FIRST-BLOCK FUNCTION EVALUATIONS

j	1	2	3	4	5
x_1^j	11.87(12)	15(15)	26.87(27)	30(30)	41.87(42)

After the third block of function evaluations placed by the golden block search, the remaining interval has a length $\frac{1}{3}(3.791)^{-2}(45) = 1.0436$. That is, this interval is only 4.36% longer than the corresponding final interval of unit length that could have obtained by the optimal block search method. ∎

In general, the more simultaneous function evaluations placed in a block, the better will be the approximation of the optimal block method by the golden block search.

EXERCISES

8.A. Write short computer programs that implement Newton's method and the secant method. Use them to find the solutions of simple equations. Compute the errors in each iteration and compare the experimental convergence rates obtained with the theoretical ones.

8.B. Show that the inequalities in (8.34) imply that the coefficient c of the quadratic term in (8.32) is positive and the predicted stationary point of ϕ is indeed a minimum.

8.C. Modify the quadratic and the bracketing methods by assuming that the first derivative of the function to be minimized is known everywhere.

8.D. Let f be a real function on R. Also let $x^0 \in R^n$, $z \in R^n$, and $\theta \in R$. Define

$$F(\theta) = f(x^0 + \theta z) \tag{8.108}$$

and suppose that we are looking for the minimum of F—that is, for the minimum of f in the direction z through the point x^0. Let $x^0 + \theta_1 z$, $x^0 + \theta_2 z$, and $x^0 + \theta_3 z$ be three points where f is evaluated. Show that the minimum predicted by applying the quadratic approximation method is at $x^0 + \theta^* z$, where [11]

$$\theta^* = \frac{[(\theta_2)^2 - (\theta_3)^2]F(\theta_1) + [(\theta_3)^2 - (\theta_1)^2]F(\theta_2) + [(\theta_1)^2 - (\theta_2)^2]F(\theta_3)}{2[(\theta_2 - \theta_3)F(\theta_1) + (\theta_3 - \theta_1)F(\theta_2) + (\theta_1 - \theta_2)F(\theta_3)]} \tag{8.109}$$

and it is indeed the minimum of the parabola passing through the above three points if

$$\frac{(\theta_2 - \theta_3)F(\theta_1) + (\theta_3 - \theta_1)F(\theta_2) + (\theta_1 - \theta_2)F(\theta_3)}{(\theta_2 - \theta_3)(\theta_3 - \theta_1)(\theta_1 - \theta_2)} < 0. \tag{8.110}$$

8.E. The following version of the bracketing and quadratic approximation methods is due to Davies, Swann, and Campey [14]. Evaluate the function f at some starting point $x^0 \in R$ and then at $x^0 + h$, where $h > 0$ is some predetermined step length. If

$$f(x^0) \geq f(x^0 + h), \tag{8.111}$$

the step length is doubled and a further value of f is computed in the direction in which the function is decreasing. The process is repeated until a function increase is obtained. If

$$f(x^0) < f(x^0 + h), \tag{8.112}$$

then we evaluate $f(x^0 - 2h)$, $f(x^0 - 4h)$, and so on, until a function increase results. The final three points form a bracket that includes a minimum of f.

The function is then evaluated at the midpoint between the last two points reached by the method. This process yields four points placed at some equal distance s. The endpoint farthest from the point with the lowest function value is discarded, and there remain three equally spaced points to start the quadratic approximation. Let $x^1 < x^2 < x^3$ be these three points and let f^i be the corresponding function values. Show that x^4, the minimum predicted by the quadratic approximation, is given by

$$x^4 = x^2 + \frac{s(f^1 - f^3)}{2(f^1 - 2f^2 + f^3)}. \tag{8.113}$$

The bracketing process is then repeated with a smaller step length. Can x^4 correspond to a maximum of the parabola passing through the three points? Draw a flow diagram of this method.

8.F. Describe in detail a search for minimization of functions defined everywhere on R, based on quadratic approximation with extrapolation (i.e., without bracketing) by using function evaluations only. Draw a flow diagram to illustrate your method.

8.G. Verify Eqs. (8.44) through (8.47) of the cubic approximation method.

8.H. Show that the length ratio of the final-to-starting interval in the golden section method with N function evaluations is given by $1/(\tau)^N$. Prove (8.86) by noting that

$$F_k = \frac{(\tau)^k - (-\tau)^{-k}}{\sqrt{5}}. \tag{8.114}$$

8.I. Using the formulas of the optimal block search method, derive the symmetric Fibonacci search technique ($\xi = 1$) and compare with the formulas given in this chapter.

8.J. Using the formulas of the optimal block search method, derive the **simultaneous** search procedure ($n = 1$) for arbitrary odd or even number of function evaluations. Compare the efficiency of the method for odd and even number of evaluations.

REFERENCES

1. AVRIEL, M., and D. J. WILDE, "Optimality Proof for the Symmetric Fibonacci Search Technique," *Fibonacci Quarterly*, **4**, 265–269 (1966).

2. AVRIEL, M., and D. J. WILDE, "Optimal Search for a Maximum with Sequences of Simultaneous Function Evaluations," *Management Science*, **12**, 722–731 (1966).

3. AVRIEL, M., and D. J. WILDE, "Golden Block Search for the Maximum of Unimodal Functions," *Management Science*, **14**, 307–319 (1968).

4. Bartle, R. G., *The Elements of Real Analysis*, 2nd ed., John Wiley & Sons, New York, 1976.

5. Beamer, J. H., and D. J. Wilde, "Minimax Optimization of Unimodal Functions by Variable Block Search," *Management Science*, **16**, 529–541 (1970).

6. Davidon, W. C., "Variable Metric Method for Minimization," AEC Research and Development Report ANL-5990 (Rev.), November 1959.

7. Goldstein, A. A., *Constructive Real Analysis*, Harper & Row, New York, 1967.

8. Karp, R. M., and W. L. Miranker, "Parallel Minimax Search for a Maximum," *J. Combinatorial Theory*, **4**, 19–35 (1968).

9. Kiefer, J., "Sequential Minimax Search for a Maximum," *Proc. Am. Math. Soc.*, **4**, 502–506 (1953).

10. Ostrowski, A. M., *Solution of Equations and Systems of Equations*, 2nd ed., Academic Press, New York, 1966.

11. Powell, M. J. D., "An Efficient Method for Finding the Minimum of a Function of Several Variables without Calculating Derivatives," *Computer J.*, **7**, 155–162 (1964).

12. Rosenbrock, H. H., "An Automatic Method for Finding the Greatest or Least Value of a Function," *Computer J.*, **3**, 175–184 (1960).

13. Slotnick, D. L., "The Fastest Computer," *Scientific American*, **224**, 2, 76–87 (1971).

14. Swann, W. H., "Report on the Development of a New Direct Search Method of Optimization," I.C.I. Central Instrument Lab. Research Note 64/3, London, 1964.

15. Wilde, D. J., *Optimum Seeking Methods*, Prentice-Hall, Englewood Cliffs, N.J., 1964.

16. Wilde, D. J., and C. S. Beightler, *Foundations of Optimization*, Prentice-Hall, Englewood Cliffs, N.J., 1967.

9 MULTIDIMENSIONAL UNCONSTRAINED OPTIMIZATION WITHOUT DERIVATIVES: EMPIRICAL AND CONJUGATE DIRECTION METHODS

The more we know about a function, the better or the more efficient algorithms we can derive for seeking its extremum. On the other hand, obtaining more information about the function may involve extensive computational effort. For example, if a function is convex, we know that every stationary point must correspond to a global minimum, but the amount of computational work required to establish convexity may be very large. Similarly, the availability of first and second derivatives of a function on R^n can greatly facilitate locating an extremum, but the number of arithmetic operations required for computing derivatives (especially for large values of n) can be so large that one may try to derive algorithms without using derivatives. In many practical problems such derivatives are simply unavailable, for sometimes even an analytic expression for the function to be minimized cannot be found. These considerations justify presenting a number of algorithms for unconstrained multidimensional minimization in which only values of the function are computed.

In later chapters more efficient algorithms that require evaluation of the function as well as its derivatives will be discussed. Some of these algorithms have been modified so that derivatives are not directly computed but only approximated by function values. We shall present these modifications, along with the algorithms for which they were derived. In each of the first three sections of this chapter we describe an "empirical" method that represents early efforts in unconstrained optimization. Although more advanced methods are available today, the simplicity of the empirical techniques and their underlying principles make them useful in certain cases. The rest of the chapter deals with methods that use conjugate directions. The notion of conjugate directions is introduced here in order to understand Powell's

method without derivatives, which is generally considered superior to the empirical methods.

9.1 THE SIMPLEX METHOD

For a long time it was known that finding minima of functions of n variables by the simplest concepts—such as setting up a grid on R^n and evaluating the function at every point of this grid, or searching for a minimum by random moves—was quite inefficient. Improved empirical methods emerged in the early 1960s, the first of which, called the simplex method, is described now. Parenthetically, it should be mentioned that the simplex method of unconstrained minimization should not be confused with the simplex method in linear programming, although the origin of the name is the same for both. A simplex is the convex hull of $n + 1$ points in R^n—for example, a line segment in R, a triangle in R^2 and so forth. The simplex method of unconstrained minimization was devised by Spendley, Hext, and Himsworth [15] and later improved by Nelder and Mead [7].

Consider the minimization of the real function $f(x)$, $x \in R^n$ and let $x^0, x^1, \ldots, x^n$ be points in R^n that form a current simplex. Let x^h and x^l be defined by

$$f(x^h) = \max\{f(x^0), f(x^1), \ldots, f(x^n)\} \tag{9.1}$$

and

$$f(x^l) = \min\{f(x^0), f(x^1), \ldots, f(x^n)\}. \tag{9.2}$$

Denote by $\bar{x}$ the centroid of all the vertices of the simplex except x^h:

$$\bar{x} = \frac{1}{n}\sum_{i=0}^{n} x^i, \qquad x^i \neq x^h. \tag{9.3}$$

The main idea of the algorithm is to replace x^h, the vertex of the current simplex that has the highest function value, by a new and better point. The replacement of this point involves three types of steps: reflection, expansion, and contraction.

In the **reflection** step we compute x^r by the formula

$$x^r = \bar{x} + \alpha(\bar{x} - x^h), \tag{9.4}$$

where α is a positive constant, called the reflection coefficient.

Let us consider three possible cases:

1. If $f(x^l) > f(x^r)$—that is, the reflection step has generated a new minimum—then we take an **expansion** step by computing

$$x^e = \bar{x} + \gamma(x^r - \bar{x}), \tag{9.5}$$

where $\gamma > 1$, the expansion coefficient, is a given constant. If $f(x^r) > f(x^e)$, then x^e replaces x^h and a new simplex is obtained. If, however, $f(x^e) \geq f(x^r)$, then the expansion step failed and x^h is replaced by x^r in the new simplex.

2. If

$$\max \{f(x^i), x^i \neq x^h\} \geq f(x^r) \geq f(x^l), \tag{9.6}$$

then x^h is replaced by x^r and the result is a new simplex.

3. If

$$f(x^r) > \max_i \{f(x^i), x^i \neq x^h\}, \tag{9.7}$$

then replacing x^h by x^r would make x^r the new x^h. In this case, we define a point $x^{h'}$ by

$$f(x^{h'}) = \min \{f(x^h), f(x^r)\} \tag{9.8}$$

and take a **contraction** step

$$x^c = \bar{x} + \beta(x^{h'} - \bar{x}), \tag{9.9}$$

where $0 < \beta < 1$ is the contraction coefficient. The point x^c replaces x^h in the new simplex unless $f(x^c) > f(x^{h'})$, in which case we replace all the x^i by the new points $\hat{x}^i$, defined as

$$\hat{x}^i = x^i + \tfrac{1}{2}(x^l - x^i), \qquad i = 0, \ldots, n. \tag{9.10}$$

That is, the distances between the vertices of the old simplex and the point with the lowest function value are halved.

The evaluation of the performance of the simplex method on several test functions indicated that the choice of $\alpha = 1$, $\beta = \frac{1}{2}$, $\gamma = 2$ gave good results [7]. A typical search for the minimum by the simplex method is shown in Figure 9.1. The numbers indicate vertices of successive simplexes. Dashed lines show moves made, some of which ended in failure. Nelder and Mead suggest that the algorithm should terminate if

$$\left\{\frac{1}{n+1} \sum_{i=0}^{n} [f(x^i) - f(\bar{x})]^2\right\}^{1/2} < \epsilon, \tag{9.11}$$

where ϵ is some predetermined positive number.

The original simplex method of Spendley, Hext, and Himsworth, based on regular simplexes without expansion and contraction steps, as well as the Nelder and Mead version just described have been successfully tested on many problems, but they were found to be considerably affected by the scale and orientation chosen for the first simplex [14]. The Nelder and Mead

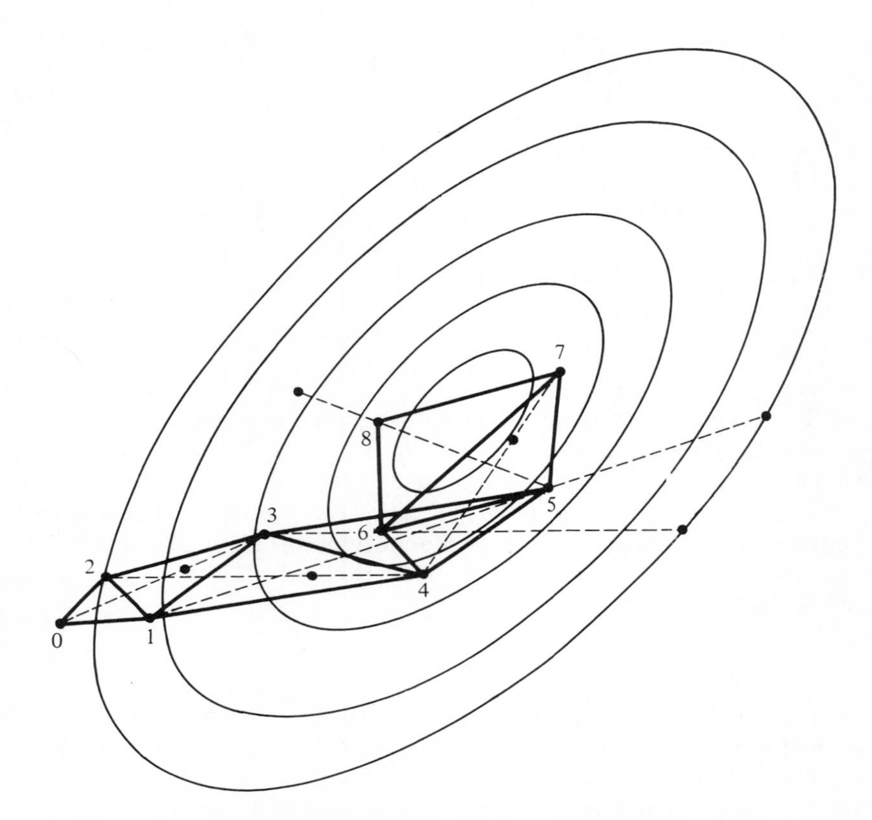

Fig. 9.1 The simplex method illustrated on a quadratic function.

version, which is usually regarded as superior to the original method was reported to be quite inefficient for problems with a large number of variables, say $n \geq 10$ [10].

9.2 PATTERN SEARCH

This simple and easily implemented algorithm is due to Hooke and Jeeves [5]. Suppose again that we are seeking the minimum of a real-valued function $f(x)$, $x \in R^n$. Given an initial point x_B^0 (it is also marked as t_0^1 for reasons that will become clear later), we perform a sequence of **exploratory moves** around x_B^0 as follows: Compute $f(x_B^0)$ and $f(x_B^0 + \Delta_1)$, where Δ_j is an n vector whose jth element is $d_j > 0$ and the rest of whose elements are zeros; that is,

$$\Delta_1 = (d_1, 0, 0, \ldots, 0)^T \tag{9.12}$$

and d_1 is some prescribed positive number (step length). Hence

$$(x_B^0 + \Delta_1) = (x_{B1}^0 + d_1, x_{B2}^0, \ldots, x_{Bn}^0). \tag{9.13}$$

If $f(x_B^0) > f(x_B^0 + \Delta_1)$, the exploratory move was successful, we set $t_1^1 = x_B^0 + \Delta_1$ and proceed to make a move in the x_2 direction; otherwise we reverse the search direction and compute $f(x_B^0 - \Delta_1)$. If $f(x_B^0 - \Delta_1) < f(x_B^0)$, then this exploratory move is declared successful and we set $t_1^1 = x_B^0 - \Delta_1$. If the last move has also failed—that is,

$$f(x_B^0 - \Delta_1) \geq f(x_B^0), \tag{9.14}$$

we set $t_1^1 = x_B^0$.

Having completed the exploratory moves in the direction of the x_1 axis, we compute $f(t_1^1 + \Delta_2)$, where $\Delta_2 = (0, d_2, 0, \ldots, 0)^T$. If $f(t_1^1 + \Delta_2) < f(t_1^1)$, we set $t_2^1 = t_1^1 + \Delta_2$; otherwise we compute $f(t_1^1 - \Delta_2)$ and compare it with $f(t_1^1)$. We set $t_2^1 = t_1^1 - \Delta_2$ if $f(t_1^1 - \Delta_2) < f(t_1^1)$ and $t_2^1 = t_1^1$ otherwise. Continuing in this way, we make exploratory moves along all n axial directions and arrive at a point t_n^1. This point is called the new base point x_B^1, and we immediately move to

$$t_0^2 = x_B^1 + (x_B^1 - x_B^0) \tag{9.15}$$

in the "promising" direction $x_B^1 - x_B^0$. This is called a **pattern move.** If $t_0^2 \neq x_B^0$, we perform a new set of exploratory moves and establish a new base point x_B^2; otherwise the exploratory moves around t_0^1 were all failures, and we reduce the d_j for additional exploratory moves around t_0^1 with smaller step lengths. If $f(x_B^2) < f(x_B^1)$, we make our next pattern move to

$$t_0^3 = x_B^2 + (x_B^2 - x_B^1) \tag{9.16}$$

and the exploratory procedure is restarted around t_0^3.

We continue in this way until we arrive at a base point x_B^k such that $f(x_B^k) \geq f(x_B^{k-1})$. In this case, we return to x_B^{k-1}, set $t_0^k = x_B^{k-1}$, and recommence the exploratory moves from there. If these exploratory moves are successful, the alternating pattern and exploratory moves are restarted; otherwise we reduce the d_j for additional exploratory moves around t_0^k with smaller step lengths. The algorithm terminates when the d_j become smaller than some predetermined values. This algorithm is illustrated on a function of two variables in Figure 9.2. The circled numbers indicate the sequence of points visited.

Although the pattern search method may require a great number of function evaluations in order to reach a point that approximates the minimum of a function, it is regarded as an easily programmed and reliable

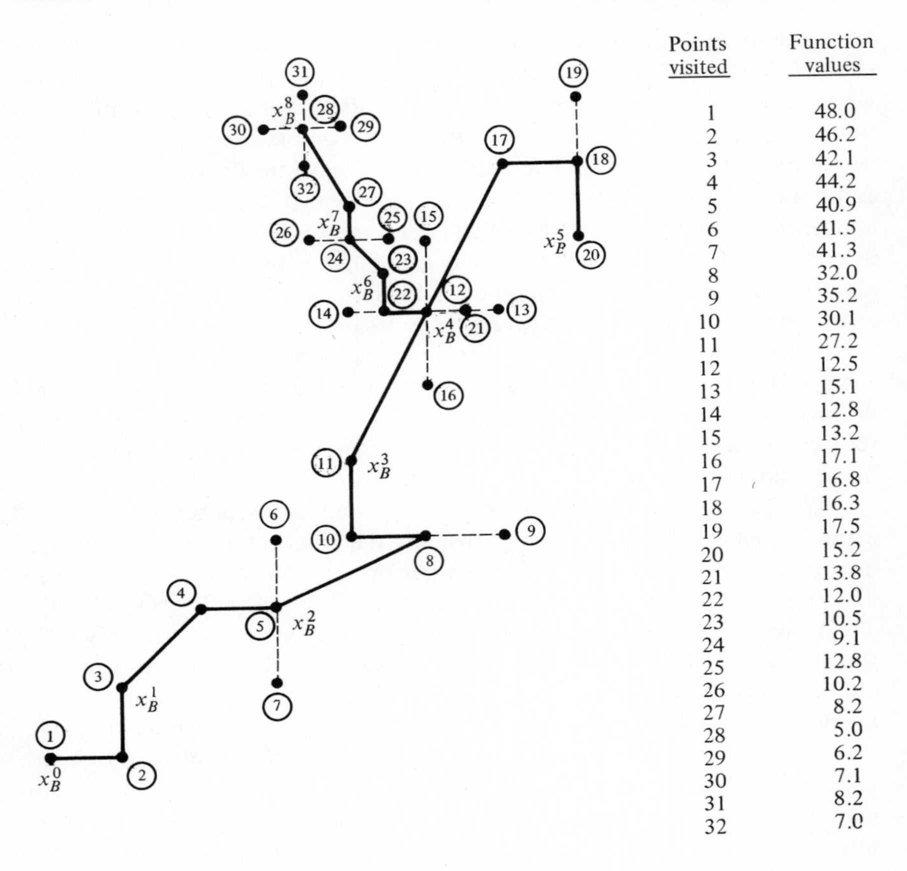

Points visited	Function values
1	48.0
2	46.2
3	42.1
4	44.2
5	40.9
6	41.5
7	41.3
8	32.0
9	35.2
10	30.1
11	27.2
12	12.5
13	15.1
14	12.8
15	13.2
16	17.1
17	16.8
18	16.3
19	17.5
20	15.2
21	13.8
22	12.0
23	10.5
24	9.1
25	12.8
26	10.2
27	8.2
28	5.0
29	6.2
30	7.1
31	8.2
32	7.0

Fig. 9.2 Pattern search on a function of two variables.

method. Its main feature consists of following "ridges" and "valleys." The pattern moves can take long steps in the assumed direction of valleys, whereas the exploratory moves find the way back to these valleys if a pattern move has climbed out of them.

9.3 THE ROTATING DIRECTIONS METHOD

The importance of following ridges or valleys is clearly reflected in Hooke and Jeeves' pattern search, as well as in Rosenbrock's rotating directions method [13], which is described next. The reason for deriving these two early techniques of unconstrained minimization is probably due to the poor performances of the two oldest algorithms—the alternating directions and the steepest descent methods—on general functions. In the alternating directions method we start from an arbitrary point and search

for a minimum in the direction of the x_1 axis by one of the one-dimensional methods described in the previous section. Starting from this minimum, we then perform a search parallel to the x_2 axis, and so on until all directions $x_1, \ldots, x_n$ are searched in turn. After completing such a cycle, the whole process is restarted. The reader should have no difficulty in drawing contours of two-dimensional functions for which this technique would approach the minimum in extremely small steps. The method of steepest descent, which we shall mention later, suffers from similar deficiencies. The main disadvantage of both methods is their inability to revise their search directions in an adaptive manner.

Let us see how the rotating directions method of Rosenbrock searches for a minimum of an unconstrained real-valued function f on R^n. In his original work Rosenbrock actually extends the method for minimization subject to certain constraints, but in this section we shall discuss only the unconstrained version.

We start at a given initial point $x_B^0 \in R^n$ and choose a set of n mutually orthonormal directions $\zeta^1, \ldots, \zeta^n$ (usually the unit vectors in the coordinate directions are chosen first). Next we perform exploratory moves in these directions as follows: Compute $f(x_B^0)$ and $f(x_B^0 + \zeta^1)$. If $f(x_B^0 + \zeta^1) \leq f(x_B^0)$, we call this step a success, replace x_B^0 by $x_B^0 + \zeta^1$, and multiply ζ^1 by a scalar $\alpha > 1$. If $f(x_B^0 + \zeta^1) > f(x_B^0)$, we call the step a failure, retain x_B^0, and multiply ζ^1 by a scalar $-\beta$, $0 < \beta < 1$. In either case, we proceed to the next direction and make an exploratory move in the same manner around the current point. This procedure is then carried out in all n directions. Unlike the pattern search method of Hooke and Jeeves, this set of exploratory moves in the mutually orthogonal directions is repeated until at least one success, followed by a failure, occurs in all directions. This part of the algorithm is called an **exploratory stage.** A typical exploratory stage, consisting of eight function evaluations, is illustrated in Figure 9.3, where the circled numbers correspond to points at which the function was sequentially evaluated. Solid and dashed lines indicate success and failure moves, respectively.

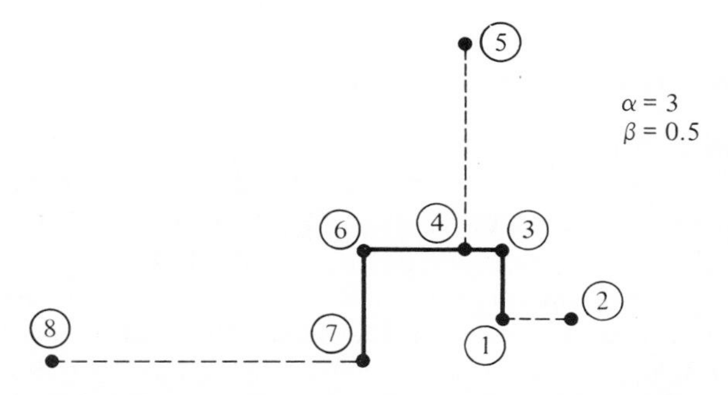

Fig. 9.3 An exploratory stage of the rotating directions method.

Having completed an exploratory stage, we turn now to the construction of a new set of orthogonal directions. Let $\alpha_j, j = 1, \ldots, n$, denote the net distance moved in the exploratory stages in the jth current direction. Then let

$$\begin{aligned} u^1 &= \alpha_1\zeta^1 + \alpha_2\zeta^2 + \cdots + \alpha_n\zeta^n \\ u^2 &= \qquad\quad \alpha_2\zeta^2 + \cdots + \alpha_n\zeta^n \\ &\vdots \\ u^n &= \qquad\qquad\qquad\qquad \alpha_n\zeta^n. \end{aligned} \tag{9.17}$$

Note that u^1 is the vector joining the initial and final points of the exploratory stage and that it is assumed to be a "promising" direction. Next we find a new set of orthonormal vectors by the Gram-Schmidt orthogonalization procedure [8]. Suppose that all the α_j in (9.17) are nonzero. Then the vectors $u^j, j = 1, \ldots, n$, are linearly independent. Let

$$w^1 = u^1 \tag{9.18}$$

$$\hat{\zeta}^1 = \frac{w^1}{\|w^1\|} \tag{9.19}$$

$$w^2 = u^2 - [(u^2)^T\hat{\zeta}^1]\hat{\zeta}^1 \tag{9.20}$$

$$\hat{\zeta}^2 = \frac{w^2}{\|w^2\|} \tag{9.21}$$

and, in general,

$$w^j = u^j - \sum_{k=1}^{j-1} [(u^j)^T\hat{\zeta}^k]\hat{\zeta}^k, \qquad j = 1, \ldots, n \tag{9.22}$$

$$\hat{\zeta}^j = \frac{w^j}{\|w^j\|}, \qquad j = 1, \ldots, n. \tag{9.23}$$

The $\hat{\zeta}$ thus obtained serves as a new set of n mutually orthonormal directions for a new exploratory stage. The first direction of search $\hat{\zeta}^1$ will be the one passing through the first and last points of the previous exploratory stage. The step lengths in the new directions are generally chosen to be the same as in the previous stage.

It is possible that some of the numbers α_j (the net distance moved in the jth direction) vanish, in which case the foregoing orthogonalization procedure must be modified [2]. A different, and somewhat more economical, orthogonalization procedure has been proposed by Palmer [9]. The algorithm can be terminated if $\|u^1\| < \epsilon$ for each of several consecutive exploratory stages, where ϵ is some small positive number.

Example 9.3.1

Consider the problem of minimizing the function

$$f(x_1, x_2) = (x_1)^2 + (x_2)^2 - 3x_1 - x_1x_2 + 3 \tag{9.24}$$

by Rosenbrock's rotating directions method.

Suppose that we start at $x_B^0 = (0, 0)$ and first perform exploratory moves in the coordinate directions; that is,

$$\zeta^1 = \begin{pmatrix} 1 \\ 0 \end{pmatrix} \qquad \zeta^2 = \begin{pmatrix} 0 \\ 1 \end{pmatrix}. \tag{9.25}$$

Also suppose that we choose $\alpha = 3$, $\beta = 0.5$. The locations and outcomes of the exploratory moves are summarized in Table 9.1.

Table 9.1 FIRST EXPLORATORY STAGE

Direction	Point Visited	Function Value	Outcome
–	(0.0, 0.0)	3.0	–
ζ^1	(1.0, 0.0)	1.0	Success
ζ^2	(1.0, 1.0)	1.0	Success
ζ^1	(4.0, 1.0)	4.0	Failure
ζ^2	(1.0, 4.0)	13.0	Failure

Having obtained a success, followed by a failure, in both search directions, we establish the new starting point at $x_B^1 = (1, 1)$ and rotate directions. Note that $\alpha_1 = 1$, $\alpha_2 = 1$. Hence by (9.17) through (9.23), we get

$$u^1 = 1\begin{pmatrix} 1 \\ 0 \end{pmatrix} + 1\begin{pmatrix} 0 \\ 1 \end{pmatrix} = \begin{pmatrix} 1 \\ 1 \end{pmatrix} \tag{9.26}$$

$$u^2 = \qquad 1\begin{pmatrix} 0 \\ 1 \end{pmatrix} = \begin{pmatrix} 0 \\ 1 \end{pmatrix} \tag{9.27}$$

$$w^1 = \begin{pmatrix} 1 \\ 1 \end{pmatrix} \qquad \hat{\zeta}^1 = \begin{pmatrix} \frac{\sqrt{2}}{2} \\ \frac{\sqrt{2}}{2} \end{pmatrix} \cong \begin{pmatrix} 0.7071 \\ 0.7071 \end{pmatrix} \tag{9.28}$$

$$w^2 = \begin{pmatrix} 0 \\ 1 \end{pmatrix} - \left[(0, 1)\begin{pmatrix} \frac{\sqrt{2}}{2} \\ \frac{\sqrt{2}}{2} \end{pmatrix}\right]\begin{pmatrix} \frac{\sqrt{2}}{2} \\ \frac{\sqrt{2}}{2} \end{pmatrix} = \begin{pmatrix} -\frac{1}{2} \\ \frac{1}{2} \end{pmatrix} \tag{9.29}$$

$$\hat{\zeta}^2 = \begin{pmatrix} -\frac{\sqrt{2}}{2} \\ \frac{\sqrt{2}}{2} \end{pmatrix} \cong \begin{pmatrix} -0.7071 \\ 0.7071 \end{pmatrix}. \tag{9.30}$$

We start a new exploratory stage from x_B^1, and the moves with their outcomes are given in Table 9.2.

Table 9.2 SECOND EXPLORATORY STAGE

Direction	Point Visited	Function Value	Outcome
–	(1.0000, 1.0000)	1.0000	–
ξ_1	(1.7071, 1.7071)	0.7929	Success
ξ_2	(1.0000, 2.4142)	4.4142	Failure
ξ_1	(3.8284, 3.8284)	6.1716	Failure
ξ_2	(2.0606, 1.3536)	0.1072	Success
ξ_1	(1.0000, 0.2929)	0.7929	Failure
ξ_2	(3.1213, 0.2929)	2.5503	Failure

The best point obtained in the second exploratory stage is $x_B^2 = (2.0606, 1.3536)$, from which a new exploratory stage begins after rotating the search directions. The exact minimum of f is at $x^* = (2.0, 1.0)$, where the function value is $f(x^*) = 0$. ∎

Davies, Swann, and Campey [16] have modified Rosenbrock's method by replacing the exploratory stages as described above by a sequence of n one-dimensional searches in the n orthogonal directions in turn. The rotation of search directions is carried out as in Rosenbrock's method. A recommended technique [2] for the one-dimensional searches is a single quadratic approximation, as described in the preceding chapter.

The Davies, Swann, and Campey method possesses an interesting property whose generalizations will be shown to be of primary importance in almost all subsequent methods of unconstrained minimization to be discussed. Suppose that the function to be minimized is quadratic and given by

$$f(x) = a + b^T x + \tfrac{1}{2} x^T Q x, \qquad x \in R^n, \tag{9.31}$$

where Q is a symmetric positive definite matrix. Then f has a global minimum at

$$x^* = -Q^{-1} b. \tag{9.32}$$

If $Q = cI$, where $c > 0$ is a real number and I is the identity matrix, the contours $f(x) =$ constant are hyperspheres with center at x^*. Starting at any point in R^n and performing a sequence of n one-dimensional searches in any n mutually orthogonal directions in turn, we can locate the minimum of f exactly. Let us illustrate this property by an example.

Example 9.3.2

Let the quadratic function f be given by

$$f(x) = 3(x_1)^2 + 3(x_2)^2 - 3x_1 + 4x_2. \tag{9.33}$$

The global minimum of this function is at $x^* = (\frac{1}{2}, -\frac{2}{3})$. Suppose that we start at $x_B^0 = (-3, 1)$. Let the two orthonormal directions be

$$\zeta^1 = \begin{pmatrix} \frac{2}{\sqrt{5}} \\ -\frac{1}{\sqrt{5}} \end{pmatrix} \qquad \zeta^2 = \begin{pmatrix} \frac{1}{\sqrt{5}} \\ \frac{2}{\sqrt{5}} \end{pmatrix}. \tag{9.34}$$

First we find the minimum of f in the ζ^1 direction:

$$\begin{aligned} f(x_B^0 + \theta_1\zeta^1) = 3\Big(-3 + \frac{2\theta_1}{\sqrt{5}}\Big)^2 + 3\Big(1 - \frac{\theta_1}{\sqrt{5}}\Big)^2 \\ - 3\Big(-3 + \frac{2\theta_1}{\sqrt{5}}\Big) + 4\Big(1 - \frac{\theta_1}{\sqrt{5}}\Big). \end{aligned} \tag{9.35}$$

Since this is a quadratic function in the variable θ_1, the second-order polynomial approximation will be exact (neglecting roundoff errors). The minimum of (9.35) is attained at

$$\theta_1^* = \frac{26\sqrt{5}}{15} \tag{9.36}$$

and the best point in the ζ^1 direction is $t_1^1 = (t_{11}^1, t_{12}^1)$, where

$$t_{11}^1 = -3 + \frac{(2)(26)}{15} = \frac{7}{15} \tag{9.37}$$

$$t_{12}^1 = 1 - \frac{26}{15} = -\frac{11}{15}. \tag{9.38}$$

Proceeding from this point in the ζ^2 direction, we again perform a one-dimensional search for the minimum of the function

$$\begin{aligned} f(t_1^1 + \theta_2\zeta^2) = 3\Big(\frac{7}{15} + \frac{\theta_2}{\sqrt{5}}\Big)^2 + 3\Big(-\frac{11}{15} + \frac{2\theta_2}{\sqrt{5}}\Big)^2 \\ - 3\Big(\frac{7}{15} + \frac{\theta_2}{\sqrt{5}}\Big) + 4\Big(-\frac{11}{15} + \frac{2\theta_2}{\sqrt{5}}\Big). \end{aligned} \tag{9.39}$$

This is once more a quadratic function of θ_2, and the minimum of (9.39) is attained at

$$\theta_2^* = \frac{\sqrt{5}}{30}. \tag{9.40}$$

Consequently,

$$t_{21}^1 = \frac{7}{15} + \frac{1}{30} = \frac{1}{2} \tag{9.41}$$

$$t_{22}^1 = -\frac{11}{15} + \frac{2}{30} = -\frac{2}{3}. \tag{9.42}$$

The point $t_2^1 = (t_{21}^1, t_{22}^1)$ is the unconstrained minimum of f, as asserted. ∎

In the case of a more general quadratic function having a positive definite matrix Q, this method usually will not locate the exact minimum by successively searching along n mutually orthogonal directions once only. As the next sections will show, conjugate directions, which can be considered generalized orthogonal directions, have the following property: by a method similar to the one just described, we can arrive at the exact minimum of a quadratic function defined on R^n by at most n one-dimensional searches. The first of a family of minimization methods based on conjugate directions will be described later in this chapter. It turns out to be one of the best minimization methods without calculating derivatives for general functions on R^n.

9.4 CONJUGATE DIRECTIONS

We now introduce an important concept for quadratic functions, one on which a whole class of unconstrained minimization methods is based.

Two vectors $x \in R^n$, $y \in R^n$ are said to be **conjugate directions** with respect to the $n \times n$ symmetric positive definite matrix A if

$$x^T A y = 0. \tag{9.43}$$

Note that the notion of conjugate directions is a generalization of orthogonality—that is, the case when A is the $n \times n$ identity matrix I. It is well known that a symmetric $n \times n$ matrix A has n orthogonal eigenvectors [8]. This set of n vectors is also mutually conjugate, for if x^1, x^2 are such eigenvectors of A, then $Ax^2 = \lambda x^2$, where λ is the corresponding eigenvalue and

$$(x^1)^T A x^2 = (x^1)^T \lambda x^2 = \lambda (x^1)^T (x^2) = 0. \tag{9.44}$$

Thus for every $n \times n$ symmetric positive definite matrix there is at least one set of n mutually conjugate directions.

We can construct n mutually conjugate directions with respect to A from a set of n linearly independent vectors $u^1, \ldots, u^n$ in R^n by a procedure similar to the Gram-Schmidt orthogonalization method [8] used earlier.

Let

$$z^1 = u^1 \tag{9.45}$$

$$z^j = u^j - \sum_{k=1}^{j-1} \frac{(u^j)^T A z^k}{(z^k)^T A z^k} z^k, \qquad j = 2, \ldots, n. \tag{9.46}$$

Then the vectors $z^1, \ldots, z^n$ are mutually conjugate with respect to A. The reader will be asked to show that the preceding relations do indeed yield a set of n conjugate vectors and that every set of nonzero vectors, $z^1, \ldots, z^k$, mutually conjugate with respect to A are linearly independent. Note that conjugate directions could have been defined without the property of positive definiteness, but most of the meaningful results depend on this property.

A geometric interpretation of conjugate vectors can be given as follows: Let f be the quadratic function given by (9.31) and assume that Q is a positive definite matrix. Let x^* be the point minimizing $f(x)$ for all $x \in R^n$. Then the surfaces $f(x) = c$ (constant) are generally ellipsoids with center at x^*. Let x^0 be a point satisfying $f(x^0) = c$. Construct the hyperplane tangent to the surface $f(x) = c$ at x^0. Then the vector joining x^0 and x^* is conjugate with respect to Q to every vector in the tangent hyperplane. The two-dimensional case is illustrated in Figure 9.4.

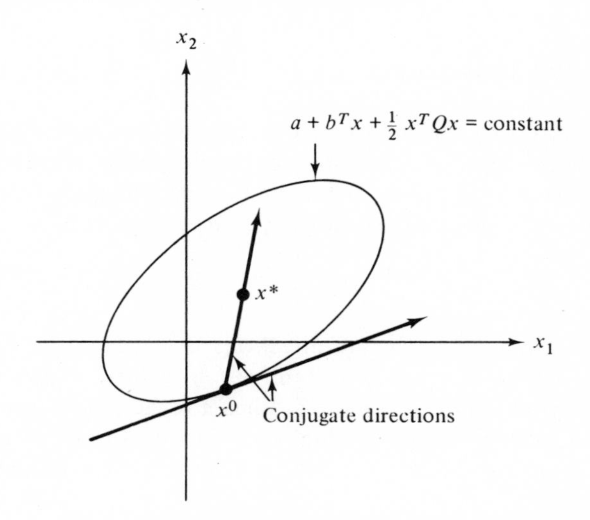

Fig. 9.4 Conjugate directions with respect to Q.

Suppose next that we have m nonzero vectors $z^1, \ldots, z^m$ in R^n, $m \leq n$, mutually conjugate with respect to a positive definite matrix Q. Then these vectors are linearly independent, and thus they span an m-dimensional subspace of R^n, given by the vectors satisfying

$$x = \sum_{j=1}^{m} \alpha_j z^j, \tag{9.47}$$

where the α_j are arbitrary real numbers. For a given point $x^0 \in R^n$, the set of vectors satisfying

$$x = x^0 + \sum_{j=1}^{m} \alpha_j z^j, \tag{9.48}$$

where the z^j are m linearly independent vectors and the α_j are arbitrary numbers, is an affine set or linear manifold (also defined in Chapter 4). It is said to be generated by the point x^0 and the vectors $z^1, \ldots, z^m$. For $m = n$, affine sets become the whole space R^n.

We can now state and prove an important result, due to Powell [11], which relates conjugate directions to unconstrained minimization of quadratic functions.

Theorem 9.1

If $z^1, \ldots, z^m$ are nonzero, mutually conjugate directions with respect to the positive definite matrix Q, then the minimum of the quadratic function

$$f(x) = a + b^T x + \tfrac{1}{2} x^T Q x, \qquad x \in R^n \tag{9.49}$$

over the affine set generated by the point $x^0 \in R^n$ and the vectors $z^1, \ldots, z^m$ will be found by searching along each of the conjugate directions once only.

Proof. Every point x in the affine set is given by (9.48). The required minimum is the point $x^0 + \alpha_1^* z^1 + \cdots + \alpha_m^* z^m$, where the α_j were chosen so as to minimize the function

$$f(x) = f\left(x^0 + \sum_{j=1}^{m} \alpha_j z^j\right) = f(x^0) + \sum_{j=1}^{m} [\alpha_j (z^j)^T (Q x^0 + b) + \tfrac{1}{2}(\alpha_j)^2 (z^j)^T Q z^j]. \tag{9.50}$$

Note that nonzero conjugate directions are linearly independent; hence $z^1, \ldots, z^m$ span an m-dimensional subspace. Since there are no $\alpha_j \alpha_k$ terms with $j \neq k$, the optimal α_j are found by the following m minimization problems:

$$\min_{\alpha_j} \{f(x^0) + \alpha_j (z^j)^T (Q x^0 + b) + \tfrac{1}{2}(\alpha_j)^2 (z^j)^T Q z^j\}, \qquad j = 1, \ldots, m. \tag{9.51}$$

But this problem is equivalent to solving

$$\min_{\alpha_j} f(x^0 + \alpha_j z^j), \qquad j = 1, \ldots, m. \tag{9.52}$$

Consequently, searching for the minimum in the m directions $z^1, \ldots, z^m$ only once will yield the optimal α_j^* and hence the minimum of f restricted to the affine set. ■

Letting $m = n$ in Theorem 9.1 implies that we can find the minimum of the quadratic function (9.49) over R^n by searching along n nonzero directions, mutually conjugate with respect to Q.

Example 9.4.1

Let us demonstrate the preceding result on a quadratic function of two variables. Let

$$f(x) = 2(x_1)^2 + 6(x_2)^2 + 2x_1x_2 + 2x_1 + 3x_2 + 3. \tag{9.53}$$

This function can be rewritten as (9.49) with $a = 3$, $b = (2, 3)^T$, and

$$Q = \begin{bmatrix} 4 & 2 \\ 2 & 12 \end{bmatrix}. \tag{9.54}$$

Suppose that we choose $(z^1)^T = (1, 0)$. A conjugate direction to z^1 with respect to Q is $(z^2)^T = (-\frac{1}{2}, 1)$. Let us find the minimum of f generated by the point $x^0 = (0, 0)$ and the vectors z^1, z^2 above. In our case, the affine set is clearly the whole space R^2. By Theorem 9.1 we must perform two independent one-dimensional searches in the z^1 and z^2 directions. We can carry out these searches by the method presented in Exercise 8.D, since the quadratic approximation in this case will be exact. Starting with the z^1 direction, we want to minimize $F_1(\theta^1) = f(x^0 + \theta^1 z^1)$. We can choose $\theta_1^1 = 0$, $\theta_2^1 = 1$, $\theta_3^1 = 2$, and then $F_1(0) = 3$, $F_1(1) = 7$, $F_1(3) = 15$. Hence

$$\theta^{1*} = \frac{(-3)(3) + (4)(7) + (-1)(15)}{2[(-1)(3) + (2)(7) + (-1)(15)]} = -\frac{1}{2}. \tag{9.55}$$

Proceeding now to the z^2 direction, we minimize $F_2(\theta^2) = f(x^0 + \theta^2 z^2)$. Again we choose $\theta_1^2 = 0$, $\theta_2^2 = 1$, $\theta_3^2 = 2$. Then $F_2(0) = 3$, $F_2(1) = \frac{21}{2}$, $F_2(2) = 29$. Hence

$$\theta^{2*} = \frac{(-3)(3) + (4)(21/2) + (-1)(29)}{2[(-1)(3) + (2)(21/2) + (-1)(29)]} = -\frac{2}{11}. \tag{9.56}$$

The minimum of f on R^2 is then given by

$$x^* = x^0 + \theta^{1*}z^1 + \theta^{2*}z^2 = \begin{pmatrix} 0 \\ 0 \end{pmatrix} + \begin{pmatrix} -\frac{1}{2} \\ 0 \end{pmatrix} + \begin{pmatrix} \frac{1}{11} \\ -\frac{2}{11} \end{pmatrix} = \begin{pmatrix} -\frac{9}{22} \\ -\frac{2}{11} \end{pmatrix}. \tag{9.57}$$

The reader can verify this result by the simple differentiation of f. ∎

Two affine sets S and T, $S \neq T$, are said to be parallel if they are generated by the same directions, $z^1, \dots, z^m$, but different points, $x(S) \in S$ and $x(T) \in T$. We have then

Theorem 9.2

Let $x^*(S)$ and $x^*(T)$ be the points minimizing $f(x)$, as given in (9.49), over two parallel affine sets S and T, respectively. Then the vector $x^*(T) - x^*(S)$ is conjugate with respect to Q to any direction that is contained in S and in T.

Proof. Let z be a direction contained in S and in T. Then

$$\frac{d}{d\alpha}[f(x^*(S) + \alpha z)] = 0 \tag{9.58}$$

for $\alpha = 0$ or, equivalently,

$$z^T[Qx^*(S) + b] = 0. \tag{9.59}$$

Similarly,

$$z^T[Qx^*(T) + b] = 0. \tag{9.60}$$

Subtracting (9.59) from (9.60) yields

$$z^T Q[x^*(T) - x^*(S)] = 0. \tag{9.61}$$

∎

These two theorems are necessary in order to prove that Powell's method, which will be described in the next section, possesses the quadratic termination property, to be defined below.

9.5 POWELL'S METHOD

The first three algorithms described so far were all derived for finding minima of general unconstrained functions, usually in an infinite number of iterations. On the other hand, if we had a method that could find the

minimum of a quadratic function on R^n by a finite number of steps, then such a method should also be efficient for minimizing a general function having continuous second derivatives. In the rest of this chapter and in the following ones we present a whole class of methods that share the common property of **quadratic termination**: if the method of finding the minimum of a quadratic function on R^n as given by (9.31) with a positive definite Q is used, it will terminate in at most n steps. The word "step" prescribed by an algorithm should be interpreted broadly as moving from one point in R^n to another, and clearly, in some cases, these moves can be carried out in an exact manner only by an infinite procedure. Such might be the case, for example, if an algorithm consists of steps that are defined by minimizing a function along a direction in R^n. Thus applying the golden section method for minimizing, say, a quadratic function along a direction would require an infinite number of function evaluations. On the other hand, using a polynomial approximation method, as described in Chapter 8 (neglecting roundoff errors), would yield the exact minimum of the same quadratic function in a finite number of arithmetic operations. Generally, however, one-dimensional minimizations are infinite procedures.

To start, Powell's method is presented as an empirical technique, similar to earlier ones in this chapter. The underlying principles, which are based on conjugate directions, will be explained later in this section.

The first and basic version of Powell's method [11] can be described as follows: Each stage of the procedure consists of $n + 1$ successive one-dimensional line searches, first along n linearly independent directions and then along the direction connecting the best point (obtained at the end of the n one-dimensional line searches) with the starting point of that stage. After these searches, one of the first n directions is replaced by the $(n + 1)$th and a new stage begins.

The kth stage of the method is given by the following steps. Let $x_B^{k-1} = t_0^k \in R^n$ be the starting point of the kth stage and suppose that n linearly independent directions $\Delta_1^k, \ldots, \Delta_n^k$ are given (for $k = 1$, the coordinate directions are usually chosen). Find numbers θ_j^* such that

$$f(t_{j-1}^k + \theta_j^* \Delta_j^k) = \min_{\theta_j} f(t_{j-1}^k + \theta_j \Delta_j^k) \tag{9.62}$$

for $j = 1, \ldots, n$ and define

$$t_j^k = t_{j-1}^k + \theta_j^* \Delta_j^k, \qquad j = 1, \ldots, n. \tag{9.63}$$

Let

$$\Delta_j^{k+1} = \Delta_{j+1}^k, \qquad j = 1, \ldots, n - 1 \tag{9.64}$$

and

$$\Delta_n^{k+1} = \Delta_{n+1}^k = t_n^k - t_0^k. \tag{9.65}$$

Find θ^*_{n+1} such that

$$f(t^k_n + \theta^*_{n+1}(t^k_n - t^k_0)) = \min_{\theta_{n+1}} f(t^k_n + \theta_{n+1}(t^k_n - t^k_0)) \tag{9.66}$$

and let

$$x^k_B = t^k_n + \theta^*_{n+1}(t^k_n - t^k_0). \tag{9.67}$$

If $||x^k_B - x^{k-1}_B|| < \epsilon$, where $\epsilon > 0$ is some predetermined number, stop, otherwise proceed to stage $k + 1$.

A few steps of this algorithm are illustrated in Figure 9.5, in which a general function of two variables is minimized.

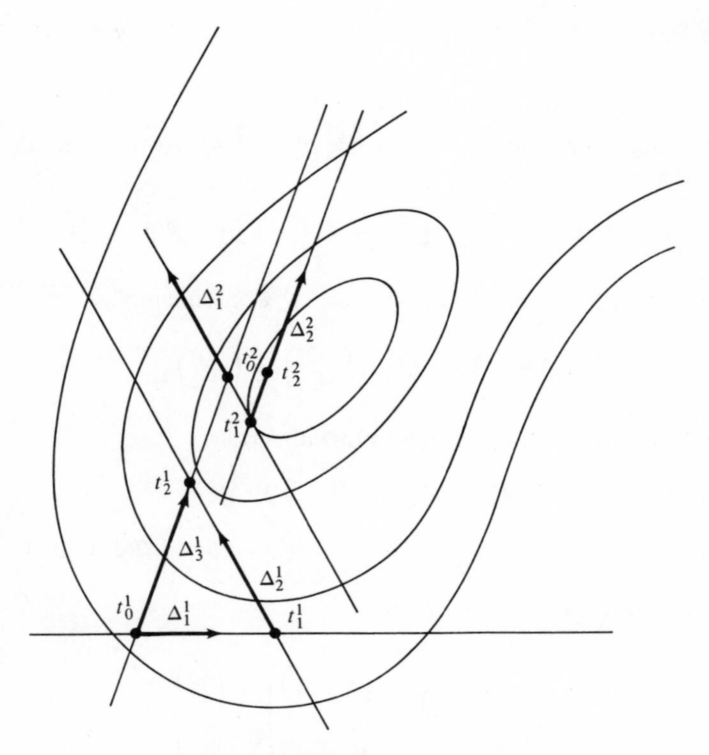

Fig. 9.5 Powell's method.

It will be shown below that Powell's algorithm can find the exact minimum of a quadratic function on R^n with a positive definite matrix Q in n stages at most, provided that the search directions $\Delta^k_1, \ldots, \Delta^k_n$ are linearly independent for $k = 1, \ldots, n$.

Example 9.5.1

Let us demonstrate Powell's method on minimizing the quadratic function in two variables

$$f(x) = \tfrac{3}{2}(x_1)^2 + \tfrac{1}{2}(x_2)^2 - x_1x_2 - 2x_1. \tag{9.68}$$

The minimum of this function is at $x^* = (1, 1)$. Suppose that we start the search at $x_B^0 = t_0^1 = (-2, 4)$ in the coordinate directions

$$\Delta_1^1 = \begin{pmatrix} 1 \\ 0 \end{pmatrix} \qquad \Delta_2^1 = \begin{pmatrix} 0 \\ 1 \end{pmatrix}. \tag{9.69}$$

The first minimization is in the Δ_1^1 direction:

$$\min_{\theta_1} f(t_0^1 + \theta_1\Delta_1^1) = \tfrac{3}{2}(-2 + \theta_1)^2 + \tfrac{1}{2}(4)^2 - 4(-2 + \theta_1) - 2(-2 + \theta_1) \tag{9.70}$$

and the optimum is at $\theta_1^* = 4$. Hence $t_1^1 = (2, 4)$. Now we minimize in the Δ_2^1 direction:

$$\min_{\theta_2} f(t_1^1 + \theta_2\Delta_2^1) = \tfrac{3}{2}(2)^2 + \tfrac{1}{2}(4 + \theta_2)^2 - 2(4 + \theta_2) - 4 \tag{9.71}$$

and obtain $\theta_2^* = -2$. Hence $t_2^1 = (2, 2)$. Consequently,

$$\Delta_3^1 = \begin{pmatrix} 2 - (-2) \\ 2 - 4 \end{pmatrix} = \begin{pmatrix} 4 \\ -2 \end{pmatrix} \tag{9.72}$$

and we perform a minimization in this direction:

$$\min_{\theta_3} f(t_2^1 + \theta_3\Delta_3^1) = \tfrac{3}{2}(2 + 4\theta_3)^2 + \tfrac{1}{2}(2 - 2\theta_3)^2 - (2 + 4\theta_3)(2 - 2\theta_3) - 2(2 + 4\theta_2). \tag{9.73}$$

The optimum is at $\theta_3^* = -\frac{2}{17}$ and

$$x_B^1 = t_0^2 = \begin{pmatrix} 2 - \dfrac{8}{17} \\ 2 + \dfrac{4}{17} \end{pmatrix} = \begin{pmatrix} \dfrac{26}{17} \\ \dfrac{38}{17} \end{pmatrix}. \tag{9.74}$$

This concludes the first iteration of the algorithm. The first two search directions of the second iteration are, by (9.64) and (9.65),

$$\Delta_1^2 = \begin{pmatrix} 0 \\ 1 \end{pmatrix} \qquad \Delta_2^2 = \begin{pmatrix} 4 \\ -2 \end{pmatrix}. \tag{9.75}$$

We solve

$$\min_{\theta_1} f(t_0^2 + \theta_1 \Delta_1^2) = \frac{3}{2}\left(\frac{26}{17}\right) + \frac{1}{2}\left(\frac{38}{17} + \theta_1\right)^2 - \frac{26}{17}\left(\frac{38}{17} + \theta_1\right) - \frac{52}{17} \tag{9.76}$$

and find $\theta_1^* = -\frac{12}{17}$, $t_1^2 = (\frac{26}{17}, \frac{26}{17})$. Next we solve

$$\min_{\theta_2} f(t_1^2 + \theta_2 \Delta_2^2) = \frac{3}{2}\left(\frac{26}{17} + 4\theta_2\right)^2 + \frac{1}{2}\left(\frac{26}{17} - 2\theta_2\right)^2 - \left(\frac{26}{17} + 4\theta_2\right)\left(\frac{26}{17} - 2\theta_2\right) - 2\left(\frac{26}{17} + 4\theta_2\right). \tag{9.77}$$

The optimum is at $\theta_2^* = -\frac{18}{289}$. Hence $t_2^2 = (\frac{370}{289}, \frac{478}{289})$.

Now we find the third direction of the second iteration

$$\Delta_3^2 = \begin{pmatrix} \frac{370}{289} - \frac{26}{17} \\ \frac{478}{289} - \frac{38}{17} \end{pmatrix} = -\begin{pmatrix} \frac{72}{289} \\ \frac{168}{289} \end{pmatrix} \tag{9.78}$$

and solve

$$\min_{\theta_3} f(t_2^2 + \theta_3 \Delta_3^2) = \frac{3}{2}\left(\frac{370}{289} - \frac{72}{289}\theta_3\right)^2 + \frac{1}{2}\left(\frac{478}{289} - \frac{168}{289}\theta_3\right)^2 - \left(\frac{370}{289} - \frac{72}{289}\theta_3\right)\left(\frac{478}{289} - \frac{168}{289}\theta_3\right) - 2\left(\frac{370}{289} - \frac{72}{289}\theta_3\right). \tag{9.79}$$

The optimal solution is at $\theta_3^* = \frac{9}{8}$ and $x_B^2 = (1, 1)$. That is, the exact minimum of the quadratic function was found in two iterations, as asserted. ∎

Note that the search directions Δ_1^k and Δ_2^k, for $k = 1, 2$, were linearly independent in this example. This condition is quite important, for examples exist that demonstrate that the first version of Powell's method, as outlined above, may fail to reach the minimum of f in n iterations and, in fact, may not reach it in any number of iterations.

Example 9.5.2 [17]

Consider the function f on R^3, given by

$$f(x_1, x_2, x_3) = (x_1 - x_2 + x_3)^2 + (-x_1 + x_2 + x_3)^2 + (x_1 + x_2 - x_3)^2. \tag{9.80}$$

This is a strictly convex quadratic function with a unique minimum at

$x^* = (0, 0, 0)$. Let $x_B^0 = (\frac{1}{2}, 1, \frac{1}{2})$ and start Powell's method with

$$\Delta_1^1 = \begin{pmatrix} 1 \\ 0 \\ 0 \end{pmatrix}, \quad \Delta_2^1 = \begin{pmatrix} 0 \\ 1 \\ 0 \end{pmatrix}, \quad \Delta_3^1 = \begin{pmatrix} 0 \\ 0 \\ 1 \end{pmatrix}. \tag{9.81}$$

The first three steps and their results are summarized in Table 9.3.

Table 9.3 FIRST THREE STEPS BY POWELL'S METHOD

j	Direction Δ_j^1	t_j^1	Function Value $f(t_j^1)$
0	–	$(\frac{1}{2}, 1, \frac{1}{2})$	2
1	Δ_1^1	$(\frac{1}{2}, 1, \frac{1}{2})$	2
2	Δ_2^1	$(\frac{1}{2}, \frac{1}{3}, \frac{1}{2})$	$\frac{2}{3}$
3	Δ_3^1	$(\frac{1}{2}, \frac{1}{3}, \frac{5}{18})$	$\frac{42}{81}$

The new direction would be

$$t_3^1 - t_0^1 = \begin{pmatrix} \frac{1}{2} \\ \frac{1}{3} \\ \frac{5}{18} \end{pmatrix} - \begin{pmatrix} \frac{1}{2} \\ 1 \\ \frac{1}{2} \end{pmatrix} = \begin{pmatrix} 0 \\ -\frac{2}{3} \\ -\frac{2}{9} \end{pmatrix} \tag{9.82}$$

and we get

$$\Delta_1^2 = \begin{pmatrix} 0 \\ 1 \\ 0 \end{pmatrix}, \quad \Delta_2^2 = \begin{pmatrix} 0 \\ 0 \\ 1 \end{pmatrix}, \quad \Delta_3^2 = \begin{pmatrix} 0 \\ -\frac{2}{3} \\ -\frac{2}{9} \end{pmatrix}. \tag{9.83}$$

Note that the first components of all three new search directions vanish. Thus the first components of all forthcoming points reached will remain $\frac{1}{2}$, and the true optimum at $x^* = (0, 0, 0)$ can never be achieved. The reason is that the vectors Δ_j^2, $j = 1, 2, 3$, are linearly dependent and do not span the whole space R^3. ∎

The reader can try to construct additional examples of the same type by starting the search at a point that is already a minimum of the function along the first search direction.

Let us now show how Theorems 9.1 and 9.2, which deal with properties of conjugate directions, can be used to prove quadratic termination of Powell's method under suitable assumptions. Suppose that we are given a quadratic function f on R^n with a positive definite Q, a starting point $x_B^0 \in R^n$, and n linearly independent directions $\Delta_1^1, \ldots, \Delta_n^1$. After performing the steps of the first stage, we have a direction $z^1 = t_n^1 - t_0^1 = \Delta_n^2$ and a new starting point $x_B^1 = t_0^2$ for the second stage. Note that if a point in R^n is optimal in n linearly independent directions, then it must be the global

optimum of the quadratic function. Suppose that $t_n^1 \neq t_0^1$; that is, the direction z^1 is nonzero. By (9.67), the point x_B^1 is a minimum in the z^1 direction; and assuming that the directions $\Delta_1^2, \ldots, \Delta_n^2$ are linearly independent, we arrive at a point t_n^2 that is also a minimum in the z^1 direction, contained in a parallel affine set. By Theorem 9.2, the direction $z^2 = t_n^2 - t_0^2$ is conjugate to z^1 with respect to Q. Suppose now that k stages of the algorithm have been completed and that k nonzero directions $z^1, \ldots, z^k$, mutually conjugate with respect to Q, were generated. If the directions $\Delta_1^k, \ldots, \Delta_{n-k}^k, z^1, \ldots, z^k$ are linearly independent, then, by Theorems 9.1 and 9.2, the direction $z^{k+1} = t_n^{k+1} - t_0^{k+1}$ is mutually conjugate to $z^1, \ldots, z^k$. After completing n such stages, all the search directions are mutually conjugate with respect to Q. And so, by Theorem 9.1, the minimum of f over R^n has been reached.

9.6 AVOIDING LINEARLY DEPENDENT SEARCH DIRECTIONS

Powell noted that, in situations that are even less extreme than in Example 9.5.2, his method may choose nearly linearly dependent search directions, especially if there are a large number of variables—a possibility that has serious consequences in terms of convergence. To avoid this difficulty, he proposed a modified version of the method. The modified version does not possess the quadratic termination property, but its performance has generally been very satisfactory.

The kth stage of the modified Powell method [11], slightly improved by Sargent [14], is given by the following rules. Let $x_B^{k-1} = t_0^k$ once more be the starting point of the kth stage and suppose that n linearly independent directions $\Delta_1^k, \ldots, \Delta_n^k$ are given. Find t_j^k for $j = 1, \ldots, n$ as in the first version. Next, find the index m such that

$$f(t_{m-1}^k) - f(t_m^k) = \max_{j=1,\ldots,n} \{f(t_{j-1}^k) - f(t_j^k)\}. \tag{9.84}$$

Set $\Delta_{n+1}^k = (t_n^k - t_0^k)$; and if $\|t_n^k - t_0^k\| \leq \epsilon$, stop. Otherwise find a number α_{n+1}^* such that

$$f(t_0^k + \alpha_{n+1}^* \Delta_{n+1}^k) = \min_{\alpha_{n+1}} f(t_0^k + \alpha_{n+1} \Delta_{n+1}^k) \tag{9.85}$$

and let $t_0^{k+1} = x_B^k = t_0^k + \alpha_{n+1}^* \Delta_{n+1}^k$. If $\|x_B^k - x_B^{k-1}\| \leq \epsilon$, stop; convergence to a minimum is assumed. Otherwise if

$$|\alpha_{n+1}^*| < \left[\frac{f(t_0^k) - f(t_0^{k+1})}{f(t_{m-1}^k) - f(t_m^k)}\right]^{1/2}, \tag{9.86}$$

set

$$\Delta_j^{k+1} = \Delta_j^k, \qquad j = 1, \ldots, n. \tag{9.87}$$

In other words, the search directions of the $(k + 1)$th stage remain the same as in the kth stage. If (9.86) does not hold, set

$$\Delta_j^{k+1} = \Delta_j^k, \qquad j = 1, \ldots, m - 1 \tag{9.88}$$

$$\Delta_j^{k+1} = \Delta_{j+1}^k, \qquad j = m, \ldots, n, \tag{9.89}$$

Proceed to stage $k + 1$.

Example 9.6.1

Let us illustrate Powell's modified method on the quadratic function of the previous example. The first steps are the same as the ones given there. We can see that the largest function decrease is obtained by going from t_1^1 to t_2^1; hence $m = 2$. We have

$$\Delta_4^1 = \begin{pmatrix} 0 \\ -\frac{2}{3} \\ -\frac{2}{9} \end{pmatrix} \tag{9.90}$$

and we find an α_4^* that minimizes $f(\frac{1}{2}, 1 - \frac{2}{3}\alpha_4, \frac{1}{2} - \frac{2}{9}\alpha_4)$. It is easy to compute that $\alpha_4^* = \frac{9}{8}$. Thus

$$t_0^2 = \begin{pmatrix} \frac{1}{2} \\ 1 \\ \frac{1}{2} \end{pmatrix} + \frac{9}{8}\begin{pmatrix} 0 \\ -\frac{2}{3} \\ -\frac{2}{9} \end{pmatrix} = \begin{pmatrix} \frac{1}{2} \\ \frac{1}{4} \\ \frac{1}{4} \end{pmatrix} \tag{9.91}$$

and $f(t_0^2) = \frac{1}{2}$. Now

$$\left[\frac{f(t_0^1) - f(t_0^2)}{f(t_1^1) - f(t_2^1)}\right]^{1/2} = \left(\frac{2 - \frac{1}{2}}{2 - \frac{2}{3}}\right)^{1/2} = \left(\frac{9}{8}\right)^{1/2}. \tag{9.92}$$

Since

$$\alpha_4^* > \left(\frac{9}{8}\right)^{1/2}, \tag{9.93}$$

we can see that (9.86) does not hold. Accordingly, the new directions will be

$$\Delta_1^2 = \begin{pmatrix} 1 \\ 0 \\ 0 \end{pmatrix}, \quad \Delta_2^2 = \begin{pmatrix} 0 \\ 0 \\ 1 \end{pmatrix}, \quad \Delta_3^2 = \begin{pmatrix} 0 \\ -\frac{2}{3} \\ -\frac{2}{9} \end{pmatrix} \tag{9.94}$$

and these vectors are linearly independent. The computation of the next stage is left for the reader. ▮

We now derive some results by which Powell's modified method can be explained. We have already seen that it is desirable to search for the minimum of a quadratic function along conjugate directions. First we show how the closeness to mutual conjugacy of vectors with respect to a positive definite symmetric matrix can be measured. Then we shall see that the search directions generated by Powell's modified method approach mutual conjugacy with respect to the Hessian matrix Q of a quadratic function whose minimum is sought and that these directions are linearly independent in each iteration. After having shown the efficiency of this method for quadratic functions, we shall prove convergence of a simplified version for a large family of functions.

We start by recalling the definition of orthogonal matrices and some of their properties.

A real matrix P such that $P^TP = PP^T = I$ (the identity matrix) is said to be an orthogonal matrix [8]. Some elementary properties of orthogonal matrices are that $P^{-1} = P^T$ and $|\det P| = 1$. Every row and every column of P are vectors of unit length (normalized). Moreover, P is orthogonal if and only if the rows and columns of P are mutually orthonormal. Given any vector x, let $y = Px$. Then $||y|| = ||x||$. Let x and y be arbitrary vectors and let $x^1 = Px$, $y^1 = Py$. Then $(x^1)^Ty^1 = x^TP^TPy = x^Ty$. Thus scalar products are invariant under orthogonal transformations. In particular, orthogonal vectors will be transformed into orthogonal vectors by a matrix P. For additional properties of orthogonal matrices and transformations, the reader is referred to standard textbooks of linear algebra. The next result, due to Powell [11], provides a suitable measure of the closeness of a set of search directions to mutual conjugacy by the value of a determinant. This result forms the basis for the modified version of Powell's method, as described above, as well as for a possible new algorithm to be outlined later in this section.

Theorem 9.3

Let G be a real $n \times n$ symmetric positive definite matrix and let $d^1, \ldots, d^n$ be n vectors normalized such that

$$(d^i)^TGd^i = 1, \qquad i = 1, \ldots, n. \tag{9.95}$$

Let D be the matrix whose columns are the vectors d^i. Then $|\det D|$ attains its maximum value if and only if the vectors d^i are mutually conjugate with respect to G.

Proof. We begin by showing that if (9.95) holds and $|\det D|$ is maximal, then the d^i must be conjugate with respect to G. Suppose that d^1 and d^2 are

normalized vectors that are not conjugate with respect to G; that is,

$$(d^1)^T G d^2 = c \neq 0. \tag{9.96}$$

Since G is positive definite, it has a "square root" matrix B; that is, $G = B^T B$. By the Cauchy-Schwarz inequality [8],

$$|(d^1)^T G d^2| = |(d^1)^T B^T B d^2| \leq [(Bd^1)^T Bd^1]^{1/2}[(Bd^2)^T Bd^2]^{1/2} \tag{9.97}$$

$$= [(d^1)^T G d^1]^{1/2}[(d^2)^T G d^2]^{1/2} = 1. \tag{9.98}$$

Hence $|c| \leq 1$ with equality holding if and only if

$$d^1 = \frac{d^2}{(d^1)^T G d^2}. \tag{9.99}$$

If (9.99) holds, $\det D = 0$, and, clearly, $|\det D|$ is not maximal. Suppose now that $0 < |c| < 1$. Define a new vector $\hat{d}^1$ by

$$\hat{d}^1 = \frac{d^1 - cd^2}{[1 - (c)^2]^{1/2}} \tag{9.100}$$

and a new matrix $\hat{D}$ by

$$\hat{D} = [\hat{d}^1, d^2, \ldots, d^n]. \tag{9.101}$$

It can easily be shown that $\hat{d}^1$ satisfies (9.95) and

$$|\det \hat{D}| = [1 - (c)^2]^{-1/2} |\det D| > |\det D|. \tag{9.102}$$

And so $|\det D|$ does not attain its maximum value.

Conversely, let $d^1, \ldots, d^n$ and $\bar{d}^1, \ldots, \bar{d}^n$ be two different sets of vectors, conjugate with respect to G and normalized by (9.95). Then there exists an $n \times n$ real matrix $P = [p_{ij}]$ such that

$$\bar{d}^i = \sum_{j=1}^{n} p_{ij} d^j, \qquad i = 1, \ldots, n. \tag{9.103}$$

Let $\bar{D}$ be the matrix whose columns are the vectors $\bar{d}^i$. Then

$$\det \bar{D} = \det D \det P. \tag{9.104}$$

Therefore we must show that $|\det P| = 1$.

Now

$$(\bar{d}^i)^T G \bar{d}^k = \sum_{j=1}^{n} \sum_{l=1}^{n} p_{ij} p_{kl} (d^j)^T G d^l \tag{9.105}$$

$$= \sum_{j=1}^{n} \sum_{l=1}^{n} p_{ij} p_{kl} \delta_{jl} = \sum_{j=1}^{n} p_{ij} p_{kj} = \delta_{ik}, \tag{9.106}$$

where $\delta_{rs} = 1$ if $r = s$ and $\delta_{rs} = 0$ if $r \neq s$. It follows from (9.106) that the rows of the matrix P constitute a set of n orthonormal vectors. Hence P is an orthogonal matrix, which implies that $|\det P| = 1$. ∎

Let us apply the last result to the modified method of Powell, whose aim in deriving this version was to avoid the possibility of linearly dependent search directions. Suppose that we have n linearly independent search directions $\Delta_1^k, \ldots, \Delta_n^k$ and also suppose that

$$\Delta_j^k = \mu_j^k d_j^k, \qquad j = 1, \ldots, n \tag{9.107}$$

so that the d_j^k satisfy (9.95). Hence

$$(\Delta_j^k)^T G \Delta_j^k = (\mu_j^k)^2. \tag{9.108}$$

Recall that

$$\Delta_{n+1}^k = (t_n^k - t_0^k) = \sum_{j=1}^{n} \theta_j^* \, \Delta_j^k = \mu_{n+1}^k \, d_{n+1}^k, \tag{9.109}$$

where μ_{n+1}^k and d_{n+1}^k were chosen so that d_{n+1}^k also satisfies the normalization condition (9.95). Let D^k be the matrix given by

$$D^k = [d_1^k, \ldots, d_n^k] \tag{9.110}$$

and denote by D^{k+1} the matrix obtained by replacing d_m^k, the mth column in D^k, by d_{n+1}^k. Now by (9.107) and (9.109),

$$d_{n+1}^k = \sum_{j=1}^{n} \frac{\theta_j^* \mu_j^k}{\mu_{n+1}^k} d_j^k. \tag{9.111}$$

Hence

$$D^{k+1} = \left[d_1^k, \ldots, \frac{\theta_1^* \mu_1^k d_1^k}{\mu_{n+1}^k} + \cdots + \frac{\theta_m^* \mu_m^k d_m^k}{\mu_{n+1}^k} + \cdots + \frac{\theta_n^* \mu_n^k d_n^k}{\mu_{n+1}^k}, \ldots, d_n^k \right]. \tag{9.112}$$

Recalling from linear algebra that the value of a determinant is unchanged if a multiple of one column is added to another column and that when multiplying the elements of a column of a matrix by a constant c, the determinant is also multiplied by c, we conclude from (9.112) that

$$\det D^{k+1} = \frac{\theta_m^* \mu_m^k}{\mu_{n+1}^k} \det D^k. \tag{9.113}$$

In order to approach a situation in which there are n mutually conjugate directions, Theorem 9.3 suggests that we choose the index m so that it

yields the largest increase in the absolute value of the ratio det D^{k+1}/det D^k; that is,

$$|\theta_m^* \mu_m^k| = \max_{j=1,\ldots,n} |\theta_j^* \mu_j^k|. \tag{9.114}$$

Note that if

$$\left|\frac{\theta_m^* \mu_m^k}{\mu_{n+1}^k}\right| < 1, \tag{9.115}$$

no increase in the value of $|\det D^{k+1}/\det D^k|$ is made if the mth search direction Δ_m^k is replaced by Δ_{n+1}^k. It can be shown that, for a quadratic function f as given by (9.31), we have

$$f(t_{j-1}^k) - f(t_j^k) = b^T(-\theta_j^* \Delta_j^k) + \tfrac{1}{2}(2t_j^k - \theta_j^* \Delta_j^k)^T Q(-\theta_j^* \Delta_j^k). \tag{9.116}$$

Since t_j^k was set so that $f(t_{j-1}^k + \theta_j \Delta_j^k)$ is minimized there, it follows that

$$(b + Qt_j^k)^T \Delta_j^k = 0, \tag{9.117}$$

and by (9.107) and (9.108), we obtain

$$f(t_{j-1}^k) - f(t_j^k) = \tfrac{1}{2}(\theta_j^* \mu_j^k)^2. \tag{9.118}$$

Thus the index m that maximizes the expression in (9.114) corresponds to that search direction of the kth stage in the algorithm that yields the largest function decrease, as given in (9.84). It is easy to show that

$$f(t_0^k) - f(t_0^{k+1}) = \tfrac{1}{2}(\alpha_{n+1}^* \mu_{n+1}^k)^2 \tag{9.119}$$

and the condition for not changing search directions, as given in (9.115), can be rewritten as

$$|\alpha_{n+1}^*| < \left[\frac{f(t_0^k) - f(t_0^{k+1})}{f(t_{m-1}^k) - f(t_m^k)}\right]^{1/2}, \tag{9.120}$$

which is precisely (9.86). It is clear that if we start with a set of linearly independent directions, the subsequent search directions will remain independent and will approach conjugacy with respect to the matrix Q in the sense of Powell's determinant measure.

If the function to be minimized is not quadratic, there is no fixed matrix G, so that we cannot normalize the directions d_j^k, and there is no guarantee that by condition (9.120), the value of $|\det D^k|$ will not decrease. Although the preceding modified Powell method is quite successful in minimizing general functions [14], the condition for ensuring linear independence of the search directions should be slightly altered. We can choose any fixed matrix

G for normalization, and the simplest choice is clearly $G = I$ (the identity matrix). Then we require for $k = 1, 2, \ldots$

$$(d_j^k)^T d_j^k = 1, \qquad j = 1, \ldots, n \tag{9.121}$$

and

$$\mu_j^k = ||\Delta_j^k||, \qquad j = 1, \ldots, n. \tag{9.122}$$

Equation (9.113) becomes

$$\det D^{k+1} = \frac{\theta_m^* ||\Delta_m^k||}{||\Delta_{n+1}^k||} \det D^k. \tag{9.123}$$

Assume now that

$$|\det D^k| \geq \epsilon, \tag{9.124}$$

where ϵ is a positive constant. Then the index m of the direction that is replaced by Δ_{n+1}^k should be such that

$$|\theta_m^*| \, ||\Delta_m^k|| = ||t_m^k - t_{m-1}^k|| = \max_{j=1,\ldots,n} ||t_j^k - t_{j-1}^k||, \tag{9.125}$$

and if

$$|\det D^{k+1}| = \frac{||t_m^k - t_{m-1}^k||}{||\Delta_{n+1}^k||} |\det D^k| < \epsilon, \tag{9.126}$$

then the search directions of stage $k + 1$ should remain the same as in stage k. The algorithm, due to Zangwill [17], can therefore be summarized as follows:

Given n linearly independent search directions $\Delta_1^1, \ldots, \Delta_n^1$, a starting point $x_B^0 \in R^n$, and a positive number ϵ such that

$$|\det D^1| \geq \epsilon, \tag{9.127}$$

the kth stage of the method then consists of the following steps:

1. Find t_j^k for $j = 1, \ldots, n + 1$ and x_B^k as in Powell's original method.
2. Find the index m for which $||t_j^k - t_{j-1}^k||$, $j = 1, \ldots, n$, is maximized.
3. If

$$|\det D^{k+1}| = \frac{||t_m^k - t_{m-1}^k||}{||\Delta_{n+1}^k||} |\det D^k| \geq \epsilon, \tag{9.128}$$

let

$$\Delta_j^{k+1} = \Delta_j^k, \qquad j = 1, \ldots, n, \qquad j \neq m \tag{9.129}$$

and

$$\Delta_m^{k+1} = \Delta_{n+1}^k. \tag{9.130}$$

If (9.128) does not hold, let

$$\Delta_j^{k+1} = \Delta_j^k, \qquad j = 1, \ldots, n \tag{9.131}$$

and

$$\det D^{k+1} = \det D^k. \tag{9.132}$$

Go to stage $k + 1$. Conditions for terminating the algorithm can be the same as in Powell's method.

We now prove that the last method converges to the minimum of continuously differentiable, strictly pseudoconvex functions. It should be noted that the collection of these functions is considerably larger than the family of quadratic functions with a positive definite Q matrix. The proof is based on Zangwill [17], who considered strictly convex functions, and on the work of Daniel [4].

First we have the following immediate result.

Lemma 9.4

The search directions $d_1^k, \ldots, d_n^k$ are linearly independent for all k.

Proof. For $k = 1$, the lemma is true by hypothesis. For $k = 2, 3, \ldots,$ we have $|\det D^k| \geq \epsilon$ by step 3 above. ■

In the following discussion let K, sometimes with a superscript (e.g., K^1), denote a subsequence of the natural numbers. The notation $K^1 \subset K$ means that K^1 is a subsequence of K. Suppose that we write $\{p^k\}$, $k \in K$, which means that we form a subsequence of the p^k by taking $k \in K$. Then $\{p^{k+1}\}$, $k \in K$, indicates the subsequence of the p^k formed by adding 1 to each $k \in K$.

Lemma 9.5

For any subsequence K there is a $K^1 \subset K$ such that

$$\lim_{k \in K^1} \{d_j^k\} = d_j^0, \qquad j = 1, \ldots, n. \tag{9.133}$$

The directions d_j^0 are linearly independent and $(d_j^0)^T d_j^0 = 1$ for $j = 1, \ldots, n$.

Proof. Since the d_j^k are normalized by (9.121), they are constrained in a compact set. Therefore there exists a $K^1 \subset K$ such that (9.133) holds and $(d_j^0)^T d_j^0 = 1$. Also, let

$$\lim_{k \in K^1} \{|\det D^k|\} = |\det [d_1^0, \ldots, d_n^0]| = |\det D^0|. \tag{9.134}$$

Since $|\det D^k| \geq \epsilon$ for all k, it follows that $|\det D^0| \geq \epsilon$ and the d_j^0 are linearly independent. ∎

Let us recall now the definitions of strongly quasiconvex and strictly pseudoconvex functions from Chapter 6. A real-valued function f on R^n is called strongly quasiconvex if

$$f(q_1 x^1 + q_2 x^2) < \max [f(x^1), f(x^2)] \tag{9.135}$$

for any two points $x^1 \neq x^2$ and for any $q_1 > 0$, $q_2 > 0$, $q_1 + q_2 = 1$. It is called strictly pseudoconvex if it is differentiable and

$$(x^1 - x^2)^T \nabla f(x^2) \geq 0 \quad \text{implies} \quad f(x^1) > f(x^2) \tag{9.136}$$

for any two points $x^1 \neq x^2$. It was also shown in Chapter 6 that a strongly quasiconvex function can attain its minimum on R^n (or over some convex subset of it) at no more than one point. In particular, this result holds for strictly pseudoconvex functions, for they are also strongly quasiconvex.

Theorem 9.6

Let f be a continuous strongly quasiconvex function on R^n that is to be minimized by the preceding method, starting with an arbitrary point x_B^0. Suppose that the level set

$$S(f, x_B^0) = \{x : x \in R^n, f(x) \leq f(x_B^0)\} \tag{9.137}$$

is bounded. Then any limit point t^0 of $\{t_j^k\}$ as $k \longrightarrow \infty$ is also a limit point of $\{t_l^k\}$, $l \neq j$, as $k \longrightarrow \infty$. For each such limit point there exist n linearly independent directions $d_1^0, \ldots, d_n^0$ such that

$$f(t^0) \leq f(t^0 + \theta_j d_j^0), \qquad j = 1, \ldots, n \tag{9.138}$$

for all θ_j.

Proof. For any subsequence K we can find a $K^1 \subset K$ such that

$$\lim_{k \in K^1} \{t_j^k\} = t_j^0, \qquad j = 0, \ldots, n. \tag{9.139}$$

We now show that $t_j^0 = t_{j+1}^0$. It follows from the algorithm that for all $k = 1, 2, \ldots$

$$f(t_{j+1}^k) \leq f(t_j^k), \qquad j = 0, \ldots, n-1 \tag{9.140}$$

and $f(t_0^{k+1}) \leq f(t_n^k)$.

Since the sequence $\{f(t_j^k)\}$ is monotonic,

$$\lim_{k \in K^1} \{f(t_{j+1}^k)\} = \lim_{k \in K^1} \{f(t_j^k)\}, \qquad j = 0, \ldots, n-1. \tag{9.141}$$

By continuity of f, we obtain

$$\lim_{k \in K^1} \{f(t_{j+1}^k)\} = f(t_{j+1}^0) \tag{9.142}$$

and

$$\lim_{k \in K^1} \{f(t_j^k)\} = f(t_j^0). \tag{9.143}$$

For all $\bar{\theta}_{j+1}$ we have

$$f(t_{j+1}^k) \leq f(t_j^k + \bar{\theta}_{j+1}\, d_{j+1}^k) \tag{9.144}$$

and again, by continuity, for all $\bar{\theta}_{j+1}$

$$f(t_{j+1}^0) = f(t_j^0) \leq f(t_j^0 + \bar{\theta}_{j+1}\, d_{j+1}^0), \qquad j = 0, \ldots, n-1. \tag{9.145}$$

It $t_{j+1}^0 \neq t_j^0$, then the function f is minimized along the direction d_{j+1}^0 at two distinct points. Since f is strongly quasiconvex, (9.135) implies

$$f(q_1 t_j^0 + q_2 t_{j+1}^0) < f(t_j^0) = f(t_{j+1}^0), \tag{9.146}$$

a contradiction. Hence $t_{j+1}^0 = t_j^0$. In fact,

$$t^0 = t_1^0 = t_2^0 = \cdots = t_n^0. \tag{9.147}$$

Consequently, for all θ_j

$$f(t^0) \leq f(t^0 + \theta_j d_j^0), \qquad j = 1, \ldots, n \tag{9.148}$$

and by Lemma 9.5, the d^0 are linearly independent. ■

Corollary 9.7

Suppose that the hypotheses of Theorem 9.6 hold and assume that f is also continuously differentiable. Then $\nabla f(t^0) = 0$ for any limit point t^0 of $\{t_j^k\}$.

Proof. From Theorem 9.6 we have for all θ_j

$$f(t^0) \leq f(t^0 + \theta_j d_j^0), \qquad j = 1, \ldots, n \tag{9.149}$$

or

$$f(t^0) \leq f(t^0) + \theta_j (d_j^0)^T \nabla f(t^0 + \lambda \theta_j d_j^0), \tag{9.150}$$

where $\lambda \in [0, 1]$. By a reasoning similar to that used in the proof of Theorem 2.3, we conclude that

$$(d_j^0)^T \nabla f(t^0) = 0, \qquad j = 1, \ldots, n. \tag{9.151}$$

Since the d_j^0 are linearly independent, we must have $\nabla f(t^0) = 0$. ■

It can be shown [4] that if, in addition to the hypotheses of the last corollary, the set $\{x : \nabla f(x) = 0\}$ contains no **continuum** (a closed set, not expressable as the union of two nonempty disjoint closed sets), then $\{t_j^k\}$ converges. Hence if $\nabla f(t^0) = 0$ implies that t^0 is the unique minimum of a continuously differentiable, strongly quasiconvex function, a condition clearly satisfied by a strictly pseudoconvex function, the convergence proof is complete. Zangwill [17] has further elaborated on this type of method and developed a similar algorithm with quadratic termination and convergence properties as above. Since Zangwill's algorithm is similar to the preceding ones described here and since, from a computational point of view, it does not seem to be superior [14], we refer the reader to [17] for further details.

9.7 FURTHER CONJUGATE DIRECTION-TYPE ALGORITHMS

Two additional algorithms for unconstrained minimization without using derivatives are given here. First we state and prove a result of Powell that deals with a novel method of approaching mutual conjugacy of search directions. This result may serve as a basis for a new minimization algorithm without derivatives. The second topic concerns a sequential-simultaneous "block"-type search method for multidimensional minimization. This method can be considered a generalization of Powell's first method for computers with parallel arithmetic processors.

The following result is due to Powell [12].

Theorem 9.8

Let $d^1, \ldots, d^n$ be any set of vectors satisfying the normalization condition (9.95) and let $P = [p_{ij}]$ be an orthogonal matrix. Define a new set of vectors by

$$\bar{d}^i = \sum_{j=1}^{n} p_{ij} d^j, \qquad i = 1, \ldots, n \tag{9.152}$$

and apply the normalization formula

$$\hat{d}^i = \frac{\bar{d}^i}{[(\bar{d}^i)^T G \bar{d}^i]^{1/2}}, \tag{9.153}$$

where $G = [g_{jk}]$ is a symmetric positive definite matrix. Let D and $\hat{D}$ be the matrices whose columns are the vectors d^i and $\hat{d}^i$, respectively. Then

$$|\det \hat{D}| \geq |\det D|. \tag{9.154}$$

Proof. From (9.152) we have $\bar{D} = DP^T$, where $\bar{D}$ is the matrix whose columns are the $\bar{d}^i$. Then

$$|\det \bar{D}| = |\det D||\det P| = |\det D| \tag{9.155}$$

and

$$|\det \hat{D}| = \frac{|\det \bar{D}|}{\Pi^{1/2}}, \tag{9.156}$$

where

$$\Pi = [(\bar{d}^1)^T G \bar{d}^1][(\bar{d}^2)^T G \bar{d}^2] \cdots [(\bar{d}^n)^T G \bar{d}^n]. \tag{9.157}$$

We must prove that $\Pi \leq 1$. Since G is positive definite, every term of Π is positive. Thus by the inequality of the arithmetic and geometric means,

$$\Pi \leq \left[\sum_{i=1}^{n} \frac{(\bar{d}^i)^T G \bar{d}^i}{n}\right]^n. \tag{9.158}$$

From (9.95), (9.152), and the orthogonality of P, we get

$$\sum_{i=1}^{n} (\bar{d}^i)^T G \bar{d}^i = \sum_{i=1}^{n} \sum_{j=1}^{n} \sum_{k=1}^{n} g_{jk} \bar{d}^i_j \bar{d}^i_k \tag{9.159}$$

$$= \sum_{i=1}^{n} \sum_{j=1}^{n} \sum_{k=1}^{n} \sum_{l=1}^{n} \sum_{m=1}^{n} g_{jk} p_{il} d^l_j p_{im} d^m_k \tag{9.160}$$

$$= \sum_{l=1}^{n} (d^l)^T G d^l = n. \tag{9.161}$$

■

The last theorem implies that if we change a current set of search directions $(d^1, \ldots, d^n)$ to a new set $(\hat{d}^1, \ldots, \hat{d}^n)$ by an orthogonal matrix P, and in accordance with (9.152) and (9.153), then the new directions will be at least as "close" to conjugacy, where "closeness" is measured by the value of the determinant at the current directions. Thus (9.152) and (9.153) can be used as a rule for updating search directions, provided that we have a suitable matrix G. It is interesting to note that, for $n = 2$, Powell found an orthogonal matrix

$$P_{(2)} = \begin{bmatrix} \cos\frac{\pi}{4} & \sin\frac{\pi}{4} \\ -\sin\frac{\pi}{4} & \cos\frac{\pi}{4} \end{bmatrix} = \begin{bmatrix} \frac{1}{\sqrt{2}} & \frac{1}{\sqrt{2}} \\ -\frac{1}{\sqrt{2}} & \frac{1}{\sqrt{2}} \end{bmatrix} \tag{9.162}$$

that transforms any two directions d^1, d^2, normalized by (9.95) through the use of a symmetric positive definite matrix G, into directions $\hat{d}^1$, $\hat{d}^2$, which are conjugate with respect to G. Note that $P_{(2)}$ is independent of the directions and the matrix involved.

Example 9.7.1

Let

$$G = \begin{bmatrix} 4 & 2 \\ 2 & 16 \end{bmatrix}, \quad d^1 = \begin{pmatrix} \frac{1}{2} \\ 0 \end{pmatrix}, \quad d^2 = \begin{pmatrix} 0 \\ \frac{1}{4} \end{pmatrix}. \tag{9.163}$$

The vectors d^1 and d^2 are normalized with respect to G, but they are not conjugate with respect to it. Now let $P_{(2)}$ be given by (9.162). Then by (9.152),

$$\bar{d}^1 = \frac{1}{\sqrt{2}} d^1 + \frac{1}{\sqrt{2}} d^2 = \begin{pmatrix} \dfrac{1}{2\sqrt{2}} \\ \dfrac{1}{4\sqrt{2}} \end{pmatrix} \tag{9.164}$$

$$d^2 = \frac{-1}{\sqrt{2}} d^1 + \frac{1}{\sqrt{2}} d^2 = \begin{pmatrix} -\dfrac{1}{2\sqrt{2}} \\ \dfrac{1}{4\sqrt{2}} \end{pmatrix} \tag{9.165}$$

and

$$(\bar{d}^1)^T G \bar{d}^1 = \tfrac{5}{4} \qquad (\bar{d}^2)^T G \bar{d}^2 = \tfrac{3}{4}. \tag{9.166}$$

And so, by (9.153),

$$\hat{d}^1 = \begin{pmatrix} \dfrac{1}{\sqrt{10}} \\ \dfrac{1}{2\sqrt{10}} \end{pmatrix}, \qquad \hat{d}^2 = \begin{pmatrix} -\dfrac{1}{\sqrt{6}} \\ \dfrac{1}{2\sqrt{6}} \end{pmatrix}, \tag{9.167}$$

which are both normalized and conjugate with respect to G. ▮

Powell [12] does not elaborate on the idea in the last theorem by precisely stating a new algorithm. We can, however, speculate on how such an algorithm could be based on Theorem 9.8.

The success of such an algorithm would seem to depend on estimating second derivatives and evaluating the denominator in (9.153). For a quadratic function, as given in (9.31), we clearly have $G = Q$. For a general function, G must be an estimate of the Hessian matrix. Suppose that we approximate

an arbitrary real function f by a quadratic function by using the Taylor expansion of f around a point $x^* \in R^n$:

$$f(x) \cong f(x^*) + (x - x^*)^T \nabla f(x^*) + \tfrac{1}{2}(x - x^*)^T \nabla^2 f(x^*)(x - x^*). \tag{9.168}$$

For a given point x and direction d, let x^* be our guess for the minimizing point of $f(x + \theta d)$; that is, $x^* = x + \theta^* d$. Then

$$d^T \nabla^2 f(x^*) d \cong \frac{2[f(x) - f(x^*)]}{(\theta^*)^2}. \tag{9.169}$$

This relation can be used in (9.153) in some iterative way. Here we may assume that if $\theta^* = 0$, then $d^T \nabla^2 f(x^*) d = 1$.

No numerical experience, based on the last theorem, is available as yet. We feel, however, that the result is important, and new methods based on it will eventually emerge.

As in the case of one-dimensional optimization, we conclude this chapter with a "block"-type search method that extends the algorithms of Powell [11] and Zangwill [17] by carrying out the one-dimensional optimizations simultaneously. The algorithm, due to Chazan and Miranker [3], was developed for use on computers with parallel arithmetic processors [6]. It actually generates a set of conjugate directions and has the quadratic termination property. The authors also prove convergence of the algorithm for strictly convex functions.

Let us describe this algorithm for the case of minimizing an unconstrained function f on R^n and then demonstrate it by an example. Let $u^1, \ldots, u^n$ be n linearly independent directions in R^n of unit length (usually the coordinate directions are taken), let $z_1^1, \ldots, z_{n-1}^1$ be $n - 1$ directions in R^n, and let an initial point $t_0^1 \in R^n$ be given. It is desirable to choose the z_j^1 to be linearly independent also. At the kth iteration, we have directions $z_1^k, \ldots, z_{n-1}^k$, a point t_0^k, and an integer r, $1 \leq r \leq n$. Let

$$z_n^k = u^r. \tag{9.170}$$

Define n points as follows:

$$p_j^k = t_0^k + \sum_{i=1}^{j} z_i^k, \qquad j = 1, \ldots, n. \tag{9.171}$$

Then perform n **simultaneous** minimizations from the points $p_1^k, \ldots, p_n^k$ in the common z_1^k direction. That is, find numbers $\theta_1^*, \ldots, \theta_n^*$ such that

$$f(p_j^k + \theta_j^* z_1^k) = \min_{\theta_j} f(p_j^k + \theta_j z_1^k), \qquad j = 1, \ldots, n. \tag{9.172}$$

The next step is to set

$$t_0^{k+1} = p_1^k + \theta_1^* z_1^k \tag{9.173}$$

$$z_j^{k+1} = z_{j+1}^k + (\theta_{j+1}^* - \theta_j^*) z_1^k, \qquad j = 1, \ldots, n-1. \tag{9.174}$$

Increase the index r by 1 if $r < n$ and set $r = 1$ otherwise. Go to iteration $k + 1$. A few steps of this method are presented in Figure 9.6.

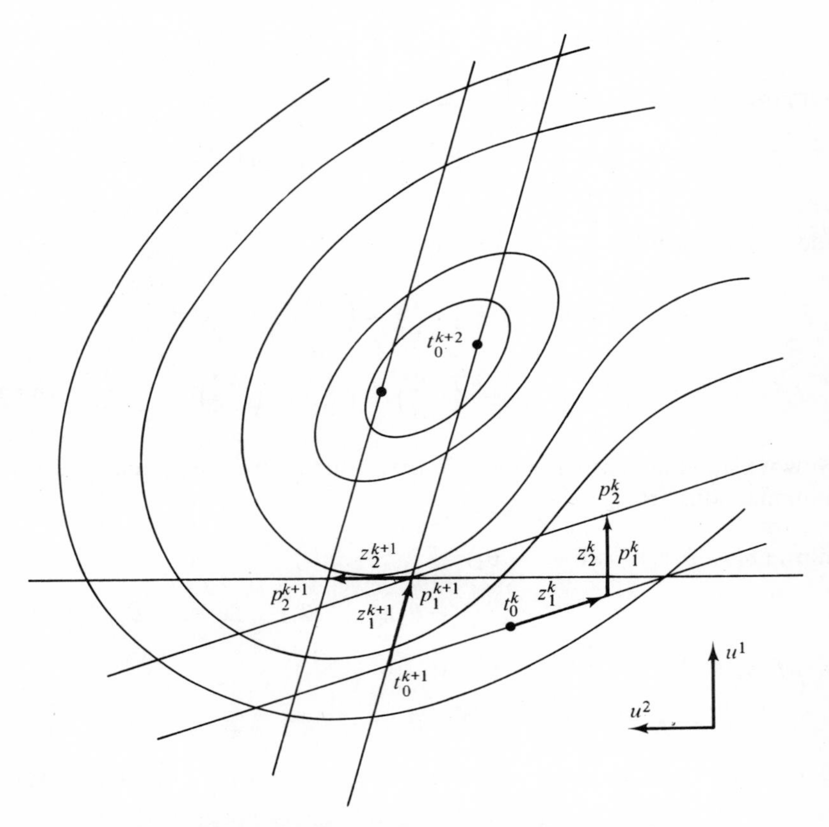

Fig. 9.6 Simultaneous-sequential nongradient algorithm.

The main purpose of this algorithm is to reduce computing time by the simultaneous line searches since the most time-consuming part of the Powell and related algorithms consists of the one-dimensional minimization steps. The number of line searches required for the minimization of a quadratic function in R^n is on the order of $(n)^2$. If each minimization takes one unit of time, then, because of the sequential nature of the algorithm, a total of $(n)^2$ units of time are required for quadratic termination. On the other

hand, the simultaneous-sequential algorithm of Chazan and Miranker, as described above, would need only n time units in order to perform the same task.

Example 9.7.2

To illustrate the Chazan-Miranker algorithm on the quadratic function of Example 9.5.1, find the minimum of

$$f(x) = \tfrac{3}{2}(x_1)^2 + \tfrac{1}{2}(x_2)^2 - x_1x_2 - 2x_1. \tag{9.175}$$

Suppose that we start at $t_0^1 = (-2, 4)$. Let

$$u^1 = \begin{pmatrix} 1 \\ 0 \end{pmatrix}, \quad u^2 = \begin{pmatrix} 0 \\ 1 \end{pmatrix}, \quad z_1^1 = \begin{pmatrix} -1 \\ 2 \end{pmatrix} \tag{9.176}$$

and $r = 1$. Then $z_2^1 = u^1$. By (9.171),

$$p_1^1 = \begin{pmatrix} -2 \\ 4 \end{pmatrix} + \begin{pmatrix} -1 \\ 2 \end{pmatrix} = \begin{pmatrix} -3 \\ 6 \end{pmatrix} \tag{9.177}$$

$$p_2^1 = \begin{pmatrix} -2 \\ 4 \end{pmatrix} + \begin{pmatrix} -1 \\ 2 \end{pmatrix} + \begin{pmatrix} 1 \\ 0 \end{pmatrix} = \begin{pmatrix} -2 \\ 6 \end{pmatrix}. \tag{9.178}$$

Now we simultaneously minimize in the z_1^1 direction from p_1^1 and p_2^1. The two minimizations are

$$\min_{\theta_1} f(p_1^1 + \theta_1 z_1^1) = \tfrac{3}{2}(-3 - \theta_1)^2 + \tfrac{1}{2}(6 + 2\theta_1)^2 - (-3 - \theta_1)(6 + 2\theta_1) - 2(-3 - \theta_1) \tag{9.179}$$

$$\min_{\theta_2} f(p_2^1 + \theta_2 z_1^1) = \tfrac{3}{2}(-2 - \theta_2)^2 + \tfrac{1}{2}(6 + 2\theta_2)^2 - (-2 - \theta_2)(6 + 2\theta_2) - 2(-2 - \theta_2) \tag{9.180}$$

and the optimal solutions are $\theta_1^* = -\tfrac{35}{11}$, $\theta_2^* = -\tfrac{30}{11}$, respectively. Hence

$$t_0^2 = \begin{pmatrix} -3 \\ 6 \end{pmatrix} - \frac{35}{11}\begin{pmatrix} -1 \\ 2 \end{pmatrix} = \begin{pmatrix} \frac{2}{11} \\ -\frac{4}{11} \end{pmatrix} \tag{9.181}$$

$$z_2^1 = \begin{pmatrix} 1 \\ 0 \end{pmatrix} + \left(-\frac{30}{11} + \frac{35}{11}\right)\begin{pmatrix} -1 \\ 2 \end{pmatrix} = \begin{pmatrix} \frac{6}{11} \\ \frac{10}{11} \end{pmatrix}. \tag{9.182}$$

Letting $r = 2$, we get $z_2^2 = u^2$. This concludes the first iteration. Continuing with the second iteration, we obtain

$$p_1^2 = \begin{pmatrix} \frac{2}{11} \\ -\frac{4}{11} \end{pmatrix} + \begin{pmatrix} \frac{6}{11} \\ \frac{10}{11} \end{pmatrix} = \begin{pmatrix} \frac{8}{11} \\ \frac{6}{11} \end{pmatrix} \tag{9.183}$$

$$p_2^2 = \begin{pmatrix} \frac{2}{11} \\ -\frac{4}{11} \end{pmatrix} + \begin{pmatrix} \frac{6}{11} \\ \frac{10}{11} \end{pmatrix} + \begin{pmatrix} 0 \\ 1 \end{pmatrix} = \begin{pmatrix} \frac{8}{11} \\ \frac{17}{11} \end{pmatrix}. \tag{9.184}$$

The two simultaneous minimizations are

$$\begin{aligned} \min_{\theta_1} f(p_1^2 + \theta_1 z_1^2) &= \frac{3}{2}\left(\frac{8}{11} + \frac{6}{11}\theta_1\right)^2 + \frac{1}{2}\left(\frac{6}{11} + \frac{10}{11}\theta_1\right)^2 \\ &\quad - \left(\frac{8}{11} + \frac{6}{11}\theta_1\right)\left(\frac{6}{11} + \frac{10}{11}\theta_1\right) - 2\left(\frac{8}{11} + \frac{6}{11}\theta_1\right) \end{aligned} \tag{9.185}$$

$$\begin{aligned} \min_{\theta_2} f(p_2^2 + \theta_2 z_1^2) &= \frac{3}{2}\left(\frac{8}{11} + \frac{6}{11}\theta_2\right)^2 + \frac{1}{2}\left(\frac{17}{11} + \frac{10}{11}\theta_2\right)^2 \\ &\quad - \left(\frac{8}{11} + \frac{6}{11}\theta_2\right)\left(\frac{17}{11} + \frac{10}{11}\theta_2\right) - 2\left(\frac{8}{11} + \frac{6}{11}\theta_2\right) \end{aligned} \tag{9.186}$$

with respective minima at $\theta_1^* = \frac{1}{2}$, $\theta_2^* = -\frac{17}{22}$. Consequently,

$$t_0^3 = \begin{pmatrix} \frac{8}{11} \\ \frac{6}{11} \end{pmatrix} + \left(\frac{1}{2}\right)\begin{pmatrix} \frac{6}{11} \\ \frac{10}{11} \end{pmatrix} = \begin{pmatrix} 1 \\ 1 \end{pmatrix}, \tag{9.187}$$

and we can see that the global optimum of f has indeed been reached in $n = 2$ iterations. The reader can verify that the directions z_1^1 and z_1^2 are conjugate with respect to the Q matrix of (9.175). ▮

Proof of quadratic termination of this algorithm will be left to the reader; convergence proof for more general functions can be found in [3].

EXERCISES

9.A. Perhaps the earliest unconstrained optimization technique in n variables is the so-called alternating directions or axial relaxation technique in which the minimum of a real function $f(x)$, $x \in R^n$, is sought by searching along the

coordinate directions in turn, starting at an arbitrary point. Draw a flow diagram of this method that employs a one-dimensional search technique from the previous chapter. Try out the algorithm on the following functions:

(i) $$f_1(x) = (x_1 - 3)^2 + 2(x_2 - 2)^2, \tag{9.188}$$

starting at $x^0 = (0, 0)$.

(ii) $$f_2(x) = (x_1)^2 + 106(x_2)^2 + 10x_1x_2 \tag{9.189}$$

starting at $x^0 = (4, 3)$.
Find numbers α and β such that the transformation

$$w_1 = x_1 + \alpha x_2, \qquad w_2 = \beta x_2 \tag{9.190}$$

yields for f_2

$$f_2(w) = (w_1)^2 + (w_2)^2. \tag{9.191}$$

Discuss the effects of such transformations on the efficiency of the alternating direction technique.

9.B. Apply the simplex method to minimizing the following functions:

(i) $$f_1(x) = (x_1 - 3)^2 + (x_2 - 2)^2 + (x_1 + x_2 - 4)^2, \tag{9.192}$$

starting with

$$x^0 = (0, 8), \quad x^1 = (0, 9), \quad x^2 = (1, 9) \tag{9.193}$$

and using $\alpha = 1$, $\beta = \frac{1}{2}$, $\gamma = 2$. Draw contours of this quadratic function and indicate on the drawing the progress of the algorithm.

(ii) $$f_2(x) = (-6 - x_1 - x_2)^2 + (2 - 3x_1 - 3x_2 - x_1x_2)^2, \tag{9.194}$$

starting with

$$x^0 = (-4, 6), \quad x^1 = (-4, 7), \quad x^2 = (-3, 6). \tag{9.195}$$

This function, having a minimum at $x^* = (0, 0)$, is a translated version of an objective function appearing in an electrical engineering optimization problem [10].
Apply the alternating directions technique to these two functions and compare the efficiencies of the two search methods on the functions tested.

9.C. Draw a flow diagram of the pattern search method and program it for a computer. Try out your program on the functions of the previous exercises. If you have no access to a computer, apply pattern search to the problem of minimizing

$$f(x) = (x_1)^2 + (x_2)^2 - 3x_1 - x_1x_2 + 3. \tag{9.196}$$

Start the search at $x_B^0 = (0, 0)$ and set $d_1 = 1$, $d_2 = 1$ initially. Complete a few exploratory and pattern stages. Draw contours of the function and illustrate the trajectory of the algorithm. Compare your results with Example 9.3.1.

9.D. Suppose that at the end of the kth exploratory stage of the rotating directions method, there are $p(> 0)$ directions for which the α_j are zero. Explain why the Gram-Schmidt orthogonalization method for finding new directions will fail in this case. One possible way to obtain new directions if $p(> 0)$ of the α_j are zero is as follows: Reorder the u^j such that the first $(n - p)$ are associated with the nonzero α_j and are defined as in (9.17). For $j = 1, \ldots, n - p$, perform the orthonormalization as suggested by Rosenbrock and obtain the new ζ^j, $j = 1, \ldots, n - p$. Define the remaining ζ^j to be the same as in the exploratory stage just completed. Show that the resulting n direction vectors are mutually orthonormal.

9.E. Suppose that at the last steps of the kth stage of the rotating directions method, applied to minimizing a function on R^2, the following results were listed:

Point Visited	Function Value
(1, 8)	25.0
(2, 9)	26.0
(0, 9)	22.5
(−0.5, 8.5)	21.0
(−3.5, 11.5)	24.8
(−2, 7)	23.6

What were $\zeta^{1,k}$ and $\zeta^{2,k}$? What will be the search directions $\zeta^{1,k+1}$, $\zeta^{2,k+1}$?

9.F. Apply the rotating directions method to the functions (9.192) and (9.196), starting at $x_B^0 = (0, 8)$ and (0, 0), respectively. First choose $\zeta^1 = (1, 0)$, $\zeta^2 = (0, 1)$. Let $\alpha = 3$, $\beta = 0.5$. Draw the trajectory of the algorithm and compare it with the simplex and pattern search methods.

9.G. Repeat Exercise 9.F with the Davies-Swann-Campey (DSC) method instead of Rosenbrock's rotating directions method.

9.H. Show that the formulas of (9.45) and (9.46) yield mutually conjugate directions that are linearly independent. Also show that if $z^1, \ldots z^k$, are nonzero vectors, mutually conjugate with respect to a symmetric positive definite matrix A, then they are linearly independent.

9.I. Find mutually conjugate directions with respect to the following matrices:

$$\text{(a)}\quad G_1 = \begin{bmatrix} 1 & 2 \\ 2 & 9 \end{bmatrix} \qquad \text{(b)}\quad G_2 = \begin{bmatrix} 2 & -2 & 0 \\ -2 & 3 & -1 \\ 0 & -1 & 6 \end{bmatrix}. \tag{9.197}$$

Suppose that z^1, z^2, and z^1, z^3 are conjugate directions with respect to a 3×3 positive definite symmetric matrix G. Will z^2, z^3 also be conjugate with respect to G?

9.J. Given a quadratic function f on R^n with a positive definite Q matrix, suppose that $\theta_1^* = \theta_2^* = \cdots = \theta_n^* = 0$, where the optimal step lengths θ_j^* are defined by

$$f(x^0 + \theta_j^* \Delta^j) = \min_{\theta_j} f(x^0 + \theta_j \Delta^j), \qquad j = 1, \ldots, n \tag{9.198}$$

and the Δ^j are orthogonal directions. Show that x^0 is the minimum of f. For which type of functions can you generalize this result?

9.K. Show that the directions Δ_3^1 and Δ_3^2 in Example 9.5.1 are conjugate with respect to the Q matrix that corresponds to the quadratic function in (9.68).

9.L. A computer program for minimizing real functions without derivatives was written to implement Powell's method as given in Section 9.5. It uses quadratic approximation line searches with extrapolation (without bracketing) and starting search directions parallel to the coordinate axes. The convergence criterion is satisfied if the difference between the endpoints of two consecutive iterations is less than some tolerance. In order to test the code, the problem

$$\min f_1(x) = 10(x_1)^2 + (x_2)^2 - 60x_1 - 8x_2 + 106 \tag{9.199}$$

was solved, starting from different points, by the preceding method. In all cases, the same optimal solution was printed out by the computer after six line searches. On the other hand, when the problem

$$\min f_2(x) = 3(x_1)^2 + (x_2)^2 - 2x_1x_2 - 4x_1 \tag{9.200}$$

was solved by the same method, the optimum was printed out only after nine line searches. Explain the difference between the number of line searches required in order to obtain the respective optima of these two problems. Assume absolute accuracy for the computer. Estimate the total number of function evaluations needed for solving these problems.

9.M. Continue Example 9.6.1 by performing two more stages. Compute the determinant of the search directions normalized by (9.95) at each stage and note if the directions approach conjugacy.

9.N. Compute three stages of Zangwill's algorithm as applied to minimizing the quadratic function of Example 9.6.1.

9.O. Let G be the 3×3 matrix in Exercise 9.I. Choose a 3×3 orthogonal matrix P and three directions in R^3. Using the matrices G and P, obtain a new set of directions that will be at least as close to mutual conjugacy (in the sense of Theorem 9.8) as the original directions. Repeat the computations by starting with a set of three directions that are mutually conjugate with respect to G.

9.P. Draw a flow diagram of the Chazan-Miranker method. Apply the algorithm to the quadratic function (9.188), starting at the origin, and plot its trajectory.

REFERENCES

1. Box, M. J., "A Comparison of Several Current Optimization Methods, and the Use of Transformations in Constrained Problems," *Computer J.*, **9**, 67–77 (1966).
2. Box, M. J., D. Davies, and W. H. Swann, *Non-Linear Optimization Techniques*, I.C.I. Monograph No. 5, Oliver & Boyd, Edinburgh, 1969.
3. Chazan, D., and W. L. Miranker, "A Nongradient and Parallel Algorithm for Unconstrained Minimization," *SIAM J. Control*, **8**, 207–217 (1970).
4. Daniel, J. W., *The Approximate Minimization of Functionals*, Prentice-Hall, Englewood Cliffs, N.J., 1971.
5. Hooke, R., and T. A. Jeeves, "Direct Search Solution of Numerical and Statistical Problems," *J. Assoc. Comp. Mach.*, **8**, 212–221 (1961).
6. Miranker, W. L., "A Survey of Parallelism in Numerical Analysis," *SIAM Review*, **13**, 524–547 (1971).
7. Nelder, J. A., and R. Mead, "A Simplex Method for Function Minimization," *Computer J.*, **7**, 308–313 (1965).
8. Noble, B., *Applied Linear Algebra*, Prentice-Hall, Englewood Cliffs, N.J., 1969.
9. Palmer, J. R., "An Improved Procedure for Orthogonalising the Search Vectors in Rosenbrock's and Swann's Direct Search Optimization Methods," *Computer J.*, **12**, 69–71 (1969).
10. Pierre, D. A., *Optimization Theory with Applications*, John Wiley & Sons, New York, 1969.
11. Powell, M. J. D., "An Efficient Method for Finding the Minimum of a Function of Several Variables without Calculating Derivatives," *Computer J.*, **7**, 155–162 (1964).
12. Powell, M. J. D., "Unconstrained Minimization Algorithms without Computation of Derivatives," Report T. P. 483, A.E.R.A. Harwell, United Kingdom, April 1972.
13. Rosenbrock, H. H., "An Automatic Method for Finding the Greatest or Least Value of a Function," *Computer J.*, **3**, 175–184 (1960).
14. Sargent, R. W. H., "Minimization without Constraints," in *Optimization and Design*, M. Avriel, M. J. Rijckaert, and D. J. Wilde (Eds.), Prentice-Hall, Englewood Cliffs, N.J., 1973.
15. Spendley, W., G. R. Hext, and F. R. Himsworth, "Sequential Application of

Simplex Designs in Optimisation and Evolutionary Operation," *Technometrics*, **4**, 441–461 (1962).

16. SWANN, W. H., "Report on the Development of a New Direct Search Method of Optimization," I.C.I. Central Instrument Lab. Research Note 64/3, 1964.
17. ZANGWILL, W. I., "Minimizing a Function without Calculating Derivatives", *Computer J.*, **10**, 293–296 (1967).

10 SECOND DERIVATIVE, STEEPEST DESCENT, AND CONJUGATE GRADIENT METHODS

In the first of two chapters on unconstrained minimization methods that use derivatives, we start our discussion with two classical algorithms—Newton's method, which is the multidimensional extension of the algorithm described in Section 8.1, and Cauchy's steepest descent method. Both methods are unsatisfactory for minimizing general nonlinear functions in many variables, since Newton's method, for example, may not converge to the minimum sought, and the convergence rate of the steepest descent method is very slow. They are important, however, for the understanding of their various modifications, which are considered the most successful methods in practice: After presenting the method of Goldstein and Price, which combines the preceding two classic methods, we turn to the method of conjugate gradients.

The underlying principle of the conjugate gradient method, which was originally developed for solving systems of linear equations, is that most general nonlinear functions can be reasonably well approximated in the neighborhood of a minimum by a quadratic function. Thus any efficient iterative minimization method should be especially efficient for minimizing quadratic functions. Here we return to the concept of conjugate directions and use it to derive the methods of conjugate gradients. The chapter concludes with a discussion of the convergence properties of this method. One of the interesting and not so well-understood features of the conjugate gradient method is that it exhibits most of the important properties of the more complex and sophisticated variable metric methods, which are considered the best unconstrained minimization algorithms. The latter will be discussed in the next chapter.

10.1 NEWTON-TYPE AND STEEPEST DESCENT METHODS

Suppose that the real-valued function f is continuously differentiable on R^n. From Chapter 2 we know that a necessary (and sometimes sufficient) condition for a minimum of f at some point $x^* \in R^n$ is

$$\nabla f(x^*) = 0. \tag{10.1}$$

If we have sufficient information about the function so that a solution of (10.1) is a minimum, then we can solve the above system of n generally nonlinear equations in n variables. Just as in Chapter 8, where we discussed the case of $n = 1$, the classic method here for finding a solution of (10.1) is also Newton's method for solving systems of equations. In order to apply Newton's method, we must assume that f is at least twice continuously differentiable. That is, in addition to the gradient vector $\nabla f(x)$, we also need the $n \times n$ Hessian matrix $\nabla^2 f(x)$ at every point $x \in R^n$. We expand (linearize) each component of ∇f around a point x^k and set the linear functions equal to zero:

$$\frac{\partial f(x^k)}{\partial x_j} + \sum_{i=1}^{n} \frac{\partial^2 f(x^k)}{\partial x_i \, \partial x_j}(x_i - x_i^k) = 0, \qquad j = 1, \ldots, n \tag{10.2}$$

or in vector notation

$$\nabla f(x^k) + \nabla^2 f(x^k)(x - x^k) = 0. \tag{10.3}$$

Assuming that $\nabla^2 f(x^k)$ is nonsingular, we can solve the preceding system of linear equations for x. Letting x^{k+1} be a solution, we get

$$x^{k+1} = x^k - [\nabla^2 f(x^k)]^{-1} \nabla f(x^k). \tag{10.4}$$

Newton's method consists of using (10.4) iteratively. At each iteration, the gradient of f and the inverse of the Hessian at x^k must be evaluated. In practice, we do not invert the Hessian matrix directly but solve the system of linear equations (10.2). Under suitable assumptions, similar to those presented for the one-dimensional case, Newton's method converges to a solution of (10.1). For a thorough discussion of convergence, the reader is referred to Goldstein [15], Ortega and Rheinboldt [25], or Ostrowski [27].

If there are a large number of variables, the function and derivative evaluations, and especially the matrix inversion process, even for a nicely behaving function that has a positive definite Hessian matrix everywhere, are time-consuming and expensive operations. The situation becomes worse if the Hessian approaches singularity. Another difficulty that frequently

arises with Newton's method as applied to general functions is that x^{k+1}, obtained from (10.4), does not ensure a decrease in the function value at each iteration. In other words, it is posible that $f(x^{k+1}) > f(x^k)$, and the sequence of points generated by iterative applications of (10.4) may converge to a saddlepoint or maximum of f.

There is another way of looking at Newton's method. Suppose that we make a quadratic approximation of f at some point x by expanding it around x^k in a Taylor series:

$$f(x) \cong f(x^k) + (x - x^k)^T \nabla f(x^k) + \tfrac{1}{2}(x - x^k)^T \nabla^2 f(x^k)(x - x^k). \qquad (10.5)$$

The point x^{k+1}, given by (10.4), is then the minimum of this quadratic approximation, provided that $\nabla^2 f(x^k)$ is positive definite. Most difficulties with Newton's method occur if $\nabla^2 f$ is not positive definite at x^k or if the point x^{k+1}, obtained by (10.4), is not close to x^k, so that the quadratic approximation of f around x^k is not valid at x^{k+1}. Partial remedies have been proposed. We can, for example, replace (10.4) by a **limited-step Newton formula**

$$x^{k+1} = x^k - \theta[\nabla^2 f(x^k)]^{-1} \nabla f(x^k), \qquad (10.6)$$

where θ is chosen so that $f(x^{k+1}) < f(x^k)$, and, in some versions of the algorithm, it is chosen to minimize $f(x)$ along the Newton direction. This step can be carried out by a line search from Chapter 8.

An interesting improvement over Newton's method for minimizing a general function has been suggested by Goldfeld, Quandt, and Trotter [14]. Their algorithm is based on minimizing, at the kth iteration, the quadratic approximation of f as given in (10.5) on a closed ball centered at x^k. The size of this ball is adjusted by the algorithm and is taken to be as large as possible, provided that the quadratic approximation of f in the ball remains reasonably good. This algorithm has been subsequently improved by minimizing the quadratic approximation over an ellipsoid instead of a spherical region. The shape and direction of the ellipsoid are modified at each iteration.

Based on a limited number of small test problems, the authors report that their algorithm is very robust, with an overall performance comparable to Powell's method without derivatives (Chapter 9). It seems, however, that (as in the original Newton method) the efficiency of this algorithm also rapidly decreases with an increase in the number of variables. The main reason is the need to invert the Hessian matrix or, equivalently, to solve a system of linear equations at each iteration.

In one case, Newton's method becomes extremely simple and converges in one iteration. This situation occurs if f is a quadratic function given by

$$f(x) = a + b^T x + \tfrac{1}{2} x^T Q x, \qquad (10.7)$$

where Q is positive definite. Note that $\nabla^2 f(x) = Q$; that is, the Hessian is a

constant matrix. Let x^0 be an arbitrary point in R^n and let x^* be the minimum of f. Then

$$\nabla f(x^0) = b + Qx^0 \tag{10.8}$$

and

$$0 = b + Qx^*. \tag{10.9}$$

Combining these equations, we get

$$x^* = x^0 - Q^{-1}\nabla f(x^0). \tag{10.10}$$

Comparing (10.10) with (10.4), it can be seen that Newton's method, starting from any point x^0, arrives at the minimum of such a quadratic function in just one step.

Example 10.1.1

Consider again the quadratic function $f(x) = (x_1 - x_2 + x_3)^2 + (-x_1 + x_2 + x_3)^2 + (x_1 + x_2 - x_3)^2$ used in Example 9.5.2. By simple algebraic manipulations, it can be transformed into the form

$$f(x) = \tfrac{1}{2}x^T Qx, \qquad x \in R^3 \tag{10.11}$$

where Q and its inverse Q^{-1} are given by

$$Q = \begin{bmatrix} 6 & -2 & -2 \\ -2 & 6 & -2 \\ -2 & -2 & 6 \end{bmatrix} \qquad Q^{-1} = \begin{bmatrix} \frac{1}{4} & \frac{1}{8} & \frac{1}{8} \\ \frac{1}{8} & \frac{1}{4} & \frac{1}{8} \\ \frac{1}{8} & \frac{1}{8} & \frac{1}{4} \end{bmatrix}. \tag{10.12}$$

Let $x^0 = (\frac{1}{2}, 1, \frac{1}{2})$. Then

$$\nabla f(x^0) = Qx^0 = \begin{bmatrix} 6 & -2 & -2 \\ -2 & 6 & -2 \\ -2 & -2 & 6 \end{bmatrix} \begin{pmatrix} \frac{1}{2} \\ 1 \\ \frac{1}{2} \end{pmatrix} = \begin{pmatrix} 0 \\ 4 \\ 0 \end{pmatrix} \tag{10.13}$$

and by (10.10) we get the exact minimum

$$x^* = \begin{pmatrix} \frac{1}{2} \\ 1 \\ \frac{1}{2} \end{pmatrix} - \begin{bmatrix} \frac{1}{4} & \frac{1}{8} & \frac{1}{8} \\ \frac{1}{8} & \frac{1}{4} & \frac{1}{8} \\ \frac{1}{8} & \frac{1}{8} & \frac{1}{4} \end{bmatrix} \begin{pmatrix} 0 \\ 4 \\ 0 \end{pmatrix} = \begin{pmatrix} 0 \\ 0 \\ 0 \end{pmatrix}. \tag{10.14}$$

∎

Newton's method uses an iterative scheme to find a point where the gradient vector of the function whose minimum is sought vanishes. A related use of the gradient vector involves finding directions along which a function

decrease occurs. In Chapter 4 we defined the directional derivative of a differentiable function f as

$$Df(x^0; y) = y^T\nabla f(x^0) = \lim_{t\to 0} \frac{f(x^0 + ty) - f(x^0)}{t}. \tag{10.15}$$

Consider now all the vectors (directions) $y \in R^n$ such that, for a given point $x^0 \in R^n$, we have

$$y^T\nabla f(x^0) < 0. \tag{10.16}$$

Then it follows from (10.15) that, for sufficiently small positive t, we obtain

$$f(x^0 + ty) < f(x^0). \tag{10.17}$$

In other words, if we are seeking the minimum of f on R^n and at some point $x^0 \in R^n$ the gradient of f does not vanish, then a sufficiently small move in a direction y that satisfies (10.16) will result in a function decrease. The directional derivative $Df(x^0; y)$ actually measures the instantaneous increase (if $Df(x^0; y) > 0$) or decrease (if $Df(x^0; y) < 0$) in the value of f at x^0 along the direction y. We can, therefore, seek among all directions y having some bounded length, say $\|y\| \leq 1$, that particular direction that yields the steepest descent in the value of f at a given point x^0 for which $\nabla f(x^0) \neq 0$. We have then the nonlinear programming problem

$$\min_y \mathrm{y}^T\nabla f(x^0) = \sum_{j=1}^{n} \frac{\partial f(x^0)}{\partial x_j} y_j \tag{10.18}$$

subject to

$$\|y\| = \left\{\sum_{j=1}^{n} (y_j)^2\right\}^{1/2} \leq 1. \tag{10.19}$$

The reader will be asked to show that the optimal solution of this problem is

$$y^* = \frac{-\nabla f(x^0)}{\|\nabla f(x^0)\|}. \tag{10.20}$$

The steepest descent in the function value is thus in the direction of the negative gradient. The **method of steepest descent**, first derived by Cauchy [3], can be described as follows: Given a point $x^0 \in R^n$, compute, for $k = 0, 1, \ldots$, the sequence of points

$$x^{k+1} = x^k - \theta_k^* \nabla f(x^k), \tag{10.21}$$

where $\theta_k^* > 0$ satisfies

$$f(x^k - \theta_k^*\nabla f(x^k)) = \min_{\theta_k \geq 0} f(x^k - \theta_k \nabla f(x^k)). \tag{10.22}$$

In Cauchy's steepest descent method, the global minimum of f is found along the negative gradient direction. Curry [6] has modified this method by choosing θ_k^* to be the **first** stationary point of f along the direction $-\nabla f(x^k)$; that is, that stationary point for which θ_k has the least positive value.

The steepest descent direction vector, as given by (10.20), is not invariant under a change in the norm constraint (10.19). Suppose that, for example, instead of the Euclidean norm condition, we require

$$y^T A y \leq 1, \tag{10.23}$$

where A is a symmetric positive definite matrix and we minimize $y^T \nabla f(x^0)$ subject to (10.23). The reader will also be asked to show that the optimal y, in this case, is given by

$$y^A = \frac{-A^{-1}\nabla f(x^0)}{[(\nabla f(x^0))^T A^{-1} \nabla f(x^0)]^{1/2}}. \tag{10.24}$$

If we let $A = \nabla^2 f(x^k)$ at each iteration, provided that the Hessian is positive definite, we get an interesting analogy between the steepest descent method that uses non-Euclidean norms and Newton's method as described earlier in this section. We shall, however, refer in the sequel to the steepest descent method that uses the Euclidean norm only. Greenstadt [17] studied the effects of using different A matrices on the relative efficiencies (defined below) of gradient methods as applied to minimizing quadratic functions. From such a study we can also try to draw conclusions about nonquadratic functions.

Suppose that we choose a search direction y^A from x^0 by (10.24) and minimize a quadratic function f along this direction. Let this optimal point be denoted by x^A. It can be shown that

$$f(x^0) - f(x^A) = \frac{[(y^A)^T \nabla f(x^0)]^2}{2(y^A)^T \nabla^2 f(x^0) y^A} = \frac{[(\nabla f(x^0))^T A^{-1} \nabla f(x^0)]^2}{2(\nabla f(x^0))^T A^{-1} \nabla^2 f(x^0) A^{-1} \nabla f(x^0)}. \tag{10.25}$$

Letting $A = \nabla^2 f(x^0)$, the search direction becomes that of Newton's method, denoted by y^N. Minimizing in the Newton direction, we arrive at a point x^N, and

$$f(x^0) - f(x^N) = \tfrac{1}{2}(\nabla f(x^0))^T [\nabla^2 f(x^0)]^{-1} \nabla f(x^0). \tag{10.26}$$

Defining the relative efficiency η by

$$\eta = \frac{f(x^0) - f(x^A)}{f(x^0) - f(x^N)} \tag{10.27}$$

and the matrix M by

$$M = A^{-1/2} \nabla^2 f(x^0) A^{-1/2}, \tag{10.28}$$

where $A^{-1/2}$ is the symmetric positive definite "square root" of A^{-1}—that is, $A^{-1/2} A^{-1/2} = A^{-1}$—we obtain

$$\frac{4\mu}{(1+\mu)^2} \leq \eta \leq 1, \tag{10.29}$$

where μ is the ratio of the largest to the smallest eigenvalue of M. It is also called the condition number of M. The upper bound of unity on η expresses the previously established fact that Newton's method, starting from any x^0 arrives at the minimum of a quadratic function with a positive definite Hessian matrix in a single step.

The inequalities of (10.29) can also shed some light on the poor performance of the steepest descent method ($A = I$) when applied to a function having elongated "valleys." For such functions, the condition number of M will be high and the progress made by the steepest descent method will be very slow. This phenomenon is called hemstitching (see Figure 10.1).

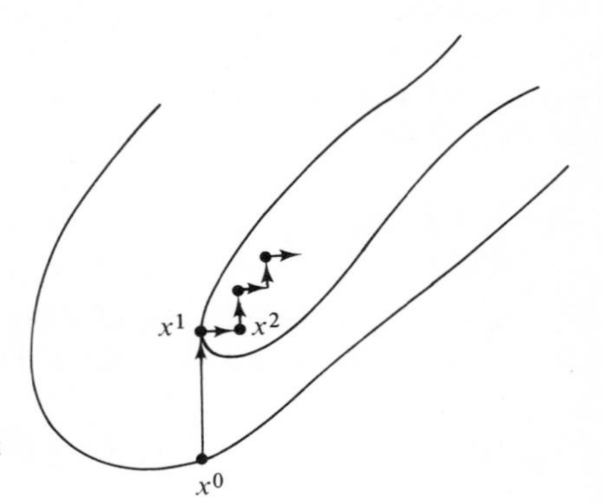

Fig. 10.1 Hemstitching by the steepest descent method.

Let us see what can be said about the convergence of the steepest descent method. We remind the reader that $\hat{x} \in R^n$ is called a cluster point of the sequence $\{x^k\} \subset R^n$ if for each positive number ϵ there exists a natural number $K(\epsilon)$, such that if $k \geq K(\epsilon)$ then $x^k \in N_\epsilon(\hat{x})$, where

$$N_\epsilon(\hat{x}) = \{x : ||x - \hat{x}|| < \epsilon\}. \tag{10.30}$$

We can state and prove now

Theorem 10.1 [15]

Let f be a real-valued function on R^n. Given $x^0 \in R^n$, $f(x^0) = \alpha$, assume that the level set

$$S(f, \alpha) = \{x : x \in R^n, f(x) \leq \alpha\} \tag{10.31}$$

is bounded and f is continuously differentiable on the convex hull of $S(f, \alpha)$. Let $\{x^k\}$ be the sequence of points generated by the steepest descent method. Then every cluster point $\hat{x}$ of $\{x^k\}$ satisfies $\nabla f(\hat{x}) = 0$.

Proof. Since $S(f, \alpha)$ is bounded, the Bolzano-Weierstrass theorem [2] implies that there exists at least one cluster point of $\{x^k\}$. Let $\hat{x}$ be such a cluster point and suppose that $\nabla f(\hat{x}) \neq 0$. Then there exists a $\hat{\theta}^* > 0$ minimizing $f(\hat{x} - \hat{\theta}\nabla f(\hat{x}))$; that is, there exists a $\delta > 0$ such that $f(\hat{x}) = f(\hat{x} - \hat{\theta}^*\nabla f(\hat{x})) + \delta$. Thus $\hat{x} - \hat{\theta}^*\nabla f(\hat{x})$ is an interior point of $S(f, \alpha)$.

For any x^k we can write, by the Mean Value Theorem,

$$f(x^k - \hat{\theta}^*\nabla f(x^k)) = f(\hat{x} - \hat{\theta}^*\nabla f(\hat{x})) + (\nabla f(\xi^k))^T[x^k - \hat{x} + \hat{\theta}^*(\nabla f(\hat{x}) - \nabla f(x^k))], \tag{10.32}$$

where $\xi^k = \hat{x} - \hat{\theta}^*\nabla f(\hat{x}) + \lambda[(x^k - \hat{x}) - \hat{\theta}^* (\nabla f(x^k) - \nabla f(\hat{x}))]$ for some $0 < \lambda < 1$. It also follows from the Bolzano-Weierstrass theorem that there is a subsequence $\{x^{k_m}\}$ of $\{x^k\}$ that converges to $\hat{x}$. It can be seen then that $\{\nabla f(\xi^{k_m})\}$ converges to $\nabla f(\hat{x} - \hat{\theta}^*\nabla f(\hat{x}))$ and $\{(x^{k_m} - \hat{x}) - \hat{\theta}^*(\nabla f(x^{k_m}) - \nabla f(\hat{x}))\}$ converges to 0. For sufficiently large k_m, the vector ξ^{k_m} belongs to the convex hull of $S(f, \alpha)$ and

$$f(x^{k_m} - \hat{\theta}^*\nabla f(x^{k_m})) \leq f(\hat{x} - \hat{\theta}^*\nabla f(\hat{x})) + \frac{\delta}{2} = f(\hat{x}) - \frac{\delta}{2}. \tag{10.33}$$

Let $\theta^*_{k_m}$ be the minimizing point of $f(x^{k_m} - \theta_{k_m}\nabla f(x^{k_m}))$. Since $\{f(x^{k_m})\}$ is monotone decreasing and converges to $f(\hat{x})$, we have

$$f(\hat{x}) < f(x^{k_m} - \theta^*_{k_m}\nabla f(x^{k_m})) \leq f(x^{k_m} - \hat{\theta}^*\nabla f(x^{k_m})) \leq f(\hat{x}) - \frac{\delta}{2}, \tag{10.34}$$

a contradiction. Hence we must have $\nabla f(\hat{x}) = 0$. ∎

This theorem can be generalized [28] by replacing (10.21) in the steepest descent method by the formula

$$x^{k+1} = x^k - \bar{\theta}_k B(x^k)\nabla f(x^k), \tag{10.35}$$

where $B(x)$ is an $n \times n$ positive definite matrix whose elements are continuous functions of x and $\bar{\theta}_k \geq 0$ is a number found by the Curry rule mentioned above. Thus "steepest descent" methods using non-Euclidean norms also converge under the hypotheses of Theorem 10.1. As noted, computational experience with the steepest descent method has generally been disappointing, and convergence, except in the case of functions having nearly spherical contours, was slow. Theoretical studies by Akaike [1], Forsythe [13], and Kantorovich and Akilov [19] show that the convergence rate of the steepest

descent method for quadratic functions is linear, and usually a higher rate for nonquadratic functions cannot be expected.

More specifically, we can assume without loss of generality that the quadratic function f is given by $f(x) = \frac{1}{2}x^T Qx$, where Q is an $n \times n$ symmetric positive definite matrix whose smallest and largest eigenvalues are λ_1 and λ_n, respectively. The global minimum of this function is, of course, at the origin. It was shown by Forsythe [13] that, starting from any $x^0 \in R^n$, the steepest descent method generates points such that there exist constants c_1 and c_2, depending on x^0, that satisfy

$$0 < c_1 \leq \frac{f(x^{k+1})}{f(x^k)} \leq c_2 \leq \left(\frac{\lambda_n - \lambda_1}{\lambda_n + \lambda_1}\right)^2 < 1, \qquad k = 0, 1, \ldots. \tag{10.36}$$

This slow rate of convergence can be heuristically explained by looking at consecutive search directions. Suppose that at iteration k the point x^k was reached, where $\nabla f(x^k) \neq 0$. Then we minimize the function $f(x^k - \theta_k \nabla f(x^k))$ with respect to θ_k and set $x^{k+1} = x^k - \theta_k^* \nabla f(x^k)$. By simple differentiation, we can show that $(\nabla f(x^k))^T \nabla f(x^{k+1}) = 0$; that is, consecutive search directions are orthogonal. In R^2, this means that we search only along directions that are parallel to the first two directions. The algorithm actually behaves in a similar way in more than two dimensions, and the search directions asymptotically approach two fixed vectors [13].

We conclude this section by describing a minimization algorithm, due to Goldstein and Price [16], which can be considered as a method that combines Newton's method with the steepest descent method and eliminates some undesirable features of both techniques. In the following discussion let δ and r be positive numbers with $\delta < \frac{1}{2}$. Let x^0 be an arbitrary point in R^n and let I_i be the ith column of the $n \times n$ identity matrix I. Also let f be a real function, defined at least on the level set $S(f, \alpha)$, given by (10.31). The kth iteration of the Goldstein-Price algorithm for minimizing f consists of the following computations. Evaluate the matrix $Q(x^k)$ whose ith column is given by

$$Q_i(x^k) = \frac{\nabla f(x^k + \beta_k I_i) - \nabla f(x^k)}{\beta_k}, \qquad i = 1, \ldots, n \tag{10.37}$$

where $\beta_0 = r$ and

$$\beta_k = r\,||\phi(x^{k-1})||, \qquad k = 1, 2, \ldots. \tag{10.38}$$

Here

$$\phi(x^k) = \nabla f(x^k) \tag{10.39}$$

for $k = 0$ or if $Q(x^k)$ is singular or if

$$(\nabla f(x^k))^T [Q(x^k)]^{-1} \nabla f(x^k) \leq 0. \tag{10.40}$$

Otherwise let

$$\phi(x^k) = [Q(x^k)]^{-1}\nabla f(x^k). \tag{10.41}$$

Define

$$g(x^k, \theta) = \frac{f(x^k) - f(x^k - \theta\phi(x^k))}{\theta(\nabla f(x^k))^T\phi(x^k)}. \tag{10.42}$$

If $g(x^k, 1) < \delta$, let θ_k satisfy

$$\delta \leq g(x^k, \theta_k) \leq 1 - \delta; \tag{10.43}$$

otherwise set $\theta_k = 1$. Let the next point x^{k+1} be given by

$$x^{k+1} = x^k - \theta_k\phi(x^k). \tag{10.44}$$

The iterations terminate if $||\nabla f(x^k)|| < \epsilon$, where ϵ is some predetermined small positive number.

Before illustrating this algorithm on a numerical example, a few remarks are in order. The matrix $Q(x^k)$ is an approximation to $\nabla^2 f(x^k)$, the Hessian matrix of f at x^k. The approximation is made by using only gradient information, and no second derivatives need be computed. The search direction in the algorithm is either the steepest descent $-\nabla f(x^k)$ or an approximate Newton direction $-[Q(x^k)]^{-1}\nabla f(x^k)$. The steepest descent direction is used at the first iteration or if $Q(x^k)$ is singular, so that it cannot be inverted (in practice, we choose the steepest descent direction whenever $Q(x^k)$ is close to singularity), or if the approximate Newton direction does not cause an instantaneous function decrease—that is, if

$$-(\nabla f(x^k))^T[Q(x^k)]^{-1}\nabla f(x^k) \geq 0. \tag{10.45}$$

If we obtain $g(x^k, 1) < \delta$ by taking $\theta_k = 1$ (the "full Newton step"), which is the case, for example, if $f(x^k - \phi(x^k)) > f(x^k)$, then a smaller step length is chosen so that $g(x^k, \theta_k)$ becomes positive and bounded away from zero. The upper and lower bounds in (10.43) can best be understood by considering a quadratic function f with a positive definite Hessian matrix. In this case, $Q(x^k)$ becomes a constant matrix, the exact Hessian of f. After the first move in the negative gradient direction, the next direction will be according to Newton's method; that is, $\phi(x^1)$ is given by (10.41). Expanding $g(x^1, \theta)$ in a Taylor series around x^1, we get, after some simple algebraic manipulations,

$$g(x^1, \theta) = 1 - \frac{\theta}{2} \tag{10.46}$$

and the full Newton step of $\theta = 1$ will satisfy (10.43).

Example 10.1.2

Consider the minimization of the quadratic function

$$f(x) = \tfrac{3}{2}(x_1)^2 + \tfrac{1}{2}(x_2)^2 - x_1 x_2 - 2x_1 \tag{10.47}$$

by the Goldstein-Price algorithm. This function was also minimized by Powell's method in Example 9.5.1. Let $r = \frac{1}{2}$, $\delta = \frac{1}{8}$ and suppose that we start at $x^0 = (-2, 4)$. The gradient of f is given by

$$\nabla f(x) = \begin{pmatrix} 3x_1 - x_2 - 2 \\ x_2 - x_1 \end{pmatrix} \tag{10.48}$$

and

$$\nabla f(x^0) = \begin{pmatrix} -12 \\ 6 \end{pmatrix}. \tag{10.49}$$

The columns of the matrix $Q(x^0)$ are given by

$$Q_1(x^0) = \begin{pmatrix} 3 \\ -1 \end{pmatrix} \qquad Q_2(x^0) = \begin{pmatrix} -1 \\ 1 \end{pmatrix}. \tag{10.50}$$

Note that since f is quadratic, the matrix Q is exactly the (constant) Hessian of f. The first search direction is given by $\phi(x^0) = \nabla f(x^0)$. Now compute $g(x^0, 1)$:

$$g(x^0, 1) = \frac{f(x^0) - f(x^0 - \nabla f(x^0))}{(\nabla f(x^0))^T \nabla f(x^0)} \tag{10.51}$$

$$= \frac{26 - 152}{180} = -\frac{126}{180} < \frac{1}{8}. \tag{10.52}$$

Let us find a θ_0 that satisfies (10.43). Trying $\theta_0 = \frac{1}{10}$, we get

$$g(x^0, \tfrac{1}{10}) = \frac{f(x^0) - f(x^0 - (\frac{1}{10})\nabla f(x^0))}{(\frac{1}{10})(\nabla f(x^0))^T \nabla f(x^0)} \tag{10.53}$$

$$= \frac{26 - 11.06}{18} = \frac{14.94}{18} \tag{10.54}$$

and

$$\frac{1}{8} < \frac{14.94}{18} < \frac{7}{8}. \tag{10.55}$$

Hence $x^1 = (-0.8, 3.4)$. Noting that $Q(x^1) = Q(x^0) = Q$, we compute

$$Q^{-1} = \begin{bmatrix} \frac{1}{2} & \frac{1}{2} \\ \frac{1}{2} & \frac{3}{2} \end{bmatrix}. \tag{10.56}$$

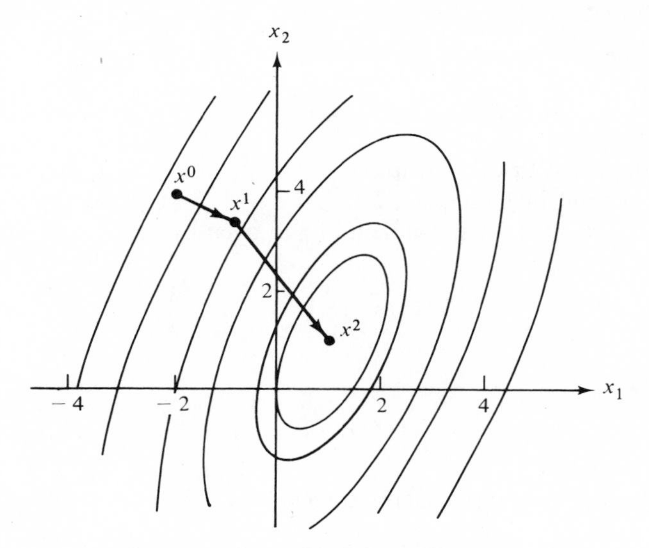

Fig. 10.2 The Goldstein-Price algorithm.

The next search direction will be the Newton direction

$$\phi(x^1) = Q^{-1}\nabla f(x^1) = \begin{bmatrix} \frac{1}{2} & \frac{1}{2} \\ \frac{1}{2} & \frac{3}{2} \end{bmatrix} \begin{pmatrix} -7.8 \\ 4.2 \end{pmatrix} = \begin{pmatrix} -1.8 \\ 2.4 \end{pmatrix} \tag{10.57}$$

and

$$g(x^1, 1) = \frac{11.06 - (-1)}{(-7.8)(-1.8) + (4.2)(2.4)} = \frac{1}{2} > \frac{1}{8}. \tag{10.58}$$

Thus $x^2 = (1, 1)$ and $\nabla f(x^2) = 0$; that is, the global minimum has been attained. The trajectory of the algorithm is illustrated in Figure 10.2. ▌

In this example the algorithm terminated after a finite number of iterations. Goldstein and Price [16] have established convergence to a global minimum at a superlinear rate for a broad class of strictly convex functions on R^n. In their limited numerical experiments they have shown that the method compares favorably with other minimization algorithms that will be described later in the chapter, and in the following one. This comparison, however, is made on the number of function and gradient evaluations only and does not take into account the extra computational effort required in order to invert the $Q(x^k)$ matrix at each iteration.

It can be shown that the matrices $Q(x^k)$ approach the true Hessian $\nabla^2 f(x^k)$ of the function to be minimized in a certain sense. So it is possible to try to derive similar algorithms in which a sequence of matrices appro-

aches the inverse Hessian, $[\nabla^2 f(x^k)]^{-1}$ in the limit without the need to invert matrices at each iteration. Such methods will be extensively discussed in the next chapter.

Other Newton-type minimization methods, using second derivatives, are described in Murray [23]. Ritter [31] proposed a Newton-type algorithm for which superlinear convergence can be proved under the same assumptions as for the Goldstein-Price method. Convergence (not necessarily superlinear) of the Ritter algorithm has also been established under the more general conditions of Theorem 10.1.

10.2 CONJUGATE GRADIENT METHODS

An iterative method for minimizing a real function f on R^n can be described by a sequence of moves from an initial point x^0 to a new point x^1, then to x^2, and so on, where the successive points are given by the relation

$$x^k = x^{k-1} + \alpha_k z^k. \tag{10.59}$$

Here x^{k-1} is the current point, z^k is the direction vector along which we move, and α_k is the step length. Suppose that the direction z^k is given and α_k is chosen so that the function f is minimized along z^k. Let

$$F(\alpha_k) = f(x^{k-1} + \alpha_k z^k) \tag{10.60}$$

and at α_k^*, the minimum of F, we have

$$\frac{dF(\alpha_k^*)}{d\alpha_k} = (z^k)^T \nabla f(x^{k-1} + \alpha_k^* z^k) = (z^k)^T \nabla f(x^k) = 0. \tag{10.61}$$

Assume that f is a quadratic function, given as before by

$$f(x) = a + b^T x + \tfrac{1}{2} x^T Q x, \tag{10.62}$$

where Q is an $n \times n$ symmetric positive definite matrix. In this case, the gradients of f at any two points are related by

$$\nabla f(x^k) = \nabla f(x^{k-1}) + Q(x^k - x^{k-1}). \tag{10.63}$$

If $x^k = x^{k-1} + \alpha_k^* z^k$, then from (10.59), (10.61), and (10.63) we obtain an explicit formula for α_k^*:

$$\alpha_k^* = -\frac{(z^k)^T \nabla f(x^{k-1})}{(z^k)^T Q z^k}. \tag{10.64}$$

The relation between the function values at two points is given by

$$f(x^k) = f(x^{k-1}) + (x^k - x^{k-1})^T \nabla f(x^{k-1}) + \tfrac{1}{2}(x^k - x^{k-1})^T Q (x^k - x^{k-1}) \tag{10.65}$$

and, by (10.59) to (10.65), we get

$$f(x^{k-1}) - f(x^k) = \frac{[(z^k)^T \nabla f(x^{k-1})]^2}{2(z^k)^T Q z^k}. \tag{10.66}$$

Since Q is assumed to be positive definite, the right-hand side of this equation is nonnegative for $z^k \neq 0$ and is positive if z^k is not orthogonal to $\nabla f(x^{k-1})$. In the latter case, the algorithm is called a **descent method**, since $f(x^{k-1}) > f(x^k)$. We only require $(z^k)^T \nabla f(x^{k-1}) \neq 0$, and so it follows from (10.64) that if $(z^k)^T \nabla f(x^{k-1}) > 0$, then $\alpha_k^* < 0$ (in our discussion of the steepest descent method we knew that the preceding scalar product is negative, which implies $\alpha_k^* > 0$).

In addition to having a descent minimization method, we would also like to have an algorithm that converges rapidly or, even better, that terminates in a finite number of steps when applied to minimizing a quadratic function. Since general nonlinear functions can be reasonably well approximated by a quadratic function in the neighborhood of a minimum, the quadratic termination property seems desirable for fast convergence in the case of general functions. From the results of the previous chapter, especially Theorem 9.1, it follows that if the search directions z^k are mutually conjugate with respect to Q for $k = 1, \ldots, n$, then the point x^n attained will be the exact minimum of the quadratic function f. The choice of the conjugate directions can be done in the following way.

Suppose that we start at a point $x^0 \in R^n$ and choose

$$z^1 = -\nabla f(x^0). \tag{10.67}$$

The next point is

$$x^1 = x^0 + \alpha_1^* z^1, \tag{10.68}$$

where α_1^* is given by (10.64). Evaluate $\nabla f(x^1)$ and set

$$z^2 = -\nabla f(x^1) + \beta_{11} z^1, \tag{10.69}$$

where β_{11} is a number chosen so that z^1 and z^2 will be conjugate with respect to Q. Hence

$$(z^1)^T Q z^2 = (z^1)^T Q(-\nabla f(x^1) + \beta_{11} z^1) = 0 \tag{10.70}$$

and

$$\beta_{11} = \frac{(z^1)^T Q \nabla f(x^1)}{(z^1)^T Q z^1}. \tag{10.71}$$

Now we move from x^1 along the direction z^2 to a new point x^2 and compute $\nabla f(x^2)$. The new direction z^3 (provided that $n \geq 3$) should be conjugate to both z^1 and z^2. It will be if we let

$$z^3 = -\nabla f(x^2) + \beta_{21} z^1 + \beta_{22} z^2, \qquad (10.72)$$

where β_{21} and β_{22} are chosen so that $(z^1)^T Q z^3 = (z^2)^T Q z^3 = 0$. In general, we get

$$z^{k+1} = -\nabla f(x^k) + \sum_{j=1}^{k} \beta_{kj} z^j, \qquad k = 0, \ldots, n-1. \qquad (10.73)$$

The difficulty with this formula is that the coefficients β_{kj} are functions of Q, and in trying to use (10.73) for a nonquadratic function, we would need to compute Hessian matrices, an undesirable operation. So we shall show how these directions can be generated without the explicit use of Q. The following result will be useful in the sequel.

Theorem 10.2

Let f be given by (10.62) *and suppose that the m nonzero vectors $z^1, \ldots, z^m$, $z^j \in R^n$, $m \leq n$, are mutually conjugate with respect to Q. Starting at a point $x^0 \in R^n$, we move to $x^1, x^2, \ldots, x^m$, along $z^1, z^2, \ldots, z^m$, respectively, such that*

$$(z^j)^T \nabla f(x^j) = 0, \qquad j = 1, \ldots, m. \qquad (10.74)$$

Then

$$(z^j)^T \nabla f(x^m) = 0, \qquad j = 1, \ldots, m. \qquad (10.75)$$

Proof. For $j = m$ the result is trivial. From (10.63) we have

$$\nabla f(x^m) = \nabla f(x^j) + Q(x^m - x^j), \qquad j = 0, \ldots, m-1 \qquad (10.76)$$

and

$$x^j = x^{j-1} + \alpha_j^* z^j, \qquad j = 1, \ldots, m \qquad (10.77)$$

where the α_j^* were chosen so that (10.74) holds. Applying (10.77) recursively, we get

$$x^m - x^j = \sum_{i=j+1}^{m} \alpha_i^* z^i, \qquad j = 0, \ldots, m-1. \qquad (10.78)$$

Substituting into (10.76) yields

$$\nabla f(x^m) = \nabla f(x^j) + \sum_{i=j+1}^{m} \alpha_i^* Q z^i, \qquad j = 0, \ldots, m-1 \qquad (10.79)$$

and

$$(z^j)^T\nabla f(x^m) = (z^j)^T\nabla f(x^j) + \sum_{i=j+1}^{m} \alpha_i^*(z^j)^T Q z^i, \qquad j = 1, \ldots, m-1. \quad (10.80)$$

The first term on the right-hand side vanishes by (10.74) and the second by conjugacy. Hence (10.75) holds. ■

Corollary 10.3

If $m = n$ in Theorem 10.2, then

$$\nabla f(x^n) = 0; \quad (10.81)$$

that is, x^n is the unconstrained minimum of f.

Proof. Since the z^j are linearly independent, the gradient of f at x^n must vanish by (10.75). ■

Denote by γ^i the difference between the gradients of f at two consecutive points x^i and x^{i-1}. That is,

$$\gamma^i = \nabla f(x^i) - \nabla f(x^{i-1}), \qquad i = 1, 2, \ldots. \quad (10.82)$$

For the quadratic function f we obtain from (10.63)

$$\gamma^i = Q(x^i - x^{i-1}). \quad (10.83)$$

The points x^i and directions z^i are so chosen that (10.59) and (10.83) hold. Thus

$$(\gamma^i)^T z^j = \alpha_i^*(z^i)^T Q z^j, \qquad i = 1, \ldots, n, \qquad j = 1, \ldots, n. \quad (10.84)$$

Assuming that $z^1, \ldots, z^k$, $k \leq n$, were chosen to be mutually conjugate with respect to Q, we get

$$(\gamma^i)^T z^j = 0, \qquad i = 1, \ldots, k, \quad j = 1, \ldots, k, \quad i \neq j. \quad (10.85)$$

Let us use this result to obtain another expression for the coefficient β_{11}:

$$(\gamma^1)^T z^2 = (\nabla f(x^1) - \nabla f(x^0))^T[-\nabla f(x^1) + \beta_{11}\nabla f(x^0)] = 0 \quad (10.86)$$

and

$$\beta_{11} = \frac{(\nabla f(x^1) - \nabla f(x^0))^T \nabla f(x^1)}{(\nabla f(x^1) - \nabla f(x^0))^T(-\nabla f(x^0))} = \frac{(\gamma^1)^T \nabla f(x^1)}{(\gamma^1)^T(-\nabla f(x^0))}. \quad (10.87)$$

Since $(z^1)^T\nabla f(x^1) = -(\nabla f(x^0))^T\nabla f(x^1) = 0$ by (10.61) and (10.67), and assuming that $\nabla f(x^0) \neq 0$, we get another expression

$$\beta_{11} = \frac{(\nabla f(x^1))^T\nabla f(x^1)}{(\nabla f(x^0))^T\nabla f(x^0)}. \tag{10.88}$$

The point x^2 is reached by minimizing along the conjugate directions z^1 and z^2. Hence by Theorem 10.2,

$$(z^1)^T\nabla f(x^2) = (z^2)^T\nabla f(x^2) = 0. \tag{10.89}$$

Substituting (10.67) and (10.69) into (10.89), we conclude that

$$(\nabla f(x^0))^T\nabla f(x^2) = 0 \tag{10.90}$$

$$(\nabla f(x^1))^T\nabla f(x^2) = 0. \tag{10.91}$$

We can now find a formula for the coefficients β_{21} and β_{22} in (10.72). Combining the equations $(y^1)^Tz^3 = (y^2)^Tz^3 = 0$ with (10.89) to (10.91), we obtain

$$\beta_{21} = 0, \qquad \beta_{22} = \frac{(\nabla f(x^2))^T\nabla f(x^2)}{(\nabla f(x^1))^T\nabla f(x^1)}. \tag{10.92}$$

Note that β_{22} (and β_{kk}, defined below) can also be expressed in several forms, just as the equivalent expressions (for quadratic functions) of β_{11} in (10.87) and (10.88). We shall return to this equivalency later.

In a similar way, we can also establish that

$$(\nabla f x^i))^T\nabla f(x^j) = 0, \qquad 0 \leq i < j \leq n \tag{10.93}$$

$$\beta_{kj} = 0, \qquad k \neq j \tag{10.94}$$

and

$$\beta_k = \beta_{kk} = \frac{(\nabla f(x^k))^T\nabla f(x^k)}{(\nabla f(x^{k-1}))^T\nabla f(x^{k-1})}, \qquad k = 1, \ldots, n. \tag{10.95}$$

Since $\beta_{kj} = 0$ for $k \neq j$, we drop the double subscript kk in the forthcoming discussions. Thus

$$z^{k+1} = -\nabla f(x^k) + \frac{(\nabla f(x^k))^T\nabla f(x^k)}{(\nabla f(x^{k-1}))^T\nabla f(x^{k-1})} z^k. \tag{10.96}$$

A complete unconstrained minimization algorithm, called the **conjugate gradient method**, resulting from these derivations is due to Fletcher and Reeves [12] and is based on a method suggested by Hestenes and Stiefel [18]

for the solution of a system of linear equations. The conjugate gradient method can be described as follows:

Choose a starting point $x^0 \in R^n$. Evaluate $\nabla f(x^0)$ and then let $z^1 = -\nabla f(x^0)$. Move to $x^1, x^2, \ldots, x^n$ by minimizing $f(x)$ along the directions $z^1, z^2, \ldots, z^n$ in turn, where the z^k are chosen by (10.96). Restart the procedure by letting x^n and $-\nabla f(x^n)$ be the new x^0 and z^1, respectively. Terminate the algorithm if

$$(\nabla f(x^k))^T \nabla f(x^k) \leq \epsilon, \tag{10.97}$$

where $\epsilon > 0$ is some predetermined small number. Note that the directions z^k are mutually conjugate with respect to Q if f is given by (10.62). Thus the conjugate gradient method has the quadratic termination property.

Example 10.2.1

Consider the problem of minimizing the quadratic function

$$f(x) = \tfrac{3}{2}(x_1)^2 + \tfrac{1}{2}(x_2)^2 - x_1x_2 - 2x_1 \tag{10.98}$$

by the conjugate gradient algorithm of Fletcher and Reeves. Rewriting f in the form of (10.62), we see that

$$b = \begin{pmatrix} -2 \\ 0 \end{pmatrix} \qquad Q = \begin{bmatrix} 3 & -1 \\ -1 & 1 \end{bmatrix}. \tag{10.99}$$

Suppose that we begin the search at $x^0 = (-2, 4)$. We get

$$\nabla f(x^0) = \begin{pmatrix} -12 \\ 6 \end{pmatrix} \quad \text{and} \quad z^1 = \begin{pmatrix} 12 \\ -6 \end{pmatrix}. \tag{10.100}$$

Minimizing $f(x^0 + \alpha_1 z^1)$ with respect to α_1, we find that $\alpha_1^* = \frac{5}{17}$. Thus

$$x^1 = \begin{pmatrix} -2 + \dfrac{(5)(12)}{17} \\ 4 - \dfrac{(5)(6)}{17} \end{pmatrix} = \begin{pmatrix} \dfrac{26}{17} \\ \dfrac{38}{17} \end{pmatrix} \qquad \nabla f(x^1) = \begin{pmatrix} \dfrac{6}{17} \\ \dfrac{12}{17} \end{pmatrix}. \tag{10.101}$$

Now we must find z^2, the second search direction, by (10.96).

$$z^2 = -\nabla f(x^1) + \frac{(\nabla f(x^1))^T \nabla f(x^1)}{(\nabla f(x^0))^T \nabla f(x^0)} z^1 \tag{10.102}$$

and

$$z^2 = \begin{pmatrix} -\dfrac{6}{17} \\ -\dfrac{12}{17} \end{pmatrix} + \frac{(\frac{6}{17})^2 + (\frac{12}{17})^2}{(-12)^2 + (6)^2} \begin{pmatrix} 12 \\ -6 \end{pmatrix} = \begin{pmatrix} -\dfrac{90}{289} \\ -\dfrac{210}{289} \end{pmatrix}. \tag{10.103}$$

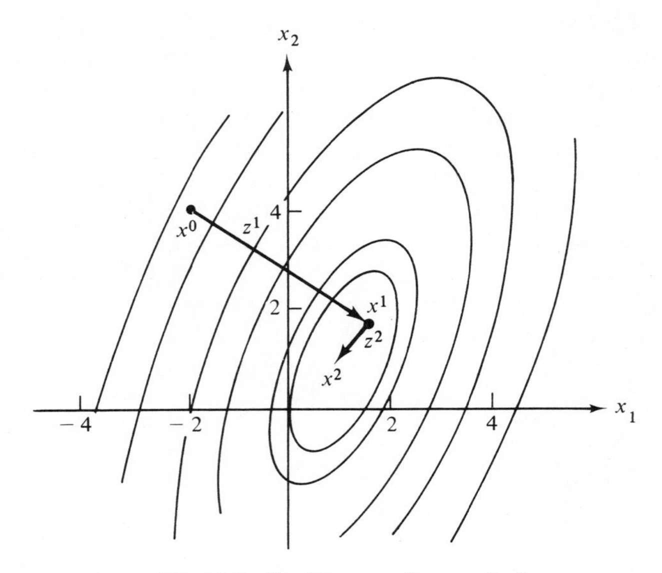

Fig. 10.3 Conjugate gradient method.

Minimizing $f(x^1 + \alpha_2 z^2)$ with respect to α_2, we obtain $\alpha_2^* = \frac{17}{10}$. Consequently,

$$x^2 = \begin{pmatrix} \frac{26}{17} - \frac{(17)(90)}{(10)(289)} \\ \frac{38}{17} - \frac{(17)(210)}{(10)(289)} \end{pmatrix} = \begin{pmatrix} 1 \\ 1 \end{pmatrix}, \tag{10.104}$$

and this is the global minimum of f, as asserted. The search directions and steps of the algorithm are illustrated in Figure 10.3. ∎

The reader will be asked to show that there are at least three more equivalent formulas (if used for minimizing quadratic functions) of the coefficient β_k as given in (10.95). They are

$$\beta_k = \frac{(z^k)^T \nabla^2 f(x^k) \nabla f(x^k)}{(z^k)^T \nabla^2 f(x^k) z^k} \tag{10.105}$$

$$\beta_k = \frac{(\gamma^k)^T \nabla f(x^k)}{(\gamma^k)^T z^k} \tag{10.106}$$

and

$$\beta_k = \frac{(\gamma^k)^T \nabla f(x^k)}{(\nabla f(x^{k-1}))^T \nabla f(x^{k-1})}. \tag{10.107}$$

These formulas are equivalent in the sense that all yield the same search directions when used in minimizing a quadratic function that has a positive definite Q matrix with a starting direction $z^1 = -\nabla f(x^0)$. Formula (10.105) was suggested by Daniel [7] and formula (10.106) by Sorenson [34] and Wolfe [35], whereas (10.107) is used in the conjugate gradient algorithms of Polak and Ribière [29] and Polyak [30]. Hestenes and Stiefel [18] found (10.106) to be less sensitive than (10.95) to roundoff error.

The choice of the starting direction $z^1 = -\nabla f(x^0)$ seems to be quite important. For example, if we choose $z^1 \neq -\nabla f(x^0)$, it is possible that the Fletcher-Reeves method will terminate on a quadratic function only after the restarting step, whereas the other versions arrive at the minimum in n steps. Let us illustrate the last assertion on a simple quadratic function [35].

Example 10.2.2

Consider the minimization of

$$f(x) = \tfrac{1}{2}(x_1)^2 + \tfrac{1}{2}(x_2)^2. \tag{10.108}$$

Suppose that $x^0 = (1, 1)$ and, instead of the negative gradient direction $-\nabla f(x^0) = (-1, -1)^T$, we take $z^1 = (-1, 0)^T$. Minimizing along z^1, we arrive at the next point, $x^1 = (0, 1)$, and $\nabla f(x^1) = (0, 1)$. Using (10.95), the next search direction is given by

$$z^2 = -\begin{pmatrix} 0 \\ 1 \end{pmatrix} + \frac{1}{2}\begin{pmatrix} -1 \\ 0 \end{pmatrix} = \begin{pmatrix} -\frac{1}{2} \\ -1 \end{pmatrix} \tag{10.109}$$

and $x^2 = (-\frac{2}{5}, \frac{1}{5})$. Without restarting, we would move in a spiral trajectory approaching the origin. On the other hand, using either (10.106) or (10.107), we obtain $z^2 = (0, -1)^T$, since $(y^1)^T \nabla f(x^1) = 0$, and the minimum in the z^2 direction is attained at $x^2 = (0, 0)$, where, of course, $\nabla f(x^2) = 0$, so the algorithm terminates. ∎

The reason for not finding the exact minimum in two steps by using (10.95) in the last example is due to the fact that this formula ensures $(z^k)^T Q z^{k+1} = 0$ at the kth iteration only if all the previous steps have been taken accurately and, in particular, if $z^1 = -\nabla f(x^0)$ was chosen. The alternative formulas presented earlier ensure that two consecutive search directions z^k and z^{k+1} will be conjugate with respect to Q even if $z^1 \neq -\nabla f(x^0)$; but mutual conjugacy of n search directions for $n > 2$ generally cannot be obtained in this case, and quadratic termination may not occur. It follows that conjugate gradient algorithms without the "standard" choice of

$z^1 = -\nabla f(x^0)$ and without periodic restarting lose much of their efficiency when applied to minimizing a quadratic function. We shall return to this point later.

10.3 CONVERGENCE OF CONJUGATE GRADIENT ALGORITHMS

As for the question of convergence, so far we only know that properly executed conjugate gradient methods have the quadratic termination property. We now prove convergence to a global minimum for a large class of functions. Convergence properties of various versions of the conjugate gradient method have been widely discussed in the literature. See, for example, Daniel [10], Elkin [11], McCormick and Ritter [21, 22], Ortega and Rheinboldt [25], Polak [28], Polak and Ribière [29], and Polyak [30]. Here we prove convergence of the Polak-Ribière-Polyak conjugate gradient method, abbreviated the PRP method, without restarting. The PRP algorithm can be described as follows: Let $x^0 \in R^n$ be given. Choose z^1 so that

$$(z^1)^T\nabla f(x^0) < 0. \tag{10.110}$$

For $k = 1, 2, \ldots$, let

$$z^{k+1} = -\nabla f(x^k) + \beta_k z^k, \tag{10.111}$$

where β_k is given by (10.107), and

$$x^k = x^{k-1} + \alpha_k^* z^k, \tag{10.112}$$

where α_k^* minimizes $f(x^{k-1} + \alpha_k z^k)$; that is, α_k^* is found by an exact line search. Our proof is based mainly on the works of Daniel [10] and McCormick and Ritter [22].

We start by proving that the PRP algorithm is a strictly descent method.

Lemma 10.4

Suppose that f is a continuously differentiable real function on an open subset of R^n. Then for the sequence of points $\{x^k\}$ generated by the PRP method,

$$f(x^k) > f(x^{k+1}). \qquad k = 0, 1, \ldots \tag{10.113}$$

whenever $\nabla f(x^k) \neq 0$.

Proof. Multiplying (10.111) by $\nabla f(x^k) \neq 0$, we obtain

$$(\nabla f(x^k))^T z^{k+1} = -(\nabla f(x^k))^T\nabla f(x^k) < 0, \qquad k = 1, 2, \ldots \tag{10.114}$$

since $(\nabla f(x^k))^T z^k = 0$ by the choice of α_k^*. So for sufficiently small α_{k+1}, it follows from (10.110) and (10.114) that

$$f(x^k) > f(x^k + \alpha_{k+1} z^{k+1}). \tag{10.115}$$

Since $x^{k+1} = x^k + \alpha_{k+1}^* z^{k+1}$, where α_{k+1}^* minimizes $f(x^k + \alpha_{k+1} z^k)$, the inequality of (10.113) holds. ∎

In the convergence proof of the PRP method given in a forthcoming theorem, we assume that f is a twice continuously differentiable real function on some open convex set $C \subset R^n$ and that its Hessian matrix $\nabla^2 f$ is positive definite (which implies, of course, that f is strictly convex). We further suppose that for every $x^0 \in C$ with $f(x^0) = \alpha$, the level set $S(f, \alpha) \subset C$, defined by

$$S(f, \alpha) = \{x : x \in C, f(x) \leq \alpha\}, \tag{10.116}$$

is compact. The compactness of the level sets is a widely used assumption in convergence theorems for minimization algorithms. For this reason, it is appropriate to consider some conditions on a function ϕ that ensure that its level sets are compact. Note that, for continuous functions ϕ, all level sets are closed, hence these sets are compact if and only if they are bounded. The compactness of all level sets can be characterized by the following

Theorem 10.5

Let ϕ be a continuous real function on an unbounded set $T \subset R^n$. Then for any $\alpha \in R$ the nonempty level set

$$S(\phi, \alpha) = \{x : x \in T, \phi(x) \leq \alpha\} \tag{10.117}$$

is compact if and only if

$$\lim_{k \to \infty} \{\phi(x^k)\} = +\infty \tag{10.118}$$

for every sequence $\{x^k\}$, such that $x^k \in T$ and

$$\lim_{k \to \infty} \{\| x^k \|\} = +\infty. \tag{10.119}$$

Proof. Assume that all level sets of ϕ are bounded. Let $\{x^k\}$ be a sequence of points in T satisfying (10.119) and suppose that there exists a real number β such that $\phi(x^k) \leq \beta$ for infinitely many k. Then there exists a subsequence $\{x^{k_m}\}$ of $\{x^k\}$ such that $\phi(x^{k_m}) \leq \beta$ for every x^{k_m}. Thus all members of $\{x^{k_m}\}$ are in the bounded set $S(\phi, \beta)$, thereby contradicting (10.119). Conversely, sup-

pose that (10.118) and (10.119) hold and $S(\phi, \hat{\alpha})$ is unbounded for some $\hat{\alpha}$. Thus there exists a sequence $\{x^k\}$ with $x^k \in S(\phi, \hat{\alpha})$ that satisfies (10.119), and $\phi(x^k) \leq \hat{\alpha}$, thereby contradicting (10.118). ■

Suppose now that f is a twice continuously differentiable function on C and that there exist positive numbers μ, η such that for every $x \in C$ and $y \in R^n$ we have

$$\mu \|y\|^2 \leq y^T \nabla^2 f(x) y \leq \eta \|y\|^2 \tag{10.120}$$

or, equivalently, the eigenvalues of $\nabla^2 f(x)$ are bounded by μ and η. These inequalities also clearly imply that the Hessian matrix $\nabla^2 f(x)$ is positive definite for every $x \in C$, and so f is strictly convex on C. Let us prove that the search directions of the PRP methods are bounded.

Lemma 10.6

Suppose that (10.120) *holds and that the* z^k, x^k; α_k *are determined by the PRP method. Then*

$$\|\nabla f(x^k)\| \leq \|z^{k+1}\| \leq \frac{\mu + \eta}{\mu} \|\nabla f(x^k)\|, \qquad k = 1, 2, \ldots. \tag{10.121}$$

Proof. We have

$$(\nabla f(x^k))^T z^k = 0, \qquad k = 1, 2, \ldots \tag{10.122}$$

$$(\nabla f(x^k))^T z^{k+1} = -\|\nabla f(x^k)\|^2, \qquad k = 1, 2, \ldots. \tag{10.123}$$

Thus β_k can be written

$$\beta_k = \frac{(y^k)^T \nabla f(x^k)}{(y^k)^T z^k}. \tag{10.124}$$

By the Mean Value Theorem,

$$\nabla f(x^k) = \nabla f(x^{k-1}) + \nabla^2 f(\xi^{k-1})(x^k - x^{k-1}) \tag{10.125}$$

$$= \nabla f(x^{k-1}) + \nabla^2 f(\xi^{k-1}) \alpha_k^* z^k, \tag{10.126}$$

where ξ^{k-1} is some point on the line joining x^k and x^{k-1}. Hence

$$y^k = \nabla^2 f(\xi^{k-1}) \alpha_k^* z^k \tag{10.127}$$

and

$$|(y^k)^T \nabla f(x^k)| = |\alpha_k^* (z^k)^T \nabla^2 f(\xi^{k-1}) \nabla f(x^k)|. \tag{10.128}$$

By the Cauchy-Schwarz inequality [2],

$$|(\gamma^k)^T \nabla f(x^k)| \leq \alpha_k^* \| \nabla f(x^k) \| \| \nabla^2 f(\xi^{k-1}) z^k \| \tag{10.129}$$

$$\leq \alpha_k^* \| \nabla f(x^k) \| \left\| \nabla^2 f(\xi^{k-1}) \frac{z^k}{\| z^k \|} \right\| \| z^k \| \tag{10.130}$$

$$\leq \alpha_k^* \| \nabla f(x^k) \| \| z^k \| \left\{ \sup_{\|z\|=1} \| \nabla^2 f(\xi^{k-1}) z^k \| \right\} \tag{10.131}$$

$$\leq \alpha_k^* \| \nabla f(x^k \| \| z^k \| \| \nabla^2 f(\xi^{k-1}) \| \tag{10.132}$$

$$\leq \alpha_k^* \| \nabla f(x^k) \| \| z^k \| \eta, \tag{10.133}$$

where (10.132) and (10.133) follow from the definition of $\| A \|$, the norm of a symmetric matrix [15], and from (10.120). As noted in Chapter 1,

$$\| A \| = \sup_{\|x\|=1} \| Ax \| = \sup_{\|x\|=1} x^T A x. \tag{10.134}$$

Similarly, it can be shown that

$$|(\gamma^k)^T z^k| \geq \alpha_k^* \mu \| z^k \|^2. \tag{10.135}$$

Consequently,

$$|\beta_k| \leq \frac{\eta}{\mu} \frac{\| \nabla f(x^k) \|}{\| z^k \|} \tag{10.136}$$

and

$$\| \beta_k z^k \| \leq \frac{\eta}{\mu} \| \nabla f(x^k) \|. \tag{10.137}$$

By (10.111), (10.137), and the triangle inequality [2], we obtain

$$\| z^{k+1} \| \leq \| \nabla f(x^k) \| + \| \beta_k z^k \| \leq \frac{\mu + \eta}{\mu} \| \nabla f(x^k) \|. \tag{10.138}$$

Now

$$\| z^{k+1} \|^2 = -(\nabla f(x^k))^T [-\nabla f(x^k) + \beta_k z^k] + \beta_k (z^k)^T [-\nabla f(x^k) + \beta_k z^k] \tag{10.139}$$

and from (10.122) it follows that

$$\| z^{k+1} \|^2 = \| \nabla f(x^k) \|^2 + (\beta_k)^2 \| z^k \|^2 \geq \| \nabla f(x^k) \|^2. \tag{10.140}$$

Hence

$$\| z^{k+1} \| \geq \| \nabla f(x^k) \|. \tag{10.141}$$

■

We now state and prove the following convergence theorem for the PRP method.

Theorem 10.7

Let f be a twice continuously differentiable real function on an open convex set $C \subset R^n$ and satisfying (10.120). *Let $x^0 \in C$, $f(x^0) = \alpha$ and suppose that $S(f, \alpha)$ is compact. Then the PRP algorithm is a strictly descent method. That is,*

$$f(x^k) > f(x^{k+1}), \qquad k = 0, 1, \ldots \tag{10.142}$$

provided that $\nabla f(x^k) \neq 0$ and either $\nabla f(x^K) = 0$ for some K or

$$\lim_{k \to \infty} \{\| \nabla f(x^k) \|\} = 0. \tag{10.143}$$

The sequence $\{x^k\}$ converges to a point $x^* \in S(f, \alpha)$, where $\nabla f(x^*) = 0$ and x^* is the strict global minimum of f.

Proof. By Lemma 10.4, the PRP algorithm is a strictly descent method and (10.142) holds. By (10.127), we have

$$(y^{k+1})^T z^{k+1} = \alpha^*_{k+1}(z^{k+1})^T \nabla^2 f(\xi^k) z^{k+1} \tag{10.144}$$

and

$$\alpha^*_{k+1} = \frac{-(\nabla f(x^k))^T z^{k+1}}{(z^{k+1})^T \nabla^2 f(\xi^k) z^{k+1}} = \frac{\| \nabla f(x^k) \|^2}{(z^{k+1})^T \nabla^2 f(\xi^k) z^{k+1}}. \tag{10.145}$$

From (10.120) and Lemma 10.6, we get

$$\alpha^*_{k+1} \geq \frac{\| \nabla f(x^k) \|^2}{\eta \| z^{k+1} \|^2} \geq \frac{(\mu)^2}{\eta(\mu + \eta)^2}. \tag{10.146}$$

For all $0 < \alpha_{k+1} \leq \alpha^*_{k+1}$, we can write, by Taylor's theorem,

$$f(x^k + \alpha_{k+1} z^{k+1}) = f(x^k) + \alpha_{k+1}(\nabla f(x^k))^T z^{k+1} + \tfrac{1}{2}(\alpha_{k+1})^2 (z^{k+1})^T \nabla^2 f(\hat{\xi}^k) z^{k+1}, \tag{10.147}$$

where $\hat{\xi}^k$ is some point on the line joining $x^k + \alpha_{k+1} z^{k+1}$ and x^k. By (10.120), (10.121), and (10,123),

$$f(x^k + \alpha_{k+1} z^{k+1}) \leq f(x^k) - \alpha_{k+1} \| \nabla f(x^k) \|^2 + \tfrac{1}{2}(\alpha_{k+1})^2 \eta \frac{(\mu + \eta)^2}{(\mu)^2} \| \nabla f(x^k) \|^2. \tag{10.148}$$

Letting

$$\alpha_{k+1} = \frac{(\mu)^2}{\eta(\mu+\eta)^2}, \tag{10.149}$$

and by the definitions of α^*_{k+1}, x^{k+1}, we get

$$f(x^{k+1}) \leq f\left(x^k + \frac{(\mu)^2}{\eta(\mu+\eta)^2} z^{k+1}\right) \tag{10.150}$$

$$\leq f(x^k) - \frac{(\mu)^2}{\eta(\mu+\eta)^2} \|\nabla f(x^k)\|^2 + \frac{1}{2}\frac{(\mu)^2}{\eta(\mu+\eta)^2}\|\nabla f(x^k)\|^2 \tag{10.151}$$

$$\leq f(x^k) - \frac{1}{2}\frac{(\mu)^2}{\eta(\mu+\eta)^2}\|\nabla f(x^k)\|^2. \tag{10.152}$$

And so

$$f(x^k) - f(x^{k+1}) \geq \frac{1}{2}\frac{(\mu)^2}{\eta(\mu+\eta)^2}\|\nabla f(x^k)\|^2. \tag{10.153}$$

Since the sequence $\{x^k\}$ converges to x^* in the compact set $S(f, \alpha)$ and the function f is bounded below on $S(f, \alpha)$, it follows from (10.153) that $\{\|\nabla f(x^k)\|\}$ must converge to zero, and so $\nabla f(x^*) = 0$. Since f is strictly convex, the point x^* satisfies the sufficient conditions for a strict global minimum of f, as given in Chapter 4. ■

Similar to the Fletcher-Reeves method, we can define the PRP method with periodic restart by letting

$$z^{k+1} = -\nabla f(x^k), \qquad k = p, 2p, 3p, \ldots \tag{10.154}$$

where p is usually chosen to be n or $n + 1$ and the rest of the procedure is as defined earlier. The reader can verify that Theorem 10.7 also holds for the PRP method with periodic restart. A weaker result for both versions of the PRP method can be stated as follows:

Theorem 10.8

Let f be a continuously differentiable real function on an open convex set $C \subset R^n$. Let $x^0 \in C$, $f(x^0) = \alpha$, and suppose that $S(f, \alpha)$ is compact. Then the PRP algorithm, with or without periodic restart, is a strictly descent method. That is,

$$f(x^k) > f(x^{k+1}), \qquad k = 0, 1, \ldots \tag{10.155}$$

provided that $\nabla f(x^k) \neq 0$ and either $\nabla f(x^K) = 0$ for some K or

$$\lim_{k\to\infty} \{\|\nabla f(x^k)\|\} = 0. \tag{10.156}$$

Every cluster point of the sequence $\{x^k\}$ *is a solution of* $\nabla f(x) = 0$; *and if it has a unique solution* $x^* \in S(f, \alpha)$, *then* x^* *is a global minimum of* f *over* C.

The proof is similar to previous ones and will be omitted.

Let us turn now to the rate of convergence of conjugate gradient methods. As we already know, these methods have the quadratic termination property if the starting direction z^1 is chosen to be $-\nabla f(x^0)$. It was shown by Crowder and Wolfe [5] that, in general, the rate of convergence of some conjugate gradient methods is not worse than that of the steepest descent method; that is, it is not worse than linear. More specifically, let $f(x) = \frac{1}{2}x^T Qx$, $x \in R^n$, where Q is a symmetric positive definite matrix. Suppose now that, at iteration k, we have a point x^{k-1} and a direction z^k. Minimizing $f(x^{k-1} + \alpha_k z^k)$ with respect to α_k, we arrive at the point x^k that satisfies $(z^k)^T \nabla f(x^k) = 0$. Now we can choose $z^{k+1} = -\nabla f(x^k)$, as in the steepest descent method, or by the conjugate gradient formula

$$z^{k+1} = -\nabla f(x^k) + \beta_k z^k, \tag{10.157}$$

where β_k is given by (10.105), (10.106), or (10.107), as, for example, in the PRP method. Next we minimize $f(x^k + \alpha_{k+1} z^{k+1})$ with respect to α_{k+1} and arrive at a point x^{k+1}. The two different types of directions z^{k+1} will generally lead to two different points x^{k+1}. In the case of the steepest descent (SD) move we obtain

$$f(x_{\mathrm{SD}}^{k+1}) = \tfrac{1}{2}(\nabla f(x^k))^T Q^{-1} \nabla f(x^k) - \tfrac{1}{2}\frac{\|\nabla f(x^k)\|^4}{(\nabla f(x^k))^T Q \nabla f(x^k)}, \tag{10.158}$$

whereas in the case of the conjugate gradient (CG) move, we get

$$(z^{k+1})^T Q z^{k+1} = (\nabla f(x^k))^T Q \nabla f(x^k) - \frac{[(z^k)^T Q \nabla f(x^k)]^2}{(z^k)^T Q z^k} \tag{10.159}$$

and

$$f({}_{\mathrm{CG}}^{k+1}) = \tfrac{1}{2}(\nabla f(x^k))^T Q^{-1} \nabla f(x^k) - \tfrac{1}{2}\frac{\|\nabla f(x^k)\|^4}{(z^{k+1})^T Q z^{k+1}}. \tag{10.160}$$

Hence by (10.158) to (10.160),

$$f(x_{\mathrm{CG}}^{k+1}) \leq f(x_{\mathrm{SD}}^{k+1}). \tag{10.161}$$

In other words, each iteration of any conjugate gradient method reduces the value of f at least as much as the steepest descent method. Since the steepest descent method has a linear rate of convergence, we conclude that conjugate gradient methods with β_k defined by (10.105) to (10.107) have convergence rates that are no worse than the linear rate. Example 10.2.2, however,

shows that, for the choice of β_k by (10.95), the inequality (10.161) does not necessarily hold, because the directions z^1 and z^2 chosen by the Fletcher-Reeves method were not conjugate. It was also shown by Crowder and Wolfe [5] that without the "standard" starting direction $z^1 = -\nabla f(x^0)$, the rate of convergence of conjugate gradient methods can be as slow as the linear rate. They use a simple quadratic function to demonstrate the linear convergence rate, as we shall see now.

Example 10.3.1

Let $f(x) = \frac{1}{2}x^T Q x$, where

$$Q = \begin{bmatrix} 0.1 & 0 & 0 \\ 0 & 1 & 0 \\ 0 & 0 & 1 \end{bmatrix}. \tag{10.162}$$

Then suppose that we start the search at $x^0 = (10/\sqrt{6}, -\sqrt{5}/\sqrt{6}, 0)$ and choose $z^1 = (-5/2\sqrt{6}, 7/2\sqrt{30}, -3/4\sqrt{5})^T$. Note that $\nabla f(x^k) = Qx^k$ and

$$(z^1)^T \nabla f(x^0) = \frac{-5}{2\sqrt{6}} \frac{1}{\sqrt{6}} + \frac{7}{2\sqrt{30}} \frac{-\sqrt{5}}{\sqrt{6}} < 0. \tag{10.163}$$

This z^1 is a descent direction. Next, suppose that z^{k+1} is given by (10.157), where β_k is chosen by formula (10.105). It can be shown that for all $k = 1, 2, \ldots$

$$\alpha_k^* = \frac{8}{5}, \qquad \beta_k = \frac{9}{25}. \tag{10.164}$$

Furthermore,

$$\nabla f(x^{k+1}) = rR\nabla f(x^k) \qquad z^{k+1} = rRz^k, \tag{10.165}$$

where $r = \frac{3}{5}$ and

$$R = \begin{bmatrix} 1 & 0 & 0 \\ 0 & -\frac{1}{5} & -\frac{2\sqrt{6}}{5} \\ 0 & -\frac{2\sqrt{6}}{5} & -\frac{1}{5} \end{bmatrix} \tag{10.166}$$

is an orthogonal matrix. Consequently,

$$\frac{\|x^{k+1}\|}{\|x^k\|} = \frac{3}{5}, \qquad \frac{\|f(x^{k+1})\|}{\|f(x^k)\|} = \frac{9}{25} \tag{10.167}$$

and the convergence rate is linear. ∎

Using the starting search direction $z^1 = -\nabla f(x^0)$ and periodic restart, it has been shown by Cohen [4], Daniel [10], McCormick and Ritter [22],

Ortega and Rheinboldt [25], Polak [28], and Polyak [30] that, under suitable assumptions on f, such as (10.120), the rate of convergence of various versions of the conjugate gradient method is "n-step superlinear"; that is,

$$\lim_{k\to\infty} \frac{\|x^{k+n} - x^*\|}{\|x^k - x^*\|} = 0, \tag{10.168}$$

where x^* is the minimum of f. Several of these works also show that the actual rate is n-step quadratic. Daniel [7, 8, 9] proved that for his conjugate gradient method, using (10.105) with the "standard" start $z^1 = -\nabla f(x^0)$ and without periodic restart, we have

$$\|x^{k+n} - x^*\| \leq q\|x^k - x^*\|^{[1+1/(4n-3)]} \tag{10.169}$$

for sufficiently large k. We can conclude that, based on theoretical convergence rates, the conjugate gradient method with periodic restart is preferable to the continued version. An interesting discussion of the convergence rates of conjugate gradient and some Newton-type methods, such as the Goldstein-Price or the Ritter method, can be found in McCormick and Ritter [21]. It seems that algorithms should not, in general, be preferred on the basis of convergence rates only but that other aspects, such as the amount of computation needed per iteration and computer storage requirements, should also be considered. For example, it should be mentioned here that the convergence of conjugate gradient methods strongly depends on accurate one-dimensional optimizations and they are quite costly. Klessig and Polak [20], however, modified the PRP method by eliminating line searches but retaining some of the convergence properties.

Ortega and Rheinboldt [24, 25, 26] use a unified approach to the convergence properties of algorithms described in this chapter. They consider continuously differentiable real functions f on R^n and iterative descent methods, defined by

$$x^k = x^{k-1} + \alpha_k z^k, \qquad z^k \neq 0. \tag{10.170}$$

In order to ensure the convergence of such methods, the step length α_k should be chosen so that a "sufficient" decrease of f occurs at each iteration. This notion is defined with the aid of a **forcing function** σ whose domain and range are the set of nonnegative numbers with the property that for any sequence $\{t^k\}$ of nonnegative numbers

$$\lim_{k\to\infty} \{\sigma(t^k)\} = 0 \quad \text{implies} \quad \lim_{k\to\infty} \{t^k\} = 0. \tag{10.171}$$

For example, $\sigma(t) = ct$, where $c > 0$ satisfies the preceding relation. An iterative descent method provides a sufficient decrease of f at any point x^k if

there exists a function $\hat{\sigma}$, independent of x^k and such that

$$f(x^{k-1}) - f(x^k) \geq \hat{\sigma} \frac{|(z^k)^T \nabla f(x^{k-1})|}{\|z^k\|}. \tag{10.172}$$

If f is bounded below, then the difference $f(x^{k-1}) - f(x^k)$ approaches zero as $k \longrightarrow \infty$ and

$$\lim_{k \to \infty} \left\{ \frac{(z^k)^T \nabla f(x^{k-1})}{\|z^k\|} \right\} = 0. \tag{10.173}$$

Now we would like to say that (10.173) implies

$$\lim_{k \to \infty} \{\nabla f(x^k)\} = 0. \tag{10.174}$$

This assertion depends on the choice of z^k. Ortega and Rheinboldt define a sequence $\{z^k\}$ of nonzero vectors in R^n as **gradient related** to the sequence $\{x^k\}$, $x^k \in R^n$, if for some forcing function σ

$$\frac{|(z^k)^T \nabla f(x^{k-1})|}{\|z^k\|} \geq \sigma(\|\nabla f(x^{k-1})\|). \tag{10.175}$$

For example, if $z^k = -\nabla f(x^{k-1})$, then this relation holds with $\sigma(t) = t$. It follows that if the sequence of points $\{x^k\}$ generated by a descent method satisfies (10.173) for a gradient-related sequence $\{z^k\}$, then (10.175) implies (10.174), and convergence to a stationary point of f has been obtained. In particular, the directions chosen by the steepest descent method, certain limited-step Newton methods, and conjugate gradient methods were shown to be gradient related. Thus convergence results similar to those presented in this chapter can be alternatively derived by using this approach. An interesting extension of the convergence results is that descent methods in which there is a periodic restart with gradient-related vectors will also converge, no matter how the intermediate descent directions are chosen.

Finally, it should be noted that many aspects of conjugate gradient methods are not fully understood at present. Such questions as how periodic restart, the choice of a starting search direction, or avoidance of exact line searches affect the performance of conjugate gradient methods are still to be answered.

EXERCISES

10.A. One "classical" test function for every unconstrained optimization algorithm is Rosenbrock's curved valley function in two variables [32]:

$$f(x) = 100[x_2 - (x_1)^2]^2 + (1 - x_1)^2, \tag{10.176}$$

which has a unique minimum at $x^* = (1, 1)$.

(a) Observe the behavior of Newton's method on this function by computing the quadratic approximations of f, as given by (10.5), at the points $(-1.2, 1)$, $(0, 0)$, and $(1.5, 1)$ and finding their minima. Compare the values of f at the minima predicted by the quadratic functions with the values of f at those points. Did you obtain a function decrease in every case?

(b) Evaluate the gradients of f at each of the foregoing three points and compare the steeepest descent directions with the directions pointing toward the minimum of f.

10.B. What is the Newton direction for the function given by (9.194), starting at $x^0 = (-4, 6)$? Compute the next two points reached by Newton's method.

10.C. Show that the problem of finding the steepest descent direction, as posed in (10.18) and (10.19), is a convex programming problem whose solution is given by (10.20). Show the same for the case in which (10.19) is replaced by (10.23) and the solution is given by (10.24).

10.D. Determine the relative efficiencies η of the steepest descent method, as given by (10.27), on the quadratic functions (9.188) and (9.189) of Exercise 9.A at their respective starting points x^0. Compute the lower bounds on η as given by (10.29) for both cases.

10.E. Suppose that we modify the steepest descent method by letting

$$x^{k+1} = x^k - \theta_k^* B(x^k)\nabla f(x^k), \tag{10.177}$$

where $B(z)$ is an $n \times n$ positive definite matrix whose elements are continuous functions of z and θ_k^* is a scalar determined by minimizing $f(x^k - \theta_k B(x^k)\nabla f(x^k))$. Prove an analog of Theorem 10.1 for this method.

10.F. Draw a flow diagram of the Goldstein-Price algorithm. Apply the algorithm to a search for the minimum of the function given by (9.194), starting at $(-4, 6)$. Perform three iterations and compare the matrices $Q(x^k)$ that are obtained with the actual Hessian matrices $\nabla^2 f(x^k)$.

10.G. Prove Equations (10.93) to (10.95).

10.H. Derive formula (10.106) from the following construction of a descent algorithm for minimizing a differentiable function on R^n. Choose a point $x^0 \in R^n$ and a search direction $z^1 = -\nabla f(x^0)$. Perform a line search along z^1 ending at x^1; that is $(x^1)^T\nabla f(x^1) = 0$. Let z^2 satisfy the following requirements:

(a) It is orthogonal to γ^1.

(b) It should be a descent direction; that is, $(z^2)^T\nabla f(x^1) < 0$.

(c) It should be a linear combination of z^1 and $\nabla f(x^1)$.

Show that z^2 can be written

$$z^2 = -\nabla f(x^1) + \beta_1 z^1, \tag{10.178}$$

where

$$\beta_1 = \frac{(\gamma^1)^T\nabla f(x^1)}{(\gamma^1)^T z^1}. \tag{10.179}$$

If all subsequent z^k are defined in a similar way, show that the corresponding β_k are given by (10.106). Note that these derivations do not assume that f is quadratic.

10.I. Show that formulas (10.95), (10.105), (10.106), and (10.107) yield the same search directions if applied to minimizing a quadratic function with a positive definite Q matrix. Under what conditions will (10.106) and (10.107) yield the same points for general functions?

10.J. Carry out three iterations of the Fletcher-Reeves algorithm on the function given by (9.194), starting at $(-4, 6)$, and letting each line search be a single quadratic approximation as described in Chapter 8. Compare the points reached with the ones obtained in Exercise 10.F.

10.K. Shah, Buehler, and Kempthorne [33] proposed the following algorithm, called the **method of parallel tangents** or, briefly, **partan,** for minimizing a differentiable function on R^n. Given x^0, let $z^1 = -\nabla f(x^0)$ and minimize f along z^1, arriving at the point x^1. Let $z^2 = -\nabla f(x^1)$ and minimize $f(x^1 + \alpha_2 z^2)$ with respect to α_2. Set $x^2 = x^1 + \alpha_2^* z^2$. Let $z^3 = x^2 - x^0$. This is called an acceleration direction. Minimize f along z^3, arriving at x^3 After arriving at x^3, move to $x^4, x^5, \ldots$ by performing successive line searches along negative gradient and acceleration directions alternately. That is, let

$$z^k = -\nabla f(x^{k-1}), \qquad k = 1, 2, 4, 6, \ldots \tag{10.180}$$

$$z^k = (x^{k-1} - x^{k-3}), \qquad k = 3, 5, 7, 9, \ldots. \tag{10.181}$$

Try this method on the quadratic function of Example 10.2.1, starting at $(-2, 4)$. Make at least four steps.

10.L. Prove the following properties of the partan method when applied to a quadratic function f on R^n with a positive definite Q matrix:

(a) The vectors $(x^1 - x^0), (x^3 - x^1), \ldots, (x^{2n-1} - x^{2n-3})$ are mutually conjugate with respect to Q.

(b) The points $x^3, x^5, \ldots, x^{2n-1}$ minimize f over affine sets generated by the vectors z^1 and $\nabla f(x^1), z^1, \nabla f(x^1)$ and $\nabla f(x^3), \ldots, z^1, \nabla f(x^1), \nabla f(x^3), \ldots$ and $\nabla f(x^{2n-3})$, respectively.

(c) The vectors $\nabla f(x^0), \nabla f(x^1), \nabla f(x^3), \ldots, \nabla f(x^{2n-3})$ are mutually orthogonal.

(d) The partan method possesses the quadratic termination property.

10.M. Derive Eqs. (10.158) through (10.161).

10.N. Establish relations (10.164) to (10.167) in Example 10.3.1.

REFERENCES

1. Akaike, H., "On a Successive Transformation of Probability Distribution and its Application to the Analysis of the Optimum Gradient Method," *Ann. Inst. Stat. Math. Tokyo*, **11**, 1–16 (1959).

2. BARTLE, R. G., *The Elements of Real Analysis*, 2nd ed., John Wiley & Sons, New York, 1976.

3. CAUCHY, A., "Méthode Générale pour la Résolution des Systéms d'Équations Simultanées," *Comp. Rend. Acad. Sci. Paris*, pp. 536–538 (1847).

4. COHEN, A., "Rate of Convergence for Root Finding and Optimization Algorithms," *SIAM J. Numer. Anal.*, **9**, 248–259 (1972).

5. CROWDER, H., and P. WOLFE, "Linear Convergence of the Conjugate Gradient Method," *IBM J. Res. & Dev.*, **16**, 431–433 (1972).

6. CURRY, H. B., "The Method of Steepest Descent for Non-Linear Minimization Problems," *Quart. Appl. Math.*, **2**, 258–261 (1944).

7. DANIEL, J. W., "The Conjugate Gradient Method for Linear and Nonlinear Operator Equations," *SIAM J. Numer. Anal.*, **4**, 10–26 (1967).

8. DANIEL, J. W., "Convergence of the Conjugate Gradient Method with Computationally Convenient Modifications," *Numerische Mathematik*, **10**, 125–131 (1967).

9. DANIEL, J. W., "A Correction Concerning the Convergence Rate for the Conjugate Gradient Method," *SIAM J. Numer. Anal.*, **7**, 277–280 (1970).

10. DANIEL, J. W., *The Approximate Minimization of Functionals*, Prentice-Hall, Englewood Cliffs, N.J., 1971.

11. ELKIN, R., "Convergence Theorems for Gauss-Seidel and Other Minimization Algorithms," Ph.D. dissertation, University of Maryland, College Park, 1968.

12. FLETCHER, R., and C. M. REEVES, "Function Minimization by Conjugate Gradients," *Computer J.*, **7**, 149–154 (1964).

13. FORSYTHE, G. E., "On the Asymptotic Directions of the s-Dimensional Optimum Gradient Method," *Numerische Mathematik*, **11**, 57–76 (1968).

14. GOLDFELD, S. M., R. E. QUANDT, and H. F. TROTTER, "Maximization by Quadratic Hill-Climbing," *Econometrica*, **34**, 541–551, (1966).

15. GOLDSTEIN, A. A., *Constructive Real Analysis*, Harper & Row, New York, 1967.

16. GOLDSTEIN, A. A., and J. F. PRICE, "An Effective Algorithm for Minimization," *Numerische Mathematik*, **10**, 184–189 (1967).

17. GREENSTADT, J., "On the Relative Efficiencies of Gradient Methods," *Math. Comp.*, **21**, 360–367 (1967).

18. HESTENES, M. R., and E. STIEFEL, "Methods of Conjugate Gradients for Solving Linear Systems," *J. Res. Nat. Bur. Stand.*, **49**, 409–436 (1952).

19. KANTOROVICH, L. V., and G. P. AKILOV, *Functional Analysis in Normed Spaces*, Pergamon Press, Oxford, 1964.

20. KLESSIG, R., and E. POLAK, "Efficient Implementations of the Polak-Ribière Conjugate Gradient Algorithm," *SIAM J. Control*, **10**, 524–549 (1972).

21. MCCORMICK, G. P., and K. RITTER, "Methods of Conjugate Directions Versus Quasi-Newton Methods," *Math. Prog.*, **3**, 101–116 (1972).

22. McCormick, G. P., and K. Ritter, "Alternative Proofs of the Convergence Properties of the Conjugate Gradient Method," *J. Optimization Theory & Appl.*, **13**, 497–518 (1974).

23. Murray, W., "Second Derivative Methods," in *Numerical Methods of Unconstrained Optimization*, W. Murray (Ed.), Academic Press, London, 1972.

24. Ortega, J. M., and W. C. Rheinboldt, "Local and Global Convergence of Generalized Linear Iterations," in *Numerical Solution of Nonlinear Problems*, J. M. Ortega, and W. C. Rheinboldt (Eds.), SIAM, Philadelphia, 1970.

25. Ortega, J. M., and W. C. Rheinboldt, *Iterative Solution of Nonlinear Equations in Several Variables*, Academic Press, New York, 1970.

26. Ortega, J. M., and W. C. Rheinboldt, "A General Convergence Result for Unconstrained Minimization Methods," *SIAM J. Numer. Anal.*, **9**, 40–43 (1972).

27. Ostrowski, A. M., *Solution of Equations and Systems of Equations*, 2nd ed., Academic Press, New York, 1966.

28. Polak, E., *Computational Methods in Optimization*, Academic Press, New York, 1971.

29. Polak, E., and G. Ribière, "Note sur la Convergence de Méthodes de Directions Conjugées," *Rev. Fr. Inform. Rech. Oper.*, **16-R1**, 35–43 (1969).

30. Polyak, B. T., "The Conjugate Gradient Method in Extremal Problems," *USSR Comp. Maths. and Math. Phys.*, **9**, No. 4, 94–112 (1969).

31. Ritter, K., "A Superlinearly Convergent Method for Unconstrained Minimization," in *Nonlinear Programming*, J. B. Rosen, O. L. Mangasarian, and K. Ritter (Eds.), Academic Press, New York, 1970.

32. Rosenbrock, H. H., "An Automatic Method for Finding the Greatest or Least Value of a Function," *Computer J.*, **3**, 175–184 (1960).

33. Shah, B. V., R. J. Buehler, and O. Kempthorne, "Some Algorithms for Minimizing a Function of Several Variables," *J. SIAM*, **12**, No. 1, 74–92 (1964).

34. Sorenson, H. W., "Comparison of Some Conjugate Direction Procedures for Function Minimization," *J. Franklin Inst.*, **288**, 421-441 (1969).

35. Wolfe, P., "The Method of Conjugate Gradients," lectures delivered at the NATO Advanced Study Institute on Mathematical Programming in Theory and Practice, Figueira da Foz, Portugal, June 1972.

11 VARIABLE METRIC ALGORITHMS

As noted earlier, minimization algorithms that are based on the assumption that a real differentiable function on R^n can be approximated in a neighborhood of a minimum by a quadratic function are usually highly successful in solving unconstrained optimization problems. One of the most important aspects of such algorithms is the estimation of second derivatives from calculated first derivatives. Most of the presently known methods that estimate the Hessian matrix of the function to be minimized, or its inverse, can be classified as variable metric algorithms, the subject of this chapter. Since the appearance of Davidon's work [11] in 1959 many additional methods have been proposed, methods that differ mainly in the way that the second derivative estimates are changed from one iteration to another and in the type and accuracy of the line searches performed. Only in the past few years have several unified approaches been developed for variable metric methods. In Section 11.1 we present such an approach by deriving a whole family of algorithms, each member of which possesses the quadratic termination property by defining search directions that are mutually conjugate if applied to minimizing a quadratic function with a positive definite Q matrix. Conjugate directions are thus the link between previous chapters and the present one. Another property of quadratic functions is embedded in an equation, called the secant relation, that characterizes the class of quasi-Newton algorithms to be discussed in Section 11.2. Variable metric methods can be quite successfully modified for the case in which only function values are calculated and even first derivatives must be approximated. Such methods are the subject of Section 11.3.

Finally, in Section 11.4 we present two other algorithms that already go one step beyond variable metric methods in the sense that they make use of

nonquadratic properties of the function to be minimized. These methods are new, and no extensive numerical experience is available yet. Their conceptual form may, however, represent a direction in which the next generations of unconstrained minimization algorithms will evolve.

This is the last chapter on unconstrained minimization—an important subject in its own right—and one that will be very useful in subsequent chapters on methods for constrained optimization.

11.1 A FAMILY OF VARIABLE METRIC ALGORITHMS

In Section 10.1 we discussed two classical optimization methods—Newton's method, in which the search direction is given by $z^{k+1} = -[\nabla^2 f(x^k)]^{-1}\nabla f(x^k)$, and the steepest descent method, in which $z^{k+1} = -\nabla f(x^k)$. We have seen that by defining a norm $x^T A x$, where A is a symmetric positive definite matrix, the direction that minimizes the directional derivative of a differentiable function f at a point x^k is given by $-A^{-1}\nabla f(x^k)$. Such a direction can be more efficient than the steepest descent direction, as can easily be seen by an example of a quadratic function with ellipsoidal contours $f(x) = \frac{1}{2}x^T Q x$ on which the norm $x^T Q x$ is defined. Here, of course, the non-Euclidean steepest descent direction is $-Q^{-1}\nabla f(x^k)$, which is also the Newton direction, since $Q = \nabla^2 f(x^k)$. Note that the norm function is a special case of a metric [3]. Since methods that do not use second derivatives are usually preferred, because of the extra work involved in such computations, the suggestion was made to approximate the inverse Hessian required for computing the Newton direction or, equivalently, to change the non-Euclidean norm (metric) at each iteration. An iterative minimization algorithm that, at the current point x^k, uses a search direction z^{k+1}, given by

$$z^{k+1} = -H_k^T \nabla f(x^k), \tag{11.1}$$

where H_k is an $n \times n$ matrix, varying from iteration to iteration, is called a **variable metric** method. Note that this definition is somewhat broader than the ones that can be found in some references, where H_k is restricted to be symmetric and positive definite. Indeed, as we shall see later, such symmetric positive definite matrices H_k form the basis of a distinguished subclass of variable metric algorithms. The similarity of (11.1) to the Newton direction also justifies calling these methods quasi-Newton algorithms, especially since we can choose the matrices H_k so that the minimization of a quadratic function $\frac{1}{2}x^T Q x$ on R^n terminates after n iterations with $H_n^T = Q^{-1}$ and, in the case of more general nonlinear functions, the H_k tend to $[\nabla^2 f(x^k)]^{-1}$, the inverse Hessian at the optimum. Unfortunately, research workers in unconstrained optimization do not use a unified terminology. The reader can find

a large variety of not necessarily coinciding definitions of "variable metric" and "quasi-Newton" methods in the literature. We shall restrict the name quasi-Newton to that subclass of variable metric methods in which the matrices H_k satisfy the so-called secant relation. These methods will be discussed in the next section.

The first and perhaps best-known variable metric algorithm was proposed by Davidon [11]. It was subsequently simplified by Fletcher and Powell [20], and we shall refer to it as the Davidon-Fletcher-Powell or DFP method. This algorithm for minimizing a differentiable function f on R^n can be described as follows: Choose an $x^0 \in R^n$ and an arbitrary symmetric $n \times n$ positive definite matrix H_0. At iteration k, we have a point x^{k-1} and a matrix H_{k-1}. Compute the search direction z^k by (11.1). Find the exact minimum of $f(x^{k-1} + \alpha_k z^k)$ with respect to α_k by one of the methods of Chapter 8 and let α_k^* be the value of α_k at this minimum. Set $x^k = x^{k-1} + \alpha_k^* z^k$ and compute the matrix H_k by the updating formula

$$H_k = H_{k-1} + \frac{p^k(p^k)^T}{(p^k)^T\gamma^k} - \frac{H_{k-1}\gamma^k(\gamma^k)^T H_{k-1}}{(\gamma^k)^T H_{k-1}\gamma^k}, \tag{11.2}$$

where

$$p^k = x^k - x^{k-1} \qquad \gamma^k = \nabla f(x^k) - \nabla f(x^{k-1}) \tag{11.3}$$

If $\|\nabla f(x)^k\| \leq \epsilon$, stop; otherwise proceed to iteration $k + 1$. Computational experience with the DFP algorithm has generally been satisfactory. If the function to be minimized is quadratic with a positive definite Q matrix, some important properties of the algorithm can be established. It has quadratic termination, the search directions are mutually conjugate with respect to Q, and at iteration n we have $H_n = Q^{-1}$. Thus the algorithm has features of conjugate directions as well as features of Newton-type algorithms. Sorenson [51] has derived several interesting properties of the DFP method and related them to some methods of conjugate gradients. We shall see below that the DFP method is only one choice in an infinite variety of variable metric methods possessing the same properties

Parenthetically, we can note here that the conjugate gradient algorithms discussed in Chapter 10 can also be viewed as variable metric methods, since, for example, the formula for the search direction of the Fletcher-Reeves method, as given by (10.96), can be rewritten

$$z^{k+1} = -H_k^T \nabla f(x^k) \tag{11.4}$$

where

$$H_k = I - \frac{\nabla f(x^k)(z^k)^T}{(\nabla f(x^{k-1}))^T \nabla f(x^{k-1})}. \tag{11.5}$$

In the following discussion we shall be concerned with the problem of modifying the matrix H_k from one iteration to another. The forthcoming derivations are mainly based on the work of Huang [30], who developed an elegant approach to a large class of variable metric algorithms.

Since quadratic termination seems to be a desirable property of minimization algorithms, and, as we have already seen, mutually conjugate search directions with exact one-dimensional minimizations are sufficient conditions for quadratic termination in the case of a function given by (10.62), we can require that for $k = 1, 2, \ldots$ all search directions should be mutually conjugate with respect to Q. That is,

$$(z^j)^T Q z^{k+1} = -(z^j)^T Q H_k^T \nabla f(x^k) = 0, \qquad j = 1, \ldots, k. \tag{11.6}$$

If we also assume that we move from point to point as outlined in Theorem 10.2, then

$$(z^j)^T \nabla f(x^k) = 0, \qquad j = 1, \ldots, k. \tag{11.7}$$

Equations (11.6) and (11.7) can be simultaneously satisfied by a matrix H_k, given by

$$H_k Q z^j = \omega z^j, \qquad j = 1, \ldots, k \tag{11.8}$$

where ω is an arbitrary real number. Suppose that we decide to update the matrices H_k at each iteration according to the relation

$$H_k = H_{k-1} + \Delta H_k, \qquad k = 1, 2, \ldots \tag{11.9}$$

where the matrix ΔH_k is called the **correction matrix**. Let us find a general form for the correction matrix. Rewriting (11.8) for $k - 1$ and subtracting from (11.8), we obtain

$$\Delta H_k Q z^j = 0, \qquad j = 1, \ldots, k - 1. \tag{11.10}$$

From (11.8) and (11.9) we get

$$\Delta H_k Q z^k = \omega z^k - H_{k-1} Q z^k. \tag{11.11}$$

Since $p^j = \alpha_j^* z^j$, $\gamma^j = Q p^j$, the last two relations become

$$\Delta H_k \gamma^j = 0, \qquad j = 1, \ldots, k - 1 \tag{11.12}$$

$$\Delta H_k \gamma^k = \omega p^k - H_{k-1} \gamma^k. \tag{11.13}$$

A simple form for the correction matrix considered by Huang is given by

$$\Delta H_k = p^k (u^k)^T + H_{k-1} \gamma^k (v^k)^T, \tag{11.14}$$

where u^k, v^k are n vectors to be specified. First we see that, in order to satisfy (11.12) and (11.13), these vectors can be chosen so that

$$(u^k)^T \gamma^j = \begin{cases} 0 & j = 1, \ldots, k-1 \\ \omega & j = k \end{cases} \tag{11.15}$$

$$(v^k)^T \gamma^j = \begin{cases} 0 & j = 1, \ldots, k-1 \\ -1 & j = k. \end{cases} \tag{11.16}$$

Another important factor in the choice of u^k and v^k is to use only information available at the point x^k and the preceding point x^{k-1}, Since we assumed that the search directions are conjugate with respect to Q, we have

$$(z^j)^T Q z^k = 0, \qquad j = 1, \ldots, k-1 \tag{11.17}$$

or

$$(\gamma^j)^T p^k = 0, \qquad j = 1, \ldots, k-1. \tag{11.18}$$

From (11.7) it follows that

$$(z^j)^T \gamma^k = 0, \qquad j = 1, \ldots, k-1 \tag{11.19}$$

and by (11.8),

$$(z^j)^T Q H_{k-1}^T \gamma^k = 0, \qquad j = 1, \ldots, k-1 \tag{11.20}$$

or

$$(\gamma^j)^T H_{k-1}^T \gamma^k = 0, \qquad j = 1, \ldots, k-1. \tag{11.21}$$

As a result of (11.18) and (11.21), we can see that u^k and v^k will satisfy (11.15) and (11.16) for $j = 1, \ldots, k-1$ if they are given by

$$u^k = a_{11}^k p^k + a_{12}^k H_{k-1}^T \gamma^k \tag{11.22}$$

$$v^k = a_{21}^k p^k + a_{22}^k H_{k-1}^T \gamma^k, \tag{11.23}$$

where the a_{il}^k are arbitrary parameters to be chosen so that (11.15) and (11.16) also hold for $j = k$.

Summarizing, if the matrices H_k are updated according to the formula

$$H_k = H_{k-1} + p^k (u^k)^T + H_{k-1} \gamma^k (v^k)^T, \tag{11.24}$$

where u^k and v^k are given by (11.22) and (11.23) and such that

$$(u^k)^T \gamma^k = \omega, \tag{11.25}$$

$$(v^k)^T \gamma^k = -1, \tag{11.26}$$

then the search directions will be mutually conjugate, provided that an exact line search is carried out at each iteration. Note that the updating formula can be also written as

$$H_k = H_{k-1} + C_k^1 A_k (C_k^2)^T, \tag{11.27}$$

where C_k^1 and C_k^2 are the $n \times 2$ matrices

$$C_k^1 = [p^k \quad H_{k-1}\gamma^k] \qquad C_k^2 = [p^k \quad H_{k-1}^T\gamma^k] \tag{11.28}$$

and A_k is a 2×2 matrix given by

$$A_k = \begin{bmatrix} a_{11}^k & a_{12}^k \\ a_{21}^k & a_{22}^k \end{bmatrix}. \tag{11.29}$$

This discussion can be further generalized by allowing the number ω to vary from iteration to iteration [48]. Since no practical implementation of this generalization has been proposed to date, we discuss only the case where ω is fixed. Let us call the collection of matrices H_k, updated by (11.24) and satisfying (11.22), (11.23), (11.25), and (11.26), **Huang's family of matrices.**

Letting $\omega = 1$ and $a_{12}^k = a_{21}^k = 0$, the reader can verify that the H_k matrix of the DFP method, as given in (11.2), is a member of Huang's family. The following results will show that in order to minimize a quadratic function with a positive definite Q matrix, the choice of the parameters in the updating formula of the H_k matrices is unimportant, since, for a given H_0 and x^0, all members of Huang's family generate the same sequence of points. For a general nonlinear function, the sequence of points generated will be shown to depend only on ω. In what follows we shall often refer to variable metric algorithms that use matrices H_k belonging to Huang's family. Such algorithms are defined precisely as the DFP method except that the general updating formula (11.24) is used instead of (11.2). The next lemma, in a slightly different form, is due to Shanno and Kettler [50].

Lemma 11.1

Suppose that f is a differentiable real function on R^n. Let x^{k-1}, x^k, and H_{k-1} be given such that $(p^k)^T\nabla f(x^{k-1}) \neq 0$ and $(p^k)^T\nabla f(x^k) = 0$. Then the direction of the vector z^{k+1}, specified by (11.1), *will be identical for all H_k matrices in Huang's family.*

Proof. By (11.1),

$$z^{k+1} = -H_k^T\nabla f(x^k) = -H_{k-1}^T\nabla f(x^k) - v^k(\gamma^k)^T H_{k-1}^T\nabla f(x^k) \tag{11.30}$$

$$= -H_{k-1}^T\nabla f(x^k) - (a_{21}^k p^k + a_{22}^k H_{k-1}^T\gamma^k)(\gamma^k)^T H_{k-1}^T\nabla f(x^k). \tag{11.31}$$

Noting that

$$H_{k-1}^T \gamma^k = H_{k-1}^T \nabla f(x^k) + \frac{p^k}{\alpha_k^*}, \tag{11.32}$$

we get, after some algebraic manipulations,

$$z^{k+1} = -\{1 + a_{22}^k (\gamma^k)^T H_{k-1}^T \nabla f(x^k)\}\left[I - \frac{p^k(\gamma^k)^T}{(p^k)^T \gamma^k}\right] H_{k-1}^T \nabla f(x^k). \tag{11.33}$$

Since the expression in the square brackets depends only on the points x^{k-1}, x^k and the gradient difference $\gamma^k = \nabla f(x^k) - \nabla f(x^{k-1})$, we conclude that the direction of the vector z^{k+1} (but not its magnitude) is independent of the parameters in the correction matrix. ■

The "nonorthogonality" condition $(p^k)^T \nabla f(x^{k-1}) \neq 0$ in Lemma 11.1 ensures that $f(x^{k-1}) > f(x^k)$ for quadratic functions (see (10.66) in the preceding chapter). The condition $(p^k)^T \nabla f(x^k) = 0$ is necessary for minimizing f along the direction z^k. It should be noted that in trying to apply the preceding lemma inductively, a choice must be made as to how the α_k^* are selected. For example, if f is a strictly convex function with bounded level sets, then α_k^* is uniquely determined by the solution of the equation

$$(z^k)^T \nabla f(x^{k-1} + \alpha_k z^k) = 0. \tag{11.34}$$

If, however, f is a general nonlinear function, more than one value of α_k^* may solve (11.34). In this case, we can decide on the choice of α_k^* by the "Curry rule"—that is, the smallest $|\alpha_k^*|$ that solves (11.34) and such that $f(x^{k-1} + \alpha_k^* z^k) < f(x^{k-1})$; see Fig. 11.1.

Let us show that the sequence of points generated by any variable metric algorithm using an H_k matrix that belongs to Huang's family will be identical in the case of minimizing a quadratic function with a positive definite Q matrix. Define the n vector q^{k+1} by

$$q^{k+1} = -\left[I - \frac{p^k(\gamma^k)^T}{(p^k)^T \gamma^k}\right] H_{k-1}^T \nabla f(x^k) \tag{11.35}$$

and the real number μ_{k+1} by

$$\mu_{k+1} = \{1 + a_{22}^k (\gamma^k)^T H_{k-1}^T \nabla f(x^k)\}. \tag{11.36}$$

The vector q^{k+1} is the common direction for all members of Huang's family as given by (11.33). That is, we can write

$$z^{k+1} = \mu_{k+1} q^{k+1}. \tag{11.37}$$

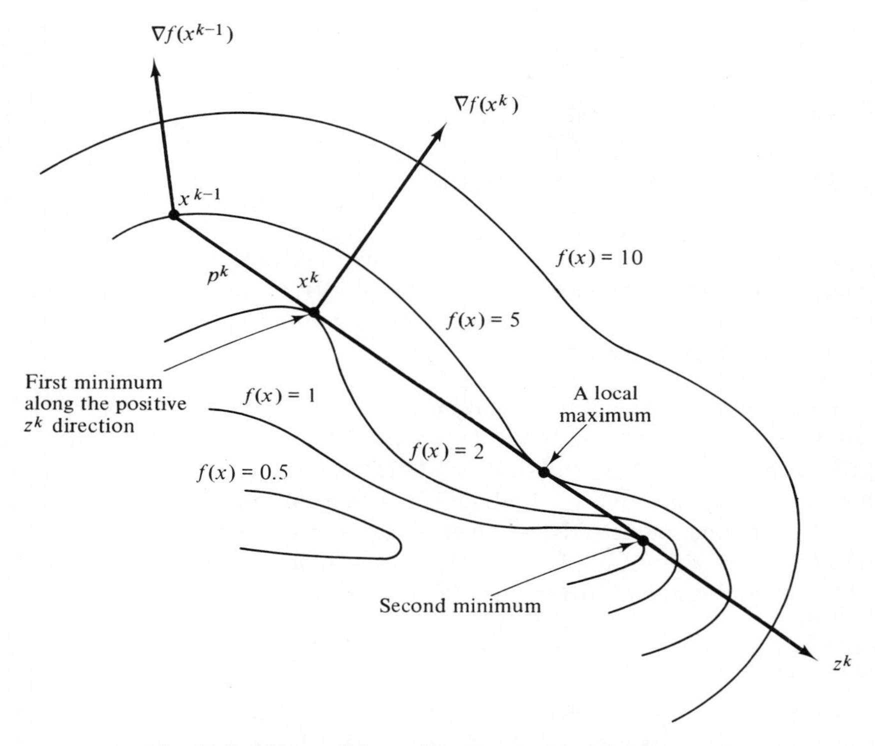

Fig. 11.1 The conditions of Lemma 11.1 and the Curry rule.

Now $p^{k+1} = \alpha^*_{k+1} z^{k+1} = \alpha^*_{k+1}\mu_{k+1} q^{k+1}$, and by (10.64), we get

$$p^{k+1} = -\frac{(q^{k+1})^T \nabla f(x^k)}{(q^{k+1})^T Q q^{k+1}} q^{k+1}. \tag{11.38}$$

Thus the next point, x^{k+1}, is independent of the parameters of the correction matrices and is the same for all members of Huang's family. In terms of the direction vector q^{k+1}, we can write (10.66) as

$$f(x^k) - f(x^{k+1}) = \frac{[(q^{k+1})^T \nabla f(x^k)]^2}{2(q^{k+1})^T Q q^{k+1}} \tag{11.39}$$

and the above difference in function values is positive if any only if

$$(q^{k+1})^T \nabla f(x^k) \neq 0. \tag{11.40}$$

From (11.34) and (11.35) we have

$$(q^{k+1})^T\nabla f(x^k) = -(\nabla f(x^k))^T H_{k-1}\left[\nabla f(x^k) - \frac{\gamma^k(p^k)^T\nabla f(x^k)}{(p^k)^T\gamma^k}\right] \tag{11.41}$$

$$= -(\nabla f(x^k))^T H_{k-1}\nabla f(x^k). \tag{11.42}$$

We can express H_{k-1} in terms of H_{k-2}, using the general updating formulas (11.22) to (11.26), and obtain

$$(\nabla f(x^k))^T H_{k-1}\nabla f(x^k) = (\nabla f(x^k))^T H_{k-2}\nabla f(x^k). \tag{11.43}$$

Continuing backward, we get

$$\nabla f(x^k))^T H_{k-1}\nabla f(x^k) = (\nabla f(x^k))^T H_0\nabla f(x^k). \tag{11.44}$$

Hence (11.40) is satisfied if the right-hand side of (11.44) is nonzero. It follows that for a descent algorithm we need

$$(\nabla f(x^k))^T H_0\nabla f(x^k) \neq 0, \qquad k = 0, \ldots, n-1 \tag{11.45}$$

and a simple condition for satisfying (11.45) with arbitrary gradient vectors is that $\frac{1}{2}(H_0 + H_0^T)$ be positive or negative definite. In particular, we can choose H_0 to be symmetric and positive definite, a choice taken in many variable metric algorithms.

Let the search directions be given by

$$q^1 = -H_0^T\nabla f(x^0) \tag{11.46}$$

$$q^{k+1} = -\left[I - \frac{p^k(\gamma^k)^T}{(p^k)^T\gamma^k}\right]H_{k-1}^T\nabla f(x^k), \qquad k = 1, \ldots. \tag{11.47}$$

The matrix H_{k-1} can again be expressed in terms of H_{k-2}, and, by continuing recursively, we finally obtain

$$q^{k+1} = -\left[I - \sum_{j=1}^{k}\frac{p^j(\gamma^j)^T}{(p^j)^T\gamma^j}\right]H_0^T\nabla f(x^k), \qquad k = 1, \ldots. \tag{11.48}$$

Starting at an arbitrary point $x^0 \in R^n$ with any starting matrix H_0 such that $\frac{1}{2}(H_0 + H_0^T)$ is positive or negative definite, we use (11.46) to find the next point x^1 by minimizing f along q^1. From x^1 we move in the direction q^2 as given in (11.48). Since q^2 depends on x^0, x^1, $\nabla f(x^0)$, $\nabla f(x^1)$, and H_0, it will be the same for all updating formulas of Huang's family. Minimizing along q^2, we arrive at x^2 and determine a new direction q^3 by using (11.48) again, and so we get a sequence of points $x^0, x^1, x^2, \ldots, x^n$. We have just proven

Theorem 11.2

For a given x^0 and H_0, let the sequence of points $x^1, \ldots, x^n$ be generated by a variable metric algorithm that uses a matrix-updating formula that belongs to Huang's family. If the function to be minimized is quadratic on R^n with a positive definite Q matrix, then the same sequence can be obtained by any other member of Huang's family.

Dixon [14, 15] extended the last result for general nonlinear functions by showing that subfamilies of variable metric algorithms based on matrices that belong to Huang's family generate identical sequences of points:

Theorem 11.3

Let f be a differentiable real function on R^n. Suppose that x^0, H_0 are given. Then the sequence of points $x^0, x^1, x^2, \ldots$, generated by a variable metric algorithm that uses matrices H_k that belong to Huang's family will depend only on the parameter ω chosen, provided that $(p^k)^T\gamma^k \neq 0$, $a_{21}^k + a_{22}^k/\alpha_k^* \neq 0$ for all k and the α_k^* are chosen uniquely so that $(z^k)^T\nabla f(x^k) = 0$.

Proof. For a given point x^0 and matrix H_0, the direction z^1 and, consequently, x^1, p^1, γ^1 are common to all variable metric algorithms. By Lemma 11.1, it follows that z^2, x^2, p^2, and γ^2 will also be common to all algorithms. Now assume that there is a common sequence of points $x^0, x^1, \ldots, x^k$ for all algorithms. We shall show that x^{k+1} will also be the same for all of them. By Lemma 11.1, we have for $k = 1, 2, \ldots$ and for all members of Huang's family

$$z^{k+1} = -\{1 + a_{22}^k(\gamma^k)^T H_{k-1}^T \nabla f(x^k)\} R_k H_{k-1}^T \nabla f(x^k), \tag{11.49}$$

where

$$R_k = I - \frac{p^k(\gamma^k)^T}{(p^k)^T\gamma^k}. \tag{11.50}$$

Note that $R_k p^k = 0$ for all k. Hence,

$$R_{k+1}p^{k+1} = -\alpha_{k+1}^*\{1 + a_{22}^k(\gamma^k)^T H_{k-1}^T \nabla f(x^k)\} R_{k+1} R_k H_{k-1}^T \nabla f(x^k) = 0. \tag{11.51}$$

Since by (11.16), (11.23), and the hypotheses,

$$1 + a_{22}^k(\gamma^k)^T H_{k-1}^T \nabla f(x^k) = -\left(a_{21}^k + \frac{a_{22}^k}{\alpha_k^*}\right)(p^k)^T\gamma^k \neq 0, \tag{11.52}$$

we must have

$$R_{k+1} R_k H_{k-1}^T \nabla f(x^k) = 0. \tag{11.53}$$

By (11.15), (11.22), (11.32), and (11.50),

$$R_k u^k = u^k - \frac{p^k(\gamma^k)^T u^k}{(p^k)^T\gamma^k} = u^k - \frac{\omega p^k}{(p^k)^T\gamma^k} = a_{12}^k R_k H_{k-1}^T \nabla f(x^k). \tag{11.54}$$

Similarly,

$$R_k v^k = v^k - \frac{p^k(\gamma^k)^T v^k}{(p^k)^T\gamma^k} = v^k + \frac{p^k}{(p^k)^T\gamma^k} = a_{22}^k R_k H_{k-1}^T \nabla f(x^k) \tag{11.55}$$

and by (11.53),

$$R_{k+1} u^k = \frac{\omega R_{k+1} p^k}{(p^k)^T\gamma^k} \tag{11.56}$$

$$R_{k+1} v^k = -\frac{R_{k+1} p^k}{(p^k)^T\gamma^k}. \tag{11.57}$$

Now

$$R_k H_{k-1}^T = R_k H_{k-2}^T + R_k u^{k-1}(p^{k-1})^T + R_k v^{k-1}(\gamma^{k-1})^T H_{k-2}^T \tag{11.58}$$

$$= R_k H_{k-2}^T + \omega R_k \frac{p^{k-1}(p^{k-1})^T}{(p^{k-1})^T\gamma^{k-1}} - R_k \frac{p^{k-1}(\gamma^{k-1})^T}{(p^{k-1})^T\gamma^{k-1}} H_{k-2}^T \tag{11.59}$$

$$= R_k R_{k-1} H_{k-2}^T + \omega R_k \frac{p^{k-1}(p^{k-1})^T}{(p^{k-1})^T\gamma^{k-1}} \tag{11.60}$$

and, by recursion, we obtain

$$R_k H_{k-1}^T = \prod_{i=1}^{k} R_i H_0^T + \omega \sum_{j=1}^{k-1} \left(\prod_{i=j+1}^{k} R_i \right) \frac{p^j(p^j)^T}{(p^j)^T\gamma^j}. \tag{11.61}$$

Substituting into (11.49), we conclude that z^{k+1}, and hence x^{k+1}, depends only on the value of the parameter ω chosen. ■

Many variable metric algorithms have been formulated since the appearance of Davidon's method, and most use correction matrices that belong to Huang's family. Let us consider a few examples.

An important subfamily of such correction matrices can be constructed by the following considerations. At iteration k, we are given points x^{k-1}, x^k, a matrix H_{k-1}, and the next matrix is sought. We expand the gradient of f

$$\nabla f(x^{k-1}) \cong \nabla f(x^k) + \nabla^2 f(x^k)(x^{k-1} - x^k) \tag{11.62}$$

or

$$\gamma^k \cong \nabla^2 f(x^k) p^k. \tag{11.63}$$

If $\nabla^2 f$ is nonsingular, we have

$$p^k \cong [\nabla^2 f(x^k)]^{-1} y^k. \tag{11.64}$$

Note that, for a quadratic function f, the Hessian $\nabla^2 f$ is a constant matrix; and if it is nonsingular, then (11.64) exactly holds for any step p^k. Since we are interested in methods that do not use inverse Hessian matrices directly but only approximate them, such as the secant method of Chapter 8, we can define a matrix H_k that approximates $[\nabla^2 f(x^k)]^{-1}$ and satisfies the so-called **secant relation**

$$p^k = H_k y^k. \tag{11.65}$$

Several properties of methods based on the secant relation will be discussed in the next section. Here we only mention that if the matrix H_k is updated by a correction matrix of Huang's family so that (11.65) holds, then it must satisfy

$$p^k = H_{k-1} y^k + p^k (u^k)^T y^k + H_{k-1} y^k (v^k)^T y^k \tag{11.66}$$

and it follows from (11.25) and (11.26) that we must choose $\omega = 1$. If we further restrict the choice of parameters so that the matrix A_k in (11.29) is symmetric—that is, $a_{12}^k = a_{21}^k$—then only one free parameter, say a_{12}^k, remains.

It can be shown that, starting with a symmetric matrix H_0, all subsequent matrices H_k will be symmetric. They are given by

$$H_k = H_{k-1} - \frac{H_{k-1} y^k (y^k)^T H_{k-1}}{(y^k)^T H_{k-1} y^k} + \frac{p^k (p^k)^T}{(p^k)^T y^k} - a_{12}^k \frac{(p^k)^T y^k}{(y^k)^T H_{k-1} y^k} w^k (w^k)^T, \tag{11.67}$$

where

$$w^k = H_{k-1} y^k - \frac{(y^k)^T H_{k-1} y^k}{(p^k)^T y^k} p^k. \tag{11.68}$$

We remark, parenthetically, that there is one unfortunate choice for a_{12}^k, in which case algorithms based on these formulas break down. The reader will be asked to show that if

$$a_{12}^k (p^k)^T y^k = \frac{(\nabla f(x^{k-1}))^T H_{k-1} \nabla f(x^{k-1})}{(\nabla f(x^k))^T H_{k-1} \nabla f(x^k)}, \tag{11.69}$$

then $z^{k+1} = 0$.

A subfamily of variable metric algorithms that are based on the correction matrices given by (11.67) and (11.68) was suggested by Broyden [7] and subsequently investigated by him [8, 9]. The correction matrix of the DFP method is clearly obtained by letting $a_{12}^k = 0$ in (11.67).

One important aspect of the Broyden updating formula (11.67) is that if H_0 is chosen to be positive definite and $a_{12}^k \leq 0$, then all H_k will also be positive definite [7]. Several good reasons exist as to why positive definite matrices H_k are preferred. The first is that, in the neighborhood of a minimum, most functions can be reasonably well approximated by a second-order Taylor expansion (quadratic function) having a positive definite Hessian matrix. Because the H_k matrices are approximations of the inverse Hessian, requiring positive definiteness of the H_k seems to be a good choice. Next, since in this section we are discussing variable metric methods that require exact line searches to find α_k^* at each iteration, these searches should be as efficient as possible. We have already seen that the direction $z^k = -H_{k-1}^T \nabla f(x^{k-1})$ is a descent direction if H_{k-1} is positive definite. Thus, in most cases, we need only search in the positive α_k direction for the minimum of f along z^k. Another reason for using positive definite matrices is to avoid singularity of the H_k. A singular matrix H_k at iteration k may cause the breakdown of most variable metric algorithms, for if $H_k y = 0$ for some nonzero vector $y \in R^n$, then all subsequent search directions z^{k+r} will be orthogonal to y and so will be restricted to an affine subset of R^n. Consequently, they do not span the whole space R^n, and the unconstrained minimum generally cannot be attained.

An updating formula that satisfies the secant relation is given by setting $a_{12}^k = a_{21}^k, \omega = 1$, and $a_{22}^k = 0$. In this case no free parameter is left in Huang's family of correction matrices, and we get a special case of (10.67) by

$$H_k = H_{k-1} + \left[1 + \frac{(y^k)^T H_{k-1} y^k}{(p^k)^T y^k}\right] \frac{p^k (p^k)^T}{(p^k)^T y^k} - \frac{p^k (y^k)^T H_{k-1}}{(p^k)^T y^k} - \frac{H_{k-1} y^k (p^k)^T}{(p^k)^T y^k}. \tag{11.70}$$

Algorithms based on this formula were suggested by Broyden [8, 9] Fletcher [19], and Shanno [49]. It should be noted that, by Theorem 11.3, all members of Broyden's subfamily of methods yield an identical sequence of points if applied to a general nonlinear function, provided that the hypotheses of this theorem are satisfied. This result has also been observed by Dixon [14, 15]. Some of the above-mentioned algorithms, however, do not perform exact line searches and then a different sequence of the x^k may result. For example, a variable metric algorithm, suggested by Fletcher [19], and the DFP method will not behave identically, since the choice of step length α_k in Fletcher's method is different from the one in the DFP algorithm. We shall return to the question of avoiding line searches later in the chapter.

Another choice of the parameters is to let $u^k = -v^k$. By (11.25) and (11.26), this also implies $\omega = 1$, and we get

$$H_k = H_{k-1} + (p^k - H_{k-1} y^k)(a_{11}^k p^k + a_{12}^k H_{k-1}^T y^k)^T. \tag{11.71}$$

Setting $a_{11}^k = 0$ and using (11.25), the last formula becomes

$$H_k = H_{k-1} + \frac{(p^k - H_{k-1}\gamma^k)(\gamma^k)^T H_{k-1}}{(\gamma^k)^T H_{k-1}\gamma^k}. \tag{11.72}$$

Similarly, by setting $a_{12}^k = 0$, we obtain

$$H_k = H_{k-1} + \frac{(p^k - H_{k-1}\gamma^k)(p^k)^T}{(p^k)^T\gamma^k}. \tag{11.73}$$

Correction matrices appearing in (11.72) and (11.73) were suggested by Pearson and McCormick, respectively [40]. Still another correction matrix can be obtained by letting $a_{11}^k = a_{22}^k = -a_{12}^k = -a_{21}^k$ and $\omega = 1$. The following formula results:

$$H_k = H_{k-1} + \frac{(p^k - H_{k-1}\gamma^k)(p^k - H_{k-1}^T\gamma^k)^T}{(p^k - H_{k-1}^T\gamma^k)^T\gamma^k}. \tag{11.74}$$

If H_0 is chosen to be symmetric, then all subsequent matrices H_k will also be symmetric and the correction matrix will have rank one. An algorithm based on (11.74) was proposed by Murtagh and Sargent [36], who, in a later work [37], also discussed computational aspects of several variable metric algorithms.

Example 11.1.1

Let us illustrate how a variable metric algorithm is used in finding the minimum of the quadratic function considered in some previous examples. We have

$$\min f(x) = \tfrac{3}{2}(x_1)^2 + \tfrac{1}{2}(x_2)^2 - x_1x_2 - 2x_1. \tag{11.75}$$

Since every algorithm belonging to Huang's family generates the same sequence of points by Theorem 11.2, the particular choice of the correction formula is unimportant. Let us choose, say, the DFP method. Suppose that we start at $x^0 = (-2, 4)$ and let H_0 be the 2×2 identity matrix. Now $\nabla f(x^0) = (-12, 6)^T$, and thus

$$z^1 = -H_0^T\nabla f(x^0) = \begin{pmatrix} 12 \\ -6 \end{pmatrix}. \tag{11.76}$$

In a comparison with Example 10.2.1, where the same function was minimized by the Fletcher-Reeves method, we see that the first directions of the DFP and the Fletcher-Reeves methods are identical. And so minimizing

$f(x^0 + \alpha_1 z^1)$ with respect to α_1 again yields

$$x^1 = \begin{pmatrix} \frac{26}{17} \\ \frac{38}{17} \end{pmatrix} \qquad \nabla f(x^1) = \begin{pmatrix} \frac{6}{17} \\ \frac{12}{17} \end{pmatrix}. \tag{11.77}$$

Consequently,

$$p^1 = \begin{pmatrix} \frac{26}{17} - (-2) \\ \frac{38}{17} - 4 \end{pmatrix} = \begin{pmatrix} \frac{60}{17} \\ -\frac{30}{17} \end{pmatrix} \qquad \gamma^1 = \begin{pmatrix} \frac{6}{17} - (-12) \\ \frac{12}{17} - 6 \end{pmatrix} = \begin{pmatrix} \frac{210}{17} \\ -\frac{90}{17} \end{pmatrix}. \tag{11.78}$$

The next search direction is given by

$$z^2 = -H_1^T \nabla f(x^1), \tag{11.79}$$

where

$$H_1 = H_0 + \frac{p^1(p^1)^T}{(p^1)^T\gamma^1} - \frac{H_0\gamma^1(\gamma^1)^T H_0}{(\gamma^1)^T H_0 \gamma^1} \tag{11.80}$$

and

$$H_1 = \begin{bmatrix} 1 & 0 \\ 0 & 1 \end{bmatrix} + \frac{1}{17}\begin{bmatrix} 4 & -2 \\ -2 & 1 \end{bmatrix} - \frac{1}{58}\begin{bmatrix} 49 & -21 \\ -21 & 9 \end{bmatrix} = \frac{1}{986}\begin{bmatrix} 385 & 241 \\ 241 & 891 \end{bmatrix}. \tag{11.81}$$

Thus

$$z^2 = -\frac{1}{986}\begin{bmatrix} 385 & 241 \\ 241 & 891 \end{bmatrix}\begin{pmatrix} \frac{6}{17} \\ \frac{12}{17} \end{pmatrix} = \begin{pmatrix} -\frac{9}{29} \\ -\frac{21}{29} \end{pmatrix}. \tag{11.82}$$

Comparing z^2 of the DFP method with that of the Fletcher-Reeves conjugate gradient algorithm, as given in (10.103), we conclude that the directions of the search vectors are identical but that their magnitudes are different. Thus x^2 of the DFP method, attained by minimizing $(x^1 + \alpha_2 x^2)$ with respect to α_2, will also be the point (1, 1), the global minimum of f. The two methods, therefore, generate the same sequence of points. This result is not surprising, since, given a vector $z^1 \in R^2$ and a matrix Q, the direction of the conjugate vector is uniquely determined. We shall, however, subsequently show that the sequence of points generated by the two algorithms is also identical for quadratic functions of higher dimensions.

The reader can also verify that the next matrix H_2 computed by the DFP formula is equal to the inverse Hessian Q^{-1}, where Q is given by (10.99). ▮

The fact that the matrix H_2 in the example is equal to Q^{-1} can be generalized. Letting $k = n$ in (11.8), we obtain

$$H_n Q z^j = \omega z^j, \qquad j = 1, \ldots, n. \tag{11.83}$$

Since the search directions z^j are linearly independent, it follows that

$$H_n Q = \omega I, \tag{11.84}$$

where I is the $n \times n$ identity matrix. Hence

$$H_n = \omega Q^{-1}. \tag{11.85}$$

Now since $\omega = 1$ in the DFP method, we get

$$H_n = Q^{-1}. \tag{11.86}$$

An explicit formula for Q^{-1} can also be obtained from a conjugate gradient method, such as the Fletcher-Reeves algorithm. Let $z^1, \ldots, z^k$ be vectors in R^n that are mutually conjugate with respect to Q. Define

$$S_k = \sum_{j=1}^{k} \frac{z^j (z^j)^T}{(z^j)^T Q z^j}. \tag{11.87}$$

Then it can be shown that [28]

$$S_k Q z^j = \begin{cases} z^j & j = 1, \ldots, k \\ 0 & j = k+1, \ldots, n. \end{cases} \tag{11.88}$$

And so

$$S_n = \sum_{j=1}^{n} \frac{z^j (z^j)^T}{(z^j)^T Q z^j} = Q^{-1}. \tag{11.89}$$

An alternative formula, which follows from (11.87), is

$$S_k = \sum_{j=1}^{k} \frac{p^j (p^j)^T}{(p^j)^T \gamma^j}. \tag{11.90}$$

Thus in order to compute the inverse of Q, we can proceed as follows: Starting at any point x^0, move to $x^1 = x^0 + \alpha_1 z^1$, where α_1 is any positive number, not necessarily the one that minimizes $f(x^0 + \alpha_1 z^1)$. Compute z^2, conjugate to z^1 with respect to Q, and move to $x^2 = x^1 + \alpha_2 z^2$, where α_2 is

again an arbitrary positive number. Similarly, find $x^3, \ldots, x^n$. Compute S_n, which is equal to Q^{-1}.

Example 11.1.2

Let us take the quadratic function of the previous example once more. There we had the conjugate directions

$$z^1 = \begin{pmatrix} 12 \\ 6 \end{pmatrix} \qquad z^2 = \begin{pmatrix} -9 \\ -21 \end{pmatrix}. \tag{11.91}$$

Starting at $x^0 = (-2, 4)$, suppose that we move to

$$x^1 = x^0 + z^1 = \begin{pmatrix} -2 \\ 4 \end{pmatrix} + \begin{pmatrix} 12 \\ 6 \end{pmatrix} = \begin{pmatrix} 10 \\ -2 \end{pmatrix} \tag{11.92}$$

by setting $\alpha_1 = 1$. Then

$$p^1 = \begin{pmatrix} 12 \\ -6 \end{pmatrix}, \quad \nabla f(x^1) = \begin{pmatrix} 30 \\ -12 \end{pmatrix}, \quad \gamma^1 = \begin{pmatrix} 42 \\ -18 \end{pmatrix}. \tag{11.93}$$

Now we move from x^1 to the next point

$$x^2 = x^1 + z^2 = \begin{pmatrix} 10 \\ -2 \end{pmatrix} + \begin{pmatrix} -9 \\ -21 \end{pmatrix} = \begin{pmatrix} 1 \\ -23 \end{pmatrix} \tag{11.94}$$

by setting $\alpha_2 = 1$. We obtain

$$p^2 = \begin{pmatrix} -9 \\ -21 \end{pmatrix}, \quad \nabla f(x^2) = \begin{pmatrix} 24 \\ -24 \end{pmatrix}, \quad \gamma^2 = \begin{pmatrix} -6 \\ -12 \end{pmatrix}. \tag{11.95}$$

Hence

$$S_2 = \frac{\begin{bmatrix} 144 & -72 \\ -72 & 36 \end{bmatrix}}{(12)(42) + (6)(18)} + \frac{\begin{bmatrix} 81 & 189 \\ 189 & 441 \end{bmatrix}}{(9)(6) + (21)(12)} = \begin{bmatrix} \frac{1}{2} & \frac{1}{2} \\ \frac{1}{2} & \frac{3}{2} \end{bmatrix} \tag{11.96}$$

and S_2 is equal to Q^{-1}, where Q is given by (10.99). ∎

Let us show now that when minimizing a quadratic function $f(x)$, $x \in R^n$, that has a positive definite Q matrix by a conjugate gradient method with the standard starting direction or by a variable metric method that uses correction matrices that belong to Huang's family, with $H_0 = I$, the sequence of points $x^1, x^2, \ldots, x^n$ will be identical, provided that they all started at the same x^0. Rewriting (11.46) and (11.48) for the common directions of all

variable metric algorithms based on Huang's family of matrices

$$q^1 = -H_0^T \nabla f(x^0) \tag{11.97}$$

$$q^{k+1} = -\left[I - \sum_{j=1}^{k} \frac{p^j(\gamma^j)^T}{(p^j)^T\gamma^j}\right] H_0^T \nabla f(x^k), \qquad k = 1, 2, \ldots \tag{11.98}$$

we can see that they can also be written as

$$q^{k+1} = -H_k^T \nabla f(x^k), \qquad k = 0, 1, \ldots \tag{11.99}$$

where

$$H_k = H_{k-1} - \frac{H_0\gamma^k(p^k)^T}{(p^k)^T\gamma^k}, \qquad k = 1, \ldots. \tag{11.100}$$

The algorithms satisfy the conditions of Theorem 10.2, and so we have by (10.75) and (11.37)

$$(q^j)^T \nabla f(x^k) = 0, \qquad j = 1, \ldots, k. \tag{11.101}$$

From (11.98) and (11.101) we obtain

$$(\nabla f(x^{j-1}))^T H_0 \nabla f(x^k) - \sum_{i=1}^{j-1} \frac{(\nabla f(x^{j-1}))^T H_0 \gamma^i (p_i)^T \nabla f(x^k)}{(p^i)^T\gamma^i} = 0, \quad j = 1, \ldots, k. \tag{11.102}$$

But $(p^i)^T \nabla f(x^k) = 0$ for $i = 1, \ldots, j-1$, hence

$$(\nabla f(x^{j-1}))^T H_0 \nabla f(x^k) = 0, \qquad j = 1, \ldots, k. \tag{11.103}$$

Supposing now that H_0 is symmetric, we simplify (11.98) to

$$q^{k+1} = -\left[I - \frac{(q^k)(\nabla f(x^k))^T}{(q^k)^T\gamma^k}\right] H_0 \nabla f(x^k). \tag{11.104}$$

Letting $H_0 = I$, the identity matrix, we have

$$q^{k+1} = -\nabla f(x^k) + \frac{(\nabla f(x^k))^T \nabla f(x^k)}{(q^k)^T\gamma^k} q^k. \tag{11.105}$$

From (11.41) through (11.44) and (11.101) it follows that $(q^k)^T\gamma^k = (\nabla f(x^{k-1}))^T \nabla f(x^{k-1})$, hence we finally obtain

$$q^{k+1} = -\nabla f(x^k) + \frac{(\nabla f(x^k))^T \nabla f(x^k)}{(\nabla f(x^{k-1}))^T \nabla f(x^{k-1})} q^k \tag{11.106}$$

and the search directions are the same as in the Fletcher-Reeves algorithm.

So far we have discussed only conjugate gradient and variable metric methods that include "exact" line searches—that is, methods that required α_k^* to be chosen by finding exact minima of one-dimensional functions. These methods, as we saw, have the quadratic termination property. It is shown in [8] that the precise number of iterations K used to minimize a quadratic function with a positive definite Q matrix by any variable metric method with correction matrices belonging to Broyden's subfamily and employing exact line searches is equal to the number of linearly independent n vectors in the sequence $\nabla f(x^0)$, $QH_0\nabla f(x^0)$, $(QH_0)^2\nabla f(x^0), \ldots,$ where H_0 is positive definite. Clearly, K can never exceed n, and, depending on the starting conditions, it can be less than n. An extreme case is one in which H_0 is chosen to be equal to Q^{-1}, which implies $K = 1$.

Powell [46, 47] extends this result to the interesting case in which an exact line search is not performed at every iteration; and he shows that if α_k is chosen in such a way that $f(x^{k-1} + \alpha_k z^k) < f(x^{k-1})$, then after at most K exact, but not necessarily consecutive, line searches, the algorithm terminates, where K is defined above. The question of performing an exact line search at every iteration is important, for these searches are often the most time-consuming parts of variable metric algorithms. As a result, there is a great interest in variable metric methods that avoid them. In fact, later on we shall see that several variable metric methods that are based on certain types of correction matrices exhibit the quadratic termination property without any line searches.

Although variable metric methods are widely used with considerable success and although they are also being extended to solve constrained optimization problems, there are few results on their convergence in the case of nonquadratic function minimization. The following convergence results are due to Powell [44]. Proofs are quite long and will be omitted.

Theorem 11.4

Let f be a twice continuously differentiable real function on R^n such that

$$y^T\nabla^2 f(x)y \geq m\,||y||^2 \tag{11.107}$$

for some $m > 0$ and for every x and y in R^n. Then, for any $x^0 \in R^n$, the set of points x satisfying $f(x) \leq f(x^0)$ is compact, and the sequence $x^0, x^1, \ldots$ generated by the DFP algorithm converges to x^*, the unique minimum of f.

Under the conditions imposed on f it follows from Theorems 11.3 and 11.4 that every variable metric algorithm using correction matrices that belong to Huang's family and satisfy $\omega = 1$ converges to the minimum of f. Note that the conditions on f in the last theorem imply that f is a strictly convex function such that the eigenvalues of its Hessian matrix at every

point $x \in R^n$ are bounded below by the positive number m. The next theorem deals with the convergence rate of the DFP algorithm.

Theorem 11.5

Suppose that the hypotheses of Theorem 11.4 hold and that for all x satisfying $f(x) \leq f(x^0)$ there exists a constant L such that

$$\left| \frac{\partial^2 f(x)}{\partial x_i \partial x_j} - \frac{\partial^2 f(x^*)}{\partial x_i \partial x_j} \right| \leq L \|x - x^*\|, \qquad i = 1, \ldots, n \quad j = 1, \ldots, n. \tag{11.108}$$

Then the sequence of points $x^0, x^1, \ldots$ generated by the DFP algorithm converges superlinearly in the sense of (8.21); *that is,*

$$\lim_{k \to \infty} \frac{\|x^{k+1} - x^*\|}{\|x^k - x^*\|} = 0. \tag{11.109}$$

It follows again that every other variable metric algorithm with $\omega = 1$ also converges superlinearly. We emphasize this point because Powell [45] conjectures that the lengthy proofs of the preceding theorems might possibly be shortened by using a different correction matrix from Huang's family, with $\omega = 1$, instead of the DFP formula. On the other hand, proofs of the last theorems can be useful for establishing convergence of other variable metric methods using $\omega \neq 1$.

Powell [45] has extended the results of Theorem 11.4 to real convex functions with bounded Hessian matrices.

Theorem 11.6

Let f be a twice continuously differentiable convex function on R^n such that

$$\|\nabla^2 f(x)\| \leq M \tag{11.110}$$

for all x. Suppose that the set of x satisfying $f(x) \leq f(x^0)$ is compact. Then every cluster point of the sequence $x^0, x^1, \ldots,$ generated by the DFP method is a minimum of f.

Goldfarb [23] has also discussed convergence of variable metric methods. He showed that every variable metric algorithm with exact line searches at each iteration will converge to the unique minimum of a strictly convex function f, that satisfies (10.120)—that is, such that the eigenvalues of $\nabla^2 f(x)$ are bounded—if the H_k matrices are also positive definite and their eigenvalues are bounded between two positive numbers. He also found sufficient conditions on certain classes of variable metric methods such that the H_k matrices will have this property.

11.2 QUASI-NEWTON METHODS

We have just derived a family of variable metric algorithms with quadratic termination properties based on conjugate directions and exact line searches. Here we discuss several additional variable metric algorithms that do not require exact line searches for their successful application. A great deal of overlap can be noticed between the matrix updating formulas of this section and the previous one, in the sense that many particular formulas can be derived either by the arguments of the previous section or by those used here. Although the preceding variable metric algorithms are especially suitable for function minimization, some of the methods of the present section were originally derived for the solution of nonlinear simultaneous equations, not necessarily resulting from setting the gradient of a function to zero. Since our main objective is optimization, we shall continue using the same notation as before. The reader interested in nonlinear equations can easily adapt all the following results.

The methods discussed here offer certain modifications to the Newton method described in Chapter 10. Suppose, again, that we wish to minimize a differentiable function f on R^n. Newton's method for finding a solution of the n simultaneous equations in n variables given by $\nabla f(x) = 0$ is governed by the iteration formula

$$x^{k+1} = x^k - [\nabla^2 f(x^k)]^{-1} \nabla f(x^k), \qquad k = 0, 1, \ldots. \tag{11.111}$$

The two main disadvantages of this method are (a) the computational effort required to evaluate the inverse Hessian matrix and (b) the fact that the method is unstable, can converge to local maxima, or may fail to converge if a bad initial estimate is given [6]. The methods described below approximate the inverse Hessian at each iteration without evaluating second partial derivatives of f. Suppose that we have an $n \times n$ nonsingular matrix B_k that approximates $\nabla^2 f(x^k)$. Then we can choose a direction z^{k+1}, given by

$$z^{k+1} = -B_k^{-1} \nabla f(x^k), \tag{11.112}$$

and move to the next point x^{k+1} by

$$x^{k+1} = x^k + \alpha_{k+1} z^{k+1}, \tag{11.113}$$

where α_{k+1} is a real number chosen in such a way that the iterations eventually converge to the required solution. Generally we can define

$$x = x^k + \theta z^{k+1} \tag{11.114}$$

and consider $\nabla f(x) = \nabla f(x^k + \theta z^{k+1})$ as a function of the single variable θ. Now

$$\frac{d(\partial f(x)/\partial x_j)}{d\theta} = \sum_{l=1}^{n} \frac{\partial^2 f(x)}{\partial x_j \partial x_l}\frac{dx_l}{d\theta} = \sum_{l=1}^{n} \frac{\partial^2 f(x)}{\partial x_j \partial x_l} z_l^{k+1}, \qquad j = 1, \ldots, n. \tag{11.115}$$

Thus if we are seeking an approximation to the Hessian matrix at x^{k+1}, the last relation implies that it can be obtained by having an approximation to $d(\partial f(x^{k+1})/\partial x_j)/d\theta$. Such an approximation can be written, using (11.115), as

$$\frac{\partial f(x^k + \theta z^{k+1})}{\partial x_j} \cong \frac{\partial f(x^{k+1})}{\partial x_j} + \sum_{l=1}^{n} \frac{\partial^2 f(x^{k+1})}{\partial x_j \partial x_l} z_l^{k+1}(\theta - \alpha_{k+1}), \qquad j = 1, \ldots, n. \tag{11.116}$$

If the values of ∇f were evaluated at two points, $x^k + \theta z^{k+1}$ and x^{k+1}, then, by (11.116), we could have written

$$\frac{\partial f(x^{k+1})}{\partial x_j} - \frac{\partial f(x^k + \theta z^{k+1})}{\partial x_j} \cong \sum_{l=1}^{n} \frac{\partial^2 f(x^{k+1})}{\partial x_j \partial x_l} z_l^{k+1}(\alpha_{k+1} - \theta), \qquad j = 1, \ldots, n \tag{11.117}$$

or

$$\nabla f(x^{k+1}) - \nabla f(x^k + \theta z^{k+1}) \cong \nabla^2 f(x^{k+1})(\alpha_{k+1} - \theta) z^{k+1}. \tag{11.118}$$

The basic idea behind the methods of this section is to choose B_{k+1}, the next approximation to the Hessian of f, so that it satisfies the equation

$$\nabla f(x^{k+1}) - \nabla f(x^k + \theta z^{k+1}) = B_{k+1}(\alpha_{k+1} - \theta) z^{k+1}. \tag{11.119}$$

A common choice of θ is to set it equal to zero, and then we obtain

$$\nabla f(x^{k+1}) - \nabla f(x^k) = \gamma^{k+1} = B_{k+1}(x^{k+1} - x^k) = B_{k+1} p^{k+1}. \tag{11.120}$$

Since we need B_{k+1}^{-1} rather than B_{k+1} itself for the iterative procedure, we define $H_k = B_k^{-1}$, and (11.120) becomes the secant relation (11.65)

$$p^{k+1} = H_{k+1} \gamma^{k+1}. \tag{11.121}$$

Variable metric methods in which the matrix H_k is updated by the secant relation are called **quasi-Newton methods** [6]. In these methods we have

$$z^{k+1} = -H_k \nabla f(x^k) \tag{11.122}$$

where we dropped the transpose sign on H_k, and x^{k+1} is given by (11.113) such that α_{k+1} does not necessarily minimize $f(x^k + \alpha z^{k+1})$.

Turning now to the question of how to update H_k, we assume, as in the previous section, that

$$H_k = H_{k-1} + \Delta H_k. \tag{11.123}$$

Then by the secant relation,

$$\Delta H_k \gamma^k = p^k - H_{k-1}\gamma^k. \tag{11.124}$$

If we let

$$\Delta H_k = \frac{(p^k - H_{k-1}\gamma^k)(y^k)^T}{(y^k)^T\gamma^k}, \tag{11.125}$$

where y^k is an arbitrary n vector chosen so that $(y^k)^T\gamma^k \neq 0$, then (11.124), and hence (11.121), holds. Moreover, the matrix given by the right-hand side of (11.125) has the least possible rank such that (11.124) is satisfied. The matrix ΔH_k in (11.125) has, of course, rank one. The general rank-one updating formula can, therefore, be written as [7, 10]

$$H_k = H_{k-1} + \frac{(p^k - H_{k-1}\gamma^k)(y^k)^T}{(y^k)^T\gamma^k}, \tag{11.126}$$

subject to $(y^k)^T\gamma^k \neq 0$.

Let us consider some particular quasi-Newton methods, that use correction matrices which are special cases of the general rank-one matrix above. In the **secant method** of Wolfe [54] and Barnes [2], the vectors y^k are chosen as follows: For $k < n$, let y^k satisfy

$$(y^k)^T\gamma^i = 0, \qquad i = 1, \ldots, k-1. \tag{11.127}$$

Since this relation does not determine the y^k uniquely, it was suggested that y^k be taken as that linear combination of $\gamma^1, \ldots, \gamma^k$ that is orthogonal to $\gamma^1, \ldots, \gamma^{k-1}$. This step can be accomplished by the Gram-Schmidt orthogonalization procedure. If $k \geq n$, then choose y^k so that

$$(y^k)^T\gamma^i = 0, \qquad i = k-n+1, \ldots, k-1. \tag{11.128}$$

We can easily show that if y^k satisfies the preceding relations, then for $k \leq n$

$$p^i = H_k\gamma^i, \qquad i = 1, \ldots, k. \tag{11.129}$$

In its original version the secant method actually approximated the Hessian instead of the inverse Hessian. The current version is more efficient, for it is unnecessary to invert matrices.

Murtagh and Sargent [36] proved the following result, which implies that the secant method has quadratic termination.

Theorem 11.7

Given steps p^i, $i = 1, \ldots, k$ such that (11.129) *holds with H_k nonsingular. Then, for a quadratic function f, a step given by*

$$p^{k+1} = -H_k \nabla f(x^k) \tag{11.130}$$

is either linearly independent of the previous steps or it attains a stationary point of f.

Proof. Suppose that p^{k+1} is linearly dependent on $p^1, \ldots, p^k$. Then there are numbers $\beta_1, \ldots, \beta_k$, not all zero, such that

$$p^{k+1} = \sum_{i=1}^{k} \beta_i p^i. \tag{11.131}$$

Since f is a quadratic function, we have $f(x) = a + b^T x + \frac{1}{2} x^T Q x$ and

$$\gamma^i = Q p^i. \tag{11.132}$$

Hence

$$\gamma^{k+1} = Q p^{k+1} = \sum_{i=1}^{k} \beta_i Q p^i = \sum_{i=1}^{k} \beta_i \gamma^i. \tag{11.133}$$

From (11.129), (11.131), and (11.133) it follows that

$$p^{k+1} = \sum_{i=1}^{k} \beta_i H_k \gamma^i = H_k \sum_{i=1}^{k} \beta_i \gamma^i = H_k \gamma^{k+1}. \tag{11.134}$$

And from (11.130) and (11.134), since H_k is nonsingular, we obtain

$$-\nabla f(x^k) = \gamma^{k+1} = \nabla f(x^{k+1}) - \nabla f(x^k). \tag{11.135}$$

That is,

$$\nabla f(x^{k+1}) = 0. \tag{11.136}$$

■

Corollary 11.6

If H_k is nonsingular and satisfies (11.129), *then a stationary point of a quadratic function is attained in at most $n + 1$ steps.*

Proof. Follows directly from the preceding theorem, since there can be n linearly independent steps at most. ■

Quadratic termination of the secant method can also be proven by showing that the matrices $H_0, H_1, \ldots, H_n$ generated by the secant method con-

verge to Q^{-1}, such that $H_n = Q^{-1}$. Thus, after n steps, the next one must give the solution, for it is the original Newton step. Note that, in the case of a quadratic function, the step length is unimportant for finite termination.

For a general nonlinear function, more than n steps of the secant method may be required to satisfy some convergence criterion. Then the y^k are chosen by (11.128), and experience has shown that consecutive search directions may become linearly dependent and some modification of (11.128) may be necessary.

A somewhat different approach is taken in **Broyden's 1965 method** [6], which is based on (11.120). He noticed that (11.120) relates the change in $\nabla f(x^k)$ to the change of position x^k in the direction p^{k+1}, and no information is available about the change in $\nabla f(x^k)$ in a direction different from p^{k+1}. Broyden suggests choosing B_{k+1} so that the change in $\nabla f(x^k)$ predicted by B_{k+1} in a direction r^{k+1}, orthogonal to p^{k+1}, is the same as would be given by B_k. Formally,

$$B_{k+1}r^{k+1} = B_k r^{k+1} \qquad (p^{k+1})^T r^{k+1} = 0. \tag{11.137}$$

The reader can easily verify that B_k, given by

$$B_k = B_{k-1} + \frac{(y^k - B_{k-1}p^k)(p^k)^T}{(p^k)^T p^k}, \tag{11.138}$$

satisfies both (11.120) and (11.137). Since we need $B_k^{-1} = H_k$, rather than B_k itself, we can use the Sherman-Morrison formula [29] for a nonsingular $n \times n$ matrix A and n vectors u, v:

$$(A + uv^T)^{-1} = A^{-1} - \frac{A^{-1}uv^T A^{-1}}{1 + v^T A^{-1} u}. \tag{11.139}$$

Letting $A = B_{k-1}$, $A^{-1} = H_{k-1}$, $u = (y^k - B_{k-1}p^k)/(p^k)^T p^k$, $v = p^k$, from the last two equations it follows that

$$H_k = H_{k-1} - \frac{(H_{k-1}y^k - p^k)(p^k)^T H_{k-1}}{(p^k)^T H_{k-1} y^k}. \tag{11.140}$$

Broyden's method thus uses search directions z^{k+1}, given by (11.122), and step lengths $p^{k+1} = x^{k+1} - x^k = \alpha_{k+1} z^{k+1}$, where α_{k+1} is usually chosen so that

$$\|\nabla f(x^{k+1})\| \leq \|\nabla f(x^k)\|. \tag{11.141}$$

There is no special reason to determine the step lengths by exact line searches, and α_{k+1} can also be set to unity. The method does not possess the quadratic termination property.

Broyden [10] notes that if B_k is a good approximation of $\nabla^2 f(x^k)$, then the algorithm with $\alpha_k = 1$ is usually stable. If, however, B_k is not a good approximation of the Hessian of f, the method becomes unstable. Convergence properties of this algorithm, which is considered superior to the secant method, were studied by Broyden [10], who showed that, under suitable assumptions, the rate of convergence of his method is superlinear when applied to minimizing a quadratic function. These results were subsequently extended by Dennis [13] to general nonlinear functions.

Example 11.2.1

Let us illustrate Broyden's 1965 method on our well-known quadratic function

$$f(x) = \tfrac{3}{2}(x_1)^2 + \tfrac{1}{2}(x_2)^2 - x_1 x_2 - 2x_1 \tag{11.142}$$

by performing a few iterations.

Suppose that we choose $H_0 = I$ and $\alpha_k = 1$ for all k. We start the search at $x^0 = (-2, 4)$, where $f(x^0) = 26$, $\nabla f(x^0) = (-12, 6)^T$. The next point is given by

$$x^1 = x^0 - H_0 \nabla f(x^0) = \begin{pmatrix} 10 \\ -2 \end{pmatrix}. \tag{11.143}$$

Here $f(x^1) = 152$ and $\nabla f(x^1) = (30, -12)^T$. Consequently,

$$\gamma^1 = \begin{pmatrix} 42 \\ -18 \end{pmatrix} \qquad p^1 = \begin{pmatrix} 12 \\ -6 \end{pmatrix} \tag{11.144}$$

and

$$H_1 = \begin{bmatrix} 1 & 0 \\ 0 & 1 \end{bmatrix} - \frac{\left[\begin{pmatrix} 42 \\ -18 \end{pmatrix} - \begin{pmatrix} 12 \\ -6 \end{pmatrix}\right](12, -6)\begin{bmatrix} 1 & 0 \\ 0 & 1 \end{bmatrix}}{(12, -6)\begin{bmatrix} 1 & 0 \\ 0 & 1 \end{bmatrix}\begin{pmatrix} 42 \\ -18 \end{pmatrix}} \tag{11.145}$$

or

$$H_1 = \begin{bmatrix} 0.41176 & 0.29412 \\ 0.23529 & 0.88235 \end{bmatrix}. \tag{11.146}$$

The next step is given by

$$p^2 = -H_1 \nabla f(x^1) = -\begin{bmatrix} 0.41176 & 0.29412 \\ 0.23529 & 0.88235 \end{bmatrix}\begin{pmatrix} 30 \\ -12 \end{pmatrix} = \begin{pmatrix} -8.82353 \\ 3.52941 \end{pmatrix}. \tag{11.147}$$

Hence

$$x^2 = \begin{pmatrix} 10 \\ -2 \end{pmatrix} + \begin{pmatrix} -8.82353 \\ 3.52941 \end{pmatrix} = \begin{pmatrix} 1.17647 \\ 1.52941 \end{pmatrix}. \tag{11.148}$$

At this point, $f(x^2) = -0.90657$ and $\nabla f(x^2) = (0, 0.35294)^T$. Thus

$$\gamma^2 = \begin{pmatrix} 0.00042 \\ 0.35286 \end{pmatrix} - \begin{pmatrix} 30 \\ -12 \end{pmatrix} = \begin{pmatrix} -30 \\ 12.3529 \end{pmatrix} \tag{11.149}$$

and

$$H_2 = H_1 - \frac{(H_1\gamma^2 - p^2)(p^2)^T H_1}{(p^2)^T H_1 \gamma^2} \tag{11.150}$$

$$= \begin{bmatrix} 0.41498 & 0.29352 \\ 0.24494 & 0.88058 \end{bmatrix}. \tag{11.151}$$

Consequently,

$$p^3 = -H_2 \nabla f(x^2) = \begin{pmatrix} -0.10359 \\ -0.31079 \end{pmatrix} \tag{11.152}$$

and

$$x^3 = \begin{pmatrix} 1.17647 \\ 1.52941 \end{pmatrix} - \begin{pmatrix} 0.10359 \\ 0.31079 \end{pmatrix} = \begin{pmatrix} 1.07288 \\ 1.21862 \end{pmatrix}. \tag{11.153}$$

At this point,

$$f(x^3) = -0.98407 \qquad \nabla f(x^3) = \begin{pmatrix} 0.00002 \\ 0.14574 \end{pmatrix}. \tag{11.154}$$

We can see that after the first step, which is taken in the steepest descent direction (since H_0 was chosen to be I), convergence is good, but there is no quadratic termination. If we compute H_3, we obtain

$$H_3 = \begin{bmatrix} 0.49560 & 0.49981 \\ 0.48683 & 1.49945 \end{bmatrix} \tag{11.155}$$

and this matrix is a good approximation to Q^{-1}, which is

$$Q^{-1} = \begin{bmatrix} \frac{1}{2} & \frac{1}{2} \\ \frac{1}{2} & \frac{3}{2} \end{bmatrix}. \tag{11.156}$$

Note that although H_0 was chosen to be symmetric, the subsequent matrices may be unsymmetric, as we can clearly see in H_1 and H_2. Of course, as the matrices H_k converge to Q^{-1}, they must again approach symmetry. ▮

This example demonstrated that the H_k matrices in Broyden's 1965 method (and also in the secant method) are not necessarily symmetric, not even in case of a symmetric H_0. Since H_k is supposed to approximate the inverse Hessian—a symmetric matrix—it is reasonable to find an updating formula such that all matrices H_k are symmetric and satisfy the secant relation.

Such an updating formula is the basis of **symmetric rank-one** algorithms, suggested by Broyden [7], Davidon [12], Murtagh and Sargent [36], and Powell [42]. The H_k matrices are given by choosing the y^k in (11.126) to be

$$y^k = p^k - H_{k-1}\gamma^k. \tag{11.157}$$

That is,

$$H_k = H_{k-1} + \frac{(p^k - H_{k-1}\gamma^k)(p^k - H_{k-1}\gamma^k)^T}{(p^k - H_{k-1}\gamma^k)^T\gamma^k}. \tag{11.158}$$

It is clear that, for any symmetric H_{k-1}, H_k will also be symmetric. Moreover, algorithms based on this matrix updating formula possess the quadratic termination without exact line searches. We can prove this result by showing that, in the case of a quadratic function with a nonsingular Q matrix, relation (11.129) holds—that is,

$$p^i = H_k\gamma^i, \qquad i = 1, \ldots, k \tag{11.159}$$

provided that none of the scalar products $(p^k - H_{k-1}\gamma^k)^T\gamma^k$ vanishes. Suppose that

$$p^i = H_{k-1}\gamma^i, \qquad i = 1, \ldots, k-1 \tag{11.160}$$

holds. Now for $i = 1, \ldots, k-1$

$$(p^k - H_{k-1}\gamma^k)^T\gamma^i = (p^k)^T\gamma^i - (H_{k-1}\gamma^k)^T\gamma^i \tag{11.161}$$

$$= (p^k)^T\gamma^i - (\gamma^k)^T p^i \tag{11.162}$$

$$= (\gamma^k)^T Q^{-1}\gamma^i - (\gamma^k)^T Q^{-1}\gamma^i = 0, \tag{11.163}$$

where $p^i = Q^{-1}\gamma^i$ holds for $i = 1, \ldots, k$. Hence by (11.158) and (11.160) to (11.163),

$$H_k\gamma^i = H_{k-1}\gamma^i = p^i, \qquad i = 1, \ldots, k-1 \tag{11.164}$$

and since H_k satisfies the secant relation, we conclude that (11.129) holds. Quadratic termination then follows by invoking Corollary 11.6. Note that the choice of the α_k determining the step p^k along the search directions z^k is unimportant in proving quadratic termination. The reader will be asked to prove the interesting result that if the α_k are chosen to minimize a quadratic function with a positive definite Q matrix, then the search directions $z^1, \ldots, z^n$ of the symmetric rank-one algorithm are actually conjugate with respect to Q.

The H_k matrix of the symmetric rank-one method, updated by (11.158), is also a member of Huang's family, since the updating formulas (11.74) and (11.158) are identical. By choosing $\omega = 1$, it follows from Theorem 11.3

that, in minimizing a differentiable function on R^n, a symmetric rank-one algorithm with exact line searches will generate the same sequence of points as Broyden's subfamily of methods obtained by restricting the A_k matrices in Huang's family to be symmetric and such that the H_k matrices satisfy the secant relation. Whereas algorithms based on general matrices belonging to Huang's family exhibit quadratic termination only if exact line searches are carried out at each iteration, the symmetric rank-one method terminates on a quadratic function without line searches. Although the computational effort required for minimizing a quadratic function by the symmetric rank-one method is considerably smaller than that of other variable metric algorithms requiring line searches, it is not clear whether this advantage will still hold for nonquadratic functions.

Broyden [10] lists three common ways of choosing the step lengths α_k and examines their effects:

1. Exact line search—that is, such that $(z^k)^T \nabla f(x^k) = 0$.
2. The step length is chosen so that $\| \nabla f \|$ is minimized or reduced along z^k.
3. The "Newton step"—that is, $\alpha_k = 1$ for all k.

Other choices of step lengths can, of course, also be devised, such as one that ensures a "sufficient function decrease" [39], as mentioned in Chapter 10, or the related choice in the method of Goldstein and Price [24]. A numerical comparison of various choices has been performed by Dixon [16], who checked, among other choices, the effects of using a single quadratic or cubic approximation instead of a complete one-dimensional minimization at each iteration.

Since convergence of quasi-Newton methods is obtained by successive approximation of the inverse Hessian, Broyden [8, 9, 10] suggested analyzing convergence properties of quasi-Newton algorithms by examining the behavior of sequences of certain matrices, each member of which indicates the closeness of the current matrix H_k to the inverse Hessian. Suppose that we minimize a quadratic function on R^n with a nonsingular Q matrix. Define

$$E_k = Q^{-1}B_k - I \tag{11.165}$$

and

$$G_k = H_kQ - I, \tag{11.166}$$

where, as before, $B_k = H_k^{-1}$. Clearly, E_k and G_k become the zero matrices if and only if $H_k = Q^{-1}$. We can approach this situation by successively reducing to zero, either the rank of the matrices E_k or G_k or some norms of them. It is shown in [10] that the secant and the symmetric rank-one methods are "rank reducing," whereas the Broyden 1965 method is "norm reducing."

So far we have been mainly concerned with rank-one quasi-Newton algorithms. Broyden's 1967 updating relation given by (11.67) and (11.68) was derived as a particular member of Huang's family, but the original formulation follows the quasi-Newton approach used in this section and requires a rank-two correction matrix ΔH_k. For convenience, let us rewrite the Broyden 1967 updating relation, which has one free parameter:

$$H_k = H_{k-1} - \frac{H_{k-1}\gamma^k(\gamma^k)^T H_{k-1}}{(\gamma^k)^T H_{k-1}\gamma^k} + \frac{p^k(p^k)^T}{(p^k)^T\gamma^k} - a_{12}^k \frac{(p^k)^T\gamma^k}{(\gamma^k)^T H_{k-1}\gamma^k} w^k(w^k)^T, \tag{11.167}$$

where

$$w^k = H_{k-1}\gamma^k - \frac{(\gamma^k)^T H_{k-1}\gamma^k}{(p^k)^T\gamma^k} p^k. \tag{11.168}$$

A particular formula, suggested by Broyden [9], can be obtained from (11.167) by letting

$$a_{12}^k = -\frac{1}{(p^k)^T\gamma^k} \tag{11.169}$$

Substituting into (11.167), we get the updating relation (11.70), discussed in the previous section,

$$H_k = H_{k-1} + \left[1 + \frac{(\gamma^k)^T H_{k-1}\gamma^k}{(p^k)^T\gamma^k}\right]\frac{p^k(p^k)^T}{(p^k)^T\gamma^k} - \frac{p^k(\gamma^k)^T H_{k-1}}{(p^k)^T\gamma^k} - \frac{H_{k-1}\gamma^k(p^k)^T}{(p^k)^T\gamma^k}. \tag{11.170}$$

As noted, the foregoing choice of a_{12}^k ensures that if applied to minimizing a quadratic function that has a positive definite Q matrix with an exact line search and such that H_0 is positive definite, then H_k, given by the last equation, will also be positive definite for all k. This updating formula is commonly called the Broyden-Fletcher-Shanno or BFS formula in the literature. Fletcher [19] shows an interesting relation between the DFP and the BFS updating formulas. Suppose that H_k is updated by the DFP formula (11.2). Letting $B_{k-1}^{-1} = H_{k-1}$, $B_k^{-1} = H_k$, we obtain

$$B_k = B_{k-1} + \left[1 + \frac{(p^k)^T B_{k-1} p^k}{(\gamma^k)^T p^k}\right]\frac{\gamma^k(\gamma^k)^T}{(\gamma^k)^T p^k} - \frac{\gamma^k(p^k)^T B_{k-1}}{(\gamma^k)^T p^k} - \frac{B_{k-1}p^k(\gamma^k)^T}{(\gamma^k)^T p^k}. \tag{11.171}$$

That is, B_k is updated according to an analog of the BFS formula obtained by substituting B_{k-1} instead of H_{k-1} and by interchanging p^k and γ^k. Conversely, if H_k is updated by the BFS formula, then B_k is updated by the

corresponding analog of the DFP formula. Because of these relations, the BFS formula is also sometimes called the complementary DFP formula. More general relations between the approximate Hessian matrices B_k, which result from various quasi-Newton algorithms, were found by Murray [35].

Another joint property of the DFP and BFS formulas involves methods based on them that abandon exact line searches but retain the positive definiteness of the H_k matrices by reducing the function value at each iteration in a certain way. Such methods converge to the minimum of a quadratic function in the "norm-reducing" way, described above.

Fletcher [19] proved this property for any convex combination of the variable metrics H_k^{DFP} and H_k^{BFS} corresponding to the two methods—that is, for matrices H_k, given by

$$H_k = q_1 H_k^{\mathrm{DFP}} + q_2 H_k^{\mathrm{BFS}}, \tag{11.172}$$

where $q_1 \geq 0$, $q_2 \geq 0$, $q_1 + q_2 = 1$. In practice, Fletcher recommended an algorithm whose variable metric is updated either by the DFP or the BFS formula, depending on the following test. If

$$(p^k)^T \gamma^k \geq (\gamma^k)^T H_{k-1} \gamma^k, \tag{11.173}$$

then update H_k by the BFS formula; otherwise use the DFP updating formula. The step-length determination is similar to that of the Goldstein-Price algorithm [24], as described in the preceding chapter. One chooses a small positive number $\delta < \frac{1}{2}$ and at each iteration computes

$$g(x^{k-1}, \alpha_k) = \frac{f(x^{k-1}) - f(x^{k-1} + \alpha_k z^k)}{\alpha_k (z^k)^T \nabla f(x^{k-1})} \tag{11.174}$$

with $\alpha_k = 1$. If $g(x^{k-1}, 1) \geq \delta$, set $x^k = x^{k-1} + z^k$; otherwise reduce α_k until $g(x^{k-1}, \hat{\alpha}_k) \geq \delta$ is satisfied and set $x^k = x^{k-1} + \hat{\alpha}_k z^k$.

Some researchers consider the positive definiteness of the approximate Hessian matrices H_k one of the essential properties of a successful variable metric algorithm. If H_k is kept positive definite at each iteration, then a reduction in the value of the function to be minimized is guaranteed. However, even in the case of algorithms that theoretically retain positive definiteness of the H_k matrices, such as the DFP algorithm, numerical difficulties may occur, thus causing these matrices to become singular or indefinite [1]. As noted earlier, it can be shown that once a singular H_k is computed, the subsequent search directions z^{k+1}, z^{k+2}, . . . will not span the whole space on which the function to be minimized is defined. Gill and Murray [21] suggest revised versions of symmetric rank-two, quasi-Newton algorithms, such as the DFP or the BFS methods, which, in theory, ensure positive definiteness

such that this property is also kept in practice. They find the search directions z^k by solving the linear equations

$$B_k z^{k+1} = -\nabla f(x^k), \tag{11.175}$$

where B_k is an approximation to the Hessian matrix of f. If B_k is a symmetric positive definite matrix, it can be written as

$$B_k = L_k D_k L_k^T \tag{11.176}$$

where $L_k = [l_{ij}^k]$ is a lower triangular matrix with unit diagonal elements and $D_k = [d_{ij}^k]$ is a diagonal matrix whose diagonal elements are positive. The right-hand side of (11.176) is called the **Cholesky factorization** of B_k [53]. Substituting (11.176) into (11.175), we obtain

$$L_k D_k L_k^T z^{k+1} = -\nabla f(x^k) \tag{11.177}$$

and z^{k+1} is conveniently found by first solving

$$L_k y^k = -\nabla f(x^k), \tag{11.178}$$

that is, $y_1^k = -\partial f(x^k)/\partial x_i$ and

$$y_i^k = -\frac{\partial f(x^k)}{\partial x_i} - \sum_{j=1}^{i-1} l_{ij}^k y_j^k, \qquad i = 2, \ldots, n \tag{11.179}$$

and then

$$L_k^T z^{k+1} = D_k^{-1} y^k, \tag{11.180}$$

that is, $Z_n^{k+1} = y_n^k / d_{nn}^k$ and

$$z_i^{k+1} = \frac{y_i^k}{d_{ii}^k} - \sum_{j=i+1}^{n} l_{ji}^k z_j^{k+1}, \qquad i = n-1, \ldots, 1. \tag{11.181}$$

Now if a symmetric rank-two updating formula for H_k, or the symmetric rank-one version, is used, it can be shown that the corresponding update for $H_k^{-1} = B_k$ will be

$$B_k = B_{k-1} + \mu_1^k y^k (y^k)^T + \mu_2^k w^k (w^k)^T, \tag{11.182}$$

where μ_1^k, μ_2^k are scalars and y^k, w^k are vectors, which can be found from the particular updating formula used for H_k—for example, the DFP or the BFS formula. The value of B_k is obtained by successive calculations of

$$\hat{L}_k \hat{D}_k \hat{L}_k^T = B_{k-1} + \mu_1^k y^k (y^k)^T \tag{11.183}$$

and

$$B_k = L_k D_k L_k^T = \hat{L}_k \hat{D}_k \hat{L}_k^T + \mu_2^k w^k (w^k)^T. \tag{11.184}$$

If the BFS formula is used, for example, then

$$\mu_1^k = \frac{1}{(p^k)^T \gamma^k} \qquad y^k = \gamma^k \tag{11.185}$$

$$\mu_2^k = \frac{1}{(z^k)^T \nabla f(x^{k-1})} \qquad w^k = \nabla f(x^{k-1}). \tag{11.186}$$

The important feature of the Gill-Murray approach of revised variable metric algorithms is that the B_k matrices are guaranteed to be positive definite, so that z^{k+1} is a descent direction. For further details of this method, the reader is referred to Gill and Murray [21], Gill, Murray, and Pitfield [22], and Murray [34]. Another interesting property of the Gill-Murray approach is that it can be easily modified for variable metric methods that do not use derivatives. Such methods are discussed next.

11.3 VARIABLE METRIC ALGORITHMS WITHOUT DERIVATIVES

Successful applications of the variable metric algorithms described in this chapter on many difficult test functions have initiated studies that extend the applicability of these methods to the minimization of functions whose derivatives are unavailable or difficult to calculate. Since all these methods use only first derivatives, a natural extension would be to replace analytical derivatives by finite-difference approximations. There are two common formulas for finite differences of derivatives: the forward difference

$$\frac{\Delta f(x)}{\Delta x_j} = \frac{f(x + h_j I_j) - f(x)}{h_j}, \qquad j = 1, \ldots, n \tag{11.187}$$

and the central difference

$$\frac{\Delta f(x)}{\Delta x_j} = \frac{f(x + h_j I_j) - f(x - h_j I_j)}{2h_j}, \qquad j = 1, \ldots, n \tag{11.188}$$

where $\Delta f(x)/\Delta x_j$ is the approximation of the partial derivative $\partial f(x)/\partial x_j$, h_j is a small number, and I_j denotes the jth column of the identity matrix. Note that if $f(x)$ is known, the forward difference formula requires n additional function evaluations for approximating $\nabla f(x)$, compared with $2n$ evaluations in the case of the usually more accurate central difference formula.

The first modification of variable metric algorithms that replace derivatives by finite differences is due to Stewart [52]. The basis of his method is to balance two, usually conflicting, errors that are involved in using finite

differences. If h_j is chosen too large, the first-order approximation of f, on which the finite-difference formulas are based, becomes inadequate (truncation error), whereas, by taking a too small value for h_j, function values may become undistinguishable on a computer with finite accuracy and, by taking differences, cancellation of significant figures may occur. Stewart derives estimates of these errors and suggests choosing that value of h_j for which the two error estimates are equal. This calculation involves the solution of a third-degree polynomial equation in which approximations to the diagonal elements of $\nabla^2 f$ also appear. These numbers are computed by inverting the H_k matrices that are obtained by the DFP formula through the use of the Sherman-Morrison modification rule given in [29], mentioned earlier in this chapter. If the computed truncation error is greater than some predetermined tolerance, Stewart suggests using the central difference formula. Otherwise the approximation of the gradient by the forward difference formula is used in a variable metric algorithm—for example, the DFP method. The line search of each iteration uses only function evaluations and is similar to the method suggested by Powell [41] in his conjugate directions algorithm without derivatives. This line search has been slightly modified by Lill [18, 32], who also compared Stewart's modification of the DFP algorithm with Powell's method without derivatives and concluded that the former needs less function evaluations, especially in problems with a large number of variables.

Gill and Murray [21] noticed that Stewart's way of estimating the diagonal elements of $\nabla^2 f$ may result in the selection of a poor value of h_j. Accordingly, they have suggested a different procedure for gradient approximations by finite differences. In contrast to Stewart's method, in which the intervals h_j along the coordinate directions vary from one iteration to another, Gill and Murray fix the size of the h_j at the start of the algorithm. The h_j are usually taken to be proportional to some assumed range of the variables x_j, $j = 1, \ldots, n$, in which the minimum of the function is known to lie. Since the forward difference formula (11.187) requires fewer function evaluations, it is used first to estimate the gradient ∇f. Suppose now that, at iteration k, we have an estimate of $\nabla f(x^k)$. First we check the magnitude of this estimate. If the forward difference approximation yields

$$\left\{\sum_{j=1}^{n}\left(\frac{\Delta f(x^k)}{\Delta x_j}\right)^2\right\}^{1/2} < \hat{\alpha}, \tag{11.189}$$

where $\hat{\alpha}$ is some preassigned constant, then this estimate of the gradient is rejected and a new estimate is calculated by the central difference formula (11.188). Having an approximation of $\nabla f(x^k)$ either by the forward or the central difference formula, we compute z^{k+1} by using the Cholesky factorization as described above and compute α^*_{k+1} so that it approximately minimizes $f(x^k + \alpha_{k+1} z^{k+1})$ by an efficient line search, outlined in [22].

Letting $x^{k+1} = x^k + \alpha^*_{k+1} z^{k+1}$, we check the step size taken. If

$$\|x^{k+1} - x^k\| = \|p^{k+1}\| < \hat{\beta}, \tag{11.190}$$

where $\hat{\beta}$ is again a predetermined constant, and the forward difference formula was used, the direction z^{k+1} is rejected and a new value is found by approximating $\nabla f(x^k)$ by central differences. Once the central difference formula has been used, it is used in every subsequent iteration until the step size becomes large—that is, until for some $i > k$

$$\|p^i\| \geq \hat{\beta}. \tag{11.191}$$

Numerical experimentations have shown [22, 27] that both versions of the Gill-Murray algorithm, with and without analytical derivatives, are highly successful, particularly when used in conjunction with the BFS updating formula. The implementation contains many practical details that cannot be repeated here. The interested reader is referred to [22].

The two methods described use finite differences to approximate first derivatives and then employ some existing variable metric updating formula. Two other algorithms can be regarded as quasi-Newton methods without computing derivatives. The first was derived and subsequently modified by Greenstadt [26, 27]. It is based on an earlier work of the same author [25] in which variable metric correction matrices ΔH_k are derived by variational principles that minimize some matrix norm, subject to satisfying the secant relation. These derivations were modified for the case in which no analytic derivatives are available. Numerical tests of Greenstadt [27] indicate that, in most cases, the Gill-Murray-Pitfield method performed better. The second algorithm is due to Fiacco and McCormick [17], and it consists of $n(n + 1)/2$ line searches to approximate the gradient and the Hessian matrix of an n-variable function. Once the estimates of first- and second-order derivatives have been computed, a Newton step is taken. An important feature of the algorithm is that the approximations are exact for quadratic functions, and the exact minimum can be found in a finite number of iterations. In the case of a general nonlinear function, after the Newton step, all previous derivative estimates are discarded and new approximations are set up.

11.4 MINIMIZATION METHODS BASED ON NONQUADRATIC FUNCTIONS

Nearly all efficient multidimensional minimization algorithms utilize quadratic approximations to the function whose minimum is sought. Although such an approximation is usually adequate in a small neighborhood of the optimum, it may not be so for points far from it. Quasi-Newton methods, for example, that guarantee positive definiteness of the H_k matrices

may yield poor approximations to the inverse Hessian of a nonquadratic, nonconvex function away from the minimum. So it seems logical to try to modify existing algorithms or to derive new ones that explicitly take into account nonquadratic functions. In this section we present two such algorithms that have been numerically tested on a very limited scale and that compared favorably with existing variable metric methods. We start with the **dominant degree method** of Biggs [4, 5]. In this method existing updating formulas, such as the DFP, BFS, or rank one, are modified so that some nonquadratic features of the function f to be minimized are reflected. Variable metric algorithms implicitly assume that f is quadratic between two successive points generated by the algorithm. Biggs' method, on the other hand, assumes that if we define ϕ by

$$\phi(\alpha) = f(x + \alpha z), \tag{11.192}$$

where x and z are given vectors, then the one-dimensional function ϕ can be adequately represented by

$$\phi(\alpha) = a\,|\alpha - \alpha^*|^p + b, \tag{11.193}$$

where $a > 0$, $p > 1$. This is a convex function with a minimum at α^*, and it is dominated by terms in $(\alpha)^p$. In order to determine the value of p, we proceed as follows: Suppose that at some point α^0 we have the values $\phi(\alpha^0)$, the first derivative $\phi'(\alpha^0)$, and an approximate second derivative $\tilde{\phi}''(\alpha^0) = \eta\phi''(\alpha^0)$, where $\eta > 0$ is some unknown constant. Now let

$$\alpha^1 = \alpha^0 - \theta\frac{\phi'(\alpha^0)}{\tilde{\phi}''(\alpha^0)} \tag{11.194}$$

and suppose that we also know $\phi(\alpha^1)$ and $\phi'(\alpha^1)$. Biggs [4] shows that, by using these data, we can compute p and η. Defining

$$\xi = \frac{\phi(\alpha^0) - \phi(\alpha^1)}{\phi'(\alpha^0)(\alpha^1 - \alpha^0)} \tag{11.195}$$

$$\beta = \frac{\phi'(\alpha^1)}{\phi'(\alpha^0)}, \tag{11.196}$$

the unknown numbers p and η are found by solving the pair of nonlinear equations

$$\frac{\eta(p-1)(1-\beta)}{\theta p} + \frac{\beta}{p} = \xi \tag{11.197}$$

$$\left|1 - \frac{\theta}{\eta(p-1)}\right|^p \left(\frac{\eta(p-1)}{\eta(p-1)-\theta}\right) = \beta. \tag{11.198}$$

Having determined the values of p and η, it is also shown in [4] that a good estimate to the second derivative of ϕ at α^1 is given by

$$\tilde{\phi}''(\alpha^1) = \frac{1}{\eta^*}\frac{\phi'(\alpha^0) - \phi'(\alpha^1)}{(\alpha^0 - \alpha^1)}, \tag{11.199}$$

where

$$\eta^* = \frac{\eta}{\theta}\left[1 - \frac{\theta}{\eta(p-1)}\right]\left(\frac{1}{\beta} - 1\right). \tag{11.200}$$

Variable metric methods assume $\eta^* = 1$, as can be seen from the forthcoming derivations. Suppose that we use the BFS updating formula (11.170). Then $B_k = H_k^{-1}$ is given by

$$B_k = B_{k-1} + \frac{\gamma^k(\gamma^k)^T}{(p^k)^T\gamma^k} - \frac{B_{k-1}p^k(p^k)^T B_{k-1}}{(p^k)^T B_{k-1}p^k}. \tag{11.201}$$

Since B_k is the approximate Hessian $\nabla^2 f$ of the function to be minimized, the second directional derivative of f along z^k is given by $(\hat{z}^k)^T B_k \hat{z}^k$, where $\hat{z}^k = z^k/\|z^k\|$. From (11.201) it follows that

$$(\hat{z}^k)^T B_k \hat{z}^k = \frac{[(\hat{z}^k)^T\gamma^k]^2}{(p^k)^T\gamma^k} = \frac{(\hat{z}^k)^T[\nabla f(x^k) - \nabla f(x^{k-1})]}{\|p^k\|} \tag{11.202}$$

and from (11.192) and (11.202) it follows that the BFS method assumes $\eta^* = 1$ in (11.199). The reader can verify the same result by using the DFP updating formula, instead of (11.170), for the H_k. If we assume a nonquadratic behavior of f along z^k, then the appropriate expression for the second derivative is

$$(\hat{z}^k)^T B_k \hat{z}^k = \frac{(\hat{z}^k)^T[\nabla f(x^k) - \nabla f(x^{k-1})]}{\eta^*\|p^k\|} \tag{11.203}$$

and it is easy to check that the modified BFS formula

$$H_k = H_{k-1} + \left[\eta^* + \frac{(\gamma^k)^T H_{k-1}\gamma^k}{(p^k)^T\gamma^k}\right]\frac{p^k(p^k)^T}{(p^k)^T\gamma^k} - \frac{p^k(\gamma^k)^T H_{k-1}}{(p^k)^T\gamma^k} - \frac{H_{k-1}\gamma^k(p^k)^T}{(p^k)^T\gamma^k} \tag{11.204}$$

indeed yields (11.203). Similarly, the modified DFP formula is given by

$$H_k = H_{k-1} + \eta^*\frac{p^k(p^k)^T}{(p^k)^T\gamma^k} - \frac{H_{k-1}\gamma^k(\gamma^k)^T H_{k-1}}{(\gamma^k)^T H_{k-1}\gamma^k}. \tag{11.205}$$

The calculation of η^*, as we have seen, requires the solution of two nonlinear equations, and so the computations are somewhat longer. A simplified version of the dominant degree method is to assume that the function f can be

represented along lines as a polynomial of at most third degree; that is, we can write

$$\phi(\alpha) = a + b\alpha + c(\alpha)^2 + d(\alpha)^3. \tag{11.206}$$

Biggs [5], using some simple algebraic manipulations, shows that η^*, in this case, is explicitly given by

$$\eta^* = \frac{\bar{\beta} - 1}{2(2\bar{\beta} + 1 - 3\bar{\xi})}, \tag{11.207}$$

where $\bar{\xi}$ and $\bar{\beta}$ are parameters yet to be determined. The complete algorithm with modified updating formulas is very similar to any other quasi-Newton method. No exact line search is involved; only a sufficient reduction in the function value is recommended. The step length $\bar{\alpha}_k$ is set to

$$\bar{\alpha}_k = \begin{cases} \min\left[0.1, \dfrac{1}{\|z^k\|}\right] & \text{if } k \leq n \\ 1 & \text{if } k > n \end{cases} \tag{11.208}$$

unless the subsequent search directions z^{k-1} and z^k are nearly parallel.

Letting

$$x^k = x^{k-1} + \bar{\alpha}_k z^k, \tag{11.209}$$

we define

$$\bar{\xi} = \frac{f(x^k) - f(x^{k-1})}{(p^k)^T \nabla f(x^{k-1})}, \tag{11.210}$$

$$\bar{\beta} = \frac{(p^k)^T \nabla f(x^k)}{(p^k)^T \nabla f(x^{k-1})}. \tag{11.211}$$

Note that if the function to be minimized is quadratic—that is, $d = 0$ in (11.206)—then (11.207) will yield $\eta^* = 1$ and the modified method reduces to an ordinary quasi-Newton algorithm.

The next algorithm is based on a certain generalization of quadratic functions. Given a quadratic function f whose unique minimum is at $x^* \in R^n$, we can reformulate f as

$$f(x) = \tfrac{1}{2}(x - x^*)^T Q(x - x^*) + f(x^*). \tag{11.212}$$

This function also satisfies

$$f(x) = \tfrac{1}{2}(x - x^*)^T \nabla f(x) + f(x^*). \tag{11.213}$$

We can generalize this relation and define a real differentiable function **x^*—homogeneous of degree m** if

$$f(x) = \frac{1}{m}(x - x^*)^T \nabla f(x) + f(x^*), \tag{11.214}$$

where m is a nonzero number. Euler's relation [3] for homogeneous functions of degree m is a special case of (11.214) with $x^* = f(x^*) = 0$. Alternatively, (11.214) is satisfied for differentiable functions f such that

$$(t)^m[f(x) - f(x^*)] = f(tx + (1 - t)x^*) - f(x^*). \tag{11.215}$$

We know that Newton's method finds the exact minimum of a quadratic function in a single step. We shall see that this result can be generalized by showing that Newton's method with a properly limited step length finds the exact minimum of an x^*-homogeneous function in just one step. Differentiating (11.214) yields

$$\nabla f(x) = \frac{1}{m}\nabla^2 f(x)(x - x^*) + \frac{1}{m}\nabla f(x). \tag{11.216}$$

Hence

$$x^* = x - (m - 1)[\nabla^2 f(x)]^{-1}\nabla f(x). \tag{11.217}$$

Example 11.4.1

Consider the $\bar{x}$-homogeneous function of degree four

$$f(x) = [\tfrac{1}{2}(x - \bar{x})^T Q(x - \bar{x})]^2, \tag{11.218}$$

where

$$\bar{x} = (1, 1) \qquad Q = \begin{bmatrix} 3 & -1 \\ -1 & 1 \end{bmatrix}. \tag{11.219}$$

We have

$$\nabla f(x) = [(x - \bar{x})^T Q(x - \bar{x})]Q(x - \bar{x}) \tag{11.220}$$

$$\nabla^2 f(x) = [(x - \bar{x})^T Q(x - \bar{x})]Q + 2Q(x - \bar{x})(x - \bar{x})^T Q. \tag{11.221}$$

Take $x^0 = (-2, 4)$. Then $\nabla f(x^0) = (-648, 324)^T$ and

$$\nabla^2 f(x^0) = \begin{bmatrix} 450 & -198 \\ -198 & 126 \end{bmatrix} \qquad [\nabla^2 f(x^0)]^{-1} = \frac{1}{972}\begin{bmatrix} 7 & 11 \\ 11 & 25 \end{bmatrix}. \tag{11.222}$$

Hence

$$x^* = \begin{pmatrix} -2 \\ 4 \end{pmatrix} - \frac{3}{972}\begin{bmatrix} 7 & 11 \\ 11 & 25 \end{bmatrix}\begin{pmatrix} -648 \\ 324 \end{pmatrix} = \begin{pmatrix} 1 \\ 1 \end{pmatrix} \tag{11.223}$$

and $x^* = \bar{x}$, as asserted. ▮

An algorithm that, similar to variable metric methods, does not compute second derivatives and that converges in a finite number of steps when applied to the minimization of an x^*-homogeneous function was developed by Jacobson and Oksman [31]. Rearranging (11.214), we obtain

$$x^T\nabla f(x) = (x^*)^T\nabla f(x) + mf(x) - mf(x^*). \tag{11.224}$$

Defining $w \in R$, $y \in R^{n+2}$, $\lambda \in R^{n+2}$ by

$$w = x^T\nabla f(x), \quad \lambda = \begin{pmatrix} x^* \\ m \\ mf(x^*) \end{pmatrix}, \quad y = \begin{pmatrix} \nabla f(x) \\ f(x) \\ -1 \end{pmatrix}, \tag{11.225}$$

we can write (11.224), evaluated at some point $x^k \in R^n$, as

$$w^k = \lambda^T y^k. \tag{11.226}$$

Suppose that we calculate $w^1, \ldots, w^{n+2}$ and $y^1, \ldots, y^{n+2}$ such that the y^k are linearly independent. Then letting

$$Y = \begin{bmatrix} (y^1)^T \\ \cdot \\ \cdot \\ \cdot \\ (y^{n+2})^T \end{bmatrix} \qquad v = \begin{bmatrix} w^1 \\ \cdot \\ \cdot \\ \cdot \\ w^{n+2} \end{bmatrix}, \tag{11.227}$$

we can write

$$\lambda = Y^{-1}v, \tag{11.228}$$

where the "optimality vector" $\lambda \in R^{n+2}$ contains all the information on the solution of the problem.

It is recommended that (11.228) be solved recursively by updating the matrix Y^{-1}. Suppose that at iteration k we have

$$\lambda^k = R_k v^k, \qquad k = 0, 1, \ldots \tag{11.229}$$

where $R_k = Y_k^{-1}$ and $\lambda^k = [x^{*k}, m^k, m^k f(x^{*k})]^T$ is the kth estimate of the optimality vector. Let $R_0 = I$ and $v^0 = \lambda^0$, a given $(n + 2)$ vector. At each iteration, we successively replace rows of R_0^{-1} by calculated vectors y^k and elements of v^0 by calculated numbers $w^k = (\lambda^{k-1})^T y^k$. That is,

$$R_k^{-1} = R_{k-1}^{-1} + I_j[(y^k)^T - I_j^T R_{k-1}^{-1}], \tag{11.230}$$

$$v^k = v^{k-1} + I_j[(w^k)^T - I_j^T v^{k-1}], \tag{11.231}$$

where I_j is the jth column of the $(n + 2) \times (n + 2)$ identity matrix and

$j = k$. Now by the rank-one inversion formula [29],

$$R_k = R_{k-1} - \frac{R_{k-1}I_j[(y^k)^T R_{k-1} - I_j^T]}{(y^k)^T R_{k-1} I_j}. \tag{11.232}$$

From (11.229), (11.231), and (11.232) we get the updating formula for λ^k

$$\lambda^k = \lambda^{k-1} + \frac{R_{k-1}I_j[w^k - (y^k)^T \lambda^{k-1}]}{(y^k)^T R_{k-1} I_j}. \tag{11.233}$$

For an x^*-homogeneous function in n variables, the vector λ^{n+2} equals the true "optimality vector" λ. Thus after $n + 2$ steps the Jacobson-Oksman algorithm (see below) finds the exact minimum, the degree of homogeneity, and the optimal value of the function to be minimized.

The complete algorithm, in a somewhat conceptual form, consists of the following steps:

1. Assume that $x^0 \in R^n$ is given. Set $m^0 = 2$, $m^0 f(x^{*0}) = 0$, $k = 0$, and let η be some lower bound on $f(x)$ for all x.
2. Set

$$x^1 = x^0 - \theta_0 \nabla f(x^0), \tag{11.234}$$

where

$$\theta_0 = \min\left\{1, \frac{m^0[\eta - f(x^0)]}{(-\nabla f(x^0))^T \nabla f(x^0)}\right\}. \tag{11.235}$$

If $f(x^0) > f(x^1)$, go to step 3; otherwise use some line search to find a θ_0 such that $f(x^0) > f(x^1)$ and then go to step 3.

3. Set $\lambda^0 = [x^1, 2, 0]^T$; $R_0 = I$, and $j = 1$.
4. Calculate y^{k+1}, w^{k+1}. Use (11.232) and (11.233) to compute R_{k+1} and λ^{k+1}.
5. Change k to $k + 1$. If $j = n + 2$, reset $j = 1$; otherwise change j to $j + 1$.
6. Set

$$x^{k+1} = x^k + \theta_k \sigma_k (x^k - x^{*k}), \tag{11.236}$$

where σ_k is either $+1$ or -1, the correct sign being set to the sign of $-(x^k - x^{*k})^T \nabla f(x^k)$, and θ_k is given by

$$\theta_k = \min\left\{1, \frac{m^k[\eta - f(x^k)]}{\sigma_k(x^k - x^{*k})^T \nabla f(x^k)}\right\}. \tag{11.237}$$

Here we assume that $m^k > 0$. Hence if $m^k \leq 0$, use $m^k = 2$ in the preceding formula. If the θ_k resulting from (11.237) has a value such that $f(x^{k+1}) <$

$f(x^k)$, go to step 4; otherwise use some line search such that a sufficient reduction in the function value is obtained and then go to step 4.

The algorithm terminates when $\|\nabla f(x^{k+1})\|$ is less than some small positive number. Throughout the algorithm, the denominators appearing in (11.232) and (11.233) are checked. If

$$|(y^{k+1})R_k I_j| < \epsilon, \tag{11.238}$$

where ϵ is a small positive number, we let x^{k+1} be the new x^0 and return to step 1.

Example 11.4.2

Let us illustrate the Jacobson-Oksman algorithm on the fourth-degree x^*-homogeneous function in two variables presented in the preceding example.

Suppose that we start again at $x^0 = (-2, 4)$ and $f(x^0) = 729$. Set $\eta = 0$. Now

$$\theta_0 = \min\left\{1, \left[\frac{2(-729)}{-(648)^2 - (324)^2}\right]\right\} \tag{11.239}$$

$$= \min\{1, 0.00278\} = 0.00278. \tag{11.240}$$

Hence

$$x^1 = \begin{pmatrix} -0.2 \\ 3.1 \end{pmatrix}, \quad f(x^1) = 47.4032, \quad \nabla f(x^1) = \begin{pmatrix} -78.489 \\ 45.441 \end{pmatrix}. \tag{11.241}$$

Next we set

$$\lambda^0 = \begin{pmatrix} -0.2 \\ 3.1 \\ 2 \\ 0 \end{pmatrix}, \quad R_0 = I, \quad j = 1, \quad k = 0 \tag{11.242}$$

$$y^1 = \begin{pmatrix} -78.4890 \\ 45.4410 \\ 47.4032 \\ -1 \end{pmatrix} \qquad w^1 = 156.565 \tag{11.243}$$

$$R_1 = \begin{bmatrix} -0.0127 & 0.5789 & 0.6039 & -0.0127 \\ 0 & 1 & 0 & 0 \\ 0 & 0 & 1 & 0 \\ 0 & 0 & 0 & 1 \end{bmatrix} \qquad \lambda^1 = \begin{pmatrix} 1.008 \\ 3.1 \\ 2 \\ 0 \end{pmatrix}. \tag{11.244}$$

Thus the predicted minimum is at $x^{*1} = (1.008, 3.1)$. We now set $k = 1$ and $j = 2$ and $(x^1 - x^{*1}) = (-1.208, 0)^T$.

The next point is at

$$x^2 = \begin{pmatrix} -0.2 \\ 3.1 \end{pmatrix} + \theta_1 \sigma_1 \begin{pmatrix} -1.208 \\ 0 \end{pmatrix}. \tag{11.245}$$

Since $(x^1 - x^{*1})^T \nabla f(x^1) > 0$, it follows that $\sigma_1 = -1$ and

$$\theta_1 = \min \left\{1, \left[\frac{2(-47.4032)}{-0.94806}\right]\right\} = 1 \tag{11.246}$$

and so $x^2 = (1.008, 3.1)^T$. At this point,

$$f(x^2) = 4.7896 < f(x^1). \tag{11.247}$$

This point is accepted and we compute

$$\nabla f(x^2) = \begin{pmatrix} -9.0881 \\ 9.1572 \end{pmatrix} \tag{11.248}$$

$$y^2 = \begin{pmatrix} -9.0881 \\ 9.1572 \\ 4.7896 \\ -1 \end{pmatrix} \qquad w^2 = 19.2275. \tag{11.249}$$

It follows that

$$R_2 = \begin{bmatrix} -0.0299 & 0.1486 & 0.7078 & 0.1187 \\ -0.0297 & 0.2567 & 0.1795 & 0.2270 \\ 0 & 0 & 1 & 0 \\ 0 & 0 & 0 & 1 \end{bmatrix} \tag{11.250}$$

and $\lambda^2 = (-0.416, 0.641, 2, 0)^T$. Thus the next predicted minimum is at $x^{*2} = (-0.416, 0.641)$. We set $k = 2$, $j = 3$, and $(x^2 - x^{*2}) = (1.424, 2.459)^T$.

The next point is at

$$x^3 = \begin{pmatrix} 1.008 \\ 3.1 \end{pmatrix} + \theta_2 \sigma_2 \begin{pmatrix} 1.424 \\ 2.459 \end{pmatrix}. \tag{11.251}$$

Since $(x^2 - x^{*2})^T \nabla f(x^2) > 0$, we set $\sigma_2 = -1$, and

$$\theta_2 = \min \left\{1, \left[\frac{2(-4.7896)}{(-9.5792)}\right]\right\} = 1. \tag{11.252}$$

Hence $x^3 = (-0.416, 0.641)$ and

$$f(x^3) = 6.5668 > f(x^2). \tag{11.253}$$

That is, a function increase is obtained. Therefore, the point x^3 is rejected and a new x^3 is sought by, say, cubic approximation. One iteration of this technique (see Chapter 8) yields a new point with a lower function value

$$x^3 = \begin{pmatrix} 0.3721 \\ 2.0017 \end{pmatrix}, \quad f(x^3) = 2.9661, \quad \nabla f(x^3) = \begin{pmatrix} -9.9392 \\ 5.6134 \end{pmatrix}. \tag{11.254}$$

Consequently,

$$y^3 = \begin{pmatrix} -9.9392 \\ 5.6134 \\ 2.9661 \\ -1 \end{pmatrix} \qquad w^3 = 7.5386 \tag{11.255}$$

$$R^3 = \begin{bmatrix} 0.0003 & 0.1403 & -0.2312 & -0.0906 \\ -0.0221 & 0.2546 & -0.0586 & 0.1739 \\ 0.0427 & -0.0118 & -0.3266 & -0.2957 \\ 0 & 0 & 0 & 1 \end{bmatrix} \tag{11.256}$$

and $\lambda^3 = (1.0, 1.0, 4.0, 0.0)^T$.
Thus the next predicted minimum is at $x^{3*} = (1.0, 1.0)$. We set $k = 3, j = 4$, and $(x^3 - x^{*3}) = (-0.6279, 1.0017)^T$. Then

$$x^4 = \begin{pmatrix} 0.3721 \\ 2.0017 \end{pmatrix} + \theta_3\sigma_3\begin{pmatrix} -0.6279 \\ 1.0017 \end{pmatrix}. \tag{11.257}$$

Since $(x^3 - x^{3*})^T\nabla f(x^3) > 0$, it follows that $\sigma^3 = -1$ and

$$\theta^3 = \min\left\{1, \left[\frac{4(-2.9661)}{-11.8644}\right]\right\} = 1. \tag{11.258}$$

Hence $x^4 = (1.0, 1.0)$, $f(x^4) = 0.0$, $\nabla f(x^4) = (0.0, 0.0)^T$, and the algorithm terminates at the exact minimum. ∎

EXERCISES

11.A. Obtain updating formulas for matrices H_k that belong to Huang's family by letting $\omega = 0$ in (11.25) and such that
(a) ΔH_k has rank two.
(b) ΔH_k is symmetric and has rank one.
Now set $\omega = -1$ in (11.25) and find updating formulas such that
(c) ΔH_k is symmetric and has rank two.
(d) ΔH_k is symmetric and has rank one.

11.B. Apply variable metric algorithms based on the updating formulas (a) and (c) of the preceding exercise to minimizing the quadratic function

$$f(x) = (x_1 - x_2 + x_3)^2 + (-x_1 + x_2 + x_3)^2 + (x_1 + x_2 - x_3)^2. \tag{11.259}$$

Starting at $x^0 = (\frac{1}{2}, 1, \frac{1}{2})$ with $H_0 = I$, perform exact line searches. Are the x^2 and x^3 generated by the algorithms identical? Did you obtain quadratic termination? Are the H_n equal to Q^{-1}?

11.C. One of the first variable metric methods for minimizing f on R^n was suggested by Zoutendijk [55], and it can be described as follows: Let x^0 be an arbitrary point in R^n, let $H_0 = I$,

$$H_k = H_{k-1} - \frac{H_{k-1}\gamma^k(\gamma^k)^T H_{k-1}}{(\gamma^k)^T H_{k-1}\gamma^k}, \tag{11.260}$$

and perform an exact line search at each iteration. Show that the H_k obtained by this updating formula belongs to Huang's family. What is the value of ω? Apply Zoutendijk's method to minimizing a quadratic function of two variables. What is H_2?

11.D. Suppose that the $n \times n$ matrix H_{k-1} is symmetric, positive definite, and is updated by the DFP formula. Show that if $(p^k)^T\gamma^k > 0$, then H_k is also symmetric and positive definite.

11.E. It is known that the DFP method is not invariant under scaling of the function to be minimized. Let $x^1, x^2, \ldots$, and $\hat{x}^1, \hat{x}^2, \ldots$ be the sequences of points generated by this method on minimizing, respectively, $f(x)$ and $\hat{f}(x) = cf(x)$, where c is a positive number, $c \neq 1$. Show that the two sequences are not identical. Can you generalize this result to Broyden's subfamily of methods and Huang's family of methods? Discuss the effects of scaling $x \in R^n$ to ax, $a \in R$, and minimizing $f(ax)$ instead of $f(x)$ on the sequence of points generated by the DFP method.

11.F. Oren [38] has shown a way to modify the DFP method so that it becomes invariant under scaling of the function or the variables, as discussed in the preceding exercise. He suggests, for example, multiplying the matrix H_{k-1} before updating, by the factor

$$\zeta_k = \frac{(p^k)^T\gamma^k}{(\gamma^k)^T H_{k-1}\gamma^k}. \tag{11.261}$$

This is equivalent to the following modified DFP updating formula

$$H_k = \left[H_{k-1} - \frac{H_{k-1}\gamma^k(\gamma^k)^T H_{k-1}}{(\gamma^k)^T H_{k-1}\gamma^k}\right]\frac{(p^k)^T\gamma^k}{(\gamma^k)^T H_{k-1}\gamma^k} + \frac{p^k(p^k)^T}{(p^k)^T\gamma^k}. \tag{11.262}$$

Show that a variable metric algorithm based on the last updating formula generates the same sequence of points $x^1, x^2, \ldots$ and $\hat{x}^1, \hat{x}^2, \ldots$ on minimizing $f(x)$ and $cf(ax)$, respectively.

11.G. Suppose that the DFP method is applied to minimizing a quadratic function on R^n with a symmetric positive definite Q matrix. Show that, in this case, the updating formula can be written as

$$H_k = P_k + C_k, \qquad k = 1, \ldots \tag{11.263}$$

where

$$P_k = P_{k-1} - \frac{P_{k-1}\gamma^k(P_{k-1}\gamma^k)^T}{(\gamma^k)^T P_{k-1}\gamma^k} \tag{11.264}$$

$$C_k = C_{k-1} + \frac{p^k(p^k)^T}{(p^k)^T\gamma^k} \tag{11.265}$$

and $P_0 = H_0$, $C_0 = 0$. (*Hint:* $H_{k-1}\gamma^k = P_{k-1}\gamma^k$.) Combine (11.89) with the preceding relations and show that $C_n = Q^{-1}$. Is $P_n = 0$?

11.H. Show that Broyden's subfamily of matrices H_k, given by (11.67) and (11.68), yield $z^{k+1} = 0$ if (11.69) holds.

11.I. Let A be a nonsingular $n \times n$ matrix and let U, V be $n \times m$ matrices, $n \geq m$. Show that $(A + UV^T)$ has an inverse if and only if $(I + V^T A^{-1} U)$ has an inverse, where I is the $n \times n$ identity matrix and

$$(A + UV^T)^{-1} = A^{-1} - A^{-1}U(I + V^T A^{-1} U)^{-1} V^T A^{-1}. \tag{11.266}$$

Compare this formula with (11.139).

11.J. Show that if a singular H_k is used in a variable metric algorithm, then all subsequent H_k will be singular (except in a very special case). What is the effect of a singular H_k on the subsequent search directions?

11.K. Prove the "complementary" relations between the DFP and BFS updating formulas. What is the updating formula for the approximate Hessian matrices if the H_k are updated by the symmetric rank-one formula?

11.L. Powell [43] has suggested a variable metric algorithm for minimizing $f(x)$ on R^n in which the approximate Hessian matrix is updated by the formula

$$B_k = B_{k-1} + \frac{(\gamma^k - B_{k-1}p^k)(p^k)^T + p^k(\gamma^k - B_{k-1}p^k)^T}{(p^k)^T p^k} - \frac{(p^k)^T(\gamma^k - B_{k-1}p^k)p^k(p^k)^T}{[(p^k)^T p^k]^2}. \tag{11.267}$$

(a) Find the updating formula for the approximate inverse Hessian $H_k = B_k^{-1}$ and show that H_k satisfies the secant relation.

(b) Suppose that f is a quadratic function having a nonsingular Q matrix. Show that

$$B_k - Q = \left[I - \frac{p^k(p^k)^T}{(p^k)^T p^k}\right](B_{k-1} - Q)\left[I - \frac{p^k(p^k)^T}{(p^k)^T p^k}\right]. \tag{11.268}$$

(c) Show that if the updating formula (11.267) is applied iteratively n times and if the vectors $p^1, p^2, \ldots, p^n$ are mutually orthogonal, then, in the case of a quadratic function, we have $B_n = Q$.

11.M. Show that if the step lengths α_k in the Murtagh-Sargent variable metric method, based on the symmetric rank-one formula (11.158), are chosen to minimize a quadratic function with a positive definite Q matrix along z^k,

then the search directions $z^1, \ldots, z^n$ are mutually conjugate with respect to Q. Do not use the fact that the H_k are members of Huang's family.

11.N. Suppose that a quasi-Newton minimization algorithm for continuously differentiable functions on R^n is given by the following general formulas:

$$x^{k+1} = x^k + \alpha_{k+1} z^{k+1}, \qquad k = 0, 1, \ldots \tag{11.269}$$

where

$$z^{k+1} = -H_k \nabla f(x^k), \qquad k = 0, 1, \ldots \tag{11.270}$$

and H_k is a symmetric positive definite matrix. It was suggested that the step length be selected by making a single quadratic approximation of f along z^{k+1} as follows:
Let

$$x^r = x^k + \alpha_r z^{k+1}, \qquad \alpha_r > 0 \tag{11.271}$$

be any point so that $f(x^r) > f(x^k)$ and choose

$$\alpha_{k+1} = \frac{1}{2} \frac{(\alpha_r)^2 (\nabla f(x^k))^T H_k \nabla f(x^k)}{(f(x^r) - f(x^k)) + \alpha_r (\nabla f(x^k))^T H_k \nabla f(x^k)}. \tag{11.272}$$

Show that x^{k+1} is the minimum of the quadratic approximation of f along z^{k+1} and $0 < \alpha_{k+1} < \alpha_r$.

11.O. Analyze variable metric algorithms based on updating formulas mentioned in this chapter with respect to the following theoretical properties:

(a) quadratic termination (with or without exact line searches)
(b) positive definiteness of the H_k matrices
(c) stability
(d) convergence of the H_k matrices to $[\nabla^2 f(x^*)]^{-1}$.

Consult the appropriate references, if necessary.

REFERENCES

1. BARD, Y., "On a Numerical Instability of Davidon-Like methods," *Math. Comp.*, **22**, 665–666 (1968).
2. BARNES, J. G. P., "An Algorithm for Solving Non-Linear Equations Based on the Secant Method," *Computer J.*, **8**, 66–72 (1965).
3. BARTLE, R. G., *The Elements of Real Analysis*, 2 nd ed. , John Wiley & Sons, New York, 1976.
4. BIGGS, M. C., "Minimization Algorithms Making Use of Non-quadratic Properties of the Objective Function," *J. Inst. Math. Appl.*, **8**, 315–327 (1971).
5 BIGGS, M. C., "A Note on Minimization Algorithms which Make Use of Non-Quadratic Properties of the Objective Function," *J. Inst. Math. Appl.*, **12**, 337–338 (1973).

6. BROYDEN, C. G., "A Class of Methods for Solving Nonlinear Simultaneous Equations," *Math. Comp.*, **19**, 577–593 (1965).

7. BROYDEN, C. G., "Quasi-Newton Methods and their Application to Function Minimisation," *Math. Comp.*, **21**, 368–381 (1967).

8. BROYDEN, C. G., "The Convergence of a Class of Double-Rank Minimization Algorithms 1. General Considerations," *J. Inst. Math. Appl.*, **6**, 76–90 (1970).

9. BROYDEN, C. G., "The Convergence of a Class of Double-Rank Minimization Algorithms 2. The New Algorithm," *J. Inst. Math. Appl.*, **6**, 222–231 (1970).

10. BROYDEN, C. G., "The Convergence of Single-Rank Quasi-Newton Methods," *Math. Comp.*, **24**, 365–382 (1970).

11. DAVIDON, W. C., "Variable Metric Method for Minimization," AEC Research and Development Report ANL-5990 (Rev.), November 1959.

12. DAVIDON, W. C., "Variance Algorithm for Minimization," *Computer J.*, **10**, 406–410 (1968).

13. DENNIS, J. E., "On the Convergence of Broyden's Method for Nonlinear Systems of Equations," *Math. Comp.*, **25**, 559–567 (1971).

14. DIXON, L. C. W., "Quasi-Newton Algorithms Generate Identical Points," *Math. Prog.*, **2**, 383–387 (1972).

15. DIXON, L. C. W., "Quasi-Newton Techniques Generate Identical Points. II. The Proofs of Four New Theorems," *Math. Prog.*, **3**, 345–358 (1972).

16. DIXON, L. C. W., "The Choice of Step Length a Crucial Factor in the Performance of Variable Metric Algorithms," in *Numerical Methods for Non-Linear Optimization*, F. Lootsma (Ed.), Academic Press, London, 1972.

17. FIACCO, A. V., and G. P. MCCORMICK, *Nonlinear Programming: Sequential Unconstrained Minimization Techniques*, John Wiley & Sons, New York, 1968.

18. FLETCHER, R., "A Review of Methods for Unconstrained Optimization," in *Optimization*, R. Fletcher (Ed.), Academic Press, London, 1969.

19. FLETCHER, R., "A New Approach to Variable Metric Algorithms," *Computer J.*, **13**, 317–322 (1970).

20. FLETCHER, R., and M. J. D. POWELL, "A Rapidly Convergent Descent Method for Minimization," *Computer J.*, **6**, 163–168 (1963).

21. GILL, P. E., and W. MURRAY, "Quasi-Newton Methods for Unconstrained Optimization," *J. Inst. Math. Appl.*, **9**, 91–108 (1972).

22. GILL, P. E., W. MURRAY, and R. A. PITFIELD, "The Implementation of Two Revised Quasi-Newton Algorithms for Unconstrained Optimization," National Physical Laboratory DNAC Report No. 11, April 1972.

23. GOLDFARB, D., "Sufficient Conditions for the Convergence of a Variable Metric Algorithm," in *Optimization*, R. Fletcher (Ed.), Academic Press, London, 1969.

24. GOLDSTEIN, A. A., and J. F. PRICE, "An Effective Algorithm for Minimization," *Numerische Mathematik*, **10**, 184–189 (1967).

25. GREENSTADT, J., "Variations on Variable-Metric Methods," *Math. Comp.*, **24**, 1–22 (1970).

26. GREENSTADT, J., "A Quasi-Newton Method with No Derivatives," *Math. Comp.*, **26**, 145–166 (1972).

27. GREENSTADT, J., "Improvements in a QNWD Method," IBM Palo Alto Scientific Center Tech. Report No. 320–3306, October 1972.

28. HESTENES, M. R., "Multiplier and Gradient Methods," *J. Optimization Theory & Appl.*, **4**, 303–320 (1969).

29. HOUSEHOLDER, A. S., *The Theory of Matrices in Numerical Analysis*, Blaisdell Publishing Co., New York, 1964.

30. HUANG, H. Y., "Unified Approach to Quadratically Convergent Algorithms for Function Minimization," *J. Optimization Theory & Appl.*, **5**, 405–423 (1970).

31. JACOBSON, D. H., and W. OKSMAN, "An Algorithm that Minimizes Homogeneous Functions of N Variables in $N + 2$ Iterations and Rapidly Minimizes General Functions," *J. Math. Anal. & Appl.*, **38**, 535–552 (1972).

32. LILL, S. A., "A Modified Davidon Method for Finding the Minimum of a Function Using Difference Approximation for Derivatives," *Computer J.*, **13**, 111–113 (1970).

33. LOOTSMA, F. A. (Ed.), *Numerical Methods for Non-Linear Optimization*, Academic Press, London, 1972.

34. MURRAY, W., "Failure, the Causes and Cures," in *Numerical Methods for Unconstrained Optimization*, W. Murray (Ed.), Academic Press, London, 1972.

35. MURRAY, W., "The Relationship between the Approximate Hessian Matrices Generated by a Class of Quasi-Newton Methods," National Physical Laboratory DNAC Report No. 12, January 1972.

36. MURTAGH, B. A., and R. W. H. SARGENT, "A Constrained Minimization Method with Quadratic Convergence," in *Optimization*, R. Fletcher (Ed.), Academic Press, London, 1969.

37. MURTAGH, B. A., and R. W. H. SARGENT, "Computational Experience with Quadratically Convergent Minimisation Methods," *Computer J.*, **13**, 185–194 (1970).

38. OREN, S. S., "On the Selection of Parameters in Self-Scaling Variable Metric Algorithms," *Math. Prog.*, **7**, 351–367 (1974).

39. ORTEGA, J. M., and W. C. RHEINBOLDT, *Iterative Solution of Nonlinear Equations in Several Variables*, Academic Press, New York, 1970.

40. PEARSON, J. D., "Variable Metric Methods of Minimisation," *Computer J.*, **12**, 171–178 (1969).

41. POWELL, M. J. D., "An Efficient Method for Finding the Minimum of a Function of Several Variables without Calculating Derivatives," *Computer J.*, **7**, 155–162 (1964).

42. POWELL, M. J. D., "Rank One Methods for Unconstrained Optimization," in *Integer and Nonlinear Programming*, J. Abadie (Ed.), North-Holland Publishing Co., Amsterdam, 1970.

43. POWELL, M. J. D., "A New Algorithm for Unconstrained Optimization," in *Nonlinear Programming*, J. B. Rosen, O. L. Mangasarian, and K. Ritter (Eds.), Academic Press, New York, 1970.

44. POWELL, M. J. D., "On the Convergence of Variable Metric Algorithms," *J. Inst. Math. Appl.*, **7**, 21–36 (1970).

45. POWELL, M. J. D., "Some Properties of the Variable Metric Algorithm," in *Numerical Methods for Non-Linear Optimization,* F. Lootsma (Ed.), Academic Press, London, 1972.

46. POWELL, M. J. D., "Quadratic Termination Properties of Minimization Algorithms I. Statement and Discussion of Results," *J. Inst. Math. Appl.*, **10**, 333–342 (1972).

47. POWELL, M. J. D., "Quadratic Termination Properties of Minimization Algorithms II. Proofs of Theorems," *J. Inst. Math. Appl.*, **10**, 343–357 (1972).

48. SARGENT, R. W. H., "Numerical Optimization Techniques," (mimeographed lecture notes), Dept. of Chemical Engineering and Chemical Technology, Imperial College, London, 1972.

49. SHANNO, D. F., "Conditioning of Quasi-Newton Methods for Function Minimization," *Math. Comp.*, **24**, 647–656 (1970).

50. SHANNO, D. F., and P. C. KETTLER, "Optimal Conditioning of Quasi-Newton Methods," *Math. Comp.*, **24**, 657–664 (1970).

51. SORENSON, H. W., "Comparison of Some Conjugate Direction Procedures for Function Minimization," *J. Franklin Inst.*, **288**, 421–441 (1969).

52. STEWART, G. W., "A Modification of Davidon's Minimization Method to Accept Difference Approximations of Derivatives," *J. Assoc. Comp. Mach.*, **14**, 72–83 (1967).

53. WILKINSON, J. H., *The Algebraic Eigenvalue Problem,* Oxford University Press, London, 1965.

54. WOLFE, P., "The Secant Method for Simultaneous Non-linear Equations," *Comm. Assoc. Comp. Mach.*, **2**, 12–13 (1959).

55. ZOUTENDIJK, G., *Methods of Feasible Directions*, Elsevier Publishing Co., Amsterdam, 1960.

12 PENALTY FUNCTION METHODS

This chapter begins the discussion of methods for solving constrained nonlinear programs. Primarily, we deal with the problem of how to transform a constrained program into one or more equivalent unconstrained programs that can be solved by techniques discussed in earlier chapters.

The intuitive idea behind all penalty function methods is simple. Suppose that we seek a minimum of a real-valued function f on a proper subset X of R^n. This is, of course, a constrained optimization problem that can be transformed into an unconstrained optimization problem after some modification of the objective function, as we shall now see. Define

$$P(x) = \begin{cases} 0 & x \in X \\ +\infty & x \notin X \end{cases} \tag{12.1}$$

and consider the unconstrained minimization of the **augmented objective function** F, given by

$$\min_{x \in R^n} F(x) = f(x) + P(x), \tag{12.2}$$

where f is assumed to be defined on R^n. A point x^* minimizes F if and only if it also minimizes f over X. The function P is called a **penalty function,** for it imposes an (infinite) penalty on points lying outside the feasible set. In practice, however, the unconstrained optimization (12.2) cannot be carried out (except perhaps in some trivial cases) because of the discontinuity in F on the boundary of X and the infinite values outside X. Replacing $+\infty$ by some "large" finite penalty will not simplify the problem, since the numerical difficulties would still remain, and without additional assumptions, the minimum of the augmented, everywhere-finite objective function may not coincide with the minimum of f over X.

The earliest idea for solving constrained problems by penalty functions involved a sequence of unconstrained minimizations in which a penalty parameter is adjusted from one minimization to another so that the sequence of unconstrained minima converges to a feasible point of the constrained problem that satisfies some necessary or sufficient optimality conditions. The initial suggestions for studying constrained problems via related unconstrained ones are probably due to Courant [9] and Frisch [22].

The first three sections deal with such methods, which differ in the way in which a solution is approached and in the selection of the parameters. These methods have been successfully implemented in several applications (see, e.g., [4, 21, 31, 51]) and are considered one of the major tools available at present for solving constrained problems.

In recent years much research effort has been devoted to improving of penalty function methods by eliminating the need to solve a sequence of unconstrained problems. Certain transformations of constrained problems were derived in which only one unconstrained optimization is needed, provided that a sufficiently large parameter is chosen. Such transformations are also closely related to Lagrangians and optimality conditions of nonlinear programs, which were analyzed in earlier chapters. These methods are described in Sections 12.4 and 12.5. Finally, in the last section some computational aspects of penalty methods are discussed.

12.1 EXTERIOR PENALTY FUNCTIONS

This section presents a class of techniques for solving the basic nonlinear program first introduced in Chapter 3:

$$\text{(P)} \qquad \min f(x) \tag{12.3}$$

subject to the constraints

$$g_i(x) \geq 0, \qquad i = 1, \ldots, m \tag{12.4}$$

$$h_j(x) = 0, \qquad j = 1, \ldots, p \tag{12.5}$$

where $f, g_1, \ldots, g_m, h_1, \ldots, h_p$ are assumed to be continuous on R^n. Let X denote the feasible set; that is,

$$X = \{x : x \in R^n, g_i(x) \geq 0, i = 1, \ldots, m; h_j(x) = 0, j = 1, \ldots, p\}. \tag{12.6}$$

Exterior penalty function methods usually solve (P) by a sequence of unconstrained minimization problems whose optimal solutions approach the solution of (P) from outside the feasible set. In the sequence of uncon-

strained optimizations, a penalty is imposed on every $x \notin X$, such that this penalty is increased from problem to problem, thereby forcing the unconstrained optima toward the feasible set. The development of the algorithm presented here is due to Zangwill [52], where further details can be found.

Define real-valued continuous functions ψ and ζ of the variable $\eta \in R$ by

$$\psi(\eta) = |\min(0, \eta)|^\alpha \tag{12.7}$$

and

$$\zeta(\eta) = |\eta|^\beta, \tag{12.8}$$

where $\alpha \geq 1$ and $\beta \geq 1$ are given constants, usually equal to 1 or 2. Let

$$s(x) = \sum_{i=1}^{m} \psi(g_i(x)) + \sum_{j=1}^{p} \zeta(h_j(x)) \tag{12.9}$$

or

$$s(x) = \sum_{i=1}^{m} |\min[0, g_i(x)]|^\alpha + \sum_{j=1}^{p} |h_j(x)|^\beta. \tag{12.10}$$

This continuous function is called a **loss function** for problem (P). Note that

$$s(x) = 0 \quad \text{if } x \in X \tag{12.11}$$

and

$$s(x) > 0 \quad \text{if } x \notin X. \tag{12.12}$$

For any positive number ρ we can define the augmented objective function for problem (P) as

$$F(x, \rho) = f(x) + \frac{1}{\rho} s(x) \tag{12.13}$$

and observe that $F(x, \rho) = f(x)$ if and only if x is feasible; otherwise $F(x, \rho) > f(x)$. The $s(x)/\rho$ term approximates the discontinuous penalty function $P(x)$ in (12.1) as $\rho \to 0$. The exterior penalty function method consists of solving a sequence of unconstrained optimizations for $k = 0, 1, 2, \ldots$, given by

$$(\text{EP}^k) \qquad \min_{x \in R^n} F(x, \rho^k) = f(x) + \frac{1}{\rho^k}\left\{\sum_{i=1}^{m} |\min[0, g_i(x)]|^\alpha + \sum_{j=1}^{p} |h_j(x)|^\beta\right\}, \tag{12.14}$$

using a strictly decreasing sequence of positive numbers ρ^k. Defining x^{k*} as the optimal solution of (EP^k), we construct a sequence of points $\{x^{k*}\}$,

which, under rather mild conditions on (P), has a subsequence converging to an optimum of (P).

The foregoing derivation of exterior penalty functions can be generalized. Let r be a continuous real-valued function of the variable $\rho \in R$, such that $\rho^1 > \rho^2 > 0$ implies $r(\rho^2) > r(\rho^1) > 0$ and every strictly decreasing sequence of positive numbers $\{\rho^k\}$ with the property that

$$\lim_{k\to\infty} \{\rho^k\} = 0 \tag{12.15}$$

implies

$$\lim_{k\to\infty} \{r(\rho^k)\} = +\infty. \tag{12.16}$$

Also let s be any continuous function satisfying (12.11) and (12.12). Then $r(\rho^k)s(x)$ is an **exterior penalty function**, and

$$(\text{EP}^k) \qquad \min_x F(x, \rho^k) = f(x) + r(\rho^k)s(x) \tag{12.17}$$

is the corresponding unconstrained optimization problem. In Figure 12.1 we illustrate this idea for a simple case in which the feasible set X is the closed interval $[a, b]$ on the real line.

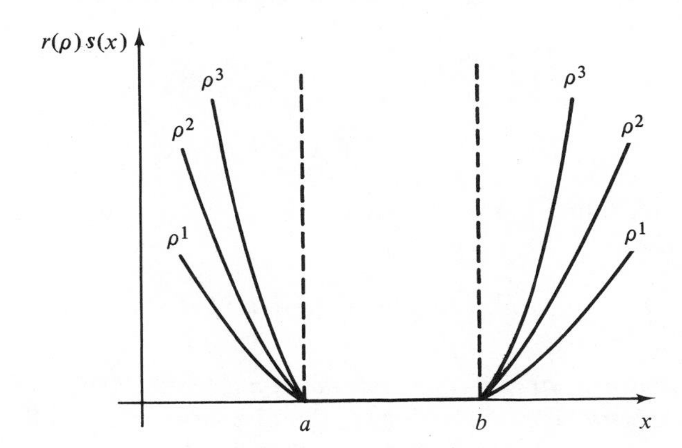

Fig. 12.1 Exterior penalty function.

Before turning to the convergence of the exterior penalty function method, consider a small example.

Example 12.1.1

We seek the minimum of $f(x) = (x)^2$, $x \in R$, subject to the constraint $x \geq 1$. The optimal solution is clearly $x^* = 1$. Let us form the augmented objective function as in (12.14) with $\alpha = 2$. We have the unconstrained

optimization problem

$$\min_{x \in R} F(x, \rho^k) = (x)^2 + \frac{1}{\rho^k}[\min(0, x - 1)]^2. \tag{12.18}$$

For any given $\rho^k > 0$, the function F is convex and its minimum is at the point

$$x^{k*} = \frac{1}{\rho^k + 1}. \tag{12.19}$$

Note that, for every $\rho^k > 0$, this point is infeasible for the original problem. As $\{\rho^k\} \longrightarrow 0$, the points x^{k*} approach x^* from outside the feasible set. Of course, in any real problem the unconstrained minimizations of $F(x, \rho^k)$ must be carried out by some numerical algorithm, such as those presented in the preceding chapters. ▮

We now state and prove some convergence results for the exterior penalty function method. First we need

Lemma 12.1

Let F be given by (12.17) *and let*

$$\rho^k \geq \rho^{k+1} > 0. \tag{12.20}$$

Assume that $F(x, \rho^k)$ *and* $F(x, \rho^{k+1})$ *attain their minima on* R^n *at* x^{k*} *and* x^{k+1*}, *respectively. Then*

$$F(x^{k+1*}, \rho^{k+1}) \geq F(x^{k*}, \rho^k) \tag{12.21}$$

$$s(x^{k*}) \geq s(x^{k+1*}) \tag{12.22}$$

and

$$f(x^{k+1*}) \geq f(x^{k*}). \tag{12.23}$$

Proof. Since $r(\rho)s(x) \geq 0$ and r is increasing as ρ is decreasing, we obtain

$$F(x^{k+1*}, \rho^{k+1}) = f(x^{k+1*}) + r(\rho^{k+1})s(x^{k+1*}) \tag{12.24}$$

$$\geq f(x^{k+1*}) + r(\rho^k)s(x^{k+1*}) \tag{12.25}$$

$$\geq f(x^{k*}) + r(\rho^k)s(x^{k*}) = F(x^{k*}, \rho^k), \tag{12.26}$$

where the last inequality follows from the fact that x^{k*} minimizes $F(x, \rho^k)$. Hence (12.21) holds. Combining (12.25) and (12.26) yields

$$f(x^{k*}) + r(\rho^k)s(x^{k*}) \leq f(x^{k+1*}) + r(\rho^k)s(x^{k+1*}). \tag{12.27}$$

Also, by the definition of x^{k+1*},

$$f(x^{k+1*}) + r(\rho^{k+1})s(x^{k+1*}) \leq f(x^{k*}) + r(\rho^{k+1})s(x^{k*}). \tag{12.28}$$

Adding the last two inequalities, we get

$$r(\rho^k)[s(x^{k*}) - s(x^{k+1*})] \leq r(\rho^{k+1})[s(x^{k*}) - s(x^{k+1*})]. \tag{12.29}$$

And since $r(\rho^k) \leq r(\rho^{k+1})$, inequality (12.22) holds. From (12.22) and (12.27) it follows that

$$f(x^{k+1*}) - f(x^{k*}) \geq r(\rho^k)[s(x^{k*}) - s(x^{k+1*})] \geq 0 \tag{12.30}$$

and (12.23) holds. ■

We can now prove a basic convergence theorem for the exterior penalty function method.

Theorem 12.2

Suppose that the feasible set X in problem (P) *is nonempty and there exists an $\epsilon > 0$ such that the set*

$$X^\epsilon = \{x : x \in R^n, g_i(x) \geq -\epsilon, i = 1, \ldots, m; |h_j(x)| \leq \epsilon, j = 1, \ldots, p\} \tag{12.31}$$

is compact. Also suppose that the $F(x, \rho^k)$ attain their unconstrained minima on R^n for all k. If $\{\rho^k\}$ is a strictly decreasing sequence of positive numbers converging to zero, then there exists a convergent subsequence $\{x^{k_i*}\}$ of the optimal solutions to (EP^k) *and the limit of any such convergent subsequence is optimal for* (P).

Proof. By Lemma 12.1, the sequence $F(x^{k*}, \rho^k)$ is an increasing sequence. Since X is compact and f is continuous, there exists at least one point $x^* \in X$ where f attains its minimum—that is, x^* is optimal for (P). Then for $k = 0, 1, \ldots,$

$$f(x^*) = f(x^*) + r(\rho^k)s(x^*) \geq F(x^{k*}, \rho^k) \tag{12.32}$$

and the sequence $\{F(x^{k*}, \rho^k)\}$ is bounded from above. Consequently, it converges to a limit F^0; see [2]. Similarly, the sequence $\{f(x^{k*})\}$ is increasing, and

$$f(x^{k*}) \leq f(x^{k*}) + r(\rho^k)s(x^{k*}) = F(x^{k*}, \rho^k). \tag{12.33}$$

By (12.32) and (12.33), we obtain

$$f(x^{k*}) \leq f(x^*) \tag{12.34}$$

and $\{f(x^{k*})\}$ converges to a limit f^0. Furthermore,

$$\lim_{k \to \infty} \{r(\rho^k)s(x^{k*})\} = F^0 - f^0. \tag{12.35}$$

By (12.16), it follows that

$$\lim_{k \to \infty} \{s(x^{k*})\} = 0. \tag{12.36}$$

From (12.11), (12.12) and (12.36) we conclude that, for every $\delta > 0$, there exists a natural number $K(\delta)$ such that $x^{k*} \in X^\delta$ for $k \geq K(\delta)$. Then for a sufficiently large $\hat{K}(\epsilon)$, the points x^{k*} will be in a compact set X^ϵ for all $k \geq \hat{K}(\epsilon)$. Hence there is a subsequence $\{x^{k_i*}\}$ that converges to a limit x^0, and it follows that $s(x^0) = 0$. Thus $x^0 \in X$. From the optimality of x^* we obtain

$$f(x^*) \leq f(x^0). \tag{12.37}$$

Now for all k_i in the convergent subsequence,

$$f(x^{k_i*}) \leq f(x^{k_i*}) + r(\rho^{k_i})s(x^{k_i*}) \leq f(x^*) + r(\rho^{k_i})s(x^*) = f(x^*). \tag{12.38}$$

Thus

$$\lim_{k_i \to \infty} \{f(x^{k_i*})\} = f(x^0) \leq f(x^*). \tag{12.39}$$

By (12.37) and (12.39), we finally obtain

$$f(x^0) = f(x^*) \tag{12.40}$$

and x^0 must be a solution for (P). ■

Zangwill [52] investigated conditions that imply that the $F(x, \rho^k)$ attain their minima, an assumption used in the convergence theorem. He found two such conditions. The first says that f is a function such that whenever $||x|| \to +\infty$, then $f(x) \to +\infty$ also. Another condition with the same implication is that X^ϵ is compact for some $\epsilon > 0$ and $F(x, \rho)$ is a convex function of x. This condition can undoubtedly be weakened to classes of generalized convex functions (Chapter 6). Note that the convergence theorem refers to a global optimum of (P), and thus it seems to be a strong result. It assumes, however, that we find a global optimum of the unconstrained program (EP^k) for each k, a rather difficult task, as we saw in earlier chapters.

For this reason, the method is mainly useful for convex programs (or their generalizations) and for those in which the augmented objective function is at least strongly quasiconvex. The question of the existence of global minima in an unconstrained program (EP^k) has been investigated by Evans and Gould [10], who derived necessary and sufficient conditions for the existence of such extrema. The verification of such conditions is, unfortunately, impractical in most problems.

Convergence to local minima of nonconvex programs by the exterior penalty function algorithm can also be obtained under slightly different assumptions. The interested reader is referred to Fiacco and McCormick [15]; here we shall only mention their main result. Suppose that the set of points A in X (the feasible set) that are local minima, with minimum value v^* of the objective function f, is a nonempty compact set satisfying a mild regularity condition. Then there exists a compact set S such that A is embedded in the interior of S, and, for sufficiently large k, the unconstrained minima of (EP^k) are attained in the interior of S. Moreover,

$$\lim_{k \to \infty} \{f(x^{k*})\} = v^* \tag{12.41}$$

and every limit point of any convergent subsequence of $\{x^{k*}\}$ is in A.

Finally, a successful application of efficient unconstrained numerical methods also requires continuous differentiability, which itself poses a problem for certain penalty functions even if all the functions appearing in (P) are continuously differentiable. This question is discussed later in the chapter. Another aspect to be covered later is that for convex programs satisfying a strong consistency, or Slater condition (see Section 4.5), a single unconstrained minimization—instead of a sequence of minimizations—can yield the sought constrained optimum.

12.2 INTERIOR PENALTY FUNCTIONS

Here inequality constrained nonlinear programs are solved through a sequence of unconstrained optimization problems whose minima are at points that strictly satisfy the constraints—that is, in the interior of the feasible set. Staying in the interior can be ensured, as we shall see, by formulating a "barrier" function by which an infinitely large penalty is imposed for crossing the boundary of the feasible set from the inside. Since the algorithm requires the interior of the feasible set to be nonempty, no equality constraints can be handled by the method described below. However, other interior-type penalty function methods have been derived that are capable of solving problems with equality constraints as in program (P) of the previous section. Such methods will be mentioned later.

Consider, therefore, the nonlinear program

$$\text{(PI)} \qquad \min f(x) \tag{12.42}$$

subject to the constraints

$$g_i(x) \geq 0, \qquad i = 1, \ldots, m. \tag{12.43}$$

As before, assume that $f, g_1, \ldots, g_m$ are continuous functions on R^n. Denote by X the feasible set for program (PI)—that is, the set of $x \in R^n$ satisfying (12.43). Let X^0 be the interior of the set X. The following regularity conditions on program (PI) are assumed to hold:

1. The set X is closed, X^0 is nonempty, and X is the closure of X^0.
2. There is a point $x^0 \in X$ with $f(x^0) = \alpha^0$ such that the set $S(f, \alpha^0) \cap X$ is compact, where $S(f, \alpha^0)$ is the level set of f at α^0.

In formulating the penalty function, let q be a real-valued function of $x \in R^n$ such that q is continuous at every point of X^0. And if $\{x^k\}$ is any sequence of points in X^0 that converges to some $\hat{x}$ on the boundary of X—that is,

$$I(\hat{x}) = \{i : g_i(\hat{x}) = 0\} \neq \emptyset, \tag{12.44}$$

then

$$\lim_{k \to \infty} \{q(x^k)\} = +\infty. \tag{12.45}$$

Also let t be a real-valued function of $\rho \in R$ such that

$$\rho^1 > \rho^2 > 0 \quad \text{implies} \quad t(\rho^1) > t(\rho^2) > 0 \tag{12.46}$$

and

$$\lim_{k \to \infty} \{\rho^k\} = 0 \quad \text{implies} \quad \lim_{k \to \infty} \{t(\rho^k)\} = 0. \tag{12.47}$$

The function $t(\rho^k)q(x)$ is called the **interior penalty** or **barrier function.**

The interior penalty method can be stated as follows: For $k = 0, 1, \ldots,$ define G to be the augmented objective function minimized in a sequence of unconstrained optimization problems (IP^k), given by

$$(\text{IP}^k) \qquad \min_x G(x, \rho^k) = f(x) + t(\rho^k)q(x). \tag{12.48}$$

Let $x^0 \in X^0$ be the starting point and assign a positive value to ρ^0. Solve problem (IP^0) by some unconstrained minimization technique, starting

at x^0, and let x^{0*} be a solution of (IP^0). Presumably $x^{0*} \in X^0$. Decrease ρ^0 to ρ^1 and solve problem (IP^1), starting at x^{0*}. Denote the optimal solution of (IP^1) by x^{1*}. In this way, continue solving (IP^k) for a strictly decreasing sequence ρ^k, starting always at x^{k-1*}. Under the already stated assumptions on (PI), this method can find an optimal solution, as we shall prove below.

The most common choices for the functions t and q are

$$t_1(\rho) = \rho \tag{12.49}$$

$$t_2(\rho) = (\rho)^2 \tag{12.50}$$

$$q_1(x) = -\sum_{i=1}^{m} \log g_i(x) \tag{12.51}$$

$$q_2(x) = \sum_{i=1}^{m} \frac{1}{g_i(x)} \tag{12.52}$$

$$q_3(x) = \sum_{i=1}^{m} \frac{1}{[g_i(x)]^2} \tag{12.53}$$

$$q_4(x) = \sum_{i=1}^{m} \frac{1}{\max\,[0,\, g_i(x)]}. \tag{12.54}$$

Example 12.2.1

Consider the following small problem in one variable

$$\min f(x) = \tfrac{1}{2}x \tag{12.55}$$

subject to

$$g(x) = x - 1 \geq 0. \tag{12.56}$$

The optimal solution is clearly at $x^* = 1$ and $f(x^*) = \frac{1}{2}$. Suppose that we choose t_2 and q_2, as given above, for the barrier function. Then

$$G(x, \rho^k) = \tfrac{1}{2}x + (\rho^k)^2(x - 1)^{-1}. \tag{12.57}$$

The reader can easily verify that the unconstrained minimum of G is given by

$$x^{k*} = 1 + \sqrt{2}\,\rho^k \tag{12.58}$$

and

$$f(x^{k*}) = \frac{1}{2} + \frac{\rho^k}{\sqrt{2}}. \tag{12.59}$$

Thus the optimal unconstrained minima are all in X^0 and converge to x^*

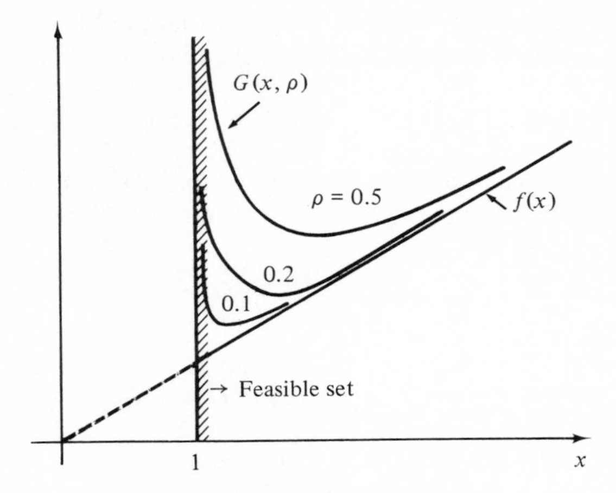

Fig. 12.2 Interior penalty method.

as the values of ρ^k are successively reduced. The original and the unconstrained problems for a few values of ρ^k are illustrated in Figure 12.2. ▮

In order to present a simple convergence proof of this method, assume that the q functions used are positive for $x \in X^0$. This assumption may not hold for the logarithmic penalty q_1, although convergence of the interior penalty method using this type of penalty can also be proved.

Theorem 12.3

Assume that regularity conditions (1) *and* (2) *stated above for program* (PI) *are satisfied and the q functions used in the augmented objective function are positive for $x \in X^0$. Suppose that the $G(x, \rho^k)$ attain their unconstrained minima in X^0 for all k. If $\{\rho^k\}$ is a strictly decreasing sequence of positive numbers converging to zero, then there exists a convergent subsequence $\{x^{k_i *}\}$ of the optimal solutions to* (IP^k), *and the limit of any such convergent subsequence is optimal for* (PI).

Proof. By condition (2), the set $S(f, \alpha^0) \cap X$ is nonempty and compact. Hence the continuous function f attains its minimum value $f(x^*)$ on X at some point $x^* \in X$. Let $G(x^{k*}, \rho^k)$ be the minimum value of the augmented objective function in program (IP^k). Then by (12.46) and (12.48),

$$G(x^{0*}, \rho^0) \geq G(x^{1*}, \rho^1) \geq \cdots \geq f(x^*). \tag{12.60}$$

Since $\{G(x^{k*}, \rho^k)\}$ is a strictly decreasing sequence bounded from below, it converges to a limit $\hat{G} \geq f(x^*)$. Suppose that $\hat{G} > f(x^*)$. From condition

(1) and the continuity of f, we conclude that there exists a number $\delta > 0$ and a spherical neighborhood $N_\delta(x^*)$ such that $X^0 \cap N_\delta(x^*) \neq \emptyset$, and

$$f(x) \leq \hat{G} - \tfrac{1}{2}[\hat{G} - f(x^*)] \tag{12.61}$$

for all $x \in N_\delta(x^*)$. Take any point $\bar{x}$ in $X^0 \cap N_\delta(x^*)$. From (12.46) and (12.47) it follows that there exists a natural number K such that for every $k \geq K$

$$t(\rho^k)q(\bar{x}) < \tfrac{1}{4}[\hat{G} - f(x^*)]. \tag{12.62}$$

Thus

$$G(x^{k*}, \rho^k) \leq f(\bar{x}) + t(\rho^k)q(\bar{x}) < \hat{G} - \tfrac{1}{4}[\hat{G} - f(x^*)] \tag{12.63}$$

for $k \geq K$, contradicting that $\{G(x^{k*}, \rho^k)\}$ monotonically converges to $\hat{G}$. Hence $\hat{G} = f(x^*)$.

By condition (2), there exists a $\hat{K}$ such that, for all $k \geq \hat{K}$, the points x^{k*} are in a compact set. And so there exists a subsequence $\{x^{k_i*}\}$ that converges to a limit $\hat{x} \in X$. Suppose that $\hat{x}$ is not optimal for (PI). Then $f(\hat{x}) > f(x^*)$, and the sequence $\{(f(x^{k_i*}) + t(\rho^{k_i})q(x^{k_i*}) - f(x^*)\}$ does not converge to zero, thereby contradicting

$$\lim_{k \to \infty} \{G(x^{k*}, \rho^k)\} = f(x^*). \tag{12.64}$$

Hence we must have $f(\hat{x}) = f(x^*)$, and $\hat{x}$ is optimal for (PI). ■

Perhaps the most important assumption in this theorem is that the $G(x, \rho^k)$ attain their minima in X^0 or, equivalently, that problems (IP^k) have optimal solutions in X^0. A sufficient condition that ensures the existence of these optima will now be given.

Program (PI) is defined to be strongly consistent if regularity condition (1) holds and if X^0, the interior of X, is given by

$$X^0 = \{x : x \in R^n, g_i(x) > 0, i = 1, \ldots, m\} \neq \emptyset. \tag{12.65}$$

We have then

Lemma 12.4

Assume that $X \subset R^n$ is compact and that program (PI) is strongly consistent. Then the $G(x, \rho^k)$ attain their unconstrained minima in X^0.

Proof. Let

$$\inf_{x \in X^0} G(x, \rho) = \alpha. \tag{12.66}$$

Then, by the definition of infimum, there exists a sequence $\{x^i\}$ of points $x^i \in X^0$ such that

$$\lim_{i\to\infty} \{G(x^i, \rho)\} = \alpha. \tag{12.67}$$

Since $\{x^i\}$ is contained in the compact set X, it has a convergent subsequence $\{x^{i_j}\}$ and

$$\lim_{i_j\to\infty} \{x^{i_j}\} = \hat{x} \in X. \tag{12.68}$$

Assume that $\hat{x} \in X^0$. Then, by the continuity of G, we have

$$\lim_{i_j\to\infty} \{G(x^{i_j}, \rho)\} = \lim_{i_j\to\infty} \{f(x^{i_j})\} + \lim_{i_j\to\infty} \{t(\rho)q(x^{i_j})\} \tag{12.69}$$

$$= f(\hat{x}) + t(\rho)q(\hat{x}) = \alpha, \tag{12.70}$$

since any subsequence of $\{G(x^i, \rho)\}$ must converge to α. Hence

$$G(\hat{x}, \rho) = \min_{x\in X^0} G(x, \rho). \tag{12.71}$$

Suppose now that $\hat{x} \notin X^0$. Then $\hat{x}$ is on the boundary of X. By (12.45), (12.69), and the positivity of $t(\rho)$, we get

$$\inf_{x\in X^0} G(x, \rho) = f(\hat{x}) + \lim_{i_j\to\infty} \{t(\rho)q(x^{i_j})\} = +\infty, \tag{12.72}$$

which is a contradiction, and $\hat{x} \in X^0$. ■

Convergence to local minima of nonconvex programs by interior penalty methods under weaker assumptions than above can also be proven (see Fiacco and McCormick [15]).

Interior penalty methods are based on an idea proposed by Carroll [6] for transforming a constrained nonlinear program into a sequence of unconstrained minimizations by using the t_1 and q_2 functions given by (12.49) and (12.52). Carroll's idea was subsequently formalized, extended, and thoroughly studied by Fiacco and McCormick [12, 15], who developed the Sequential Unconstrained Minimization Technique (SUMT), perhaps the best-known penalty function method to date.

The SUMT method has some interesting primal-dual features in cases where program (PI) is a standard convex program and the $f, g_1, \ldots, g_m$ are continuously differentiable. Consider (IP^k), the kth unconstrained minimization problem:

(IP^k)
$$\min_x G(x, \rho^k) = f(x) + \rho^k \sum_{i=1}^{m} \frac{1}{g_i(x)} \tag{12.73}$$

If $x^{k*} \in X^0$ minimizes (IP^k), then

$$\nabla G(x^{k*}, \rho^k) = \nabla f(x^{k*}) - \rho^k \sum_{i=1}^{m} \frac{\nabla g_i(x^{k*})}{[g_i(x^{k*})]^2} = 0. \tag{12.74}$$

Taking the Lagrangian of (PI)

$$L(x, \lambda) = f(x) - \sum_{i=1}^{m} \lambda_i g_i(x) \tag{12.75}$$

suggests defining Lagrange multipliers λ_i^k as

$$\lambda_i^k = \frac{\rho^k}{[g_i(x^{k*})]^2}, \qquad i = 1, \ldots, m. \tag{12.76}$$

With this choice of the λ_i^k, we have

$$\nabla G(x^{k*}, \rho^k) = \nabla L(x^{k*}, \lambda^k) = 0. \tag{12.77}$$

Note that although $\lambda_i^k \geq 0$, the Kuhn-Tucker necessary conditions for convex programs as stated in Theorem 4.41 do not hold for (PI) at x^{k*}, since

$$\lambda_i^k g_i(x^{k*}) = \frac{\rho^k}{g_i(x^{k*})} \neq 0, \qquad i = 1, \ldots, m. \tag{12.78}$$

It can be shown, however, that the vector $\lambda^k = (\lambda_1^k, \ldots, \lambda_m^k)^T$, where the λ_i^k are defined by (12.76), is feasible for the dual of program (PI). This dual program is given by (see Section 5.4)

$$\text{(DPI)} \qquad \max_{\lambda \geq 0} \{\inf_x f(x) - \sum_{i=1}^{m} \lambda_i g_i(x)\} = \max_{\lambda \geq 0} \{\inf_x L(x, \lambda)\}. \tag{12.79}$$

It follows by the convexity of f, the concavity of g_i, the definition of λ_i^k, and (12.77) that

$$L(x^{k*}, \lambda^k) = \inf_x L(x, \lambda^k) \tag{12.80}$$

and λ^k is feasible for (DPI). Consequently, by Theorem 5.5, the value $L(x^{k*}, \lambda^k)$ is a lower bound on the minimal value of f in program (PI). As the x^{k*} approach x^*, the optimal solution of (PI), the difference between $f(x^{k*})$ and $L(x^{k*}, \lambda^k)$ becomes smaller and smaller until equality is attained. Thus every successful solution of (IP^k) also provides an estimate on how far $f(x^{k*})$ is from its minimal value. This information can be useful in terminating the sequence of unconstrained minimizations.

As noted, equality constrained nonlinear programs cannot be solved by interior penalty methods. We can, however, use a **mixed penalty method**

to solve an equality-inequality constrained problem, such as the one given by (12.3) to (12.5). The augmented objective function is defined as

$$H(x, \rho, \eta) = f(x) + t(\rho)q(x) + r(\eta)s(x), \tag{12.81}$$

where the t, q, r, and s functions were defined above for the "pure" exterior and interior penalty methods but $q(x)$ and $s(x)$ are applied only to the inequality and equality constraints, respectively. For example, we can take

$$H(x, \rho, \eta) = f(x) - \rho \sum_{i=1}^{m} \log g_i(x) + \frac{1}{\eta} \sum_{j=1}^{m} (h_j(x))^2. \tag{12.82}$$

The mixed penalty method consists of solving a sequence of unconstrained minimization problems

$$(\text{MP}^k) \qquad \min_x H(x, \rho^k, \eta^k) = f(x) + t(\rho^k)q(x) + r(\eta^k)s(x). \tag{12.83}$$

Under suitable assumptions, the sequence $\{H(x^{k*}, \rho^k, \eta^k)\}$ converges to $f(x^*)$, the optimal solution value of (P), and a subsequence of $\{x^{k*}\}$ converges to x^* [13, 15].

12.3 PARAMETER-FREE PENALTY METHODS

A still-open question on the implementation of penalty function methods concerns the choice of the parameters ρ. It is necessary to decide on ρ^0, the initial value of the parameter, and on a rule to modify the value of ρ in order to obtain a monotonically decreasing sequence that converges to zero. The dilemma of choosing the parameters can be avoided by modifying penalty function methods so that the parameters are automatically chosen or, equivalently, the methods are modified so that they become parameter free. Both exterior and interior penalty function methods can be modified, as we shall now see. Starting with exterior penalty methods, suppose that we have a lower estimate ω^0 on the value of the objective function f at its global minimum x^* over the feasible set X of program (P) as given by (12.6)—that is,

$$\omega^0 \leq f(x^*). \tag{12.84}$$

Also suppose that we solve the unconstrained optimization

$$(\text{EPF}^0) \quad \min_{x \in R^n} \hat{F}(x, \omega^0) = \psi(\omega^0 - f(x)) + \sum_{i=1}^{m} \psi(g_i(x)) + \sum_{j=1}^{p} \zeta(h_j(x)), \tag{12.85}$$

where ψ and ζ are defined by (12.7) and (12.8), respectively. Let x^{0*} be the optimal solution of (12.85). If x^*, the optimal solution of (P), happens to be

the unconstrained minimum of f over R^n, then $x^{0*} = x^*$ and the algorithm terminates. Otherwise

$$\omega^0 \leq f(x^{0*}) \leq f(x^*), \tag{12.86}$$

provided that regularity conditions similar to those stated in Section 12.1 are satisfied, and we proceed as follows:

Let

$$(\mathrm{EPF}^k) \quad \min_{x \in R^n} \hat{F}(x, \omega^k) = \psi(\omega^k - f(x)) + \sum_{i=1}^{m} \psi(g_i(x)) + \sum_{j=1}^{p} \zeta(h_j(x)), \tag{12.87}$$

where $\{\omega^k\}$ is a strictly increasing sequence whose elements ω^k are computed from ω^{k-1} and the optimal solutions of (EPF^{k-1}). Solving (12.87) for $k = 1, 2, \ldots$, we obtain a sequence of points $\{x^{k*}\}$ as in Section 12.1 that contains a subsequence that converges to the optimal solution of (P). For further details see [29, 33, 36].

Consider problem (PI), given by (12.42) and (12.43), and the regularity conditions stated for the family of interior penalty methods. Parameter-free interior penalty methods are based on solving a sequence of unconstrained optimization problems, such as

$$(\mathrm{IPF}^k) \qquad \min_{x \in R^n} \hat{G}(x, x^{k-1*}) = \frac{1}{f(x^{k-1*}) - f(x)} + \sum_{i=1}^{m} \frac{1}{g_i(x)}, \tag{12.88}$$

for $k = 1, 2, \ldots$, where x^{0*} is an arbitrary point in X^0 and x^{k-1*} is the optimal solution of (IPF^{k-1}). This method is called **SUMT without parameters** [14]; in it the interior penalty method SUMT has been modified. Taking q_1 as in (12.51), we can formulate another parameter-free interior penalty method that is based on solving a sequence of problems

$$\min_{x \in R^n} \hat{G}_1(x, x^{k-1*}) = -\log [f(x^{k-1*}) - f(x)] - \sum_{i=1}^{m} \log g_i(x), \tag{12.89}$$

which is equivalent to

$$\max_{x} [f(x^{k-1*}) - f(x)] \prod_{i=1}^{m} g_i(x), \tag{12.90}$$

where x^{k-1*} is defined above. This latter method, called the **method of centres,** was developed by Huard [26]. Again, under conditions similar to those stated in Section 12.2, parameter-free interior methods converge to an optimal solution of (PI). It can be shown that parameter-free methods are actually equivalent to ordinary penalty function methods with a particular choice of parameter adjustment. For example, the relationship between SUMT and

its parameter-free version can be explicitly stated as follows: If x^{k*} solves (IPF^k), as given by (12.88), then x^{k*} also solves (IP^k), given by (12.73), for

$$\rho^k = [f(x^{k-1*}) - f(x^{k*})]^2, \tag{12.91}$$

provided that certain regularity conditions on (PI) are satisfied [15]. Similar relationships can also be derived for other parameter-free methods. Lootsma [33] has studied the rate of convergence of these methods and found that they had no particular advantage over previously discussed exterior and interior penalty algorithms. In fact, convergence of parameter-free methods is generally quite slow, and most of them have linear rates of convergence. Several modifications were suggested to improve the rate of convergence, such as assigning a small weight to the first term on the right-hand side of (12.87), which contains the objective function f; see [29, 36].

Example 12.3.1

Let us solve the problem presented in Example 12.2.1 by SUMT without parameters. We have

$$\min f(x) = \tfrac{1}{2}x \tag{12.92}$$

subject to

$$g(x) = x - 1 \geq 0. \tag{12.93}$$

At iteration k, we solve the unconstrained optimization problem

$$\min_{x \in R} [f(x^{k-1*}) - f(x)]^{-1} + (x - 1)^{-1}. \tag{12.94}$$

It is easy to show that the optimal solution x^{k*} is given by

$$x^{k*} = \frac{1 + (x^{k-1*}/\sqrt{2})}{1 + (1/\sqrt{2})}. \tag{12.95}$$

Since $x^* = 1$, we obtain

$$\frac{x^{k*} - x^*}{x^{k-1*} - x^*} = \frac{1}{1 + \sqrt{2}} \tag{12.96}$$

and the rate of convergence is linear. It is interesting to note that for the interior penalty method used in Example 12.2.1 we have

$$\frac{x^{k*} - x^*}{x^{k-1*} - x^*} = \frac{\rho^k}{\rho^{k-1}}. \tag{12.97}$$

Thus convergence can be accelerated by a proper choice of $\{\rho^k\}$. ∎

12.4 EXACT PENALTY FUNCTIONS

In all penalty methods discussed so far, the solution was obtained by solving a sequence of unconstrained optimization problems. It is natural to ask whether constrained optimization problems can perhaps be solved by performing a single unconstrained optimization. Indeed, it would be advantageous if, given a constrained nonlinear program, we could find a real function with the property that its unconstrained minima are solutions to the constrained problem. It turns out that, in certain cases, it is possible to find such functions and that they are related to penalty functions already discussed in this chapter.

Given a nonlinear program such as program (P) of Section 12.1, suppose that there exists a real function with the property that a single unconstrained minimization of it yields an optimal solution of (P). Since such a function often belongs to some particular type of penalty function with an appropriate choice of the parameters, it is called an **exact penalty function.**

One of the first works dealing with exact penalty functions is due to Zangwill [52]. He showed that, for a convex program satisfying the strong consistency condition, there exists an exact penalty function in the form of an exterior penalty function that contains a sufficiently small parameter ρ. Here we extend Zangwill's results in view of the generalizations of convexity presented in Chapter 6.

Consider the following generalized convex program, in which we use a notation consistent with that of Chapter 6 (readers unfamiliar with the special symbols introduced there for (h, ϕ)-convex functions may regard them as the ordinary algebraic operations, and the results will then apply to ordinary convex programs). Let

(GCP) $$\min f(x) \tag{12.98}$$

subject to

$$g_i(x) \geq 0_\phi, \qquad i = 1, \ldots, m \tag{12.99}$$

where f is (h, ϕ)-convex and the g_i are (h, ϕ)-concave functions, defined on an arcwise-connected set $S \subset R^n$. As before, let $X \subset S$ be the set of points $x \in R^n$ satisfying (12.99) and let $X^0 \subset X$ be the set

$$X^0 = \{x : x \in R^n, g_i(x) > 0_\phi, i = 1, \ldots, m\}. \tag{12.100}$$

Assume that X^0 is nonempty and that there exists an $x^* \in X$ that is the optimal solution of (GCP). Define

$$P(x, \rho) = f(x)\,[-]\,\frac{1}{\rho}\,[\cdot]\left\{\left[\sum_{i=1}^{m}\right] \min\,(0_\phi, g_i(x))\right\}, \tag{12.101}$$

where

$$a\,[-]\,b = a\,[+]\,((-1)\,[\cdot]\,b) = \phi^{-1}[\phi(a) - \phi(b)] \tag{12.102}$$

and

$$\left[\sum_{i=1}^{m}\right] a_i = a_1\,[+]\,a_2\,[+]\cdots[+]\,a_m. \tag{12.103}$$

If f is convex and the g_i are concave, we obtain

$$P(x, \rho) = f(x) - \frac{1}{\rho}\sum_{i=1}^{m} \min\,(0, g_i(x)). \tag{12.104}$$

Note that this form of $P(x, \rho)$ is equivalent to using ψ, defined by (12.7) with $\alpha = 1$.

We now state and prove

Theorem 12.5

There exists a $\rho^* > 0$ such that, for all $\rho^* \geq \rho > 0$, the unconstrained minimum of $P(x, \rho)$ coincides with x^*, the optimal solution of (GCP).

Proof. Let $\hat{x} \in X^0$. Define

$$\alpha = \phi(\min_i\,(g_i(\hat{x}))) > 0 \tag{12.105}$$

and

$$\beta = \phi f(\hat{x}) - \phi f(x^*) \geq 0. \tag{12.106}$$

Let

$$\frac{1}{\rho^*} = \frac{\beta + \epsilon}{\alpha} > 0, \tag{12.107}$$

where $\epsilon > 0$ is arbitrary.

Next, take any point $w \in S \subset R^n$ such that $w \notin X$. We shall show that for each w there is a point $\bar{x} \in X$ satisfying

$$P(\bar{x}, \rho^*) < P(w, \rho^*). \tag{12.108}$$

Since $\bar{x}$ is feasible, $f(\bar{x}) = P(\bar{x}, \rho^*)$. And since the optimal solution of (GCP) is attained by hypothesis, $P(x, \rho^*)$ must also attain its minimum at a point $x^0 \in X$ that is optimal for (GCP). Let $\bar{x}$ be a point on the arc $h_{\hat{x},w}$ connecting $\hat{x}$ and w—that is, $\bar{x} = h_{\hat{x},w}(\bar{\theta})$, $0 < \bar{\theta} < 1$, such that

$$I(\bar{x}) = \{i : g_i(\bar{x}) = 0_\phi,\, i = 1, \ldots, m\} \neq \emptyset. \tag{12.109}$$

In other words, $\bar{x}$ is on the boundary of X. Define an auxiliary function

$$\tau(x) = f(x)\,[-]\,\frac{1}{\rho^*}\,[\cdot]\left\{\left[\sum_{i \in I(\bar{x})}\right] g_i(x)\right\}. \tag{12.110}$$

Clearly, $\tau(\bar{x}) = P(\bar{x}, \rho^*) = f(\bar{x})$ and

$$\left[\sum_{i \in I(\bar{x})}\right] g_i(w) = \left[\sum_{i \in I(\bar{x})}\right] \min\,(0_\phi, g_i(w)) \geq \left[\sum_{i=1}^{m}\right] \min\,(0_\phi, g_i(w)). \tag{12.111}$$

Hence

$$\tau(w) = f(w)\,[-]\,\frac{1}{\rho^*}\,[\cdot]\left\{\left[\sum_{i \in I(\bar{x})}\right] g_i(w)\right\} \tag{12.112}$$

$$\leq f(w)\,[-]\,\frac{1}{\rho^*}\,[\cdot]\left\{\left[\sum_{i=1}^{m}\right] \min\,(0_\phi, g_i(w))\right\} = P(w, \rho^*). \tag{12.113}$$

We complete the proof by showing that $\tau(\bar{x}) < \tau(w)$. First we have

$$\tau(\hat{x}) = f(\hat{x})\,[-]\,\frac{\beta + \epsilon}{\alpha}\,[\cdot]\left\{\left[\sum_{i \in I(\bar{x})}\right] g_i(\hat{x})\right\} \tag{12.114}$$

$$\leq f(\hat{x})\,[-]\,\frac{\beta + \epsilon}{\alpha}\,[\cdot]\left\{\min_i g_i(\hat{x})\right\} \tag{12.115}$$

$$= f(\hat{x})\,[-]\,\phi^{-1}\left\{\frac{\beta + \epsilon}{\alpha}\,\phi\left(\min_i g_i(\hat{x})\right)\right\} \tag{12.116}$$

$$= f(\hat{x})\,[-]\,\phi^{-1}(\beta + \epsilon) \tag{12.117}$$

$$= \phi^{-1}(\phi f(x^*) - \epsilon) < f(\bar{x}) = \tau(\bar{x}). \tag{12.118}$$

Next, since τ is (h, ϕ)-convex and $\bar{x} = h_{\hat{x}, w}(\bar{\theta})$ for some $0 < \bar{\theta} < 1$, it follows that

$$\tau(\bar{x}) \leq ((1 - \bar{\theta})\,[\cdot]\,\tau(\hat{x}))\,[+]\,(\bar{\theta}\,[\cdot]\,\tau(w)) \tag{12.119}$$

$$< ((1 - \bar{\theta})\,[\cdot]\,\tau(\bar{x}))\,[+]\,(\bar{\theta}\,[\cdot]\,\tau(w)) \tag{12.120}$$

and

$$\tau(\bar{x}) = P(\bar{x}, \rho^*) < \tau(w) \leq P(w, \rho^*). \tag{12.121}$$

Hence (12.108) holds. ■

Although the existence of the critical parameter value ρ^* has been demonstrated in Theorem 12.5, in practice, it can be computed only after solving the problem itself, for ρ^* depends on $f(x^*)$. We can, however, estimate ρ^* if a lower bound on $f(x^*)$ can be found, as in many practical

situations. Let f^e be such a lower bound. Define

$$\beta^e = \phi f(\hat{x}) - \phi f^e \geq \beta \tag{12.122}$$

and

$$\frac{1}{\rho^e} = \frac{\beta^e + \epsilon}{\alpha} \geq \frac{1}{\rho^*}. \tag{12.123}$$

Then ρ^e can be used instead of ρ^* in the exact penalty function.

Example 12.4.1

The following simple example illustrates the idea of exact penalty functions:

$$\min f(x) = (x)^{1/2} \tag{12.124}$$

subject to

$$g(x) = \tfrac{1}{4}x \geq 1. \tag{12.125}$$

It follows from Chapters 6 and 7 that f and g are (log, log)-convex and (log, log)-concave functions, respectively. In other words, we have an ordinary geometric programming problem.

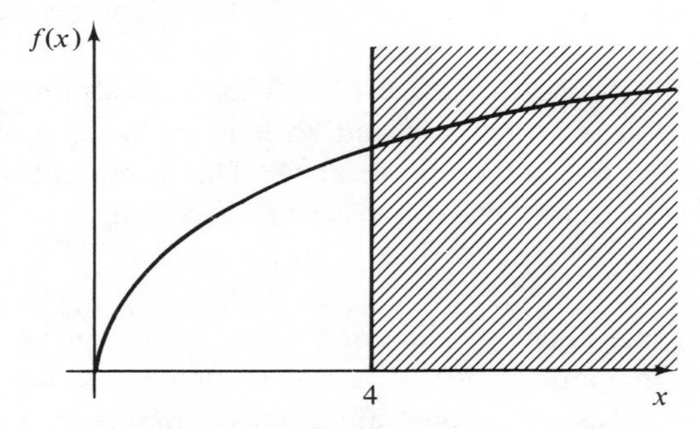

Fig. 12.3 Constrained optimization problem.

The feasible set and the objective function are illustrated in Figure 12.3. Clearly, the optimal solution is at $x^* = 4$. We begin constructing the exact penalty function by finding the critical parameter ρ^* (using our knowledge of the precise location of x^*). We take $\hat{x} = 4e$. This point is feasible. By (12.105) to (12.107) and by letting $\epsilon = \frac{1}{2}$, we obtain

$$\alpha = \log e = 1 \tag{12.126}$$

$$\beta = \log (4e)^{1/2} - \log (4)^{1/2} = \tfrac{1}{2} \tag{12.127}$$

$$\frac{1}{\rho^*} = 1. \tag{12.128}$$

Hence by (12.101),

$$P(x, \rho^*) = (x)^{1/2} [-] \frac{1}{\rho^*} [\cdot] \min\left(1, \frac{x}{4}\right) \tag{12.129}$$

$$= \exp\left\{\frac{1}{2} \log x - \min\left(0, \log \frac{x}{4}\right)\right\} \tag{12.130}$$

and

$$P(x, \rho^*) = \begin{cases} \dfrac{4}{(x)^{1/2}} & \text{if } x \leq 4 \\ (x)^{1/2} & \text{if } x \geq 4. \end{cases} \tag{12.131}$$

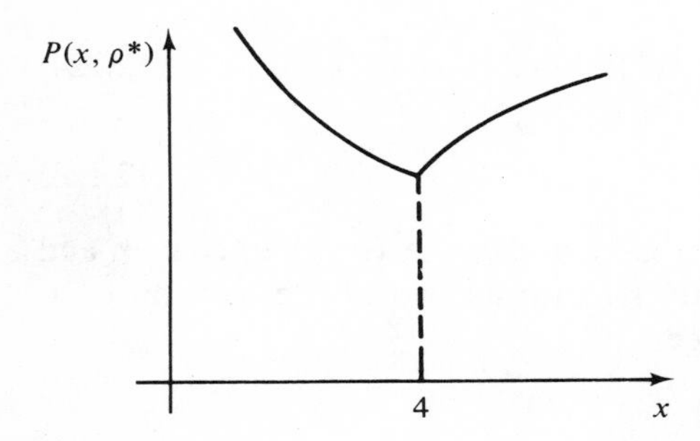

Fig. 12.4 Exact penalty function.

This function is illustrated in Figure 12.4. The unconstrained minimum of this penalty function is at $x^* = 4$, and so it is an exact penalty function. Note that $P(x, \rho^*)$ is not differentiable at x^*. This computationally undesirable property is inherent in the construction of certain exact penalty functions. ▮

Using an exact penalty function $P(x, \rho)$ as given in (12.104), Pietrzykowski [46] has obtained results that relate constrained local minima of nonconvex programs to unconstrained local minima of $P(x, \rho)$. One of the main disadvantages of the Zangwill-Pietrzykowski exact penalty function method is, as we just saw in the example, that $P(x, \rho)$ is not differentiable at an $\hat{x} \in R^n$ for which $I(\hat{x}) \neq \emptyset$—that is, $g_i(\hat{x}) = 0$ for some i—even if f and the g_i are continuously differentiable. Consequently, efficient unconstrained minimization techniques like the ones in Chapters 10 and 11 cannot be used directly. Special optimization methods for this type of piecewise-differentiable exact penalty functions were suggested by Conn [7] and Conn and Pietrzykowski [8].

In a recent work Evans, Gould, and Tolle [11] developed a quite general theory for a large class of piecewise-differentiable penalty functions. Their theory can also be regarded as a connecting link between penalty functions and Lagrangians. (This relationship will be discussed in more detail in the

next section.) Corresponding to each inequality constraint $g_i(x) \geq 0$, Evans, Gould, and Tolle define a **multiplier function** γ_i of a number β_i and a vector $\lambda \in A$, where A is a subset of R^k, such that the range of γ_i is $R \cup \{-\infty\}$. It is also assumed that, for each fixed $\lambda \in R^k$, the γ_i are continuous and nondecreasing functions of β_i. Then define the **extended Lagrangian** $P(x, \lambda)$ associated with an inequality constrained nonlinear program, such as program (PI), given by (12.42) and (12.43), as

$$P(x, \lambda) = f(x) - \sum_{i=1}^{m} \gamma_i(g_i(x), \lambda). \tag{12.132}$$

The ordinary Lagrangian associated with (PI) becomes a special case by taking $k = m$, and $A = \{\lambda : \lambda \in R^m, \lambda \geq 0\}$, and

$$\gamma_i(\beta_i, \lambda) = \lambda_i \beta_i, \qquad i = 1, \ldots, m. \tag{12.133}$$

Hence

$$P(x, \lambda) = f(x) - \sum_{i=1}^{m} \lambda_i g_i(x), \tag{12.134}$$

which is the Lagrangian introduced in Chapter 3. Actually, for convex programs (CP) defined in Chapter 4 and satisfying the strong consistency (Slater) condition, the Lagrangian $L(x, \lambda^*)$ is an exact penalty function. That is, if the vector of "optimal" Lagrange multipliers λ^* is known, then the unconstrained minimum of $L(x, \lambda^*)$ coincides with the constrained optimum of (CP), as was proven in Theorem 4.41. Another special case of (12.132) can be obtained by taking $k = 1$, $A = \{\lambda : \lambda \in R, \lambda > 0\}$, and

$$\gamma_i(\beta_i, \lambda) = \begin{cases} -\dfrac{\lambda}{\beta_i}, & \beta_i > 0, \\ -\infty, & \beta_i \leq 0, \end{cases} \qquad i = 1, \ldots, m. \tag{12.135}$$

Consequently,

$$P(x, \lambda) = \begin{cases} f(x) + \lambda \displaystyle\sum_{i=1}^{m} \frac{1}{g_i(x)}, & x \in X \\ +\infty, & x \notin X \end{cases} \tag{12.136}$$

where X is, as before, the feasible set of (PI). The function appearing in (12.136) is the interior penalty function of the SUMT method and is a member of "inside-out" class of penalty functions studied by Gould [23]. Finally, by letting $k = 1$ and $A = \{\lambda : \lambda \in R, \lambda > 0\}$ and defining

$$\gamma_i(\beta_i, \lambda) = \lambda \min(0, \beta_i), \qquad i = 1, \ldots, m \tag{12.137}$$

we obtain

$$P(x, \lambda) = f(x) - \lambda \sum_{i=1}^{m} \min\,[0, g_i(x)] \tag{12.138}$$

which is the exact penalty function of Zangwill for convex programs, mentioned above.

Assume for the following discussion that, for every $\lambda \in A$,

$$\gamma_i(0, \lambda) = 0, \qquad i = 1, \ldots, m. \tag{12.139}$$

Note that this condition eliminates (12.135) from being a special case of multiplier functions. We can now define exact penalty functions in terms of extended Lagrangians by saying that $P(x, \lambda)$ is an exact penalty function for a program

(PI) $$\min f(x) \tag{12.140}$$

subject to

$$g_i(x) \geq 0, \qquad i = 1, \ldots, m \tag{12.141}$$

if there exists a $\lambda^* \in A$ such that

$$\min_{x \in R^n} P(x, \lambda^*) = f(x^*), \tag{12.142}$$

where x^* is the optimal solution of (PI). Since γ_i is nondecreasing for each fixed λ, it follows from (12.139) that

$$\sum_{i=1}^{m} \gamma_i(g_i(x), \lambda^*) \geq 0 \tag{12.143}$$

for every $x \in R^n$ satisfying (12.141). Suppose that x^* solves (PI). Then

$$P(x^*, \lambda^*) = f(x^*) - \sum_{i=1}^{m} \gamma_i(g_i(x^*), \lambda^*) \leq f(x^*) = \min_{x \in R^n} P(x, \lambda^*). \tag{12.144}$$

That is, x^* is an unconstrained minimum of $P(x, \lambda^*)$. Conversely, let $\hat{x} \in R^n$ satisfy (12.141), such that

$$f(\hat{x}) = P(\hat{x}, \lambda^*) = \min_{x \in R^n} P(x, \lambda^*) \tag{12.145}$$

for some $\lambda^* \in A$. Then for every feasible x

$$f(\hat{x}) \leq P(x, \lambda^*) \leq f(x). \tag{12.146}$$

In other words, $\hat{x}$ is optimal for (PI).

Evans, Gould, and Tolle [11] have studied some specific multiplier functions for which a stronger result holds: Under certain mild assumptions on f and g_i, the set of optimal solutions of program (PI) is identical to the set of unconstrained global minima of the corresponding exact penalty functions $P(x, \lambda)$ for sufficiently large λ. For instance, one of these multiplier functions is

$$\gamma_i(\beta_i, \lambda) = \min(0, e^{\lambda\beta_i} - 1), \qquad i = 1, \ldots, m \tag{12.147}$$

where $\lambda \in A = \{\lambda : \lambda \in R, \lambda > 0\}$, and the corresponding extended Lagrangian is given by

$$P(x, \lambda) = f(x) - \sum_{i=1}^{m} \min(0, e^{\lambda g_i(x)} - 1). \tag{12.148}$$

The most important results of [11] are characterization theorems on the optima of (PI) in terms of unconstrained minima of corresponding extended Lagrangians. Since these Lagrangians are only piecewise differentiable, the usefulness of this type of exact penalty function method is still doubtful. As yet we do not have really efficient numerical techniques to minimize nondifferentiable functions. For this reason, we leave the subject and conclude by discussing Fletcher's method of exact penalty functions which are sufficiently smooth for the effective application of gradient-type unconstrained minimization algorithms.

In a series of publications Fletcher [16, 17] and Fletcher and Lill [19] first derived an elegant method to define differentiable exact penalty functions for nonlinear programs containing equality constraints only. This method was subsequently extended by Fletcher for inequality constrained problems as well [18]. Let us describe Fletcher's approach, beginning with the equality constrained case. Consider the program

(PE) $$\min f(x) \tag{12.149}$$

subject to

$$h_j(x) = 0, \qquad j = 1, \ldots, p \tag{12.150}$$

where $f, h_1, \ldots, h_p$ are twice continuously differentiable functions of $x \in R^n$ and $p \leq n$. Let $J_p = \{1, \ldots, p\}$, and for some subset $J_q \subset J_p$, containing q elements, we define the $n \times q$ matrix N_q at each $x \in R^n$ by

$$N_q(x) = [\nabla h_j(x), j \in J_q]. \tag{12.151}$$

The generalized inverse [44] of N_q, if it exists, is defined as

$$N_q^+ = (N_q^T N_q)^{-1} N_q^T \tag{12.152}$$

where, for brevity, the explicit dependence on x is omitted in the notation. Also define the projection matrix $\hat{P}_q$ by

$$\hat{P}_q = N_q N_q^+ = (N_q^+)^T N_q^T, \tag{12.153}$$

which projects vectors into the subspace spanned by the $\nabla h_j, j \in J_q$, and the complementary projection matrix P_q, given by

$$P_q = I - \hat{P}_q, \tag{12.154}$$

where I is the identity matrix. The matrix P_q projects vectors tangential to the manifold formed by the intersection of q constraints. It is understood that P_q and $\hat{P}_q$ are also functions of x. Further details on projection matrices are given in Section 13.2.

Fletcher has found a whole class of exact penalty functions for program (PE), a representative member of which is given by

$$\phi(x) = f(x) - (h(x))^T N_p^+ \nabla f(x) + \rho(h(x))^T N_p^+ (N_p^+)^T h(x), \tag{12.155}$$

where $h(x) = (h_1(x), \ldots, h_p(x))^T$. Let us check the behavior of ϕ at a constrained local minimum of (PE). The reader will be asked to show that, in terms of the symbols defined above, necessary conditions for such a local minimum at a point x^l are

$$P_q \nabla f(x^l) = 0 \tag{12.156}$$

$$h(x^l) = 0 \tag{12.157}$$

and

$$\nabla \phi(x^l) = 0. \tag{12.158}$$

That is, ϕ has a stationary point at x^l. But we know from Chapter 2 that the Lagrangian associated with (PE) must also have a stationary point at x^l; that is,

$$\nabla L(x^l, \mu) = \nabla f(x^l) - N_p \mu = 0. \tag{12.159}$$

So by (12.158) and (12.159), the vector of Lagrange multipliers can be written

$$\mu = N_p^+ \nabla f(x^l). \tag{12.160}$$

We could have derived ϕ by the Lagrangian approach. Letting μ be defined for every x as in (12.160), we obtain a function

$$\psi(x) = f(x) - (h(x))^T N_p^+ \nabla f(x). \tag{12.161}$$

This function is stationary at every local optimum of (PE), but it is not

necessarily equal to the ordinary Lagrangian (in which μ is independent of x) at points away from x^l. Comparing ϕ and ψ, we see that the difference between them is in the last term of ϕ, whose role is to ensure that, for sufficiently large ρ, the function ϕ has a minimum at a solution of (PE) and not just a stationary point. In fact, Fletcher [16] proves the following result.

Theorem 12.6

Let $x^* \in R^n$ be a point satisfying sufficient conditions for a strict local minimum of (PE) *as given by Theorem 2.8. Then $\nabla\phi(x^*) = 0$ and there exists a $\rho^* > 0$ such that, for every $\rho \geq \rho^*$, the matrix $\nabla^2\phi(x^*)$ is positive definite.*

The only problem remaining before we can apply this exact penalty function method is to assign a value to ρ so that a single unconstrained minimization will yield the sought optimum of (PE). For nonlinear constraint functions, the computation of the critical parameter ρ^* is by no means simple. Fletcher and Lill [19] suggest taking

$$\rho = 10\,||\nabla^2 f(x^0)|| + r, \tag{12.162}$$

where x^0 is the starting point for the unconstrained minimization and $r = 1$ is taken initially. However, this value of ρ is only an approximation of ρ^* and must be adjusted by increasing r if, during the course of minimization, a negative definite $\nabla^2\phi(x)$ is obtained. Note that if $\nabla^2 f$ is not available, a finite-difference approximation of this Hessian is used. An arbitrary and unnecessarily high value of ρ may result in numerical difficulties in the minimization algorithm and so should be avoided.

Functions such as ϕ, given by (12.155) and developed from conventional Lagrangians or their approximations by the addition of extra terms to ensure positive definiteness of their Hessian at a constrained optimum, are also called augmented Lagrangians and will be more extensively studied in the next section.

Another way of relating ϕ to a Lagrangian-type function is by writing

$$\phi(x) = f(x) - (\gamma(x))^T h(x), \tag{12.163}$$

where $\gamma(x) \in R^p$ is given by

$$\gamma(x) = N_p^+\,\nabla f(x) - \rho N_p^+(N_p^+)^T h(x). \tag{12.164}$$

Here $\gamma(x)$ is a function of x, whereas in the conventional Lagrangian the vector μ is independent of x. An interesting and eventually useful result can be derived by considering, for any given $\hat{x} \in R^n$, the quadratic program

$$\text{(QP)} \qquad \min Q(z) = z^T\nabla f(\hat{x}) + \tfrac{1}{2}\rho z^T z \tag{12.165}$$

subject to

$$l(z) = N_p^T z + h(\hat{x}) = 0. \tag{12.166}$$

We have then

Theorem 12.7

Let the Lagrangian associated with (QP) *be defined by*

$$L(z, \mu_Q) = Q(z) - \mu_Q^T l(z). \tag{12.167}$$

If N_p has full rank, then

$$\mu_Q = \gamma(\hat{x}), \tag{12.168}$$

where $\gamma(\hat{x})$ is given by (12.164).

The proof is an exercise in linear algebra and will be left for the reader. Note that the quadratic program is defined by the functions appearing in (PE) evaluated at $\hat{x}$, so that $l(z)$ in (12.166) is a linear approximation of h at $\hat{x}$ and $Q(z)$ is a kind of "quadratic approximation" of f at $\hat{x}$ in which $\nabla^2 f(\hat{x})$ is replaced by ρI.

The solution of such quadratic programs is an essential part of Fletcher's exact penalty function method for inequality constrained programs [18], as we shall now see.

Consider, therefore, problem (P), defined by (12.3) to (12.5). Also define a corresponding quadratic program

$$\text{(QP1)} \qquad \min Q(z) = z^T \nabla f(\hat{x}) + \tfrac{1}{2}\rho z^T z \tag{12.169}$$

subject to

$$l_g(z) = (N_m^g)^T z + g(\hat{x}) \geq 0 \tag{12.170}$$

$$l_h(z) = (N_p^h)^T z + h(\hat{x}) = 0, \tag{12.171}$$

where $g(x) = (g_1(x), \ldots, g_m(x))^T$, $h(x) = (h_1(x), \ldots, h_p(x))^T$, and N_m^g, N_p^h are matrices whose columns are the $\nabla g_i(\hat{x})$ and $\nabla h_j(\hat{x})$, respectively. Once a value of ρ is assigned and a starting point $x^0 \in R^n$ is chosen, (QP1) can be solved by some quadratic programming algorithm that also yields vectors of Lagrange multipliers $\lambda_Q(\hat{x})$, $\mu_Q(\hat{x})$. It is assumed that the active inequality constraints at the solution of (QP1) are the ones that are active at the solution of (P). Moreover, it is also assumed that the Lagrange multipliers of (P) and (QP1) are the same. Now we have the penalty function

$$\phi(x) = f(x) - (\lambda_Q(\hat{x}))^T g(x) - (\mu_Q(\hat{x}))^T h(x), \tag{12.172}$$

which is to be minimized by one of the methods of Chapter 10 or 11, using first and possibly second derivatives. Note that each function evaluation of (12.172) requires the solution of the quadratic programming problem (QP1). However, because of its structure, the solution of the quadratic program is not difficult; and Fletcher's limited tests [18] show that once the correct set of active inequality constraints at a local minimum of (P) is determined, convergence is attained in a remarkably small number of iterations. Algorithms for solving quadratic programs will be discussed in subsequent chapters.

12.5 MULTIPLIER AND LAGRANGIAN METHODS

One of the difficulties in implementing the methods described in the first two sections is inherent in their nature. As the parameters ρ^k or the $g_i(x^{k*})$ approach zero, numerical computations become more difficult because of discontinuities on the boundary of the feasible set and because the Hessian matrix of the penalty function becomes ill-conditioned. Thus it could be useful to derive methods in which the parameters would need to assume moderate values only. The techniques presented here are based on this fact, but as we shall see below, they are also closely related to exact penalty functions and Lagrangians.

Consider first a nonlinear program with equality constraints

$$\text{(PE)} \qquad \min f(x) \tag{12.173}$$

subject to

$$h_j(x) = 0, \qquad j = 1, \ldots, p. \tag{12.174}$$

Hestenes [25] has suggested the following interesting penalty function method for the solution of this program.

Suppose that $f, h_1, \ldots, h_p$ are twice continuously differentiable functions and that $x^* \in R^n$ is a local solution of (PE), such that the sufficient conditions for a strict local minimum as given in Theorem 2.8 are satisfied. That is, there exist multipliers $\mu_1^*, \ldots, \mu_p^*$ such that

$$\nabla f(x^*) - \sum_{j=1}^{p} \mu_j^* \nabla h_j(x^*) = 0 \tag{12.175}$$

$$h_j(x^*) = 0, \qquad j = 1, \ldots, p \tag{12.176}$$

and for every $z \neq 0$ satisfying

$$z^T \nabla h_j(x^*) = 0, \qquad j = 1, \ldots, p \tag{12.177}$$

we have

$$z^T \nabla_x^2 L(x^*, \mu^*) z > 0. \tag{12.178}$$

Let us show that it is possible to find a positive number c^* such that x^* is an unconstrained minimum of the **augmented Lagrangian** $M(x, \mu^*)$, where

$$M(x, \mu^*) = f(x) - \sum_{j=1}^{p} \mu_j^* h_j(x) + \tfrac{1}{2}c^* \sum_{j=1}^{p} (h_j(x))^2 \tag{12.179}$$

$$= L(x, \mu^*) + \tfrac{1}{2}c^* \sum_{j=1}^{p} (h_j(x))^2. \tag{12.180}$$

We begin with the following result, which is due to Arrow, Gould, and Howe [1].

Lemma 12.8

Let u and v be continuous real functions on a compact set $K \subset R^n$ such that $v(z) \geq 0$ for all $z \in K$. Then $u(z) > 0$ whenever $v(z) \leq 0$ if and only if there exists a number c^* such that, for all $c \geq c^*$, it follows that

$$u(z) + cv(z) > 0 \tag{12.181}$$

for all $z \in K$.

Proof. Suppose that (12.181) holds for all $z \in K$ and $c \geq c^*$. Then $v(z) \leq 0$ clearly implies $u(z) > 0$. Conversely, let K' be the subset of K on which $u(z) \leq 0$. If $K' = \emptyset$, we are finished; so suppose that K' is nonempty. Then

$$u(z) + cv(z) \geq u(z) > 0 \tag{12.182}$$

for all $c \geq 0$ and $z \in K$ such that $z \notin K'$. Since K' is compact, the functions u and v attain their respective minima on K'. Accordingly, denote by u^* and v^* the lowest values of these functions on K'. Then

$$v^* = v(\hat{z}) \geq 0 \tag{12.183}$$

for some $\hat{z} \in K'$. If $v(\hat{z}) = 0$, then $u(\hat{z}) > 0$ by hypothesis, thus contradicting the definition of K'. Hence

$$v^* = v(\hat{z}) > 0 \tag{12.184}$$

and for all $c \geq c^* > -u^*/v^*$ and $z \in K'$, we have

$$u(z) + cv(z) \geq u^* + cv^* > 0. \tag{12.185}$$

■

Corollary 12.9

Let A be an $n \times n$ matrix and let B be an $m \times n$ matrix. Then $z^TAz > 0$ for every $z \neq 0$ satisfying $Bz = 0$ if and only if there exists a number $c^* > 0$ such that, for all $c \geq c^*$, it follows that

$$z^T(A + cB^TB)z > 0 \tag{12.186}$$

for all $z \neq 0$.

Proof. Let $u(z) = z^TAz$ and $v(z) = (Bz)^TBz$ in the previous lemma. Also let

$$K = \{z : z \in R^n, z^Tz = 1\}. \tag{12.187}$$

Then the corollary follows from Lemma 12.8 for all z such that $z^Tz = 1$ and hence for all $z \neq 0$. ■

We have now

Theorem 12.10

Suppose that x^* and μ^* satisfy the sufficiency conditions of optimality for x^* to be a strict local minimum of (PE), *as given in Theorem 2.8. Then there exists a number $c^* > 0$ such that, for all $c \geq c^*$, the point x^* is a local unconstrained minimum of $M(x, \mu^*)$. Conversely, if $h_j(x^0) = 0$, $j = 1, \ldots, p$, and x^0 is an unconstrained minimum of $M(x, \mu^0)$ for some μ^0, then x^0 solves* (PE).

The proof follows from the preceding results and is left for the reader.

We can see from the last theorem that if the values of the Lagrange multipliers μ_j^* and a sufficiently large constant c are available, then an unconstrained minimum of M yields an optimal solution of (PE). The difficulty, of course, lies in determining the correct values of the μ_j and c.

Hestenes' **method of multipliers** consists of updating the Lagrange multipliers μ_j at each iteration and sometimes the constant c as well. Suppose that we choose a sufficiently large value of c and that at iteration k we have some estimate μ^k of the vector of Lagrange multipliers. We minimize $M(x, \mu^k)$ and let the optimal solution be x^{k*}. Observe that

$$\nabla_x M(x^{k*}, \mu^k) = \nabla f(x^{k*}) - \sum_{j=1}^{p} [\mu_j^k - ch_j(x^{k*})] \nabla h_j(x^{k*}) = 0. \tag{12.188}$$

The Lagrange multipliers are updated by the formula

$$\mu_j^{k+1} = \mu_j^k - ch_j(x^{k*}), \qquad j = 1, \ldots, p \tag{12.189}$$

and we proceed to the next iteration in which $M(x, \mu^{k+1})$ is minimized.

Example 12.5.1

Consider the quadratic program

$$\min f(x) = \tfrac{1}{2}x^T Qx + \alpha b^T x, \qquad x \in R^n \tag{12.190}$$

subject to

$$h(x) = \sum_{i=1}^{n} b_i x_i = 0, \tag{12.191}$$

where Q is a nonsingular symmetric matrix such that $x^T Qx > 0$ for all $x \neq 0$ satisfying (12.191). The vector $b \neq 0$ and the number $\alpha > 0$ are given. The reader can verify that the optimal solution is $x^* = 0$ and that $\mu^* = \alpha$ is the corresponding Lagrange multiplier. Suppose that we select a constant c such that

$$x^T Qx + c(b^T x)^2 > 0 \tag{12.192}$$

for all $x \neq 0$. Then

$$M(x, \mu^k) = \tfrac{1}{2}x^T Qx + \alpha b^T x - \mu^k b^T x + \tfrac{1}{2}c(b^T x)^2 \tag{12.193}$$

and the unconstrained minimum x^{k*} of M must satisfy the equation

$$Qx^{k*} + \alpha b - (\mu^k - cb^T x^{k*})b = 0 \tag{12.194}$$

from which we get

$$x^{k*} = \frac{(\mu^k - \alpha)Q^{-1}b}{1 + cb^T Q^{-1}b}. \tag{12.195}$$

If we start with $\mu^0 = 0$, then it can be shown that

$$x^{k*} = \frac{-\alpha Q^{-1}b}{(1 + cb^T Q^{-1}b)^{k+1}} \tag{12.196}$$

and the sequence $\{x^{k*}\}$ converges to $x^* = 0$ if

$$|1 + cb^T Q^{-1}b| > 1. \tag{12.197}$$

∎

Independently, Powell [48] has also derived a method of multipliers for equality constrained problems that is very similar to Hestenes' work. He considers the unconstrained minimization of the two-parameter augmented function

$$M(x, \sigma, \alpha) = f(x) + \sum_{j=1}^{p} \sigma_j[h_j(x) + \alpha_j]^2. \tag{12.198}$$

Note that by letting $\sigma_j = c/2$ and $\alpha_j = -\mu_j/c$, we obtain Hestenes' augmented function as given by (12.179) once more, except for a term that is independent of x. Powell's method of multipliers is based on the result that if $x^*(\sigma, \alpha)$ is an unconstrained minimum of (12.198) for some values of the parameters σ and α, then $x^*(\sigma, \alpha)$ minimizes f, subject to the constraints

$$h_j(x) = h_j[x^*(\sigma, \alpha)], \qquad j = 1, \ldots, p. \tag{12.199}$$

It follows that the parameters σ and α must be adjusted in some iterative way so that

$$\lim_{k\to\infty} \{h_j[x(\sigma^k, \alpha^k)]\} = 0, \qquad j = 1, \ldots, p. \tag{12.200}$$

Actually, Powell suggests choosing sufficiently large σ_j, $j = 1, \ldots, p$, keeping them fixed, and adjusting the α_j so that

$$\alpha_j^{k+1} = \alpha_j^k + h_j[x(\sigma, \alpha^k)], \qquad j = 1, \ldots, p. \tag{12.201}$$

This correction of the parameters α_j continues unless $\max |h_j(x(\sigma, \alpha))|$ either fails to converge or converges to zero at a too slow rate and then the σ_j are increased to higher values.

Convergence of this method to the solution of (PE) can be shown under rather mild conditions. We present the following convergence result of Powell, proof of which can be found in [48].

Theorem 12.11

Suppose that $x^* \in R^n$ *is the unique solution of* (PE) *and that for every neighborhood* $N_\delta(x^*)$ *of* x^* *there exists an* $\epsilon(\delta) > 0$ *such that*

$$X_\alpha = \{x : h_j(x) = \alpha_j, j = 1, \ldots, p\} \neq \emptyset \tag{12.202}$$

for every α_j *satisfying*

$$|\alpha_j| \leq \epsilon(\delta), \qquad j = 1, \ldots, p. \tag{12.203}$$

Also suppose that $f, h_1, \ldots, h_p$ *are continuous functions and that there exists a natural number* K *such that, for every* $k \geq K$, *the unconstrained minima of* $M(x, \sigma^k, \alpha^k)$ *lie in a compact set. Then the sequence of these minima converges to* x^*.

Additional results on the convergence of this method indicate that the rate of convergence is linear. Recently, however, it was shown that this convergence rate can be improved. Another method of multipliers, similar to the

two methods already described, was also suggested by Haarhoff and Buys [24]. Also, Hestenes' method of multipliers was extensively studied from a computational point of view by Miele and coworkers [39, 40, 41, 42].

Methods of multipliers for solving equality constrained nonlinear programs were extended to inequality constrained problems by Bertsekas [3], Buys [5], Kort and Bertsekas [27, 28], Pierre [45], and Rockafellar [49, 50]. We shall briefly describe here Rockafellar's work.

Suppose that we are interested in solving the equality-inequality constrained problem (P), as given by (12.3) to (12.5). Define

$$\theta(t) = \max\{t, 0\}. \tag{12.204}$$

Then for any positive number c we have

$$\frac{1}{4c}\{[\theta(\lambda_i - 2cg_i(x))]^2 - (\lambda_i)^2\} = \begin{cases} c(g_i(x))^2 - \lambda_i g_i(x) & \text{if } g_i(x) \leq \dfrac{\lambda_i}{2c} \\ -\dfrac{(\lambda_i)^2}{4c} & \text{if } g_i(x) \geq \dfrac{\lambda_i}{2c} \end{cases} \tag{12.205}$$

and an augmented Lagrangian $M(x, \lambda, \mu)$ can be defined as

$$M(x, \lambda, \mu) = f(x) + \frac{1}{4c}\sum_{i=1}^{m}\{[\theta(\lambda_i - 2cg_i(x))]^2 - (\lambda_i)^2\} - \sum_{j=1}^{p}\mu_j h_j(x) + \frac{c}{2}\sum_{j=1}^{p}[h_j(x)]^2. \tag{12.206}$$

The reader will be asked in the Exercise 12. L. to derive (12.206) from previous formulas. Note that, in contrast to results in Chapter 3, the Lagrange multipliers λ_i appearing in (12.206) and corresponding to the inequality constraints of (P) are not required to be nonnegative. The multiplier method of Hestenes and Powell can now be easily generalized. For a given $c > 0$ and multipliers (λ^k, μ^k), let x^{k*} be the optimal solution of the unconstrained optimization

$$\min_x M(x. \lambda^k, \mu^k). \tag{12.207}$$

Observe again that

$$\nabla_x M(x^{k*}, \lambda^k, \mu^k) = \nabla f(x^{k*}) - \sum_{i=1}^{m}\theta(\lambda_i^k - 2cg_i(x^{k*}))\,\nabla g_i(x^{k*}) - \sum_{j=1}^{p}(\mu_j - ch_j(x^{k*}))\,\nabla h_j(x^{k*}) = 0. \tag{12.208}$$

So it is reasonable to set

$$\lambda_i^{k+1} = \theta(\lambda_i^k - 2cg_i(x^{k*})), \qquad i = 1, \ldots, m \tag{12.209}$$

$$\mu_j^{k+1} = \mu_j^k - ch_j(x^{k*}), \qquad j = 1, \ldots, p. \tag{12.210}$$

Convergence of the method to an optimal solution of (P) can be shown under conditions similar to those of Theorem 12.11. In addition, Rockafellar has also derived another procedure that converges to a solution of a convex programming problem (CP), as given by (4.158) to (4.160), that satisfies a strong consistency (Slater) condition. This procedure does not require an exact minimization of M at every iteration, and the value of c can remain fixed. For details, the reader is referred to [49].

A possible computational disadvantage of the methods described so far is that the augmented Lagrangian M is usually differentiable only once. In the rest of this section we shall unify, generalize, and, if possible, improve the works mentioned here. The purpose of such a unifying approach is to associate with the optimal solutions of a general inequality-equality constrained problem some corresponding unconstrained stationary points of a class of generalized Lagrangians or, even better, unconstrained minima of these functions. The following results on this subject are due to Mangasarian [37].

Consider again the general nonlinear program (P), given by (12.3) to (12.5). Define a generalized Lagrangian

$$GL\,(x, y, w) = f(x) - \sum_{i=1}^{m} \phi(g_i(x), y_i) - \sum_{j=1}^{p} \psi(h_j(x), w_j), \tag{12.211}$$

where $\phi(z_1, z_2)$ and $\psi(z_1, z_2)$ are real differentiable functions defined on R^2. Let us list certain conditions on ϕ and ψ:

1. The equations

$$\frac{\partial \phi(0, \eta)}{\partial z_1} = \beta \qquad \frac{\partial \phi(0, \eta)}{\partial z_2} = 0 \tag{12.212}$$

have a solution η for each $\beta \geq 0$, and

$$\frac{\partial \phi(\xi, \eta)}{\partial z_1} = 0 \qquad \frac{\partial \phi(\xi, \eta)}{\partial z_2} = 0 \tag{12.213}$$

have a solution η for each $\xi > 0$.

2. The equations

$$\frac{\partial \psi(0, \eta)}{\partial z_1} = \beta \qquad \frac{\partial \psi(0, \eta)}{\partial z_2} = 0 \tag{12.214}$$

have a solution η for each β.

3. The relations

$$\frac{\partial \phi(0, \eta)}{\partial z_1} \geq 0 \qquad \frac{\partial \phi(\xi, 0)}{\partial z_1} = 0 \tag{12.215}$$

are satisfied for $\xi > 0$ and if

$$\frac{\partial \phi(\xi, \eta)}{\partial z_2} = 0, \tag{12.216}$$

then $\xi \geq 0$ and $\xi\eta = 0$.

4. If

$$\frac{\partial \psi(\xi, \eta)}{\partial z_2} = 0, \tag{12.217}$$

then $\xi = 0$.

Before reading the next result, the reader is advised to review the Kuhn-Tucker necessary conditions of optimality for a mathematical program, as presented in Theorem 3.8. We have then

Theorem 12.12

Let $f, g_1, \ldots, g_m, h_1, \ldots, h_p$ be differentiable at x^*. If x^* is a local or global solution of (P) *such that condition* (3.71) *holds, or if x^* and some $\lambda^* \in R^m$, $\mu^* \in R^p$ satisfy* (3.72) *to* (3.74)*—that is, the generalized Kuhn-Tucker necessary conditions for optimality in* (P) *hold—then x^* and some $y^* \in R^m$, $w^* \in R^p$ form a stationary point of GL. That is*

$$\nabla GL\,(x^*, y^*, w^*) = 0 \tag{12.218}$$

provided that conditions (1) *and* (2) *above are satisfied. Conversely, if* (x^*, y^*, w^*) *satisfies* (12.218), *then the vector* (x^*, λ^*, μ^*), *where*

$$\lambda_i^* = \frac{\partial \phi(g_i(x^*), y_i^*)}{\partial z_1}, \qquad i = 1, \ldots, m \tag{12.219}$$

$$\mu_j^* = \frac{\partial \psi(h_j(x^*), w_j^*)}{\partial z_1}, \qquad j = 1, \ldots, p \tag{12.220}$$

satisfies the generalized Kuhn-Tucker necessary conditions for optimality in (P), *provided that conditions* (3) *and* (4) *hold.*

Proof. If x^* is a local or global solution of (P) and (3.71) holds, then the Kuhn-Tucker conditions (3.72) to (3.74) are satisfied for x^* and some λ^*, μ^*. Let ϕ and ψ satisfy conditions (1) and (2), let $I(x^*)$ be the set of active constraints, and define y^*, w^* by

$$\frac{\partial\phi(0, y_i^*)}{\partial z_1} = \lambda_i^* \qquad \frac{\partial\phi(0, y_i^*)}{\partial z_2} = 0, \qquad i \in I(x^*) \tag{12.221}$$

$$\frac{\partial\phi(g_i(x^*), y_i^*)}{\partial z_1} = 0 \qquad \frac{\partial\phi(g_i(x^*), y_i^*)}{\partial z_2} = 0, \qquad i \notin I(x^*) \tag{12.222}$$

$$\frac{\partial\psi(0, w_j^*)}{\partial z_1} = \mu_j^* \qquad \frac{\partial\psi(0, w_j^*)}{\partial z_2} = 0 \qquad j = 1, \ldots, p. \tag{12.223}$$

Then (3.72) to (3.74) and (12.221) to (12.223) yield

$$\nabla_x GL\,(x^*, y^*, w^*) = \nabla f(x^*) - \sum_{i=1}^{m} \frac{\partial\phi(g_i(x^*), y_i^*)}{\partial z_1} \nabla g_i(x^*) - \sum_{j=1}^{p} \frac{\partial\psi(h_j(x^*), w_j^*)}{\partial z_1} \nabla h_j(x^*) = 0 \tag{12.224}$$

$$\frac{\partial\, GL\,(x^*, y^*, w^*)}{\partial y_i} = -\frac{\partial\phi(g_i(x^*), y_i^*)}{\partial z_2} = 0, \qquad i = 1, \ldots, m \tag{12.225}$$

$$\frac{\partial\, GL\,(x^*, y^*, w^*)}{\partial w_j} = -\frac{\partial\psi(h_j(x^*), w_j^*)}{\partial z_2} = 0, \qquad j = 1, \ldots, p. \tag{12.226}$$

Hence

$$\nabla\, GL\,(x^*, y^*, w^*) = 0. \tag{12.227}$$

Conversely, suppose that (x^*, y^*, w^*) satisfies (12.218) and let conditions (3) and (4) hold. Define λ^* and μ^* by (12.219) and (12.220), respectively. From (12.224) we get

$$\nabla f(x^*) - \sum_{i=1}^{m} \lambda_i^* \nabla g_i(x^*) - \sum_{j=1}^{p} \mu_j^* \nabla h_j(x^*) = 0. \tag{12.228}$$

From (12.215) in condition (3) and (12.219) we obtain

$$\lambda_i^* \geq 0, \qquad i \in I(x^*). \tag{12.229}$$

From (12.216) in condition (3) and (12.225) it follows that

$$g_i(x^*) \geq 0, \qquad i = 1, \ldots, m \tag{12.230}$$

and

$$y_i^* g_i(x^*) = 0, \qquad i = 1, \ldots, m. \tag{12.231}$$

From (12.215), (12.219), and (12.231) we obtain

$$\lambda_i^* = \frac{\partial \phi(g_i(x^*), y_i^*)}{\partial z_1} = \frac{\partial \phi(g_i(x^*), 0)}{\partial z_1} = 0, \qquad i \notin I(x^*). \tag{12.232}$$

Finally, from (12.217) and (12.226) we have

$$h_j(x^*) = 0, \qquad j = 1, \ldots, p. \tag{12.233}$$

Combining (12.229), (12.230) and (12.232) into equivalent expressions

$$\lambda_i^* g_i(x^*) = 0, \qquad i = 1, \ldots, m \tag{12.234}$$

$$\lambda^* \geq 0, \tag{12.235}$$

we conclude that (12.228), (12.230), and (12.233) to (12.235) are the Kuhn-Tucker conditions as given in Theorem 3.8. ■

If further conditions are imposed on ϕ and ψ, it is guaranteed that, for sufficiently large values of a parameter α, a stationary point (x^*, y^*, w^*) of GL, as given in the first part of the theorem, corresponds to a strict unconstrained local minimum of $GL\,(x, y^*, w^*)$. And so previous results in exact penalty functions and augmented Lagrangians are generalized.

Consider some examples of functions ϕ and ψ that satisfy conditions (1) to (4) above, required for the last theorem, as well as conditions ensuring positive definiteness of the Hessian matrix of GL at (x^*, y^*, w^*). Define

$$\Gamma(t)_+ = \begin{cases} \Gamma(t) & \text{if } t \geq 0 \\ 0 & \text{if } t < 0 \end{cases} \tag{12.236}$$

and let

$$\phi(\xi, \eta) = \Gamma(\eta) - \Gamma(\eta - \alpha\xi)_+ \tag{12.237}$$

$$\psi(\xi, \eta) = \Gamma(\eta) - \Gamma(\eta - \alpha\xi), \tag{12.238}$$

where Γ is a real differentiable function on R and $\alpha \in R$ is a parameter. The Γ function itself must satisfy certain conditions: (a) The derivative Γ' is a strictly increasing function mapping R onto R, $\Gamma'(t)_+$ maps $[0, \infty)$ onto itself, and $\Gamma'(0) = 0$. (b) Γ is a convex nonnegative function on R such that $\Gamma(0) = 0$. (c) Γ is twice differentiable on R, $\Gamma''(t) > 0$ for $t \neq 0$, and

$\Gamma''(0) = 0$. Examples of Γ functions that satisfy conditions (a), (b), and (c) are

$$\Gamma_1(t) = \frac{1}{\alpha\beta}(t)^\beta, \qquad \alpha > 0, \quad \beta = 4, 6, \ldots \tag{12.239}$$

$$\Gamma_2(t) = \frac{1}{2}[e^t + e^{-t} - (t)^2] - 1 \tag{12.240}$$

$$\Gamma_3(t) = \frac{1}{2}\left[\frac{1}{2}(e^t + e^{-t}) - 1\right]^2. \tag{12.241}$$

Corresponding to these functions, we obtain

$$\phi_1(\xi, \eta) = \frac{1}{\alpha\beta}[(\eta)^\beta - (\eta - \alpha\xi)_+^\beta], \qquad \alpha > 0, \quad \beta = 4, 6, \ldots \tag{12.242}$$

$$\psi_1(\xi, \eta) = \frac{1}{\alpha\beta}[(\eta)^\beta - (\eta - \alpha\xi)^\beta], \qquad \alpha > 0, \quad \beta = 4, 6, \ldots \tag{12.243}$$

and so on. Letting $\beta = 4$, we can, therefore, write a generalized Lagrangian for program (P) as

$$GL_1\,(x, y, w) = f(x) - \frac{1}{4\alpha}\sum_{i=1}^{m}[(y_i)^4 - (y_i - \alpha g_i(x))_+^4]$$
$$- \frac{1}{4\alpha}\sum_{j=1}^{p}[(w_j)^4 - (w_j - \alpha h_j(x))^4]. \tag{12.244}$$

An example of a function satisfying (a), (b), and (c) except for the condition $\Gamma''(0) = 0$, is given by

$$\Gamma_4(t) = \frac{1}{2\alpha}(t)^2, \qquad \alpha > 0 \tag{12.245}$$

from which we obtain

$$\phi_4(\xi, \eta) = \frac{1}{2\alpha}[(\eta)^2 - (\eta - \alpha\xi)_+^2] \tag{12.246}$$

$$\psi_4(\xi, \eta) = \frac{1}{2\alpha}[(\eta)^2 - (\eta - \alpha\xi)^2] \tag{12.247}$$

and

$$GL_4\,(x, y, w) = f(x) - \frac{1}{2\alpha}\sum_{i=1}^{m}[(y_i)^2 - (y_i - \alpha g_i(x))_+^2]$$
$$- \frac{1}{2\alpha}\sum_{j=1}^{p}[(w_j)^2 - (w_j - \alpha h_j(x))^2]. \tag{12.248}$$

Letting $c = 2$, $\lambda_i = y_i$, and $\mu_j = w_j$, the reader can easily verify that GL_4 is exactly Rockafellar's augmented Lagrangian, given by (12.206), which is not twice differentiable everywhere, in contrast to generalized Lagrangians based on Γ functions that satisfy all the conditions given in (a), (b), and (c).

Mangasarian [37] has also obtained duality results based on generalized Lagrangians and a superlinearly (or quadratically) convergent algorithm for minimizing GL. This algorithm is essentially Newton's method applied to solving the set of nonlinear equations $\nabla GL(x, y, w) = 0$. In a related development, Mangasarian has extended the method of multipliers to inequality constraints and to more general Lagrangians and obtained a linearly convergent algorithm.

12.6 SOME COMPUTATIONAL ASPECTS OF PENALTY FUNCTION METHODS

The penalty function methods introduced in the first three sections, which consist of sequential unconstrained minimizations, are widely used to solve nonlinear optimization problems. The methods of Sections 12.4 and 12.5 are more recent and so computational experience is quite limited. We believe, however, that they can be very efficient and that they may become one of the primary all-purpose tools used in obtaining numerical solutions to highly nonlinear problems.

Penalty function methods—or methods that transform constrained problems into unconstrained ones by incorporating the constraint functions in the objective function of the unconstrained problem—are expected to be particularly efficient if the constraint functions are nonlinear. Since linear constraints can be efficiently handled by various other methods, as we shall see in subsequent chapters, their incorporation in the penalty function is not particularly useful. Easy modifications of exterior and interior penalty function methods exist [15] that treat linear constraints separately in combined "unconstrained-simplicial" algorithms. The real challenge in solving nonlinear programs lies in the handling of nonlinear constraints. Penalty function methods possess some definite advantages for such problems. The most important, perhaps, is that, in contrast to other methods that explicitly treat nonlinear constraints, penalty methods avoid the time-consuming and often difficult task of moving along nonlinear boundaries of the feasible set, or trying not to cross them. Another advantage is that, in most cases, no specific assumptions on the functions appearing in a problem, such as convexity or its generalizations, are needed in order to apply the method. Later on we shall see that several other algorithms are defined only for particular families of functions and thus cannot be universally used [38]. Although many algorithms converge to solutions of a problem satisfying

the Kuhn-Tucker first-order necessary conditions as given in Theorem 3.8, sequential unconstrained minimization methods generally converge to points that satisfy second-order necessary conditions as stated in Theorem 3.10. Consequently, at these points, satisfaction of second-order sufficient conditions can also be expected. It can also be shown that, near a point where the sufficient conditions hold, iterations of the sequential penalty methods often consist of unconstrained minimization of strictly convex functions; see McCormick [38] and Polyak [47].

The efficient solving of constrained problems by unconstrained minimizations depends heavily on the numerical method used to perform the unconstrained optimization. The most efficient of the unconstrained methods belong to the group discussed in Chapters 10 and 11; such methods use information supplied by evaluating the gradient and sometimes the Hessian of the function to be minimized. Ironically, these methods are the most vulnerable to the main disadvantage of sequential penalty methods: the Hessian matrix of the penalty function becomes successively more ill-conditioned as the parameter of the penalty method approaches its limit. Consider, for example, a mixed penalty function $H(x, \rho)$, such as (12.82) for solving problem (P), having equality and inequality constraints. Let

$$H(x, \rho^k) = f(x) - \rho^k \sum_{i=1}^{m} \log g_i(x) + \frac{1}{\rho^k} \sum_{j=1}^{p} (h_j(x))^2, \qquad (12.249)$$

where $\rho^k > 0$ and a sequence of unconstrained minimizations of H is performed for $\{\rho^k\} \longrightarrow 0$. The Hessian matrix of H is given by

$$\begin{aligned} \nabla^2 H(x, \rho^k) = \nabla^2 f(x) &- \sum_{i=1}^{m} \frac{\rho^k}{g_i(x)} \nabla^2 g_i(x) + \sum_{j=1}^{p} \frac{2h_j(x)}{\rho^k} \nabla^2 h_j(x) \\ &+ \sum_{i=1}^{m} \frac{\rho^k}{(g_i(x))^2} \nabla g_i(x)(\nabla g_i(x))^T + \sum_{j=1}^{p} \frac{2}{\rho^k} \nabla h_j(x)(\nabla h_j(x))^T. \end{aligned} \qquad (12.250)$$

It can be shown [15] that the sequences $\{\rho^k/g_i(x^{k*})\}$ and $\{2h_j(x^{k*})/\rho^k\}$ converge, under mild assumptions, to the corresponding Lagrange multipliers λ_i^* and μ_j^*, respectively. Thus the first three terms of $\nabla^2 H$ converge to the Hessian of the Lagrangian, which has finite eigenvalues. The remaining two terms, however, contain the sequences $\{\rho^k/(g_i(x))^2\}$ and $\{2/\rho^k\}$, which, in most cases, tend to increase without bound and cause the ill-conditioning. This problem has received some attention recently, and several modifications of penalty methods have been suggested. The reader is referred to Fletcher and McCann [20], Lootsma [34, 36], and Murray [43]. Analysis of the problem by Lootsma [34] showed that ill-conditioning of the Hessian matrix is more rapid for the function $H(x, \rho^k)$ given by (12.249), which uses the q function of (12.51), than for penalty functions using (12.52) or (12.53).

On the other hand, convergence with the q function of (12.51) is usually faster than with the other two functions. A related difficulty, which occurs mainly when variable metric algorithms are used to solve the sequence of unconstrained minimizations, is that the inverse Hessians of the penalty function tend to singularity. It is expected that most of the disadvantages of penalty function methods associated with the Hessian matrix will be eliminated by turning to exact penalty or multiplier methods.

Lootsma [35] has tested the performance of several unconstrained minimization techniques, described in preceding chapters, when applied to minimizing penalty functions. On the basis of a limited number of test problems, he recommends the use of the BFS method or a modified Newton method [15]. A special unconstrained technique for penalty methods and a line search for barrier functions were developed, respectively, by Lasdon [30] and Lasdon, Fox, and Ratner [32].

An interesting and useful feature in penalty function methods consists of accelerating the convergence of the method by extrapolation, as proposed by Fiacco and McCormick [15]. The idea is to estimate successive points in building the sequence of optima $\{x^{k*}\}$ by series expansions of already-existing members in this sequence. The information obtained can be used to start the unconstrained optimization with the next value of the penalty parameter. Conceivably, such acceleration techniques would also be useful in other types of algorithms, but this subject has not yet received proper attention.

EXERCISES

12.A. Draw a flow diagram of the exterior penalty function method for solving problem (P). Assume that the unconstrained minimizations are carried out by (a) Powell's method, described in Section 9.5, or (b) a variable metric alogrithm, such as the BFS method. Discuss the possible computational advantages and disadvantages of both cases and enumerate the difficulties expected to arise in implementing the methods. Also discuss the behavior of the exterior penalty function method if the optimal solution of an inequality constrained problem lies in the interior of the feasible set, and illustrate this behavior by a numerical example.

12.B. Show that the set X_1, given by

$$X_1 = \{x : x \in R^2, (x_2)^2 - (x_1x_2)^2 + (x_2(x_2 - 1))^2 \geq 0\}, \qquad (12.251)$$

does not satisfy regularity condition (1) assumed for (PI) in Section 12.2.

Let

$$X_2 = \{x : x \in R^n, g(x) \geq 0\}. \qquad (12.252)$$

Find a real function g such that X_2^0, the interior of X_2, is nonempty, but $g(x) = 0$ for every $x \in X_2^0$.

12.C. Show convergence of the parameter-free exterior method under the conditions stated in Section 12.1.

12.D. Solve the following constrained problems [15, 40] by the penalty function methods discussed in this chapter. Use numerical algorithms for unconstrained minimization, developed in previous chapters, which are available in your computer.

(a)
$$\min\,(x_1)^3 - 6(x_1)^2 + 11x_1 + x_3 \tag{12.253}$$

subject to

$$(x_1)^2 + (x_2)^2 - (x_3)^2 \leq 0 \tag{12.254}$$

$$(x_1)^2 + (x_2)^2 + (x_3)^2 \geq 4 \tag{12.255}$$

$$x_3 \leq 5 \tag{12.256}$$

$$x_1 \geq 0, \quad x_2 \geq 0, \quad x_3 \geq 0. \tag{12.257}$$

An optimum is at $x^* = (0, \sqrt{2}, \sqrt{2})$.

(b)
$$\min\,(x_1 - x_2)^2 + (x_2 - x_3)^4 \tag{12.258}$$

subject to

$$x_1 + x_1(x_2)^2 + (x_3)^4 = 3. \tag{12.259}$$

An optimal solution is at $x^* = (1, 1, 1)$.

12.E. Solve the quadratic program of Example 12.5.1 by an exterior penalty method, using (12.10) and (12.14). Let $1/\rho^k = (k + 1)/2$ and $\beta = 2$. Show that the sequence of minima $\{x^{k*}\}$ is defined by

$$x^{k*} = \frac{-\alpha Q^{-1}b}{1 + (k + 1)b^T Q^{-1} b}. \tag{12.260}$$

Compare the convergence of this sequence with that of the multiplier method presented in the example.

12.F. Prove the following statement concerning exact penalty functions. Let

$$\gamma_i(\beta, \lambda) = 0 \qquad i = 1, \ldots, m \tag{12.261}$$

for $\beta \geq 0$ and for all $\lambda \in A \subset R^m$. Suppose that $\lambda \in A$ and $\hat{x}$ is feasible for (PI) such that it is a local minimum of (12.132). Then $\hat{x}$ is also a local minimum of problem (PI).

12.G. Show that the Kuhn-Tucker necessary conditions of optimality, as stated in Chapters 2 and 3, for the equality constrained program (PE) at $\hat{x}$ are

$$P\nabla f(\hat{x}) = 0 \tag{12.262}$$

$$h(\hat{x}) = 0, \tag{12.263}$$

where P is given by (12.154). Give a geometric interpretation of (12.262) and show that

$$\nabla\phi(\hat{x}) = 0, \tag{12.264}$$

where ϕ is defined by (12.155).

12.H. Prove Theorem 12.7.

12.I. Suppose that Q is positive definite in the quadratic program stated in Example 12.5.1. Solve this program by Powell's method of multipliers with $\sigma = 1$ and $\alpha^0 = 0$. Will convergence to $x^* = 0$ be obtained?

12.J. Prove Theorem 12.10.

12.K. An equality-inequality constrained nonlinear program, such as program (P) presented in the beginning of this chapter, can be converted into an equivalent program by the introduction of slack variables. Thus program (P) can be rewritten as

$$\text{(P}'\text{)} \qquad \min f(x) \tag{12.265}$$

subject to

$$g_i(x) - z_i = 0, \qquad i = 1, \ldots, m \tag{12.266}$$

$$h_j(x) = 0, \qquad j = 1, \ldots, p \tag{12.267}$$

$$z_i \geq 0, \qquad i = 1, \ldots, m. \tag{12.268}$$

Solution of this program by Hestenes' method of multipliers would call for finding an $x \in R^n$ and a $z \in R^m$, $z \geq 0$, that minimize

$$M(x, z, \lambda, \mu) = f(x) - \sum_{i=1}^{m} \lambda_i(g_i(x) - z_i) + \frac{c}{2}\sum_{i=1}^{m}(g_i(x) - z_i)^2 - \sum_{j=1}^{p} \mu_j h_j(x) + \frac{c}{2}\sum_{j=1}^{p}(h_j(x))^2. \tag{12.269}$$

Noting that the minimization with respect to $z \geq 0$ can be explicitly carried out, obtain the formula (12.206) for $M(x, \lambda, \mu)$.

12.L. Suppose that (SP) is a standard primal convex program as defined in Section 5.2—that is, we have

$$\text{(SP)} \qquad \min f(x) \tag{12.270}$$

subject to

$$g_i(x) \geq 0, \qquad i = 1, \ldots, m. \tag{12.271}$$

(a) Derive the augmented Lagrangian $M(x, \lambda)$, corresponding to (SP), as given by (12.206) from (5.120) and from the following perturbed problem:

(QSP)
$$\min f(x) + c \sum_{i=1}^{m} (w_i)^2 \tag{12.272}$$

subject to

$$g_i(x) \geq w_i, \qquad i = 1, \ldots, m \tag{12.273}$$

where $c > 0$ is a parameter.

(b) Show that the perturbation function Φ_Q, corresponding to (QSP), is given by

$$\Phi_Q(w) = \Phi_P(w) + c \sum_{i=1}^{m} (w_i)^2, \tag{12.274}$$

where Φ_P is the perturbation function that corresponds to (PSP), defined by (5.111) and (5.112).

12.M. Illustrate Theorem 12.12 by the problem presented in Example 3.1.4. Take $x^* = (0, 0)$ and choose $\alpha = 1$, $y^* = \frac{1}{2}$, $w^* = 0$. Define GL by (12.244). Is $(0, 0, \frac{1}{2}, 0)$ a minimum of the generalized Lagrangian for $\alpha \geq 1$?

12.N. Show that Γ_1, Γ_2, and Γ_3, given by (12.239) to (12.241), satisfy conditions (a), (b), and (c), as asserted.

REFERENCES

1. ARROW, K. J., F. J. GOULD, and S. M. HOWE, "A General Saddle Point Result for Constrained Optimization," *Math. Prog.*, **5**, 225–234 (1973).

2. BARTLE, R. G., *The Elements of Real Analysis*, 2nd ed., John Wiley & Sons, New York, 1976.

3. BERTSEKAS, D. P., "On Penalty and Multiplier Methods for Constrained Minimization," Engineering-Economic Systems Department Working Paper, Stanford University, August 1973, *SIAM J. Control,* to appear.

4. BRACKEN J., and G. P. MCCORMICK, *Selected Applications of Nonlinear Programming*, John Wiley & Sons, New York, 1968.

5. BUYS, J. D., "Dual Algorithms for Constrained Optimization Problems," Doctoral dissertation, University of Lèiden, The Netherlands, 1972.

6. CARROLL, C. W., "The Created Response Surface Technique for Optimizing Nonlinear Restrained Systems," *Operations Research*, **9**, 169–184 (1961).

7. CONN, A. R., "Constrained Optimization Using a Nondifferentiable Penalty Function," *SIAM J. Numer. Anal.*, **10**, 760–784 (1973).

8. CONN, A. R., and T. PIETRZYKOWSKI, "A Penalty Function Method Converging Directly to a Constrained Optimum," Research Report 73-11, Department of Combinatorics and Optimization, University of Waterloo, Ontario, Canada, May 1973, *SIAM J. Numer. Anal.*, to appear.

9. COURANT, R., "Variational Methods for the Solution of Problems of Equilibrium and Vibrations," *Bull. Am. Math. Soc.*, **49**, 1–23 (1943).

10. EVANS, J. P., and F. J. GOULD, "An Existence Theorem for Penalty Function Theory," *SIAM J. Control*, **12**, 509–516 (1974).

11. EVANS, J. P., F. J. GOULD, and J. W. TOLLE, "Exact Penalty Functions in Nonlinear Programming," *Math. Prog.*, **4**, 72–97 (1973).

12. FIACCO, A. V., and G. P. MCCORMICK, "The Sequential Unconstrained Minimization Technique for Nonlinear Programming: A Primal-Dual Method," *Management Science*, **10**, 360–366 (1964).

13. FIACCO, A. V., and G. P. MCCORMICK, "Extensions of SUMT for Nonlinear Programming: Equality Constraints and Extrapolation," *Management Science*, **12**, 816–829 (1966).

14. FIACCO, A. V., and G. P. MCCORMICK, "The Sequential Unconstrained Minimization Technique without Parameters," *Operations Research*, **15**, 820–827 (1967).

15. FIACCO, A. V., and G. P. MCCORMICK, *Nonlinear Programming: Sequential Unconstrained Minimization Techniques*, John Wiley & Sons, New York, 1968.

16. FLETCHER, R., "A Class of Methods for Nonlinear Programming with Termination and Convergence Properties," in *Integer and Nonlinear Programming*, J. Abadie (Ed.), North-Holland Publishing Co., Amsterdam, 1970.

17. FLETCHER, R., "A Class of Methods for Nonlinear Programming III, Rates of Convergence," in *Numerical Methods for Non-linear Optimization*, F. A. Lootsma (Ed.), Academic Press, London, 1972.

18. FLETCHER, R., "An Exact Penalty Function for Nonlinear Programming with Inequalities," *Math. Prog.*, **5**, 129–150 (1973).

19. FLETCHER, R., and S. A. LILL, "A Class of Methods for Nonlinear Programming II, Computational Experience," in *Nonlinear Programming*, J. B. Rosen, O. L. Mangasarian, and K. Ritter (Eds.), Academic Press, New York, 1970.

20. FLETCHER, R., and A. P. MCCANN, "Acceleration Techniques for Nonlinear Programming," in *Optimization*, R. Fletcher (Ed.), Academic Press, London, 1969.

21. FOX, R. L., "Structural and Mechanical Design Optimization," in *Optimization and Design*, M. Avriel, M. J. Rijckaert, and D. J. Wilde (Eds.), Prentice-Hall, Englewood Cliffs, N.J., 1973.

22. FRISCH, K. R., "The Logarithmic Potential Method of Convex Programming," University Institute of Economics, Oslo, Norway, Memorandum, May 13, 1955.

23. GOULD, F. J., "A Class of Inside-out Algorithms for General Programs," *Management Science*, **16**, 350–356 (1970).

24. HAARHOFF, P. C., and J. D. BUYS, "A New Method for the Optimization of a Nonlinear Function Subject to Nonlinear Constraints," *Computer J.*, **13**, 178–184 (1970).

25. HESTENES, M. R., "Multiplier and Gradient Methods," *J. Optimization Theory & Appl.*, **4**, 303–320 (1969).

26. HUARD, P., "Resolution of Mathematical Programming with Nonlinear Constraints by the Method of Centres," in *Nonlinear Programming*, J. Abadie (Ed.), North-Holland Publishing Co., Amsterdam, 1967.

27. KORT, B. W., and D. P. BERTSEKAS, "A New Penalty Function Method for Constrained Minimization," in *Proceedings of 1972 IEEE Conference on Decision and Control*, New Orleans, December 1972.

28. KORT, B. W., and D. P. BERTSEKAS, "Combined Primal-Dual and Penalty Methods for Convex Programming," Engineering-Economic Systems Dept. Working Paper, Stanford University, August 1973, *SIAM J. Control*, to appear.

29. KOWALIK, J., M. R. OSBORNE, and D. M. RYAN, "A New Method for Constrained Optimization Problems," *Operations Research*, **17**, 973–983 (1969).

30. LASDON, L. S., "An Efficient Algorithm for Minimizing Barrier and Penalty Functions," *Math. Prog.*, **2**, 65–106 (1972).

31. LASDON, L. S., "Penalty and Barrier Function Design of Electrical Networks and Sonar Transducer Arrays," in *Optimization and Design*, M. Avriel, M. J. Rijckaert, and D. J. Wilde (Eds.), Prentice-Hall, Englewood Cliffs, N.J., 1973.

32. LASDON, L. S., R. L. FOX, and M. W. RATNER, "An Efficient One-Dimensional Search Procedure for Barrier Functions," *Math. Prog.*, **4**, 279–296 (1973).

33. LOOTSMA, F. A., "Constrained Optimization via Parameter-Free Penalty Functions," *Philips Res. Repts.*, **23**, 424–437 (1968).

34. LOOTSMA, F. A., "Hessian Matrices of Penalty Functions for Solving Constrained Optimization Problems," *Philips Res. Repts.*, **24**, 322–330 (1969).

35. LOOTSMA, F. A., "Penalty Function Performance of Several Unconstrained Minimization Techniques," *Philips Res. Repts.*, **27**, 358–385 (1972).

36. LOOTSMA, F. A., "A Survey of Methods for Solving Constrained Minimization Problems via Unconstrained Minimization," in *Numerical Methods for Nonlinear Optimization*, F. A. Lootsma (Ed.), Academic Press, London, 1972.

37. MANGASARIAN, O. L., "Unconstrained Lagrangians in Nonlinear Programming," *SIAM J. Control*, **13**, 772–791 (1975).

38. MCCORMICK, G. P., "Penalty Function versus Non-penalty Function Methods for Constrained Nonlinear Programming Problems," *Math. Prog.*, **1**, 217–238 (1971).

39. MIELE, A., E. E. CRAGG, R. R. IYER, and A. V. LEVY, "Use of the Augmented Penalty Function in Mathematical Programming Problems, Part 1," *J. Optimization & Theory & Appl.*, **8**, 115–130 (1971).

40. MIELE, A., E. E. CRAGG, and A. V. LEVY, "Use of the Augmented Penalty Function in Mathematical Programming Problems, Part 2," *J. Optimization Theory & Appl.*, **8**, 131–153 (1971).

41. MIELE, A., P. E. MOSELEY, and E. E. CRAGG, "A Modification of the Method of Multipliers for Mathematical Programming Problems," in *Techniques of Optimization*, A. V. Balakrishnan (Ed.), Academic Press, New York, 1972.

42. MIELE, A., P. E. MOSELEY, A. V. LEVY, and G. M. COGGINS, "On the Method of Multipliers for Mathematical Programming Problems," *J. Optimization Theory & Appl.*, **10**, 1–33 (1972).

43. MURRAY, W., "Analytical Expressions for the Eigenvalues and Eigenvectors of the Hessian Matrices of Barrier and Penalty Functions," *J. Optimization Theory & Appl.*, **7**, 189–196 (1971).

44. NOBLE, B., *Applied Linear Algebra*, Prentice-Hall, Englewood Cliffs, N.J., 1969.

45. PIERRE, D. A., "Multiplier Algorithms for Nonlinear Programs," paper presented at the Eighth Int. Symp. on Math. Prog., Stanford University, Stanford, Ca., August 1973.

46. PIETRZYKOWSKI, T., "An Exact Potential Method for Constrained Maxima," *SIAM J. Numer. Anal.*, **6**, 299–304 (1969).

47. POLYAK, B. T., "The Convergence Rate of the Penalty Function Method," *USSR Comp. Math. & Math. Phys.*, **11**, 1, 1–12 (1971).

48. POWELL, M. J. D., "A Method for Nonlinear Constraints in Minimization Problems," in *Optimization*, R. Fletcher (Ed.), Academic Press, London, 1969.

49. ROCKAFELLAR, R. T., "The Multiplier Method of Hestenes and Powell Applied to Convex Programming," *J. Optimization Theory & Appl.*, **12**, 555–562 (1973).

50. ROCKAFELLAR, R. T., "A Dual Approach to Solving Nonlinear Programming Problems by Unconstrained Optimization," *Math. Prog.*, **5**, 354–373 (1973).

51. RUSH, B. C., J. BRACKEN, and G. P. MCCORMICK, "A Nonlinear Programming Model for Launch Vehicle Design and Costing," *Operations Research*, **15**, 185–210 (1967).

52. ZANGWILL, W. I., "Non-linear Programming via Penalty Functions," *Management Science*, **13**, 344–358 (1967).

13 SOLUTION OF CONSTRAINED PROBLEMS BY EXTENSIONS OF UNCONSTRAINED OPTIMIZATION TECHNIQUES

In contrast to the preceding chapter, where constraints were handled indirectly by incorporating them into the objective function, here methods that handle constraints in a direct way are discussed. The methods presented are extensions of techniques derived for unconstrained optimization problems in which extra steps, which take constraints into account, were introduced. These steps first find a descent direction that points toward a constrained minimum, and then a line search, restricted to the feasible set, is carried out. It will be shown that, for linearly constrained problems, these extra steps are not too difficult to implement and that some efficient unconstrained techniques, such as the variable metric methods, can also be quite efficient in the presence of linear constraints. Because of the success of such algorithms, it is generally agreed that even if penalty function methods are used, linear constraints should be handled directly and should not be transformed into various augmented objective functions. No such agreement on general nonlinear constraints exists, but it would seem that, with the present state of numerical techniques, penalty methods are preferable for this type of problem. In the direct approach, the process of following an active nonlinear constraint, such as an equality constraint, is usually a difficult and time-consuming operation that should be avoided. A possible alternative could be some successive linearizations of nonlinear constraints, which result, in general, in a method that consists of solving a sequence of linear programs. This approach is discussed in Chapter 14.

Section 13.1 presents simple constrained extensions of empirical methods for unconstrained optimization without using derivatives. The unconstrained versions of these methods were discussed in Chapter 9. Gradient projection or "deflected" gradient methods that are extensions of steepest descent and

variable metric algorithms are the subject of Section 13.2. An algorithm for quadratic programming that appears in Section 13.3 illustrates a special case of these methods. One of the first numerical techniques proposed for the solution of problems in which a general function is minimized subject to constraints, is a method of feasible directions. This method is described in Section 13.4 for linear constraints. Finally, the chapter concludes with the direct handling of nonlinear constraints by projection and feasible direction methods, as well as by several new approaches that reduce problems with nonlinear constraints to linearly constrained ones.

13.1 EXTENSIONS OF EMPIRICAL METHODS

Every empirical method given in Sections 9.1 to 9.3 has been extended to handle constrained problems. We begin a discussion of such extensions with a constrained version of the simplex method.

Suppose that the problem under consideration is given by finding a minimum $x^* \in R^n$ of a real function f, subject to inequality constraints (expressed in a notationally convenient form)

$$l_j \leq x_j \leq u_j, \qquad j = 1, \ldots, m \tag{13.1}$$

where $x = (x_1, \ldots, x_n)^T$ is the vector of independent variables and

$$x_{n+i} = g_i(x), \qquad i = 1, \ldots, m - n. \tag{13.2}$$

The upper and lower bounds u_j and l_j are assumed to be either given numbers or functions of x. Box [5] developed an extension of the simplex method to solve this problem. Appropriately, he called the constrained simplex method the **complex method**. Originally applied to the Spendley, Hext, and Himsworth [38] version, it is also applicable to the improved simplex technique of Nelder and Mead [23].

It is assumed that an initial feasible point x^0, satisfying all m constraints of (13.1), is given. Instead of generating n more points of the first simplex, Box suggests generating $N \geq n$ additional points, where a typical value of N is $2n$. These points are chosen one at a time so that their jth components satisfy

$$x_j^k = l_j + r_j^k(u_j - l_j), \qquad j = 1, \ldots, n, \quad k = 1, \ldots, N \tag{13.3}$$

where the r_j^k are pseudorandom numbers, rectangularly distributed over the open interval (0, 1). The points defined by (13.3) will satisfy the first n constraints in (13.1) but may violate the "implicit" constraints $n + 1, \ldots, m$.

In this case, the infeasible trial points are retracted halfway toward the centroid of the feasible points already selected, until, ultimately, all N points become feasible.

The function values are determined next, and relations (9.1) through (9.4) are evaluated, with N replacing n everywhere. If x^r, obtained from the reflection step, is feasible, the computations follow the steps of the unconstrained simplex method until an infeasible point is reached. If a point violates some of the first n constraints in (13.1), then the violated component is reset just inside the appropriate boundary so as to give a further trial point. If, however, some of the "implicit" constraints are violated, a further trial point is generated by retracting halfway toward the centroid around which the reflection was made, until the point becomes feasible.

Mitchell and Kaplan [21] suggest an alternative procedure to obtain the first complex without using random numbers. Their initial complex consists of $2n + 1$ points and is usually larger than the one constructed by Box's method; thus it increases the chance of finding a global minimum rather than a local one.

More than $n + 1$ points are used in a complex because, as a result of the retractions to the feasible side of the constraints, simplexes consisting of $n + 1$ points tend to collapse and cease to span the whole space. Consequently, true minima are easily missed. The general performance of the complex method is similar to the unconstrained version. It is a rather slow but reliable technique whose success in solving a given problem greatly depends on the starting point and whose efficiency rapidly declines with an increasing number of variables.

Glass and Cooper [13] have extended the pattern search method of Section 9.2 to constrained optimization. In their method the search commences at a feasible point and continues until one or more constraints are reached, so that no exploratory moves to continue the method are possible because of constraint violations. In this case, an alternate routine determines a new direction by solving a certain linear programming problem that is formulated by linear approximations of the objective and constraint functions. Since the method is considerably more complicated than either the complex method or the other direct techniques to be described next—while being much less efficient than several more sophisticated algorithms that will be discussed later in the chapter—its use is not widespread. The details of the technique and some numerical experience can be found in [13].

As noted in Section 9.3, Rosenbrock's rotating direction method [37] was derived for both constrained and unconstrained minimization problems. To illustrate the constrained version, suppose again that we have m constraints ($m \geq n$), explicit and implicit, as defined by (13.1) and (13.2). We define "boundary zones" associated with (13.1) as follows:

Any component x_j, $j = 1, \ldots, m$, is in a lower boundary zone if

$$l_j \leq x_j \leq l_j + \epsilon(u_j - l_j). \tag{13.4}$$

Similarly, it is in an upper boundary zone if

$$u_j \geq x_j \geq u_j - \epsilon(u_j - l_j), \tag{13.5}$$

where ϵ is a small constant. For a lower boundary zone we define the fractional depth of penetration as

$$\eta_j = \frac{l_j + \epsilon(u_j - l_j) - x_j}{\epsilon(u_j - l_j)} \tag{13.6}$$

and for an upper boundary zone the analogous expression is

$$\eta_j = \frac{x_j + \epsilon(u_j - l_j) - u_j}{\epsilon(u_j - l_j)}. \tag{13.7}$$

Denote by $\hat{f}$ the current lowest value of the objective function f computed by the method at feasible points—that is, at points satisfying (13.1)—and by f^+ the current lowest value of f at feasible points that do not lie in boundary zones. Clearly, $f^+ \geq \hat{f}$. Suppose that a starting point x_B^0, which does not lie in any boundary zone, is given. Initially, set $f^+ = \hat{f} = f(x_B^0)$. Carry out the search as in the unconstrained version, except that for each function evaluation, some additional computations must be performed. Let $\tilde{x}$ denote the next point at which f is to be evaluated according to the unconstrained method. Then,

1. If $\tilde{x}$ is infeasible, call the move to $\tilde{x}$ a failure and continue in the same manner as in the unconstrained version.

2. If $\tilde{x}$ is feasible and does not lie in any boundary zone, follow the unconstrained version.

3. If $\tilde{x}$ lies in some boundary zones, modify the function value $f(\tilde{x})$ as follows: Set j equal to the first component of $\tilde{x}$, that lies in a boundary zone, say j_1, and reset $f(\tilde{x})$ to $\bar{f}(\tilde{x})$ by the formula

$$\bar{f}(\tilde{x}) = f(\tilde{x}) - [f(\tilde{x}) - f^+][3\eta_{j_1} - 4(\eta_{j_1})^2 + 2(\eta_{j_1})^3]. \tag{13.8}$$

Increase j to the second component of $\tilde{x}$ that lies in a boundary zone and recompute $\bar{f}$ by using $\bar{f}(\tilde{x})$ from the previous j on the right-hand side of (13.8). Repeat for all such j.

The procedure suggested for case (3) is based on the assumption that if the unconstrained algorithm leads the search into a boundary zone, then the unconstrained minimum of the function is probably attained at an infeasible point. In order to confine the search to the boundary zone, an artificial minimum is constructed at a feasible point by the function modifications indicated above. Assume that the objective function is decreasing along a line that enters the boundary zone from the feasible side. From (13.8) we can see that $\bar{f}(\bar{x}) = f(\bar{x})$ for $\eta = 0$, and for $\eta = 1$—that is, at the boundary of the feasible set—we have $\bar{f}(\bar{x}) = f^+ \geq f(\bar{x})$. Thus the function value has been artificially raised to create the desired effect. Notice that this aspect of Rosenbrock's method has a strong resemblance to the basic idea of interior penalty function methods discussed in the preceding chapter.

The Davies, Swann, and Campey modification of the rotating directions method has also been extended to constrained problems. However, similar and more efficient techniques will be presented in later sections so we shall not discuss the method here. Interested readers are referred to Davies [6] and Davies and Swann [7] for details.

In summary, the extensions of empirical methods for unconstrained minimization to constrained problems are quite reliable but rather slow techniques and can be recommended only when derivatives are not available. Very few new ideas on constrained empirical techniques appeared in the literature in recent years. Much work remains to be done if such methods are to compete with penalty function methods or some of the techniques that are presented next.

13.2 GRADIENT PROJECTION ALGORITHMS FOR LINEAR CONSTRAINTS

The method of steepest descent is one of the basic but, as we have seen, not the most efficient numerical method for unconstrained minimization. Its extension to constrained problems by Rosen [33, 34] can be regarded as one of the fundamental works in developing the class of **gradient projection algorithms**. It should be clear to the reader from Chapters 10 and 11 that conjugate gradient or variable metric algorithms are much more efficient for unconstrained minimization than the method of steepest descent. Therefore, instead of the slowly converging steepest descent technique, we shall describe the more recent work of Goldfarb [14] (see also Goldfarb and Lapidus [16]), which is similar to Rosen's but which has the advantage of being an extension of conjugate gradient and variable metric methods to linear constraints.

Consider the linearly constrained nonlinear programming problem

$$\text{(LCP)} \qquad \min f(x) \tag{13.9}$$

subject to

$$\sum_{j=1}^{n} a_{ij}x_j - b_i = 0, \qquad i = 1, \ldots, m \tag{13.10}$$

$$\sum_{j=1}^{n} a_{ij}x_j - b_i \geq 0, \qquad i = m+1, \ldots, p \tag{13.11}$$

where f is assumed to be a continuously differentiable real function defined on a subset S of R^n containing X, the set of x satisfying (13.10) and (13.11). The numbers a_{ij} and b_i are given, and it is assumed that

$$\sum_{j=1}^{n} (a_{ij})^2 = 1, \qquad i = 1, \ldots, p. \tag{13.12}$$

Let

$$a^i = (a_{i1}, \ldots, a_{in})^T, \qquad i = 1, \ldots, p. \tag{13.13}$$

It follows then from (13.12) that

$$(a^i)^T a^i = 1, \qquad i = 1, \ldots, p \tag{13.14}$$

that is, the vectors a^i have unit length.

Denote by L_i the hyperplane defined by the ith constraint—that is,

$$L_i = \{x : x \in R^n, (a^i)^T x = b_i\}, \qquad i = 1, \ldots, p \tag{13.15}$$

and by L_i^c the closed half space defined by the inequality constraints

$$L_i^c = \{x : x \in R^n, (a^i)^T x \geq b_i\}, \qquad i = m+1, \ldots, p. \tag{13.16}$$

Then the feasible set X is given by

$$X = \left(\bigcap_{i=1}^{m} L_i\right) \cap \left(\bigcap_{i=m+1}^{p} L_i^c\right). \tag{13.17}$$

Note that the vectors a^i are orthogonal to the hyperplanes L_i and that for $i = m+1, \ldots, p$ they point into the half spaces L_i^c—that is, into the feasible set.

A set of hyperplanes is said to be linearly independent if the corresponding vectors a^i, also called the normal vectors, are linearly independent. The intersection of q linearly independent hyperplanes is an affine set of R^n and will be denoted by F^q. The $(n - q)$-dimensional subspace (affine set containing the origin) that is parallel to F^q is denoted by M^q. If F^q is itself a subspace, then, clearly, $F^q = M^q$. Let Q be a positive definite $n \times n$ matrix. The q-

dimensional subspace spanned by the vectors $Q^{-1}a^i$, where the a^i correspond to F^q, is denoted by $\hat{M}^q$. Since R^n can be expressed as the direct sum of M^q and $\hat{M}^q$, every vector $x \in R^n$ can be written as $x = u + v$, where $u \in M^q$ and $v \in \hat{M}^q$, see [24].

Let us find projection mappings of a vector $x \in R^n$ into M^q and $\hat{M}^q$. Denote by A_q the $n \times q$ matrix whose columns are q linearly independent vectors a^i belonging to I_q, a subset of $I_p = \{1, 2, \ldots, p\}$. Without loss of generality, assume that

$$I_q = \{1, 2, \ldots, q\}. \tag{13.18}$$

The matrix A_q has rank q. Hence $A_q^T Q^{-1} A_q$ has an inverse. Letting $\alpha = (\alpha_1, \ldots, \alpha_q)^T$, we can write for a vector $v \in \hat{M}^q$

$$v = Q^{-1} A_q \alpha \tag{13.19}$$

and for any $x \in R^n$

$$A_q^T x = A_q^T u + A_q^T v = A_q^T Q^{-1} A_q \alpha \tag{13.20}$$

since $A_q^T u = 0$. Hence

$$\alpha = (A_q^T Q^{-1} A_q)^{-1} A_q^T x \tag{13.21}$$

and

$$v = Q^{-1} A_q (A_q^T Q^{-1} A_q)^{-1} A_q^T x. \tag{13.22}$$

Define the $n \times n$ matrix $\hat{P}_q$ by

$$\hat{P}_q = Q^{-1} A_q (A_q^T Q^{-1} A_q)^{-1} A_q^T. \tag{13.23}$$

Then

$$\hat{P}_q v = Q^{-1} A_q (A_q^T Q^{-1} A_q)^{-1} A_q^T Q^{-1} A_q \alpha \tag{13.24}$$

$$= Q^{-1} A_q \alpha = v \tag{13.25}$$

for any $v \in \hat{M}^q$ and

$$\hat{P}_q u = Q^{-1} A_q (A_q^T Q^{-1} A_q)^{-1} A_q^T u = 0 \tag{13.26}$$

for any $u \in M^q$. Thus $\hat{P}_q$ is a **projection matrix** [24] that maps R^n onto $\hat{M}^q$. Similarly, the complementary projection matrix P_q, defined by

$$P_q = I - \hat{P}_q, \tag{13.27}$$

projects R^n onto the subspace M^q. In the special case where $Q = I$, the subspaces M^q and $\hat{M}^q$ are complementary orthogonal subspaces and the P_q, $\hat{P}_q$

are then symmetric orthogonal projection matrices. So

$$\hat{P}_q = A_q(A_q^T A_q)^{-1} A_q^T = (A_q^+)^T A_q^T, \tag{13.28}$$

where A_q^+ is the generalized inverse of A_q.

Example 13.2.1

To illustrate the preceding concepts, let $Q = I$ and

$$F^1 = \{x : x \in R^2, -\tfrac{3}{5}x_1 - \tfrac{4}{5}x_2 + \tfrac{8}{5} = 0\}. \tag{13.29}$$

Then

$$M^1 = \{x : x \in R^2, -\tfrac{3}{5}x_1 - \tfrac{4}{5}x_2 = 0\} \tag{13.30}$$

and

$$\hat{M}^1 = \{x : x \in R^2, x_2 = \tfrac{4}{3}x_1\}. \tag{13.31}$$

We have $A_1^T = (-\frac{3}{5}, -\frac{4}{5})$ and

$$\hat{P}_1 = A_1(A_1^T A_1)^{-1} A_1^T = \begin{pmatrix} -\frac{3}{5} \\ -\frac{4}{5} \end{pmatrix} \left(-\frac{3}{5}, -\frac{4}{5}\right) \tag{13.32}$$

$$= \frac{1}{25}\begin{bmatrix} 9 & 12 \\ 12 & 16 \end{bmatrix}. \tag{13.33}$$

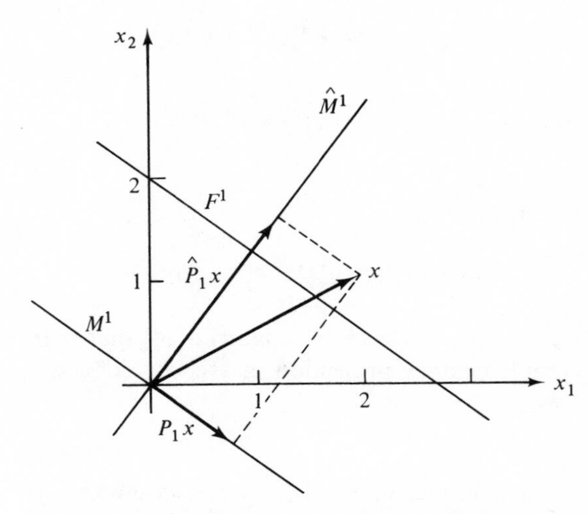

Fig. 13.1 Projection into orthogonal subspaces.

Consequently,

$$P_1 = I - \hat{P}_1 = \frac{1}{25}\begin{bmatrix} 16 & -12 \\ -12 & 9 \end{bmatrix}. \tag{13.34}$$

Now take $x = (2, 1)^T$. Then $\hat{P}_1 x = (\frac{6}{5}, \frac{8}{5})^T \in \hat{M}^1$ and $P_1 x = (\frac{4}{5}, -\frac{3}{5})^T \in M^1$, see Figure 13.1. ∎

During the course of the computations it is necessary to obtain projection matrices for various subspaces that are parallel to intersections of hyperplanes and that correspond to some current set of **active** inequality constraints [an inequality constraint $g(x) \geq 0$ is said to be active at $\hat{x}$ if $g(\hat{x}) = 0$]. Each new subspace differs from the previous one either by dropping an active constraint from the intersection or by adding a new one to it. For these computations, it is important to develop recursion relations so that $(A_q^T A_q)^{-1}$ can be easily computed from $(A_{q-1}^T A_{q-1})^{-1}$ and vice versa. Suppose that the $q \times q$ nonsingular matrix $(A_q^T A_q)^{-1}$ is partitioned as

$$(A_q^T A_q)^{-1} = \begin{bmatrix} B_1 & B_2 \\ B_3 & B_4 \end{bmatrix}, \tag{13.35}$$

where B_1, B_2, B_3, and B_4 are, respectively, $(q-1) \times (q-1)$, $(q-1) \times 1$, $1 \times (q-1)$, and 1×1 matrices. When a^q, the qth (last) vector, corresponding to hyperplane L_q, is dropped from the current matrix A_q, we can compute the $(q-1) \times (q-1)$ matrix $(A_{q-1}^T A_{q-1})^{-1}$ by

$$(A_{q-1}^T A_{q-1})^{-1} = B_1 - B_2 B_4^{-1} B_3. \tag{13.36}$$

If hyperplane L_h for $h < q$, rather than L_q, is to be dropped, the same formula can be used, provided that the hth and qth rows and columns of $(A_q^T A_q)^{-1}$ are interchanged before (13.36) is applied.

The opposite procedure of calculating $(A_q^T A_q)^{-1}$ from $(A_{q-1}^T A_{q-1})^{-1}$ by adding the vector a^q, corresponding to a hyperplane L_q, is as follows: Compute

$$r_{q-1} = (A_{q-1}^T A_{q-1})^{-1}(A_{q-1}^T a^q) \tag{13.37}$$

and

$$P_{q-1} a^q = a^q - A_{q-1} r_{q-1}. \tag{13.38}$$

Then,

$$A_0 = (P_{q-1} a^q)^T (P_{q-1} a^q) \tag{13.39}$$

$$B_1 = (A_{q-1}^T A_{q-1})^{-1} + A_0^{-1} r_{q-1} r_{q-1}^T \tag{13.40}$$

$$B_2 = B_3^T = -A_0^{-1} r_{q-1} \tag{13.41}$$

$$B_4 = A_0^{-1} > 0, \tag{13.42}$$

where the positivity of B_4 follows from the fact that $P_{q-1}a^q \neq 0$. The matrix $(A_q^T A_q)^{-1}$ is then given by (13.35). Another useful relation is a recursion formula for projection matrices

$$P_q = P_{q-1} - \frac{P_{q-1}a^q(a^q)^T P_{q-1}}{(a^q)^T P_{q-1} a^q}. \tag{13.43}$$

In addition, it can be shown that, for any vector $x \in R^n$,

$$\| P_q x \| \leq \| P_{q-1} x \| \leq \| x \|. \tag{13.44}$$

That is, projections do not increase the length of a vector.

Bartels, Golub, and Saunders [1] suggest an alternative procedure for updating the matrices $(A_q^T A_q)^{-1}$ that uses the Householder transformation [17]. The result is a numerically more stable process, sometimes at the expense of lengthier calculations.

Example 13.2.2

Suppose that we have two equality constraints in R^3

$$\tfrac{2}{3}x_1 - \tfrac{2}{3}x_2 - \tfrac{1}{3}x_3 = -\tfrac{1}{3} \tag{13.45}$$

$$\tfrac{2}{3}x_1 + \tfrac{1}{3}x_2 + \tfrac{2}{3}x_3 = \tfrac{2}{3} \tag{13.46}$$

and we wish to compute the projection matrices P_1 and P_2. From (13.43) we obtain

$$P_1 = P_0 - \frac{P_0 a^1 (a^1)^T P_0}{(a^1)^T P_0 a^1}. \tag{13.47}$$

Now let $P_0 = I$ and $a^1 = (\tfrac{2}{3}, -\tfrac{2}{3}, -\tfrac{1}{3})^T$. Hence

$$P_1 = \begin{bmatrix} 1 & 0 & 0 \\ 0 & 1 & 0 \\ 0 & 0 & 1 \end{bmatrix} - \frac{1}{9}\begin{bmatrix} 4 & -4 & -2 \\ -4 & 4 & 2 \\ -2 & 2 & 1 \end{bmatrix} = \frac{1}{9}\begin{bmatrix} 5 & 4 & 2 \\ 4 & 5 & -2 \\ 2 & -2 & 8 \end{bmatrix}. \tag{13.48}$$

Similarly, $a^2 = (\tfrac{2}{3}, \tfrac{1}{3}, \tfrac{2}{3})^T$ and

$$P_2 = P_1 - \frac{P_1 a^2 (a^2)^T P_1}{(a^2)^T P_1 a^2}. \tag{13.49}$$

Substituting the numerical values into (13.49), we obtain

$$P_2 = \frac{1}{9}\begin{bmatrix} 5 & 4 & 2 \\ 4 & 5 & -2 \\ 2 & -2 & 8 \end{bmatrix} - \frac{1}{9}\begin{bmatrix} 4 & 2 & 4 \\ 2 & 1 & 2 \\ 4 & 2 & 4 \end{bmatrix} = \frac{1}{9}\begin{bmatrix} 1 & 2 & -2 \\ 2 & 4 & -4 \\ -2 & -4 & 4 \end{bmatrix}. \tag{13.50}$$

In order to check these projection matrices, take an arbitrary vector in R^3, say $x = (1, 1, 1)^T$. Then

$$P_1 x = \begin{pmatrix} \frac{11}{9} \\ \frac{7}{9} \\ \frac{8}{9} \end{pmatrix} \quad \text{and} \quad P_2 x = \begin{pmatrix} \frac{1}{9} \\ \frac{2}{9} \\ -\frac{2}{9} \end{pmatrix}. \tag{13.51}$$

The reader can easily verify that $P_1 x \in M^1$ and $P_2 x \in M^2$, where

$$M^1 = \{x : x \in R^3, \tfrac{2}{3}x_1 - \tfrac{2}{3}x_2 - \tfrac{1}{3}x_3 = 0\} \tag{13.52}$$

and

$$M^2 = M^1 \cap \{x : x \in R^3, \tfrac{2}{3}x_1 + \tfrac{1}{3}x_2 + \tfrac{2}{3}x_3 = 0\}. \tag{13.53}$$

Also, $||x|| = \sqrt{3}$, $||P_1 x|| = \sqrt{\frac{26}{9}}$, and $||P_2 x|| = \frac{1}{3}$. Thus (13.44) holds. ∎

It can also be shown that linearly dependent hyperplanes can easily be detected by the recursion formula (13.43). If, for example, it is found that $P_{l-1}(a^l) = 0$ for some l, then the constraint corresponding to a^l can be dropped from problem (LCP) without the solution being affected since it is linearly dependent on some other constraint.

Let us describe now the Goldfarb conjugate-gradient projection algorithm. The method will be presented without proofs. Interested readers should consult [14] and [16] for further details and proofs.

Suppose that an initial feasible point $x^0 \in X$ is given. Clearly, if equality constraints are present, then $x^0 \in L_i$, $i = 1, \ldots, m$. Therefore $x^0 \in F^q$ for some $q \geq m$, and, without loss of generality, assume that $A_q = (a^1, \ldots, a^m, \ldots, a^q)$. Also suppose that H_0^0, an $n \times n$ symmetric positive definite matrix, is given. Compute H_q^0 by the recursive relation

$$H_i^0 = H_{i-1}^0 - \frac{H_{i-1}^0 a^i (a^i)^T H_{i-1}^0}{(a^i)^T H_{i-1}^0 a^i}, \qquad i = 1, \ldots, q \tag{13.54}$$

where the a^i, $i = 1, \ldots, m$, corresponding to the equality constraints (13.10), are added first. At iteration $k = 0, 1, \ldots,$ the following steps are carried out:

1. Given a feasible x^k, $\nabla f(x^k)$, and H_q^k, compute $H_q^k \nabla f(x^k)$. For $q \geq 1$, compute

$$\lambda^k = (A_q^T A_q)^{-1} A_q^T \nabla f(x^k). \tag{13.55}$$

If x^k does not lie in hyperplane L_i, corresponding to inequality constraint i, set $\lambda_i^k = 0$. If

$$H_q^k \nabla f(x^k) = 0 \tag{13.56}$$

and

$$\lambda_i^k \geq 0, \qquad i = m + 1, \ldots, p \tag{13.57}$$

then x^k satisfies first-order necessary conditions for optimality in (LCP) and the algorithm terminates.

2. If (13.56) and (13.57) are not satisfied simultaneously, then either

$$||H_q^k \nabla f(x^k)|| > \max\{0, -\tfrac{1}{2}\lambda_q^k b_{qq}^{-1/2}\} \tag{13.58}$$

or

$$||H_q^k \nabla f(x^k)|| \leq -\tfrac{1}{2}\lambda_q^k b_{qq}^{-1/2}, \qquad \lambda_q^k < 0 \tag{13.59}$$

where b_{ii} is the ith diagonal element of $(A_q^T A_q)^{-1}$ and, for convenience, it is assumed that

$$\lambda_q^k b_{qq}^{-1/2} \leq \lambda_q^k b_{ii}^{-1/2}, \qquad i = m+1, \ldots, q. \tag{13.60}$$

Note that each $b_{ii} > 0$. If (13.58) holds, proceed to step 3. Otherwise we drop constraint q from the active set of constraints by computing $(A_{q-1}^T A_{q-1})^{-1}$ from $(A_q^T A_q)^{-1}$, as outlined earlier, and obtaining

$$P_{q-1} = I - A_{q-1}(A_{q-1}^T A_{q-1})^{-1} A_{q-1}^T. \tag{13.61}$$

Next we compute

$$H_{q-1}^k = H_q^k - \frac{P_{q-1} a^q (a^q)^T P_{q-1}}{(a^q)^T P_{q-1} a^q}. \tag{13.62}$$

Let $q = q - 1$ and return to step 1.

3. Set $z^{k+1} = -H_q^k \nabla f(x^k)$. (Note that in Chapter 11 we defined z^{k+1} by multiplying $\nabla f(x^k)$ by a transpose of a matrix H. In order to simplify notation, we drop the transpose sign in the description of this algorithm.) Compute β_i^{k+1}, the maximum step length along z^{k+1} without violating the ith, currently inactive, constraint.

$$\beta_i^{k+1} = -\frac{(a^i)^T x^k - b^i}{(a^i)^T z^{k+1}}, \qquad i = q+1, \ldots p. \tag{13.63}$$

Hence the maximum step length that can be taken along z^{k+1} in the feasible region is given by

$$\beta_{\max}^{k+1} = \min_i \{\beta_i^{k+1} : \beta_i^{k+1} > 0, i = q+1, \ldots, p\}. \tag{13.64}$$

Carry out a line search to find α_{k+1}^*, $0 \leq \alpha_{k+1}^* \leq \beta_{\max}^{k+1}$, which minimizes $f(x^k + \alpha_{k+1} z^{k+1})$. It is possible that $\alpha_{k+1}^* \longrightarrow +\infty$, which indicates that f is unbounded in the direction of z^{k+1}. If $\alpha_{k+1}^* < +\infty$, set

$$x^{k+1} = x^k + \alpha_{k+1}^* z^{k+1} \tag{13.65}$$

and compute $\nabla f(x^{k+1})$.

4. If $\alpha^*_{k+1} < \beta^{k+1}_{\max}$, proceed to step 5. Otherwise add the constraint corresponding to $\beta^{k+1}_{\max}$ [say, the $(q+1)$th constraint] to the active set and compute

$$H^{k+1}_{q+1} = H^k_q - \frac{H^k_q a^{q+1}(a^{q+1})^T H^k_q}{(a^{q+1})^T H^k_q a^{q+1}}. \tag{13.66}$$

Set $q = q + 1$ and $k = k + 1$. Return to step 1.

5. In this case, the active set of constraints remains unchanged. Update H^k_q by one of the rank-one or rank-two variable metric formulas of Chapter 11, say the BFS formula:

$$H^{k+1}_q = H^k_q + \left[1 + \frac{(\gamma^{k+1})^T H^k_q \gamma^{k+1}}{(p^{k+1})^T \gamma^{k+1}}\right]\frac{p^{k+1}(p^{k+1})^T}{(p^{k+1})^T \gamma^{k+1}} - \frac{p^{k+1}(\gamma^{k+1})^T H^k_q}{(p^{k+1})^T \gamma^{k+1}} - \frac{H^k_q \gamma^{k+1}(p^{k+1})^T}{(p^{k+1})^T \gamma^{k+1}}, \tag{13.67}$$

where, as before, $p^{k+1} = x^{k+1} - x^k$ and $\gamma^{k+1} = \nabla f(x^{k+1}) - \nabla f(x^k)$. Set $k = k + 1$ and return to step 1.

This completes the description of the algorithm.

Example 13.2.3

Consider the problem

$$\min f(x) = (x_1)^2 + 4(x_2)^2 \tag{13.68}$$

subject to

$$\tfrac{3}{5}x_1 + \tfrac{4}{5}x_2 \geq \tfrac{13}{5}. \tag{13.69}$$

We shall illustrate the Goldfarb algorithm by solving this quadratic program. Suppose that $x^0 = (0, 5)$ with $f(x^0) = 100$ and that we choose

$$H^0_0 = \begin{bmatrix} 1 & 0 \\ 0 & 1 \end{bmatrix}. \tag{13.70}$$

The gradient of the objective function is given by

$$\nabla f(x) = \begin{pmatrix} 2x \\ 8x \end{pmatrix} \tag{13.71}$$

and at x^0 we have

$$\nabla f(x^0) = \begin{pmatrix} 0 \\ 40 \end{pmatrix}. \tag{13.72}$$

The point x^0 satisfies (13.69) with a strict inequality. Hence $\lambda^0_1 = 0$ and

$\| H_0^0 \nabla f(x^0) \| = 40$. We proceed to step 3 and compute the maximum distance that we can move in the direction

$$z^1 = -H_0^0 \nabla f(x^k) = \begin{pmatrix} 0 \\ -40 \end{pmatrix} \tag{13.73}$$

without violating the constraint (13.69):

$$\beta_1^1 = -\frac{(3/5)(0) + (4/5)(5) - (13/5)}{(3/5)(0) + (4/5)(-40)} = \frac{7}{160}. \tag{13.74}$$

The reader can easily verify that in carrying out a line search in the interval $0 \leq \alpha_1 \leq \frac{7}{160}$ along z^1, we find $\alpha_1^* = \beta_{\max}^1 = \frac{7}{160}$ and

$$x^1 = \begin{pmatrix} 0 \\ 5 \end{pmatrix} + \frac{7}{160}\begin{pmatrix} 0 \\ -40 \end{pmatrix} = \begin{pmatrix} 0 \\ \frac{13}{4} \end{pmatrix}, \tag{13.75}$$

where $f(x^1) = \frac{168}{4}$. The constraint (13.69) now becomes active, and, by (13.66), we compute

$$H_1^1 = H_0^0 - \frac{H_0^0 a^1 (a^1)^T H_0^0}{(a^1)^T H_0^0 a^1} \tag{13.76}$$

$$= \frac{1}{25}\begin{bmatrix} 16 & -12 \\ -12 & 9 \end{bmatrix}. \tag{13.77}$$

We also set $q = 1$, $k = 1$ and return to step 1 for an optimality test. We have $(A_1^T A_1)^{-1} = 1$, $\nabla f(x^1) = (0, 26)^T$, and $\lambda_1^1 = \frac{104}{5}$. Now

$$H_1^1 \nabla f(x^1) = \begin{pmatrix} -\frac{312}{25} \\ \frac{234}{25} \end{pmatrix}. \tag{13.78}$$

Thus x^1 is not optimal. Since $\| H_1^1 \nabla f(x^1) \| = \frac{78}{5}$ and (13.58) is satisfied, the next search direction coincides with the negative gradient projected on the hyperplane that corresponds to the constraint; that is,

$$z^2 = -H_1^1 \nabla f(x^1) = \begin{pmatrix} \frac{312}{25} \\ -\frac{234}{25} \end{pmatrix}. \tag{13.79}$$

Since there are no other constraints, we have $\beta_{\max}^2 = +\infty$. By carrying out a line search, we find that $\alpha_2^* = \frac{25}{104}$. Hence

$$x^2 = \begin{pmatrix} 0 \\ \frac{13}{4} \end{pmatrix} + \frac{25}{104}\begin{pmatrix} \frac{312}{25} \\ -\frac{234}{25} \end{pmatrix} = \begin{pmatrix} 3 \\ 1 \end{pmatrix} \tag{13.80}$$

and

$$\nabla f(x^2) = \begin{pmatrix} 6 \\ 8 \end{pmatrix}. \tag{13.81}$$

The function value is $f(x^2) = 13$, and we can see that the method is a descent one. Now we update H_1^1 by (13.67). First we compute

$$p^2 = \begin{pmatrix} 3 \\ 1 \end{pmatrix} - \begin{pmatrix} 0 \\ \frac{13}{4} \end{pmatrix} = \begin{pmatrix} 3 \\ -\frac{9}{4} \end{pmatrix} \tag{13.82}$$

and

$$\gamma^2 = \begin{pmatrix} 6 \\ 8 \end{pmatrix} - \begin{pmatrix} 0 \\ 26 \end{pmatrix} = \begin{pmatrix} 6 \\ -18 \end{pmatrix}. \tag{13.83}$$

Substituting (13.77), (13.82), and (13.83) into (13.67) yields

$$H_1^2 = \frac{1}{104}\begin{bmatrix} 16 & -12 \\ -12 & 9 \end{bmatrix}. \tag{13.84}$$

Consequently,

$$H_1^2\nabla f(x^2) = \frac{1}{104}\begin{bmatrix} 16 & -12 \\ -12 & 9 \end{bmatrix}\begin{pmatrix} 6 \\ 8 \end{pmatrix} = \begin{pmatrix} 0 \\ 0 \end{pmatrix} \tag{13.85}$$

and

$$\lambda_1 = (\tfrac{3}{5})(6) + (\tfrac{4}{5})(8) = 10. \tag{13.86}$$

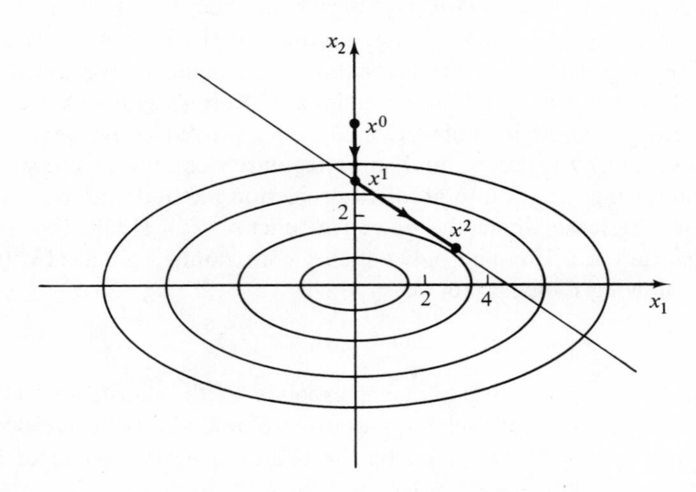

Fig. 13.2 Goldfarb's algorithm.

The point $x^2 = (3, 1)$ satisfies both (13.56) and (13.57), and so it is optimal. The trajectory of the optimization is shown in Fig 13.2. ∎

Goldfarb [14] shows that the preceding algorithm is generally a descent method, in other words, the value of the objective function is decreased at each iteration. Optimality conditions can also be formulated in terms of the H_q matrices. These conditions are given in

Theorem 13.1

Let f be a real continuously differentiable function and let $x^* \in X$ be a boundary point of X lying in F^q, the intersection of q linearly independent hyperplanes, defined by the constraints (13.10) *and* (13.11). *If x^* is a local optimum of problem* (LCP), *then*

$$H_q \nabla f(x^*) = 0 \tag{13.87}$$

and

$$\lambda_i^* \geq 0, \qquad i = m + 1, \ldots, q \tag{13.88}$$

where

$$\lambda^* = (\lambda_1^*, \ldots, \lambda_q^*)^T = (A_q^T A_q)^{-1} A_q^T \nabla f(x^*). \tag{13.89}$$

If f is also convex, then (13.87) *to* (13.89) *are sufficient for a constrained global minimum.*

Reading the proof of this theorem can also help the reader in understanding the steps of the algorithm given above. For the proof, as well as for many other details of gradient projection algorithms that cannot be covered here, see Rosen [33] and Goldfarb [14].

Numerical experience with Goldfarb's method has been good, but theoretical convergence to an optimal solution of (LCP) is not guaranteed for the foregoing version of the algorithm, even under convexity assumptions. Somewhat more complicated versions of Rosen's gradient projection method were proposed by Polak [25, 26], who proved convergence to an optimal solution if f is convex and some regularity conditions are satisfied.

An interesting insight into gradient projection methods can be obtained by analyzing the case where the objective function f in (LCP) is a strictly convex quadratic function and only equality constraints, such as (13.10), are present. Following Fletcher [10], let

$$f(x) = a + c^T x + \tfrac{1}{2} x^T Q x. \tag{13.90}$$

Let A_m be the $n \times m$ matrix that corresponds to the constraints (13.10). The optimal solution x^* of such a quadratic program and the corresponding Lagrange multipliers λ^* are given by the solution of the system of linear equations

$$\begin{bmatrix} Q & -A_m \\ -A_m^T & 0 \end{bmatrix} \begin{pmatrix} x \\ \lambda \end{pmatrix} + \begin{pmatrix} c \\ b \end{pmatrix} = 0. \tag{13.91}$$

Noting that

$$\begin{bmatrix} Q & -A_m \\ -A_m^T & 0 \end{bmatrix}^{-1}$$

$$= \begin{bmatrix} Q^{-1} - Q^{-1}A_m(A_m^TQ^{-1}A_m)^{-1}A_m^TQ^{-1} & -Q^{-1}A_m(A_m^TQ^{-1}A_m)^{-1} \\ -(A_m^TQ^{-1}A_m)^{-1}A_m^TQ^{-1} & -(A_m^TQ^{-1}A_m)^{-1} \end{bmatrix} \tag{13.92}$$

$$= \begin{bmatrix} H_m^* & -A_m^{*T} \\ -A_m^* & -(A_m^TQ^{-1}A_m)^{-1} \end{bmatrix}, \tag{13.93}$$

where H_m^* and A_m^* are just a shorthand, we can rewrite (13.91) as

$$x^* = x^0 - H_m^*\nabla f(x^0) \tag{13.94}$$

and

$$\lambda^* = A_m^*\nabla f(x^0), \tag{13.95}$$

where x^0 is any feasible point.

We can also write

$$H_m^* = [I - Q^{-1}A_m(A_m^TQ^{-1}A_m)^{-1}A_m^T]Q^{-1} \tag{13.96}$$

and, by (13.23) and (13.27), we have

$$H_m^* = P_mQ^{-1}. \tag{13.97}$$

Hence H_m^* can be regarded as the projected inverse Hessian of f. Letting $Q = I$, we get $H_m^* = P_m$, and thus Rosen's gradient projection method is recovered.

In the case of general nonlinear functions, (13.94) is not exact, and so an iterative process is required in order to find an optimum x^* by, say

$$x^{k+1} = x^k - \alpha_{k+1}^* H_k^* \nabla f(x^k), \tag{13.98}$$

where H_k^* is the approximate projected inverse Hessian—to be updated at each iteration as in Goldfarb's method.

A gradient projection method based on (13.96) and (13.98) was developed by Murtagh and Sargent [22]; it uses the search direction

$$z^{k+1} = -H_k^{*T}\nabla f(x^k) = -[H_k^T - H_k^TA_m(A_m^TH_k^TA_m)^{-1}A_m^TH_k]\nabla f(x^k), \tag{13.99}$$

where H_k^T is the current approximation of the inverse Hessian. Updating of this matrix is carried out by a rank-one correction matrix (see Chapter 11). Similar to Goldfarb's method, the Murtagh-Sargent algorithm also uses an

"active set strategy" to add or delete constraints, if necessary, as it moves from one feasible point to another during the iterations. Both the Goldfarb and the Murtagh-Sargent methods possess the quadratic termination property, appropriately modified to take constraints into account. For the Goldfarb method, we have the following result, which is due to Powell [27, 28].

Theorem 13.2

If a quadratic function f having an $n \times n$ positive definite Q matrix is minimized, subject to linear constraints (13.10) *and* (13.11), *by Goldfarb's method and if the active constraint set is changed l times during the calculations, then the exact minimum of f can be found by performing not more than $n + l$ exact line searches.*

Powell [29] extends this theorem by showing that the Murtagh-Sargent method can be considered a special case of a generalized Goldfarb algorithm and vice versa. Suppose that the H_k matrices used in (13.99) are updated by

$$H_k = H_k^{\text{BFS}} + \beta^k w^k (w^k)^T, \tag{13.100}$$

where H_k^{BFS} is obtained by the updating formula of the Broyden-Fletcher-Shanno quasi-Newton method, as given by (11.170), and w^k is defined by (11.169). The parameter β^k is related to the a_{12}^k that appears in (11.167) in a complementary way, since $\beta^k = 0$ for the BFS method while $a_{12}^k = 0$ for the DFP algorithm. Suppose next that H_k is given by (13.100) and let

$$H_q^k = H_k - H_k A_q (A_q^T H_k A_q)^{-1} A_q^T H_k. \tag{13.101}$$

The matrix H_q^k is then a suitable projection matrix for Goldfarb's method, since $H_q^k A_q = 0$. Also suppose that a quadratic function with a positive definite Q matrix is minimized by a Murtagh-Sargent-type method using a general updating formula (13.100) with an arbitrary β^0, $x^0 \in R^n$, and nonsingular H_0, where the parameters β^k and α_k (step length) are chosen arbitrarily on every iteration. Apply now Goldfarb's method, starting at the same $x^0 \in R^n$ and with H_q^0 given by (13.101), where the same value of α_k is used on every iteration and

$$H_q^k = H_q^{(\text{BFS})k} + \hat{\beta}^k w^k (w^k)^T, \tag{13.102}$$

where $\hat{\beta}^k$ is related to β^k by

$$\hat{\beta}^k = \frac{\beta^k}{1 + \beta^k (\gamma^k)^T H_k A_q (A_q^T H_k A_q)^{-1} A_q^T H_k \gamma^k} \tag{13.103}$$

and $H_q^{(\text{BFS})k}$ is obtained by the updating formula of the BFS method. It can be shown that the two methods generate an identical sequence of points.

Adding or deleting constraints has no effect on generating the same points. Consequently, the Murtagh-Sargent and related methods have the same quadratic termination properties as stated in Theorem 13.2.

13.3 A QUADRATIC PROGRAMMING ALGORITHM

A special case of projected gradient methods for solving linearly constrained problems is discussed next. Suppose that the objective function to be minimized is a strictly convex quadratic function and that the constraints are given by linear equations and inequalities. For simplicity, we shall write only inequality constraints, for they are the important ones in the active set strategy. The reader can easily modify the results if equality constraints are present as well. The presentation will follow the works of Fletcher [9] and Goldfarb [14]—the first of which is more general, since it can also be applied to nonconvex quadratic functions. For a detailed description of the more general method, as well as for comparisons with other existing methods of solving quadratic programs, the reader is referred to [9]. Quadratic programs were mentioned in Chapter 7, and in addition to being useful in applications, they serve as subproblems in a number of algorithms to be described here and in the next chapter.

Consider the problem

(SCQP) $$\min f(x) = a + c^T x + \tfrac{1}{2} x^T Q x \tag{13.104}$$

subject to

$$A^T x \geq b, \tag{13.105}$$

where $c \in R^n$, $x \in R^n$, Q is a symmetric positive definite real matrix, A is an $n \times m$ real matrix, and $b \in R^m$. We shall solve (SCQP) by generating a sequence of quadratic programs, constrained by equalities, that represent the active constraints. Members of this sequence differ only in the constraints that are added or dropped from the active set. Such an equality constrained problem is written as

(ECQP) $$\min f(x) = a + c^T x + \tfrac{1}{2} x^T Q x \tag{13.106}$$

subject to

$$A_q^T x = b^q, \tag{13.107}$$

where the $n \times q$ matrix A_q is, for simplicity, the matrix obtained by taking the first q columns of A and $b^q = (b_1, \ldots, b_q)^T$. Let us assume that A_q has

rank q. The solution of this problem and the corresponding Lagrange multipliers are explicitly given by (13.94) and (13.95). That is,

$$x^* = x^0 - H_q^* \nabla f(x^0) \tag{13.108}$$

$$\lambda^* = A_q^* \nabla f(x^0), \tag{13.109}$$

where x^0 is any point satisfying (13.107) and

$$H_q^* = [I - Q^{-1}A_q(A_q^T Q^{-1} A_q)^{-1} A_q^T] Q^{-1} \tag{13.110}$$

$$A_q^* = (A_q^T Q^{-1} A_q)^{-1} A_q^T Q^{-1}. \tag{13.111}$$

The efficiency of the algorithm depends on recurrence relations for the matrices H_q^*, A_q^*. Suppose first that we add constraint $q + 1$ to the active set. This step corresponds to adding a^{q+1} (column $q + 1$ of A) to A_q. Then

$$H_{q+1}^* = H_q^* - \frac{H_q^* a^{q+1}(a^{q+1})^T H_q^{*T}}{(a^{q+1})^T H_q^* a^{q+1}} \tag{13.112}$$

$$A_{q+1}^* = \begin{bmatrix} A_q^* \\ 0, \ldots, 0 \end{bmatrix} + \begin{pmatrix} -A_q^* a^{q+1} \\ 1 \end{pmatrix} \frac{(a^{q+1})^T H_q^{*T}}{(a^{q+1})^T H_q^{*T} a^{q+1}}. \tag{13.113}$$

Next, suppose that we wish to drop the constraint that corresponds to a^q from the active set. Let $(a^{q*})^T$ be the qth row of A_q^*. Then we have the relations

$$H_{q-1}^* = H_q^* + \frac{a^{q*}(a^{q*})^T}{(a^{q*})^T Q a^{q*}} \tag{13.114}$$

$$\begin{bmatrix} A_{q-1}^* \\ 0, \ldots, 0 \end{bmatrix} = A_q^* - \frac{A_q^* Q a^{q*}(a^{q*})^T}{(a^{q*})^T Q a^{q*}}. \tag{13.115}$$

If any constraint, say $p < q$, is to be dropped, these formulas still apply, provided that columns p and q of A_q^* are interchanged before (13.114) and (13.115) are used. Note that for n active constraints, $H_n^* = 0$ (the null matrix) and $A_n^* = A^{-1}$.

The preceding two types of matrix updating rules can be combined to so-called exchange formulas in which column q in A_q is replaced by another vector, say a^t, currently not in A_q. Denote by H_q^{**} and A_q^{**} the new matrices obtained from H_q^* and A_q^* as a result of such an exchange operation. Then

$$H_q^{**} = H_q^* + \frac{a^{q*}u^T}{y} - \frac{H_q^* a^t w^T}{y} \tag{13.116}$$

and

$$A_q^{**} = A_q^* - \frac{(A_q^* a^t - e^q)w^T}{y} - \frac{A_q^* Q a^{q*} u^T}{y}, \tag{13.117}$$

where

$$u = a^{q*}(a^t)^T H_q^* a^t - H_q^* a^t (a^t)^T a^{q*} \tag{13.118}$$

$$w = H_q^* a^t (a^{q*})^T Q a^{q*} + a^{q*}(a^t)^T a^{q*} \tag{13.119}$$

$$y = ((a^t)^T a^t)^2 + (a^{q*})^T Q a^{q*} (a^t)^T H_q^* a^t \tag{13.120}$$

$$e^q = (0, \ldots, 0, 1)^T \in R^q. \tag{13.121}$$

Such exchange formulas are very useful in cases in which the Q matrix is ill-conditioned or the H_q^* approach singularity and, although the optimum solution is well defined, the algorithm may run into numerical difficulties during the computations. Fletcher [9] showed that an efficient overall strategy of choosing among the updating formulas of adding, dropping, and exchanging constraints can be used, so that numerical difficulties are avoided and the algorithm can be also extended to problems that involve semidefinite or indefinite Q.

In order to initiate the algorithm, a feasible point is required. Such a point can be found either by a linear program, as will be shown in the next section, or by some other method, such as Fletcher's procedure [8], in which a point in the intersection of n active constraints (if one exists) is located and $A_n^* = A^{-1}$, $H_n^* = 0$ are also computed. The steps of the algorithm can now be stated:

1. Given a feasible x^0 lying in the intersection of q active constraints, compute H_q^*, A_q^*, $H_q^* \nabla f(x^0)$. Set $k = 0$.
2. Compute $z^{k+1} = -H_q^* \nabla f(x^k)$. If $z^{k+1} = 0$, go to step 4.
3. If $z^{k+1} \neq 0$, compute $x^{k+1} = x^k + \alpha_{k+1}^* z^{k+1}$ and $\nabla f(x^{k+1})$, where

$$\alpha_{k+1}^* = \min\{1, \alpha_{k+1}\} \tag{13.122}$$

and

$$\alpha_{k+1} = \min_i \left\{ -\frac{(a^i)^T x^k - b_i}{(a^i)^T z^{k+1}} : (a^i)^T z^{k+1} < 0, i = q + 1, \ldots, m \right\}. \tag{13.123}$$

Suppose, for simplicity, that the minimum is achieved for $i = q + 1$. If $\alpha_{k+1}^* < 1$, add constraint $q + 1$ to the active set and update H_q^*, A_q^* by (13.112) and (13.113). Set $k = k + 1$ and $q = q + 1$ and go to step 2. If $\alpha_{k+1}^* = 1$, set $k = k + 1$ and go to step 4.

4. Compute the vector of (approximate) Lagrange multipliers

$$\lambda^k = A_q^* \nabla f(x^k) \tag{13.124}$$

and its smallest element

$$\lambda_r^k = \min_i \{\lambda_i^k : i = 1, \dots, q\}. \tag{13.125}$$

Suppose, for simplicity, that $r = q$. If $\lambda_q^k \geq 0$, stop; an optimal solution has been reached. If $\lambda_q^k < 0$, drop constraint q from the active set and update H_q^*, A_q^* by (13.114) and (13.115). Set $q = q - 1$ and go to step 2.

Example 13.3.1

Let us apply the preceding algorithm to solve a quadratic programming problem, taken from Beale [2] and given by

$$\min f(x) = 9 - 8x_1 - 6x_2 - 4x_3 + 2(x_1)^2 + 2(x_2)^2 + (x_3)^2 + 2x_1x_2 + 2x_1x_3 \tag{13.126}$$

subject to

$$(a^1)^T x = x_1 \geq 0 \tag{13.127}$$

$$(a^2)^T x = x_2 \geq 0 \tag{13.128}$$

$$(a^3)^T x = x_3 \geq 0 \tag{13.129}$$

$$(a^4)^T x = -x_1 - x_2 - x_3 \geq -3. \tag{13.130}$$

Suppose that we start at $x^0 = (0, 0, 0)$. Thus the first three constraints are active. It follows that $H_3^* = 0$, $A_3^* = A_3^{-1} = I$, and $\nabla f(x^0) = (-8, -6, -4)^T$. We have

$$z^1 = -H_3^* \nabla f(x^0) = \begin{pmatrix} 0 \\ 0 \\ 0 \end{pmatrix} \tag{13.131}$$

and we go to step 4.

Computing the Lagrange multipliers, we observe that $\lambda^0 = (-8, -6, -4)^T$ and $\lambda_1^0 = \min \{\lambda_i^0\}$. Thus we drop the first constraint from the active set. Since $r = 1$, we permute rows 1 and 3 in A_3^* and obtain a new matrix

$$A_3^* = \begin{bmatrix} 0 & 0 & 1 \\ 0 & 1 & 0 \\ 1 & 0 & 0 \end{bmatrix}. \tag{13.132}$$

From (13.114) and (13.115) we compute

$$H_2^* = \begin{bmatrix} \frac{1}{4} & 0 & 0 \\ 0 & 0 & 0 \\ 0 & 0 & 0 \end{bmatrix} \qquad A_2^* = \begin{bmatrix} -\frac{1}{2} & 0 & 1 \\ -\frac{1}{2} & 1 & 0 \end{bmatrix} \tag{13.133}$$

and return to step 2. Next we compute

$$z^1 = -H_2^* \nabla f(x^0) = \begin{pmatrix} 2 \\ 0 \\ 0 \end{pmatrix} \tag{13.134}$$

and since $z^1 \neq 0$, we proceed to step 3. Since $(a^4)^T z^1 < 0$ and $(a^i)^T z^1 \geq 0$ for $i \neq 4$, we have

$$\alpha_1 = \frac{-(a^4)^T x^0 - b_4}{(a^4)^T z^1} = \frac{3}{2} > 1 \qquad \alpha_1^* = 1. \tag{13.135}$$

It follows that $x^1 = (2, 0, 0)$, $\nabla f(x^1) = (0, -2, 0)^T$, and we go to step 4. We compute

$$\lambda^1 = \begin{bmatrix} -\frac{1}{2} & 0 & 1 \\ -\frac{1}{2} & 1 & 0 \end{bmatrix} \begin{pmatrix} 0 \\ -2 \\ 0 \end{pmatrix} = \begin{pmatrix} 0 \\ -2 \end{pmatrix}. \tag{13.136}$$

Since $\lambda_2^1 < 0$, we drop the second constraint from the active set and, again, by (13.114) and (13.115),

$$H_1^* = \begin{bmatrix} \frac{1}{3} & -\frac{1}{6} & 0 \\ -\frac{1}{6} & \frac{1}{3} & 0 \\ 0 & 0 & 0 \end{bmatrix} \qquad A_1^* = [-\tfrac{2}{3} \quad \tfrac{1}{3} \quad 1]. \tag{13.137}$$

Returning to step 2, we compute

$$z^2 = -H_1^* \nabla f(x^1) = \begin{pmatrix} -\frac{1}{3} \\ \frac{2}{3} \\ 0 \end{pmatrix} \tag{13.138}$$

and since $z^2 \neq 0$, we continue to step 3. We find that $(a^i)^T z^2 < 0$ for $i = 1, 4$ and

$$\alpha_2 = \min\{6, 3\} > 1 \qquad \alpha_2^* = 1. \tag{13.139}$$

Hence $x^2 = (\frac{5}{3}, \frac{2}{3}, 0)$ and $\nabla f(x^2) = (0, 0, -\frac{2}{3})^T$.

Computing new Lagrange multipliers, we find

$$\lambda^2 = A_1^* \nabla f(x^0) = -\tfrac{2}{3} < 0 \tag{13.140}$$

and we drop the third constraint from the active set. Updating H_q^* and A_q^*, we obtain

$$H_0^* = \begin{bmatrix} 1 & -\frac{1}{2} & -1 \\ -\frac{1}{2} & \frac{1}{2} & \frac{1}{2} \\ -1 & \frac{1}{2} & \frac{3}{2} \end{bmatrix} \tag{13.141}$$

and A_0^* is, of course, empty. Returning to step 2, we compute

$$z^3 = -H_0^* \nabla f(x^2) = \begin{pmatrix} -\frac{2}{3} \\ \frac{1}{3} \\ 1 \end{pmatrix} \tag{13.142}$$

and in step 3 we get $\alpha_3^* = \frac{2}{5}$. Hence $x^3 = (\frac{7}{5}, \frac{4}{5}, \frac{2}{5})$ and $\nabla f(x^3) = (0, 0, -\frac{2}{5})^T$. Since α_3^* corresponds to the fourth constraint, we add it to the (empty) active set and obtain new matrices by (13.112) and (13.113) as follows:

$$H_1^* = \frac{1}{18}\begin{bmatrix} 9 & -3 & -3 \\ -3 & 5 & -1 \\ -3 & -1 & 2 \end{bmatrix} \qquad A_1^* = [\tfrac{1}{3} \quad -\tfrac{2}{9} \quad -\tfrac{5}{9}]. \tag{13.143}$$

The next search direction is given by

$$z^4 = -H_1^* \nabla f(x^3) = \frac{1}{45}\begin{pmatrix} -3 \\ -1 \\ 2 \end{pmatrix} \tag{13.144}$$

and from (13.122) and (13.123) we find that $\alpha_4^* = 1$. Consequently, $x^4 = (\frac{4}{3}, \frac{7}{9}, \frac{4}{9})$ and $\nabla f(x^4) = (0, 0, 0)^T$, that is, the optimal solution has been found. ▮

If the Q matrix of the quadratic function is not positive definite, special steps have been derived by Fletcher [9] in order to ensure the applicability of his algorithm to the general quadratic programming problem. Some refinements of this method can also be found in [30].

13.4 FEASIBLE DIRECTION METHODS

Let us review a quite general approach to solving a large class of mathematical programming problems. Suppose that we have a problem

$$\min f(x) \tag{13.145}$$

subject to

$$x \in X \subset R^n, \tag{13.146}$$

where X is a closed connected set and f is a continuously differentiable function that has bounded level sets. An algorithm for solving this problem is said to be a **feasible direction method** if it can be described in the following way.

For $k = 0, 1, \ldots$, given a point $x^k \in X$, find a direction vector z^{k+1} such that

$$(z^{k+1})^T \nabla f(x^k) < 0 \tag{13.147}$$

and such that there exists an $\bar{\alpha}_{k+1} > 0$ satisfying $x^k + \alpha_{k+1} z^{k+1} \in X$ for every $0 \leq \alpha_{k+1} \leq \bar{\alpha}_{k+1}$; that is, z^{k+1} is a **feasible descent**, also called "usable," direction. Determine an α^*_{k+1} by some criterion and such that

$$x^{k+1} = x^k + \alpha^*_{k+1} z^{k+1} \in X. \tag{13.148}$$

Continue in this way until some stopping condition—hopefully, one of convergence—is satisfied.

Methods of feasible directions have been derived by Zoutendijk [41] and can be found in a number of his papers; see, for example, [42, 43, 44]. The reader can easily see that most of the unconstrained algorithms described earlier, as well as the methods dealing with constrained problems given in this chapter, are special cases of feasible direction methods. Moreover, the simplex method of linear programming and some of the techniques to be discussed in Chapter 14 can also be viewed as feasible direction methods. It should be mentioned that because of this generality of feasible direction methods, it is questionable whether the ideas and results described in this section are indeed extensions of unconstrained optimization techniques and hence belong to this chapter. We restrict our discussion of feasible direction methods of this section to the case of the linearly constrained nonlinear program

$$\text{(LCP)} \qquad \min f(x) \tag{13.149}$$

subject to

$$\sum_{j=1}^{n} a_{ij} x_j - b_i = 0, \qquad i = 1, \ldots, m \tag{13.150}$$

$$\sum_{j=1}^{n} a_{ij} x_j - b_i \geq 0, \qquad i = m+1, \ldots, p, \tag{13.151}$$

defined in a preceding section. Two major decisions are required for each iteration of a feasible direction method:

1. selection of a feasible descent direction.
2. selection of a step length along the feasible direction.

Beginning with the first, the reader will be asked to show that, given a point $x^k \in X$—that is, satisfying (13.150) and (13.151)—a feasible descent direction z from x^k must be in $Z^1(x^k) \cap Z^2(x^k)$, which is defined in a more general sense in Chapter 3. Here

$$Z^1(x^k) = \left\{z : \sum_{j=1}^{n} a_{ij}z_j = 0, j = 1, \ldots, m; \sum_{j=1}^{n} a_{ij}z_j \geq 0, i \in I(x^k)\right\} \tag{13.152}$$

$$Z^2(x^k) = \{z : z^T\nabla f(x^k) < 0\}, \tag{13.153}$$

where

$$I(x^k) = \left\{i : \sum_{j=1}^{n} a_{ij}x_j^k = b_i, i = m + 1, \ldots, p\right\}. \tag{13.154}$$

Generally there are many feasible descent directions from a given point $x^k \in X$, and so the **steepest feasible descent direction** can be chosen from among them by solving the optimization problem

$$\text{(SFD)} \qquad \min z^T\nabla f(x^k) \tag{13.155}$$

subject to $z \in Z^1(x^k)$, and

$$||z|| \leq 1. \tag{13.156}$$

Since $Z^1(x^k)$ is a cone, it is necessary to bound $||z||$ in order to obtain a finite solution. Note that $||z||$, the Euclidean norm of z, is a nonlinear function, hence (13.156) is a nonlinear constraint. We know from the Farkas lemma and Theorem 3.2 that $Z^1(x^k) \cap Z^2(x^k) = \emptyset$—that is, there is no feasible descent direction from x^k—if and only if there exist vectors $\lambda \geq 0$ and $\mu \in R^m$ such that

$$\nabla f(x^k) = \sum_{i=1}^{m} \mu_i a^i + \sum_{i \in I(x^k)} \lambda_i a^i, \tag{13.157}$$

where $a^i = (a_{i1}, \ldots, a_{in})^T$. In other words, it is possible to find a feasible descent direction from x^k if and only if $\nabla f(x^k)$ is not in the cone spanned by the vectors a^i that correspond to the active constraints. Suppose that the latter condition holds. Let t^{k+1} be the point of the cone that is closest to $\nabla f(x^k)$—that is,

$$(t^{k+1})^T[t^{k+1} - \nabla f(x^k)] = 0. \tag{13.158}$$

We now prove that the steepest feasible descent vector that solves (SFD) is given by

$$z^{k+1} = \frac{t^{k+1} - \nabla f(x^k)}{||t^{k+1} - \nabla f(x^k)||}. \tag{13.159}$$

Theorem 13.3 [30]

Let z^{k+1} be given by (13.159). *Then it is optimal for* (SFD), *the problem of finding the steepest feasible descent direction.*

Proof. We can write

$$t^{k+1} = \sum_{i=1}^{m} \mu_i a^i + \sum_{i \in I(x^k)} \lambda_i a^i \tag{13.160}$$

for some μ and $\lambda \geq 0$. Let z be any feasible direction satisfying (13.156). Then

$$(t^{k+1})^T z = \sum_{i=1}^{m} \mu_i (a^i)^T z + \sum_{i \in I(x^k)} \lambda_i (a^i)^T z \geq 0. \tag{13.161}$$

From (13.158) and (13.159) it follows that

$$(z^{k+1})^T \nabla f(x^k) = \frac{(t^{k+1} - \nabla f(x^k))^T \nabla f(x^k)}{||t^{k+1} - \nabla f(x^k)||} = -||t^{k+1} - \nabla f(x^k)||. \tag{13.162}$$

Hence by (13.159) and (13.161),

$$z^T \nabla f(x^k) \geq -z^T(||t^{k+1} - \nabla f(x^k)||\, z^{k+1}). \tag{13.163}$$

Then by the Schwarz inequality [24] and (13.162), it follows that

$$z^T \nabla f(x^k) \geq -||z||\,||t^{k+1} - \nabla f(x^k)||\,||z^{k+1}|| \tag{13.164}$$

$$\geq -||t^{k+1} - \nabla f(x^k)|| \tag{13.165}$$

$$\geq (z^{k+1})^T \nabla f(x^k). \tag{13.166}$$

■

The idea of steepest feasible directions is illustrated in Figure 13.3. The drawing in part (a) shows a feasible set defined by two linear inequality constraints. The point x^k where both constraints are active is also shown, along with the gradient of f at this point. Part (b) shows $Z^1(x^k)$, the cone spanned by the vectors a^1 and a^2. The construction of z^{k+1} is also illustrated.

The geometric interpretation of z^{k+1} and Theorem 13.3 have an immediate application. The problem of finding the steepest-feasible-descent direc-

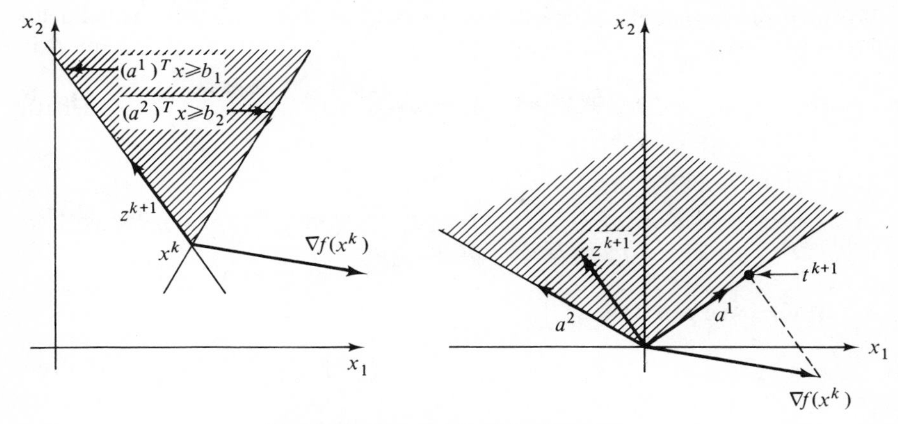

Fig. 13.3 Steepest feasible direction.

tion can be solved by a very special type of quadratic program in which the least distance from the cone spanned by the active constraints to the point $\nabla f(x^k)$ is sought. This program is given by

$$\min_{\lambda, \mu} \left\| \nabla f(x^k) - \sum_{i=1}^{m} \mu_i a^i - \sum_{i \in I(x^k)} \lambda_i a^i \right\|^2 \tag{13.167}$$

subject to

$$\lambda_i \geq 0, \qquad i \in I(x^k). \tag{13.168}$$

We note here that the solution of this quadratic program is not at all trivial, but efficient solution methods are available. See, for example, [2, 9, 11].

The need to solve nonlinear (quadratic) programs in the selection of a feasible descent direction can be avoided by replacing the Euclidean-norm constraint (13.156) by some other constraints that bound the components of the direction vector. For example, we could bound the L_∞ (Chebyshev) norm of the vector z by writing

$$\max_j |z_j| \leq 1 \tag{13.169}$$

or, equivalently,

$$-1 \leq z_j \leq 1, \qquad j = 1, \ldots, n. \tag{13.170}$$

Replacing (13.156) by (13.170) results in an another steepest-feasible direction-finding problem that is a linear program having a special structure. Additional bounding constraints of this type can be found in [44].

It is possible to extend the direction-selecting step outlined above to a method that resembles constrained variable metric algorithms if, instead of using $\nabla f(x^k)$ in the formulas, we premultiply it by a matrix H_k that

approximates the inverse Hessian of f. Alternatively, instead of Euclidean norms we could use non-Euclidean norms, such as $(z^T B z)^{1/2}$, where B is positive definite and can vary from one iteration to another.

Having found a feasible descent direction, we must select a step length α_{k+1}^*. Consequently, define

$$\alpha_{k+1}^* = \min\{\alpha'_{k+1}, \alpha''_{k+1}\} \tag{13.171}$$

where

$$\alpha'_{k+1} = \max\{\alpha_{k+1} : (z^{k+1})^T \nabla f(x^k + \alpha_{k+1} z^{k+1}) \leq 0\} \tag{13.172}$$

$$\alpha''_{k+1} = \max\{\alpha_{k+1} : (x^k + \alpha_{k+1} z^{k+1}) \in X\}. \tag{13.173}$$

For $\alpha'_{k+1} < +\infty$, we have $\nabla f(x^k + \alpha_{k+1} z^{k+1}) = 0$. Setting $x^{k+1} = x^k + \alpha_{k+1}^* z^{k+1}$, we return to the direction-selecting step, and it seems that, continuing in this way, we have a complete algorithm, provided that a starting feasible point is given. Unfortunately, such a method would not only have a very slow linear convergence rate—just like the steepest descent method in unconstrained optimization—but the constraints can also prevent convergence to an optimum because of so-called zigzagging [41] or jamming [40]. This phenomenon can be illustrated by an example, due to Wolfe [39].

Example 13.4.1

Suppose that we have

$$\min f(x) = \tfrac{4}{3}[(x_1)^2 - x_1 x_2 + (x_2)^2]^{3/4} - x_3 \tag{13.174}$$

subject to $x \geq 0$. The function f is convex; hence we have a convex programming problem. Let us use the preceding algorithm to solve this problem, starting at $x^0 = (0, \frac{1}{4}, \frac{1}{2})$. At this point, constraint $x_1 \geq 0$ is active and

$$-\nabla f(x^0) = ((\tfrac{1}{4})^{1/2}, -2(\tfrac{1}{4})^{1/2}, 1)^T. \tag{13.175}$$

Since $-\nabla f(x^0)$ is a feasible descent direction, we move along it and arrive at

$$x^1 = (\tfrac{1}{2}\tfrac{1}{4}, 0, \tfrac{1}{2} + \tfrac{1}{2}(\tfrac{1}{4})^{1/2}) \tag{13.176}$$

and here the $x_2 \geq 0$ constraint has become active. Continuing in this way, we obtain

$$x^2 = (0, \tfrac{1}{4}\tfrac{1}{4}, \tfrac{1}{2} + \tfrac{1}{2}(\tfrac{1}{4})^{1/2} + \tfrac{1}{2}(\tfrac{1}{2}\tfrac{1}{4})^{1/2}) \tag{13.177}$$

and again $x_1 \geq 0$ is active. The points generated by this method are zigzagging between the two active constraints $x_1 \geq 0$ and $x_2 \geq 0$ and finally con-

verging (jamming) to $\hat{x} = (0, 0, \hat{x}_3)$, where

$$\hat{x}_3 = \frac{1}{2} + \left(1 + \frac{\sqrt{2}}{2}\right)\left(\frac{1}{4}\right)^{1/2}. \tag{13.178}$$

The reader will be asked to show that $\hat{x}$ is not the optimum! ∎

Feasible direction algorithms, including antizigzagging procedures, have been suggested by McCormick [19], Polak [26], Zangwill [40], Zoutendijk [41], and others. We conclude this section by a method due to Zoutendijk in which zigzagging is prevented by not leaving an active constraint once it has become active unless forced to do so.

1. Suppose that $x^0 \in X$. Compute $\nabla f(x^0)$. Let $I_0 = K_0 = \emptyset$; set $k = 0$ and select a small number $\epsilon > 0$.
2. Use any feasible-descent-direction-selecting method to find a z^{k+1} having some bounded norm and satisfying

(i) $$(a^i)^T z^{k+1} = 0, \qquad i = 1, \ldots, m \tag{13.179}$$

$$(a^i)^T z^{k+1} \begin{cases} = 0, & i \in I_k \\ \geq 0, & i \in I(x^k), i \notin I_k. \end{cases} \tag{13.180}$$

(ii) $$(\nabla f(x^{h+1}) - \nabla f(x^h))^T z^{k+1} = 0, \qquad h \in K_k. \tag{13.181}$$

(iii) $$(z^{k+1})^T \nabla f(x^k) \leq -\epsilon. \tag{13.182}$$

3. If the direction-finding problem is inconsistent—that is, there is no z^{k+1} satisfying (i) to (iii)—then
   (i) If $I_k \cup K_k \neq \emptyset$, drop the oldest index from either I_k or K_k and go to step 2.
   (ii) If $I_k \cup K_k = \emptyset$, stop; an optimal solution has been obtained.
4. Determine an α^*_{k+1} as given by (13.171). If $\alpha^*_{k+1} = +\infty$, stop; the solution is unbounded.
5. Compute

$$x^{k+1} = x^k + \alpha^*_{k+1} z^{k+1} \tag{13.183}$$

and $\nabla f(x^{k+1})$.

6. (i) If $\alpha^*_{k+1} = \alpha'_{k+1}$, then let $I_{k+1} = I_k$ and $K_{k+1} = K_k \cup \{k\}$.
   (ii) If $\alpha^*_{k+1} = \alpha''_{k+1}$, then let

$$I_{k+1} = I_k \cup \{i : (a^i)^T x^{k+1} = 0 \text{ and } (a^i)^T x^k > 0, i = m+1, \ldots, p\} \tag{13.184}$$

   and let $K_{k+1} = \emptyset$.
   (iii) If $\alpha'_{k+1} = \alpha''_{k+1}$, proceed as in 6. (ii) but let $K_{k+1} = \{k\}$.
7. Set $k = k + 1$ and go to step 2.

Note that (13.181) has been added to obtain conjugate directions (see Chapter 10) until either an optimum solution is found—that is,

$$(z)^T \nabla f(x^k) > -\epsilon \tag{13.185}$$

for all feasible z—or the set of active constraints is changed. In practice, ϵ is gradually reduced to a very small positive number, thus ensuring almost-optimality. For quadratic objective functions, $\epsilon = 0$ at each iteration.

If no initial feasible point is available, we can use a "phase 1" procedure of linear programming in which we solve an auxiliary linear program

$$\min \sum_{i=1}^{p} \xi_i \tag{13.186}$$

subject to

$$\sum_{j=1}^{n} a_{ij} x_j + \xi_i = b_i, \qquad i = 1, \ldots, m \tag{13.187}$$

$$\sum_{j=1}^{n} a_{ij} x_j + \xi_i \geq b_i, \qquad i = m+1, \ldots, p \tag{13.188}$$

$$\xi \geq 0. \tag{13.189}$$

If the optimal solution of this linear program is (x^*, ξ^*) with $\xi^* \doteq 0$, then, by setting $x^0 = x^*$, the feasible direction algorithm with a nonlinear objective function can be initiated.

Best and Ritter [4] have also used condition (13.181) to attain nearly conjugate directions. Their direction-selecting step is based on a similar portion of the Frank-Wolfe [12] method, where a linear approximation of the objective function is minimized, subject to constraints that ensure feasibility of the search directions.

Finally, two other feasible direction methods for solving linearly constrained nonlinear programs that are extensions of unconstrained minimization techniques can be mentioned. The first is a method derived by Ritter [31], which is convergent for a large class of problems and has a superlinear rate of convergence for certain strictly convex functions. It is an extension of the same author's algorithm for unconstrained problems using first derivatives only, which was mentioned in Chapter 10. The second method, due to McCormick [20], uses first as well as second derivatives to obtain convergence to points satisfying second-order necessary conditions of optimality.

13.5 PROJECTION AND FEASIBLE DIRECTION METHODS FOR NONLINEAR CONSTRAINTS

Most of the gradient projection and feasible direction algorithms in Sections 13.3 and 13.4 have been extended to mathematical programs with nonlinear constraints. Solving such problems by these methods, however,

is much more difficult than for linearly constrained ones. If constraints are linear, the search directions along which the objective function is minimized lie either in the interior of the feasible set or in the intersection of hyperplanes corresponding to the active constraints. In both cases, there is no danger of leaving the feasible set by moving a small distance from the current point along a search direction. The difficulty arising from nonlinear constraints by extensions of projection or feasible direction methods is that the search directions are confined to the intersection of tangent hyperplanes of active nonlinear constraints. In this case, any movement along the search direction may lead to infeasible points, and extra steps are required to return to the feasible set. Repeated movements along a search direction and returns to the feasible set result in the "hemstitching" phenomenon shown in Figure 13.4.

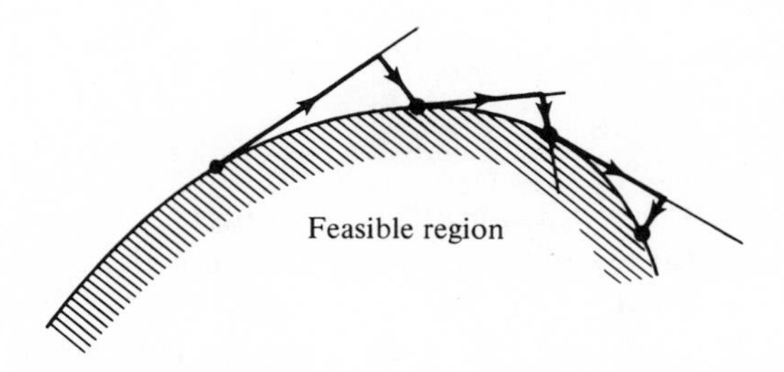

Fig. 13.4 Hemstitching by following an active nonlinear constraint.

Such a direct handling of active constraints causes these algorithms to slow down considerably. For this reason, it seems that indirect handling of nonlinear constraints by penalty function methods is more advantageous, unless the nonlinear constraints are of some special type or structure, such as in geometric programming. Methods given in the next chapter can also be recommended for solving certain nonlinearly constrained problems, although the convergence rate of the algorithms is generally slow. Nevertheless, we cannot conclude this chapter without briefly reviewing some extensions of unconstrained optimization techniques to nonlinearly constrained mathematical programming problems.

Consider the general nonlinear program

$$\min f(x) \tag{13.190}$$

subject to

$$g_i(x) \geq 0, \qquad i = 1, \ldots, m \tag{13.191}$$

$$h_j(x) = 0, \qquad j = 1, \ldots, p \tag{13.192}$$

where f, g_i, h_j are continuously differentiable functions and at least one of the constraints is nonlinear. A method for solving this problem is an extension of the Goldfarb algorithm, which is due to Davies [6].

An inequality constraint $g_i(x) \geq 0$, linearized at a point x^k, is given by

$$g_i(x^k) + (x - x^k)^T \nabla g_i(x^k) \geq 0 \tag{13.193}$$

and, similarly, a linearized equality constraint $h_j(x) = 0$ is

$$h_j(x^k) + (x - x^k)^T \nabla h_j(x^k) = 0. \tag{13.194}$$

If $g_i(x^k) = 0$, the linearized constraint (13.193) defines a closed half space that is generated by the hyperplane tangent to the surface $g_i(x) = 0$ at x^k. If $h_j(x^k) = 0$, then the points satisfying (13.194) lie in the hyperplane tangent to $h_j(x) = 0$ at x^k.

Throughout the calculations, all equality constraints must be kept "active." A nonlinear constraint is termed active at a point x^k if the corresponding linear constraint is active at x^k. If any of the g_i or h_j are linear, they are handled in the same way as in the Goldfarb algorithm. Given a feasible x^k, we determine the set of active constraints at x^k and compute z^{k+1}, the search direction, which is appropriately projected into the intersection of the hyperplanes that correspond to the active constraints. The next task is to determine a bracket $\bar{\alpha}_{k+1}$ on the step length along z^{k+1}. The distance to the nearest linearized constraint along z^{k+1} is evaluated. Let this distance be θ. Then we set

$$\bar{\alpha}_{k+1} = \min \{c, \theta\}, \tag{13.195}$$

where c is some positive constant, say $c = 2$. Such an $\bar{\alpha}_{k+1}$ will limit the step length and avoid moving far away from the feasible set. Now we let

$$x^{k,1} = x^k + \bar{\alpha}_{k+1} z^{k+1} \tag{13.196}$$

and evaluate all the constraints at $x^{k,1}$. If any constraint is violated, a return direction vector r^k is determined that points back into the feasible set. A move along this direction is taken and an interior point $x^{k,2}$ is found, followed by an interpolation scheme that locates a point $x^{k,3}$ on the boundary of the feasible set. An interesting interpolation method for this purpose was developed by Beamer and Wilde [3]. A minimum of f is then sought along the line segment joining x^k and $x^{k,3}$. If such a minimum, say at $x^{k,4}$, is found and if the feasible set is convex, then $x^{k,4}$ is in the interior of the feasible set, and we must move again to the boundary along, say, the steepest descent direction, arriving at a new point x^{k+1}. If no minimum is found between x^k and $x^{k,3}$, then a new search commences from $x^{k,3}$ along z^{k+1}. A typical situation is shown in Figure 13.5.

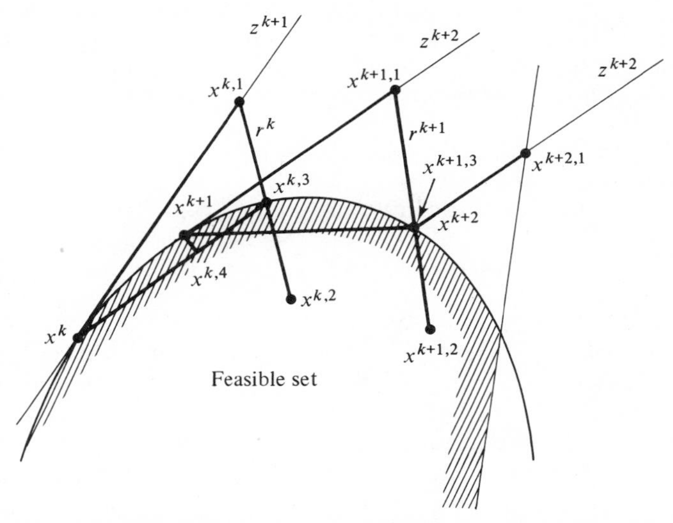

Fig. 13.5 Following active nonlinear constraints by Davies' method.

Many technical details are important for implementing Davies' method in a usable algorithm. For details, the reader is referred to Davies [6] and Rosen [34].

Let us turn now to a feasible direction algorithm for nonlinear constraints, derived by Zoutendijk [41, 42]. Consider the problem

$$\min f(x) \tag{13.197}$$

subject to

$$g_i(x) \geq 0, \qquad i = 1, \ldots, m \tag{13.198}$$

where the f and g_i are continuously differentiable. For any feasible x and $\epsilon \geq 0$, define

$$X_\epsilon(x) = \{i : g_i(x) \leq \epsilon\} \tag{13.199}$$

and let X be the feasible set. At iteration k, a feasible point x^k and $\epsilon^k > 0$ are given. We solve the direction-selecting linear program

$$\min \sigma \tag{13.200}$$

subject to

$$\sigma \geq z^T \nabla f(x^k) \tag{13.201}$$

$$z^T g_i(x^k) \geq \sigma, \qquad i \in X_{\epsilon^k}(x^k) \tag{13.202}$$

$$|z_j| \leq 1, \qquad j = 1, \ldots, n. \tag{13.203}$$

Let the optimal solution be $\sigma^{k+1} \in R$, $z^{k+1} \in R^n$. If $\sigma^{k+1} \leq -\epsilon^k$, set $\epsilon^{k+1} = \epsilon^k$; otherwise set $\epsilon^{k+1} = \frac{1}{2}\epsilon^k$. Compute α_{k+1}^* as given by (13.171) to (13.173), and let $x^{k+1} = x^k + \alpha_{k+1}^* z^{k+1}$. Zangwill [40] shows that this method converges to a solution of the problem given by (13.197) and (13.198), where a point x^* is called a solution if no feasible descent direction can be found from x^*. Of course, the computation of α_{k+1}^* can be quite complicated, and several iteration devices, such as those outlined in Davies' method, may be required.

Successful implementations of penalty function methods and algorithms for linearly constrained nonlinear programs have motivated two recent works in which nonlinearly constrained programs are solved by a sequence of linearly constrained programs, using a penalty function-Lagrangian approach.

Consider again the inequality constrained problem

$$\min f(x) \tag{13.204}$$

$$g_i(x) \geq 0, \qquad i = 1, \ldots, m \tag{13.205}$$

where f and g_i are continuously differentiable, convex and concave functions of $x \in R^n$, respectively. The method of Rosen and Kreuser [35, 36] can be described as follows: Start with an arbitrary $x^0 \in R^n$. At iteration k, we have a set of indices $I(x^k)$, defined by

$$I(x^k) = \{i : g_i(x^k) \leq 0, i = 1, \ldots, m\}. \tag{13.206}$$

We compute λ^k by

$$\lambda^k = J^+(x^k)\nabla f(x^k), \tag{13.207}$$

where

$$J^+(x^k) = [(J(x^k))^T J(x^k)]^{-1}(J(x^k))^T \tag{13.208}$$

and $J(x^k)$ is the Jacobian matrix of the g_i at x^k for $i \in I(x^k)$—that is, the matrix whose columns are $\nabla g_i(x^k)$, $i \in I(x^k)$. Then $J^+(x^k)$ is the generalized inverse of $J(x^k)$. The matrix $(J(x^k))^T J(x^k)$ is assumed to be nonsingular. Next we solve the linearly constrained problem

$$\min \left\{f(x) - \sum_{\lambda_i^k > 0} \lambda_i^k [g_i(x) - g_i(x^k) - (x - x^k)^T \nabla g_i(x^k)]\right\} \tag{13.209}$$

subject to

$$g_i(x^k) + (x - x^k)^T \nabla g_i(x^k) \geq 0, \qquad i = 1, \ldots, m \tag{13.210}$$

by, say, Goldfarb's method. Let the solution be x^{k+1}. If x^{k+1} satisfies the (first-order) Kuhn-Tucker conditions for problem (13.204) and (13.205) within some tolerance, stop; otherwise set $k = k + 1$ and repeat the preceding steps. This method can be shown to have a quadratic rate of convergence,

provided that certain regularity assumptions are satisfied. Note that linear constraints never appear in (13.209), but they are included in (13.210).

In a similar method, Robinson [32] suggests solving (13.204) and (13.205) with additional equality constraints

$$h_j(x) = 0, \qquad j = 1, \ldots, p \tag{13.211}$$

where the h_j, as well as f and the g_i, are twice continuously differentiable, as follows: At iteration k, we have a vector $x^k \in R^n$ and vectors $\lambda^k \in R^m$, $\lambda^k \geq 0$, $\mu^k \in R^p$. We solve the linearly constrained problem

$$\min_x \left\{ f(x) - \sum_{i=1}^{m} \lambda_i^k[g_i(x) - \tilde{g}_i(x^k, x)] - \sum_{j=1}^{p} \mu_j^k[h_j(x) - \tilde{h}_j(x^k, x)] \right\} \tag{13.212}$$

subject to

$$\tilde{g}_i(x^k, x) = g_i(x^k) + (x - x^k)^T \nabla g_i(x^k) \geq 0, \qquad i = 1, \ldots, m \tag{13.213}$$

$$\tilde{h}_j(x^k, x) = h_j(x^k) + (x - x^k)^T \nabla h_j(x^k) = 0, \qquad j = 1, \ldots, p. \tag{13.214}$$

Let the solution be x^{k+1}. Obtain the corresponding Lagrange multipliers $\lambda^{k+1} \geq 0$, μ^{k+1}, of this linearly constrained program. If there is more than one such triple $(x^{k+1}, \lambda^{k+1}, \mu^{k+1})$, choose from among them the one that is closest to (x^k, λ^k, μ^k) in some norm. If $(x^{k+1}, \lambda^{k+1}, \mu^{k+1})$ satisfies the (first-order) Kuhn-Tucker necessary conditions for optimality in the program defined by (13.204), (13.205), and (13.211) within some tolerance, stop. Otherwise set $k = k + 1$ and resolve (13.212) to (13.214). The point x^0, starting the algorithm, need not be feasible. Again, under suitable assumptions on the functions defining the program, the algorithm converges quadratically to an optimum.

EXERCISES

13.A. Suppose that the set of $x \in R^n$, defined by (13.1) and (13.2), is convex. Show that the moves of the complex method are all well defined. Give an example of a problem in R^2 where some moves may fail. (*Hint:* Try curved valleys.) Show that the unconstrained simplex method can be used to find x^0, the initial feasible point for the complex method. Will failure of the simplex method to find such a point prove that the feasible set is empty?

13.B. Carry out two exploratory stages of the constrained version of Rosenbrock's rotating directions method as applied to Example 9.3.1. and constrained by

$$-2x_1 + x_2 + 2 \geq 0. \tag{13.215}$$

Compare the best point x_B^2 obtained by the method with the true constrained optimum. Use the same parameters as in Example 9.3.1 and let $\epsilon = 10^{-4}$.

13.C. Continue Example 13.2.2 as follows: First drop constraint (13.45) and compute the new projection matrix P_1 from P_2 as given by (13.50). Then add the constraint

$$\tfrac{3}{5}x_2 - \tfrac{4}{5}x_3 = 2 \tag{13.216}$$

and compute the new P_2 matrix. Use (13.27), (13.28), and the recursion relations (13.35) to (13.43).

13.D. Solve the following problem by the Goldfarb method for general linearly constrained nonlinear programs:

$$\min f(x) = \tfrac{3}{2}(x_1)^2 + \tfrac{1}{2}(x_2)^2 - x_1x_2 - 2x_1 \tag{13.217}$$

subject to

$$x_1 + x_2 = 5 \tag{13.218}$$

$$6x_1 - x_2 \geq -12 \tag{13.219}$$

$$-x_1 + x_2 \geq 3. \tag{13.220}$$

Start the algorithm at $x^0 = (-1, 6)$. Note that the constraint coefficients are not normalized.

13.E. Verify relations (13.92), (13.94), and (13.95).

13.F. Consult the work of Murtagh and Sargent [22] and draw a flow diagram of their gradient projection algorithm. Compare the points generated by their method with the Goldfarb algorithm for the following problem:

$$\min (x_1)^2 + 4(x_2)^2 \tag{13.221}$$

subject to

$$x_1 + 2x_2 \geq 1 \tag{13.222}$$

$$x_1 - x_2 \geq 0 \tag{13.223}$$

$$x_1 \geq 0, \tag{13.224}$$

starting at $x^0 = (8, 8)$.

13.G. Resolve the quadratic programs given in Exercises 13.D and 13.F by the algorithm of Section 13.3 and compare the points visited with the corresponding points obtained in the exercises.

13.H. Prove that $z^{k+1} \in R^n$ is a feasible descent direction vector from a point x^k satisfying (3.110) and (3.111) if and only if it is an element of $Z^1(x^k) \cap Z^2(x^k)$, given by (13.152) and (13.153). Give an example of a linearly constrained nonlinear program in which lack of feasible descent direction vectors from a point x^k does not imply optimality of x^k and find a class of problems for which such optimality is ensured.

13.I. Noting that (SFD), the problem of finding the steepest-feasible-descent direction is a convex program, derive its dual programming problem by the results of Chapter 5. Compare with (13.167) and (13.168).

13.J. Show that the objective function in Example 13.4.1 is convex. Also show that $\hat{x}$ is not optimal. What is the optimal solution in this problem?

13.K. Levitin and Polyak [18] suggested a gradient projection-feasible direction algorithm for convex programming problems in which at iteration k, given a feasible x^k, the direction $x^k - \nabla f(x^k)$ is projected on X, the feasible set, by solving the quadratic program

$$\min_{y \in X} (y - (x^k - \nabla f(x^k))^T(y - (x^k - \nabla f(x^k)). \tag{13.225}$$

If y^k is the optimal solution to this program, then the search direction z^{k+1} is given by

$$z^{k+1} = y^k - x^k. \tag{13.226}$$

(a) Suppose that

$$X = \{x : Ax = b\}, \tag{13.227}$$

where A is an $m \times n$ matrix having rank m. Show that y^k can be explicitly written as a function of A, x^k, and $\nabla f(x^k)$. [*Hint:* Use (13.94).]

(b) Suppose now that

$$X = \{x : a_j \leq x_j \leq b_j, j = 1, \ldots, n\} \tag{13.228}$$

where the a_j and b_j are given bounds. Find explicit expressions for y^k. Hint: Consider three different cases:

(i)
$$x_j^k - \frac{\partial f(x^k)}{\partial x_j} \leq a_j \tag{13.229}$$

(ii)
$$a_j < x_j^k - \frac{\partial f(x^k)}{\partial x_j} < b_j \tag{13.230}$$

(iii)
$$b_j \leq x_j^k - \frac{\partial f(x^k)}{\partial x_j}. \tag{13.231}$$

13.L. Given the nonlinear program

$$\min (x_1 - 4)^2 + (x_2 - 4)^2 \tag{13.232}$$

subject to

$$3(x_1)^2 + (x_2)^2 - 2x_1x_2 - 4x_1 \leq 12 \tag{13.233}$$

$$3x_1 + 4x_2 \leq 28. \tag{13.234}$$

(a) Apply Davies' method to solve this problem and carry out a few steps of the algorithm, starting from $x^0 = (2, 0)$. Set up a rule for determining

the return directions r^k and compare with Davies' suggestion by consulting his work [6].

(b) Try Zoutendijk's method on the same problem, starting from the same x^0.

(c) Apply the Rosen-Kreuser and Robinson methods to solve this problem and carry out a few steps.

Draw the trajectory of points obtained by each method. Use the Goldfarb algorithm for the linearly constrained subproblems.

REFERENCES

1. BARTELS, R. H., G. H. GOLUB, and M. A. SAUNDERS, "Numerical Techniques in Mathematical Programming," in *Nonlinear Programming*, J. B. Rosen, O. L. Mangasarian, and K. Ritter (Eds.), Academic Press, New York, 1970.
2. BEALE, E. M. L., "Numerical Methods," in *Nonlinear Programming*, J. Abadie (Ed.), North-Holland Publishing Co., Amsterdam, 1967.
3. BEAMER, J. H., and D. J. WILDE, "A Minimax Search Plan for Constrained Optimization Problems," *J. Optimization Theory & Appl.*, **12**, 439–446 (1973).
4. BEST, M. J., and K. RITTER, "An Accelerated Conjugate Direction Method to Solve Linearly Constrained Minimization Problems," Research Report CORR 73-16, Department of Combinatorics and Optimization, University of Waterloo, Ontario, Canada, August 1973.
5. BOX, M. J., "A New Method of Constrained Optimization and a Comparison with Other Methods," *Computer J.*, **8**, 42–52 (1965).
6. DAVIES, D., "Some Practical Methods of Optimization," in *Integer and Nonlinear Programming*, J. Abadie (Ed.), North-Holland Publishing Co., Amsterdam, 1970.
7. DAVIES, D., and SWANN, W. H., "Review of Constrained Optimization," in *Optimization*, R. Fletcher (Ed.), Academic Press, London, 1969.
8. FLETCHER, R., "The Calculation of Feasible Points for Linearly Constrained Optimization Problems," Report R6354, A.E.R.E. Harwell, United Kingdom, 1970.
9. FLETCHER, R., "A General Quadratic Programming Algorithm," *J. Inst. Math. Appl.*, **7**, 76–91 (1971).
10. FLETCHER, R., "Minimizing General Functions Subject to Linear Constraints," in *Numerical Methods for Non-linear Optimization*, F. A. Lootsma (Ed.), Academic Press, London, 1972.
11. FLETCHER, R., and M. P. JACKSON, "Minimization of a Quadratic Function of Many Variables Subject Only to Lower and Upper Bounds," Report T.P. 528, A.E.R.E. Harwell, United Kingdom, May 1973.
12. FRANK, M., and P. WOLFE, "An Algorithm for Quadratic Programming," *Naval Res. Log. Quart.*, **3**, 95–110 (1956).

13. GLASS, H., and L. COOPER, "Sequential Search: A Method for Solving Constrained Optimization Problems," *J. Assoc. Comp. Mach.*, **12**, 71–82 (1965).

14. GOLDFARB, D., "Extension of Davidon's Variable Metric Method to Maximization under Linear Inequality and Equality Constraints," *SIAM J. Appl. Math.*, **17**, 739–764 (1969).

15. GOLDFARB, D., "Extensions of Newton's Method and Simplex Methods for Solving Quadratic Programs," in *Numerical Methods for Non-linear Optimization*, F. A. Lootsma (Ed.), Academic Press, London, 1972.

16. GOLDFARB, D., and L. LAPIDUS, "Conjugate Gradient Method for Nonlinear Programming Problems with Linear Constraints," *Ind. & Eng. Chem. Fund.*, **7**, 142–151 (1968).

17. HOUSEHOLDER, A. S., "Unitary Triangularization of a Nonsymmetric Matrix," *J. Assoc. Comp. Mach.*, **5**, 339–342 (1958).

18. LEVITIN, E. S., and B. T. POLYAK, "Constrained Minimization Methods," *USSR Comp. Maths. & Math. Phys.*, **6**, 5, 1–50 (1966).

19. MCCORMICK, G. P., "Anti-zigzagging by Bending," *Management Science*, **15**, 315–320 (1969).

20. MCCORMICK, G. P., "A Second Order Method for the Linearly Constrained Nonlinear Programming Problem," in *Nonlinear Programming*, J. B. Rosen, O. L. Mangasarian, and K. Ritter (Eds.), Academic Press, New York, 1970.

21. MITCHELL, R. A., and J. L. KAPLAN, "Nonlinear Constrained Optimization by a Nonrandom Complex Method," *J. Research of the NBS-C. Eng. and Instr.*, **72C**, 249–258 (1968).

22. MURTAGH, B. A., and R. W. H. SARGENT, "A Constrained Minimization Method with Quadratic Convergence," in *Optimization*, R. Fletcher (Ed.), Academic Press, London, 1969.

23. NELDER, J. A., and R. MEAD, "A Simplex Method for Function Minimization," *Computer J.*, **7**, 308–313 (1965).

24. NOBLE, B., *Applied Linear Algebra*, Prentice-Hall, Englewood Cliffs, N.J., 1969.

25. POLAK, E., "On the Convergence of Optimization Algorithms," *Rev. Inform. Rech. Oper.*, **16**, 17–34 (1969).

26. POLAK, E., *Computational Methods in Optimization*, Academic Press, New York, 1971.

27. POWELL, M. J. D., "Quadratic Termination Properties of Minimization Algorithms I. Statement and Discussion of Results," *J. Inst. Math. Appl.*, **10**, 333–342 (1972).

28. POWELL, M. J. D., "Quadratic Termination Properties of Minimization Algorithms II. Proofs of Theorems," *J. Inst. Math. Appl.*, **10**, 343–357 (1972).

29. POWELL, M. J. D., "Unconstrained Minimization and Extensions for Constraints," Report T.P. 495, A.E.R.E. Harwell, United Kingdom, August 1972.

30. Powell, M. J. D., "Introduction to Constrained Optimization," Report C.S.S.1, A.E.R.E. Harwell, United Kingdom, January 1974.

31. Ritter, K., "A Superlinearly Convergent Method for Minimization Problems with Linear Inequality Constraints," *Math. Prog.*, **4**, 44–71 (1973).

32. Robinson, S. M., "A Quadratically Convergent Algorithm for General Nonlinear Programming Problems," *Math. Prog.*, **3**, 145–156 (1972).

33. Rosen, J. B., "The Gradient Projection Method for Nonlinear Programming, Part I. Linear Constraints," *J. SIAM*, **8**, 181–217 (1960).

34. Rosen, J. B., "The Gradient Projection Method for Nonlinear Programming, Part II. Nonlinear Constraints," *J. SIAM*, **9**, 514–532 (1961).

35. Rosen, J. B., and J. L. Kreuser, "A Gradient Projection Algorithm for Nonlinear Constraints," in *Numerical Methods for Non-linear Optimization*, F. A. Lootsma (Ed.), Academic Press, London, 1972.

36. Rosen, J. B., and J. L. Kreuser, "A Quadratically Convergent Lagrangian Algorithm for Nonlinear Constraints," Technical Report No. 166, Computer Science Department, The University of Wisconsin, Madison, November 1972.

37. Rosenbrock, H. H., "An Automatic Method for Finding the Greatest or Least Value of a Function," *Computer J.*, **3**, 175–184 (1960).

38. Spendley, W., G. R. Hext, and F. R. Himsworth, "Sequential Application of Simplex Designs in Optimisation and Evolutionary Operation," *Technometrics*, **4**, 441–461 (1962).

39. Wolfe, P., "On the Convergence of Gradient Methods under Constraint," *IBM J. Res. Dev.*, **16**, 407–411 (1972).

40. Zangwill, W. I., *Nonlinear Programming: A Unified Approach*, Prentice-Hall, Englewood Cliffs, N.J., 1969.

41. Zoutendijk, G., *Methods of Feasible Directions*, Elsevier Publishing Co., Amsterdam, 1960.

42. Zoutendijk, G., "Nonlinear Programming: A Numerical Survey," *SIAM J. Control*, **4**, 194–210 (1966).

43. Zoutendijk, G., "Nonlinear Programming, Computational Methods," in *Integer and Nonlinear Programming*, J. Abadie (Ed.), North-Holland Publishing Co., Amsterdam, 1970.

44. Zoutendijk, G., "Some Algorithms Based on the Principle of Feasible Directions," in *Nonlinear Programming*, J. B. Rosen, O. L. Mangasarian, and K. Ritter (Eds.), Academic Press, New York, 1970.

14 APPROXIMATION-TYPE ALGORITHMS

The algorithms presented in Chapters 12 and 13 are clearly unconstrained optimization oriented. That is, they either transform a constrained problem into an unconstrained one or extend unconstrained methods to constrained problems by the appropriate adjustment of search directions and step lengths. This chapter describes constrained optimization methods that are linear programming oriented in the sense that most consist of steps in which linear programming subproblems are solved or, if the problem under consideration happens to be a linear program, the algorithm becomes some version of the well-known simplex method. Linear programming is such an extensive and important branch of optimization theory that it is usually taught as a separate subject in most curricula and cannot be covered in this book. Knowledge of linear programming, however, is, not necessary in order to understand the topics covered here. The reader need only know that the simplex method is an extremely powerful tool for solving linear programs. For a basic study of linear programs and some of its extensions, the reader is referred to Dantzig [12]. where the solution of certain nonlinear programming problems by variants of the simplex method is also discussed. In fact, the success of the simplex method in solving complicated and large linear programs led many researchers to believe that most nonlinear programs can and should be solved by methods that can be classified as belonging to this chapter. We shall not describe all the methods proposed along these lines. Supplementary material on the methods given here can be found, for example, in the reviews of Beale [7, 8] and the references mentioned there.

Section 14.1 begins with an algorithm that is based on successive linearizations of the objective function and the constraints of a nonlinear program and then presents a method that applies to linearly constrained problems in

which the objective function is successively approximated by a quadratic one. Reduced gradient methods, in which the variables are divided into basic (dependent) and nonbasic (independent) ones, similar to the simplex method, are discussed in Section 14.2. Cutting plane algorithms, the topic of Section 14.3, differ from previous algorithms in the sense that they are only applicable to convex (or generalized convex) problems, and the optimum is approached via a nonfeasible trajectory that is obtained by solving a sequence of linear programs. Finally, the last section deals with a method for solving a family of nonconvex programs in which some constraints describe nonconvex complements of convex sets. A detailed convergence proof is given that also illustrates a quite general approach to convergence results developed by Zangwill [40].

14.1 METHODS OF APPROXIMATION PROGRAMMING

The simplest approach in solving a general nonlinear programming problem by a sequence of linear programs is to approximate every nonlinear function appearing in the problem by a linear (affine) function and to solve the resulting linear program. An immediate objection is that linear approximations are usually valid only locally—that is, at the points around which the nonlinear functions are approximated or at most in some small neighborhood of the approximating points. Therefore in order to develop a useful algorithm, we must restrict movement from a current point, around which the functions are approximated, so that only small steps are made. For this reason, methods described in this section are also called "small-step" gradient methods, in contrast to "large-step" methods, which were presented in Chapter 13.

The first approximation programming method given here was derived by Griffith and Stewart [19]. Consider the general nonlinear program with bounded variables

$$\text{(BP)} \qquad \min f(x) \tag{14.1}$$

subject to

$$g_i(x) \geq 0, \qquad i = 1, \ldots, m \tag{14.2}$$

$$h_l(x) = 0, \qquad l = 1, \ldots, p \tag{14.3}$$

$$u_j \geq x_j \geq 0, \qquad j = 1, \ldots, n \tag{14.4}$$

where f, g_i, h_l are differentiable functions and the upper bounds u_j are given.

If lower bounds other than zero are imposed, they can be easily transformed to the preceding form by a simple change of variables.

Suppose that $\bar{x}$ is a feasible solution to (BP). Linearizing each nonlinear function around $\bar{x}$ yields

$$f(x) \cong \tilde{f}(x, \bar{x}) = f(\bar{x}) + (x - \bar{x})^T \nabla f(\bar{x}) \tag{14.5}$$

$$g_i(x) \cong \tilde{g}_i(x, \bar{x}) = g_i(\bar{x}) + (x - \bar{x})^T \nabla g_i(\bar{x}), \qquad i = 1, \ldots, m \tag{14.6}$$

$$h_l(x) \cong \tilde{h}_l(x, \bar{x}) = h_l(\bar{x}) + (x - \bar{x})^T \nabla h_l(\bar{x}), \qquad l = 1, \ldots, p. \tag{14.7}$$

Let us introduce a new vector variable y, defined by

$$y = x - \bar{x}. \tag{14.8}$$

Since y is generally unrestricted in sign, we employ the formula

$$y_j = y_j^+ - y_j^-, \qquad j = 1, \ldots, n \tag{14.9}$$

where

$$y_j^+ \geq 0 \qquad y_j^- \geq 0, \qquad j = 1, \ldots, n \tag{14.10}$$

and

$$y_j^+ = \begin{cases} y_j & \text{if } y_j \geq 0 \\ 0 & \text{if } y_j < 0 \end{cases} \tag{14.11}$$

$$y_j^- = \begin{cases} -y_j & \text{if } y_j \leq 0 \\ 0 & \text{if } y_j > 0. \end{cases} \tag{14.12}$$

The quantity $|y_j|$ represents the distance of any x_j from $\bar{x}_j$, the corresponding component of the approximating point $\bar{x}$. We wish to bound this distance by the inequality

$$|y_j| \leq m_j, \qquad j = 1, \ldots, n \tag{14.13}$$

where the m_j are given constants. Since the vector x is also bounded, we can combine (14.4) and (14.8) through (14.13) into

$$0 \leq y_j^+ \leq \min \{m_j, u_j - \bar{x}_j\} \tag{14.14}$$

$$0 \leq y_j^- \leq \min \{m_j, \bar{x}_j\}. \tag{14.15}$$

The linear program approximating (BP) is given by

$$\min \tilde{f}(x, \bar{x}) = f(\bar{x}) + (y^+)^T \nabla f(\bar{x}) - (y^-)^T \nabla f(\bar{x}) \tag{14.16}$$

subject to

$$\tilde{g}_i(x, \bar{x}) = g_i(\bar{x}) + (y^+)^T \nabla g_i(\bar{x}) - (y^-)^T \nabla g_i(\bar{x}) \geq 0, \qquad i = 1, \ldots, m \quad (14.17)$$

$$\tilde{h}_l(x, \bar{x}) = h_l(\bar{x}) + (y^+)^T \nabla h_l(\bar{x}) - (y^-)^T \nabla h_l(\bar{x}) = 0. \qquad l = 1, \ldots, p \quad (14.18)$$

and also subject to (14.14) and (14.15).

Griffith and Stewart's algorithm consists of an iterative scheme in which, starting from a point $\bar{x}$, the foregoing linear program is solved. If the optimal solution, say x_L^*, of this approximating program is feasible for (BP), then it is used as the new approximating point for the next linear program, with the same bounds as in the previous one being used. If, however, x_L^* is infeasible, it is not accepted, and the linear program is repeated with reduced values of the m_j. A point is termed optimal for (BP) if it is feasible within some tolerance and if some convergence criteria on the successive values of y and $f(x)$ are met.

The values of the bounds m_j greatly influence the success of the algorithm. If they are too small, the steps taken are also small and progress is rather slow. On the other hand, too large values of m_j may result in infeasible solutions, and then successive reductions of the bounds are needed. In moving from one approximating point to another, all the nonlinear functions are reapproximated, and generally a completely new linear program must be solved. The method is not restricted to convex programs or their generalizations. Convergence to a local optimum, however, has not been proven even for convex programs, although, in practice, the method usually converges in the convex case. The Griffith-Stewart method is primarily useful in nonlinear optimization problems having many variables and possibly lacking convexity assumptions, when there are only a few nonlinear constraints in addition to linear ones. In such problems it is often necessary to improve some current solution, and theoretical convergence considerations are unimportant. It is assumed, of course, that the user of this method has efficient linear programming codes available.

Example 14.1.1

To illustrate this algorithm, suppose that the nonlinear program to be solved is

$$\min f(x) = -2x_1 - x_2 \quad (14.19)$$

subject to

$$g_1(x) = 25 - (x_1)^2 - (x_2)^2 \geq 0 \quad (14.20)$$

$$g_2(x) = 7 - (x_1)^2 + (x_2)^2 \geq 0 \quad (14.21)$$

$$5 \geq x_1 \geq 0, \qquad 10 \geq x_2 \geq 0. \quad (14.22)$$

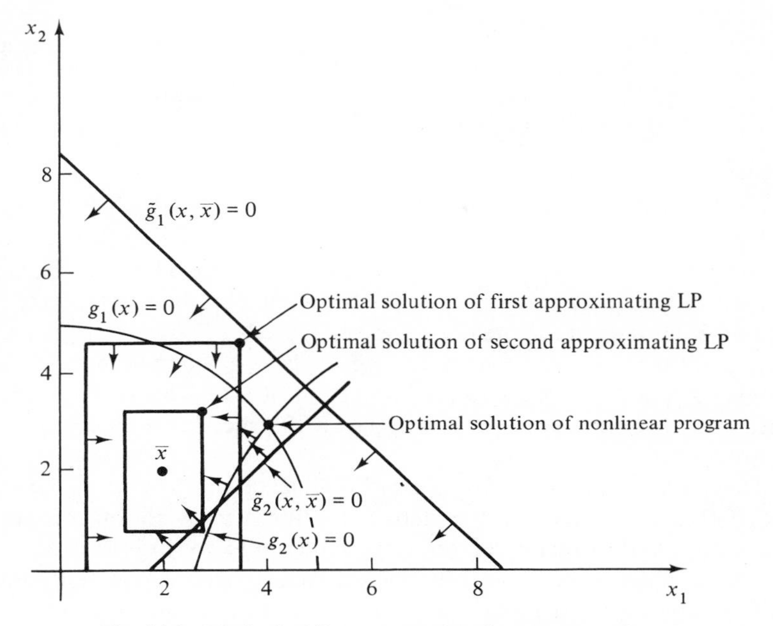

Fig. 14.1 Method of linear approximation programming.

We start at $\bar{x} = (2, 2)$ and take $m_1 = 1.5$, $m_2 = 2.5$ for the bounds m_j. The nonlinear program and the first approximating linear program (in terms of the original variables) are shown in Figure 14.1. The linear program is

$$\min \tilde{f}(x, \bar{x}) = -6 - 2y_1^+ + 2y_1^- - y_2^+ + y_2^- \tag{14.23}$$

subject to

$$\tilde{g}_1(x, \bar{x}) = 17 - 4y_1^+ + 4y_1^- - 4y_2^+ + 4y_2^- \geq 0 \tag{14.24}$$

$$\tilde{g}_2(x, \bar{x}) = \ \ 7 - 4y_1^+ + 4y_1^- + 4y_2^+ - 4y_2^- \geq 0 \tag{14.25}$$

$$1.5 \geq y_1^+ \geq 0 \tag{14.26}$$

$$1.5 \geq y_1^- \geq 0 \tag{14.27}$$

$$2.5 \geq y_2^+ \geq 0 \tag{14.28}$$

$$2 \geq y_2^- \geq 0. \tag{14.29}$$

The optimal solution is $y_1^+ = 1.5$, $y_2^+ = 2.5$, $y_1^- = y_2^- = 0$, or, in the original variables, $x_{L,1}^* = 3.5$, $x_{L,2}^* = 4.5$. This point does not satisfy (14.20) and so a reduction of the m_j is necessary. Suppose that we halve the bounds and

take $m_1 = 0.75$, $m_2 = 1.25$. The new optimal solution is at $x^*_{L,1} = 2.75$, $x^*_{L,2} = 3.25$. This point is feasible for the original program and serves as the new approximating point. ▮

We might mention that nonlinear programs in which the functions are separable, such as a program having an objective function

$$f(x) = f_1(x_1) + f_2(x_2) + \cdots + f_n(x_n) \tag{14.30}$$

and linear (or separable) constraints, can be conveniently solved by linear approximations that use some modification of the simplex method of linear programming. For details, the reader is referred to Beale [7] and Dantzig [12].

The successful implementation of approximation programming methods depends on the efficient solution of the approximating subproblems, such as the linear programs in the Griffith-Stewart method. Efficient quadratic programming algorithms, plus the fact that nonlinear functions can be more accurately approximated by quadratic ones, resulted in the development of more advanced methods of approximation programming. A representative of this class of algorithms is Fletcher's **method of hypercubes** [17], which is described next.

The method is designed to solve linearly constrained or possibly unconstrained problems. Fletcher did not consider nonlinearly constrained problems, perhaps because such constraints could be better handled indirectly—that is, by transforming them into the objective function, as we saw in Chapter 12. Consider, therefore, the problem

$$\min f(x) \tag{14.31}$$

subject to

$$A^T x \geq b, \tag{14.32}$$

where f is a differentiable real function, A is an $n \times m$ matrix, and b is an m vector. Let X denote the set of $x \in R^n$ satisfying (14.32). On each iteration of the method, a quadratic approximation $f_q(x, x^k)$ is made at x^k that agrees with f at x^k and also $\nabla f_q(x^k) = \nabla f(x^k)$. The Hessian of f_q is prescribed by a given symmetric matrix B_k. Each subproblem consists of solving a quadratic program in which f_q is minimized subject to (14.32) and subject to a constraint that limits movement from the current point x^k within a hypercube—in other words, by bounding the Chebyshev norm

$$\|x - x^k\|_\infty = \max_{1 \leq j \leq n} |x_j - x_j^k| \leq h^k. \tag{14.33}$$

As we shall see below, an important feature of the method is a systematic

adjustment of the h^k so that (14.33) defines a hypercube for which f_q is, in some sense, a valid approximation to f.

Another important feature is the updating of the B_k matrix after each iteration. As in some quasi-Newton algorithms for unconstrained optimization, B_k is updated by a rank-two correction such that a secant relation of the form

$$B_k p^k = \gamma^k \tag{14.34}$$

where $p^k = x^k - x^{k-1}$ and $\gamma^k = \nabla f(x^k) - \nabla f(x^{k-1})$ is satisfied. The suggested updating formula is due to Powell [27] and is given by

$$B_k = B_{k-1} + \frac{(\gamma^k - B_{k-1}p^k)(p^k)^T + p^k(\gamma^k - B_{k-1}p^k)^T}{(p^k)^T p^k} - \frac{(p^k)^T(\gamma^k - B_{k-1}p^k)p^k(p^k)^T}{[(p^k)^T p^k]^2}. \tag{14.35}$$

Some properties of this updating formula and the corresponding unconstrained variable metric method were examined in Exercise 11.L.

The updating formula of (14.35) does not guarantee that B_k will be positive definite if B_{k-1} is positive definite. Consequently, the quadratic subproblems may have points satisfying the Kuhn-Tucker necessary optimality conditions that are not global minima. Although Fletcher's quadratic programming algorithm [16] can efficiently handle such cases, it may be more desirable to restrict the updating formula so that, starting with a positive definite B_0, all subsequent B_k will also be positive definite. This step can be accomplished by multiplying the second and third terms on the right-hand side of (14.35) by a properly chosen constant θ, $0 \leq \theta \leq 1$, so that [26]

$$\det B_k \geq \alpha \det B_{k-1}, \tag{14.36}$$

where, say, $\alpha = 0.1$. Fletcher's method of hypercubes, which combines a sound theoretical basis with empirically determined constants so as to ensure successful application, can now be stated:

1. Suppose that an $x^0 \in X$ is given. Compute $f(x^0)$, $\nabla f(x^0)$ and choose constants $\bar{h}$, h^0 such that $\bar{h} \geq h^0 > 0$. Set $k = 0$.

2. If x^k satisfies (within some tolerance) the Kuhn-Tucker necessary conditions

$$\nabla f(x^k) - \sum_{i=1}^{m} \lambda_i a^i = 0 \tag{14.37}$$

$$\lambda_i[(a^i)^T x^k - b_i] = 0, \qquad i = 1, \ldots, m \tag{14.38}$$

$$\lambda \geq 0 \tag{14.39}$$

stop; otherwise let B_k be defined by (14.35), or by some modification of the updating formula that ensures positive definiteness, where B_0 is an arbitrary, symmetric (positive definite) $n \times n$ matrix.

3. Solve the quadratic subproblem

$$\text{(QSUB)} \qquad \min f_q(x, x^k) = f(x^k) + (x - x^k)^T \nabla f(x^k) + \tfrac{1}{2}(x - x^k)^T B_k (x - x^k) \tag{14.40}$$

subject to

$$A^T x \geq b \tag{14.41}$$

$$\|x - x^k\|_\infty \leq h^k \tag{14.42}$$

and let the solution be x^{k*}. Compute $f(x^{k*})$, $\nabla f(x^{k*})$, and

$$\rho^k = \frac{f(x^k) - f(x^{k*})}{f_q(x^k, x^k) - f_q(x^{k*}, x^k)}. \tag{14.43}$$

If $\rho^k \geq 0.5$, set $x^{k+1} = x^{k*}$ and go to step 4; otherwise go to step 5.

4. If any one of the following conditions

$$\|x^{k*} - x^k\|_\infty < h^k \tag{14.44}$$

$$I(x^{k*}) \not\subset I(x^k) \tag{14.45}$$

holds, where $I(x^{k*})$ and $I(x^k)$ are the index sets of active constraints (14.41) at x^{k*} and x^k, respectively, set $k = k + 1$ and go to step 2. Otherwise solve the one-dimensional minimization problem

$$\min_\alpha f_q(x^k + \alpha(x^{k*} - x^k)) \tag{14.46}$$

subject to

$$|\alpha| \leq 2 \tag{14.47}$$

and let α^* be the solution. Compute $\bar{x} = x^k + \alpha^*(x^{k*} - x^k)$. If

$$f_q(\bar{x}, x^k) \leq f(x^k) - \frac{4|f(x^{k*}) - f_q(x^{k*}, x^k)|}{1 - \rho^k}, \tag{14.48}$$

then let

$$h^{k+1} = \min\{\bar{h}, 2h^k\}; \tag{14.49}$$

otherwise let $h^{k+1} = h^k$. Set $k = k + 1$ and go to step 2.

5. If $\rho^k < 0.1$, decrease h^k to half its current value and return to step 3. If $0.1 \leq \rho^k < 0.5$, set $x^{k+1} = x^{k*}$, $h^{k+1} = \frac{1}{2}h^k$, $k = k + 1$, and go to step 2.

Fletcher has conducted some numerical experiments and obtained encouraging results. For details of this algorithm, as well as for some convergence results, the reader is referred to [17]. Convergence properties of the method of hypercubes without constraints, together with several other related algorithms, have also been studied by Powell [28].

In concluding this section, we should mention some earlier methods, which are based on ideas similar to those of the method of hypercubes. These methods were originally derived for unconstrained problems.

Suppose that we want to solve a system of nonlinear equations

$$f_i(x) = 0, \qquad i = 1, \ldots, m \tag{14.50}$$

in the variables $x_1, \ldots, x_n$, where $m > n$. One approach is to minimize the deviations in the least-squares sense, where we solve

$$\min_x f(x) = \sum_{i=1}^{m} [f_i(x)]^2. \tag{14.51}$$

We can linearize each f_i around a point x^k and obtain the quadratic approximation program

$$\min_x \sum_{i=1}^{m} [f_i(x^k) + (x - x^k)^T \nabla f_i(x^k)]^2. \tag{14.52}$$

Let $J(x^k)$ be the $n \times m$ Jacobian matrix whose columns are the $\nabla f_i(x^k)$. A stationary point of the quadratic function in the variable $p^k = x - x^k$ can be found by solving the system of linear equations

$$J(x^k)(J(x^k))^T p^k = -J(x^k)[f_1(x^k), \ldots, f_m(x^k)]^T. \tag{14.53}$$

An algorithm by which (14.53) can be solved iteratively is the Gauss method. This method often diverges, since the quadratic approximation of f is valid only in a small neighborhood of the approximating point. Levenberg [21] and Marquardt [23] have modified the Gauss method by adding a "damping term" to (14.52) and solving the problem

$$\min_x \sum_{i=1}^{m} [f_i(x^k) + (x - x^k)^T \nabla f_i(x^k)]^2 + \tfrac{1}{2}\lambda^k \| x - x^k \|^2, \tag{14.54}$$

where $\lambda^k \geq 0$ is a parameter, to be adjusted from one iteration to another. In a sense, this problem is equivalent to a constrained optimization problem in which (14.52) is the objective function and a constraint $\| x - x^k \|^2 \leq h$ is imposed. The Goldfeld, Quandt, and Trotter method [18], briefly described in Chapter 10, is also a similar small-step, quadratic approximation programming method. Finally, the method of Wilson [36] is closely related to

the method of hypercubes, with the difference that the matrix B_k is set equal to $\nabla^2 f(x^k)$, the exact Hessian matrix of f. A discussion of Wilson's method can be found in Beale [7].

14.2 REDUCED-GRADIENT ALGORITHMS

The discussions in this section focus on solving constrained nonlinear programs by methods that resemble the solution approach of the simplex method for linear programming. First we review the reduced-gradient method for linear constraints and then its generalizations for problems with nonlinear constraints. Consider the linearly constrained problem

$$\text{(LEP)} \qquad \min f(x) \tag{14.55}$$

subject to

$$Ax = b \tag{14.56}$$

$$x \geq 0, \tag{14.57}$$

where f is a continuously differentiable real function, A is an $m \times n$ matrix, b is an m vector, and $m \leq n$. The vector of variables x can be partitioned into two subvectors $x = (x^B, x^N)^T$, where $x^B = (x_1^B, \ldots, x_m^B)^T$ is the vector of **basic**, or dependent, variables and $x^N = (x_1^N, \ldots, x_{n-m}^N)$ is the vector of **nonbasic**, or independent, variables. Accordingly, the matrix A is also partitioned into $A = [B, C]$, where we assume, without loss of generality, that the first m columns of A correspond to the basic variables. Further assume that B, the $m \times m$ submatrix of A that corresponds to the components of the vector x^B, is nonsingular. Then we can write

$$Bx^B + Cx^N = b \tag{14.58}$$

and

$$x^B = B^{-1}b - B^{-1}Cx^N. \tag{14.59}$$

In addition, we also assume that the basic variables are nondegenerate; that is, $x^B > 0$. The nonbasic variables are called independent, since by assigning some numerical values to them, we obtain a unique solution of (14.58).

The basic idea of reduced-gradient methods is to eliminate x^B (as a function of x^N) via (14.59) and consider the optimization problem only in terms of x^N. This idea was used in the differential algorithms derived by Wilde [33] and Wilde and Beightler [34] through the notion of "constrained derivatives," in the reduced-gradient method of Wolfe [37, 38], and in the convex-simplex method of Zangwill [41].

From (14.59) we obtain the **reduced gradient** $r \in R^{n-m}$ by the formula

$$r(x^N) = \nabla_{x^N} f(x^B(x^N), x^N) - (B^{-1}C)^T \nabla_{x^B} f(x^B(x^N), x^N). \tag{14.60}$$

Now if we could make a small move from a current value of x^N in the direction of the negative reduced gradient without violating the nonnegativity constraints on the vector x, a decrease in the function value of f would occur. This step is accomplished as follows: Given a feasible x^k, compute for $i = 1, \ldots, n - m$

$$z_i^{N,k+1} = \begin{cases} 0 & \text{if } x_i^{N,k} = 0 \text{ and } r_i(x^{N,k}) > 0 \\ -r_i(x^{N,k}) & \text{otherwise} \end{cases} \tag{14.61}$$

and let

$$z^{B,k+1} = -B^{-1}Cz^{N,k+1}. \tag{14.62}$$

Then $z^{k+1} = (z^{B,k+1}, z^{N,k+1})^T$. The next point, $x^{k+1} = (x^{B,k+1}, x^{N,k+1})^T$, is given by

$$x^{k+1} = x^k + \alpha_{k+1}^* z^{k+1}, \tag{14.63}$$

where α_{k+1}^* is computed from the relations

$$\alpha_{k+1}^1 = \max \{\alpha_{k+1} : x^{B,k} + \alpha_{k+1} z^{B,k+1} \geq 0\} \tag{14.64}$$

$$\alpha_{k+1}^2 = \max \{\alpha_{k+1} : x^{N,k} + \alpha_{k+1} z^{N,k+1} \geq 0\} \tag{14.65}$$

and

$$f(x^k + \alpha_{k+1}^* z^{k+1}) = \min_{\alpha_{k+1}} \{f(x^k + \alpha_{k+1} z^{k+1}) : 0 \leq \alpha_{k+1} \leq \min (\alpha_{k+1}^1, \alpha_{k+1}^2)\}. \tag{14.66}$$

If $\alpha_{k+1}^* < \alpha_{k+1}^1$, let x^{k+1} be defined by (14.63). Otherwise

$$x_l^{B,k} + \alpha_{k+1}^1 z_l^{B,k+1} = 0 \tag{14.67}$$

for some l, and x_l^B is dropped from the vector of basic variables in exchange for the largest positive nonbasic variable. The algorithm terminates if $\| z^{k+1} \| \leq \epsilon$, where $\epsilon > 0$ is some small predetermined number.

Example 14.2.1

Let us illustrate the reduced-gradient method on the problem

$$\min f(x) = (x_1)^2 + 4(x_2)^2 \tag{14.68}$$

subject to

$$x_1 + 2x_2 - x_3 \qquad = 1 \tag{14.69}$$

$$-x_1 + x_2 \qquad + x_4 = 0 \tag{14.70}$$

$$x \geq 0 \tag{14.71}$$

by carrying out a few iterations. Suppose that we start at $x^0 = (2, 1, 3, 1)$ and let $x^{B,0} = (x_1, x_4)^T$, $x^{N,0} = (x_2, x_3)^T$. Then

$$B_0 = \begin{bmatrix} 1 & 0 \\ -1 & 1 \end{bmatrix}, \quad (B_0)^{-1} = \begin{bmatrix} 1 & 0 \\ 1 & 1 \end{bmatrix}, \quad C_0 = \begin{bmatrix} 2 & -1 \\ 1 & 0 \end{bmatrix} \tag{14.72}$$

and $\nabla f(x^0) = (4, 8, 0, 0)^T$. Let us compute the reduced gradient

$$r(x^{N,0}) = \begin{pmatrix} 8 \\ 0 \end{pmatrix} - \begin{bmatrix} 2 & 1 \\ -1 & 0 \end{bmatrix} \begin{bmatrix} 1 & 1 \\ 0 & 1 \end{bmatrix} \begin{pmatrix} 4 \\ 0 \end{pmatrix} = \begin{pmatrix} 0 \\ 4 \end{pmatrix}. \tag{14.73}$$

Hence $z^{N,1} = (0, -4)^T$ and

$$z^{B,1} = -\begin{bmatrix} 1 & 0 \\ 1 & 1 \end{bmatrix} \begin{bmatrix} 2 & -1 \\ 1 & 0 \end{bmatrix} \begin{pmatrix} 0 \\ -4 \end{pmatrix} = \begin{pmatrix} -4 \\ -4 \end{pmatrix}. \tag{14.74}$$

Now we compute the step length along z^1. First we find α_1^1 from (14.64). It is the largest α_1 satisfying

$$\begin{pmatrix} 2 - 4\alpha_1 \\ 1 - 4\alpha_1 \end{pmatrix} \geq 0 \tag{14.75}$$

and we obtain $\alpha_1^1 = \frac{1}{4}$. Similarly, we find α_1^2 from

$$\begin{pmatrix} 0 + 0\alpha_2 \\ 3 - 4\alpha_2 \end{pmatrix} \geq 0 \tag{14.76}$$

and $\alpha_1^2 = \frac{3}{4}$. Hence $\alpha_1^1 < \alpha_1^2$. Minimizing $f(x^0 + \alpha_1 z^1)$ with respect to α_1, we obtain from (14.66) $\alpha_1^* = \frac{1}{4}$. The new point is then given by $x^1 = x^0 + \alpha_1^* z^1$ and $x^1 = (1, 1, 2, 0)$. At this point, $\nabla f(x^1) = (2, 8, 0, 0)^T$. Since $\alpha_1^* = \alpha_1^1$, we change basis and x_3 enters the basis, replacing x_4. Hence

$$B_1 = \begin{bmatrix} 1 & -1 \\ -1 & 1 \end{bmatrix}, \quad (B_1)^{-1} = \begin{bmatrix} 0 & -1 \\ -1 & -1 \end{bmatrix}, \quad C_1 = \begin{bmatrix} 2 & 0 \\ 1 & 1 \end{bmatrix}. \tag{14.77}$$

The new reduced gradient is given by

$$r(x^{N,1}) = \begin{pmatrix} 8 \\ 0 \end{pmatrix} - \begin{bmatrix} 2 & 1 \\ 0 & 1 \end{bmatrix} \begin{bmatrix} 0 & -1 \\ -1 & -1 \end{bmatrix} \begin{pmatrix} 2 \\ 0 \end{pmatrix} = \begin{pmatrix} 10 \\ 2 \end{pmatrix}, \tag{14.78}$$

and $z^{N,2} = (-10, 0)^T$ by (14.61). Consequently,

$$z^{B,2} = -\begin{bmatrix} 0 & -1 \\ -1 & -1 \end{bmatrix} \begin{bmatrix} 2 & 0 \\ 1 & 1 \end{bmatrix} \begin{pmatrix} -10 \\ 0 \end{pmatrix} = \begin{pmatrix} -10 \\ -30 \end{pmatrix}. \tag{14.79}$$

Next we compute the step length along z^2. The inequalities

$$\begin{pmatrix} 1 - 10\alpha_2 \\ 2 - 30\alpha_2 \end{pmatrix} \geq 0 \tag{14.80}$$

and $1 - 10\alpha_2 \geq 0$ must hold, yielding $\alpha_2^1 = \frac{1}{15}$, $\alpha_2^2 = \frac{1}{10}$, and $\alpha_2^1 < \alpha_2^2$. The unrestricted step length minimizing $f(x^1 + \alpha_2 z^2)$ with respect to α_2 is $\alpha_2 = \frac{1}{10} > \alpha_2^1$. Thus we set $\alpha_2^* = \frac{1}{15}$.

The next point is given by $x^2 = (\frac{1}{3}, \frac{1}{3}, 0, 0)$ and $\nabla f(x^2) = (\frac{2}{3}, \frac{8}{3}, 0, 0)^T$. We change the basis and x_2 becomes a basic variable instead of x_3. Now $x^{B,2} = (x_1, x_2)^T$, $x^{N,2} = (x_3, x_4)^T$. The new partitions of the A matrix are

$$B_2 = \begin{bmatrix} 1 & 2 \\ -1 & 1 \end{bmatrix}, \quad (B_2)^{-1} = \begin{bmatrix} \frac{1}{3} & -\frac{2}{3} \\ \frac{1}{3} & \frac{1}{3} \end{bmatrix}, \quad C_2 = \begin{bmatrix} -1 & 0 \\ 0 & 1 \end{bmatrix}. \tag{14.81}$$

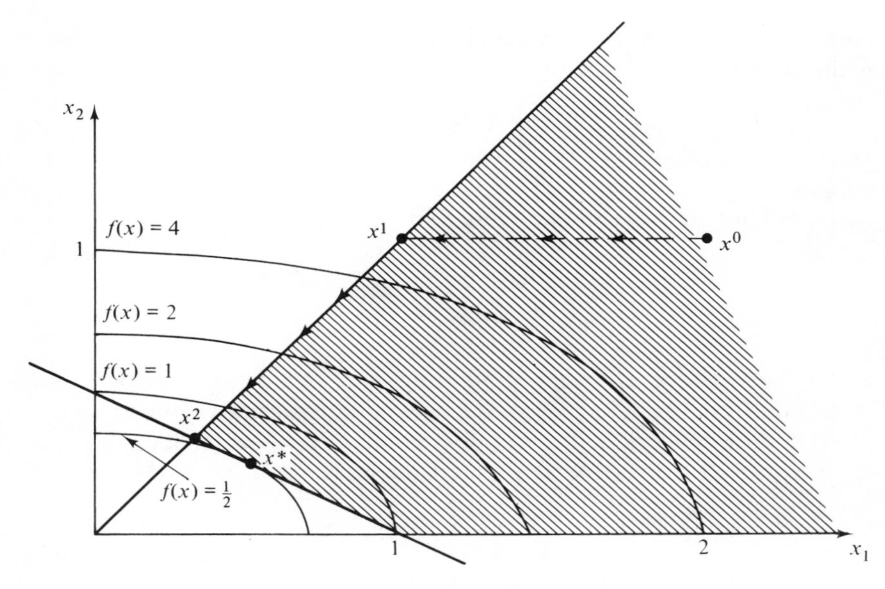

Fig. 14.2 The reduced-gradient method.

The reduced gradient at this point is given by

$$r(x^{N,2}) = \begin{pmatrix} 0 \\ 0 \end{pmatrix} - \begin{bmatrix} -1 & 0 \\ 0 & 1 \end{bmatrix} \begin{bmatrix} \frac{1}{3} & \frac{1}{3} \\ -\frac{2}{3} & \frac{1}{3} \end{bmatrix} \begin{pmatrix} \frac{2}{3} \\ \frac{8}{3} \end{pmatrix} = \begin{pmatrix} \frac{10}{9} \\ -\frac{4}{9} \end{pmatrix} \tag{14.82}$$

and $z^{N,3} = (0, \frac{4}{9})^T$. Thus

$$z^{B,3} = -\begin{bmatrix} \frac{1}{3} & -\frac{2}{3} \\ \frac{1}{3} & \frac{1}{3} \end{bmatrix} \begin{bmatrix} -1 & 0 \\ 0 & 1 \end{bmatrix} \begin{pmatrix} 0 \\ \frac{4}{9} \end{pmatrix} = \begin{pmatrix} \frac{8}{27} \\ -\frac{4}{27} \end{pmatrix}. \tag{14.83}$$

The reader is invited to continue the computations and observe convergence to the optimum at $x^* = (\frac{1}{2}, \frac{1}{4}, 0, \frac{1}{4})$. The progress of the algorithm (in x_1, x_2 space) is illustrated in Figure 14.2. ∎

The reduced-gradient method, as described above, can fail to converge to a point satisfying the Kuhn-Tucker conditions because of zigzagging. Several "antizigzagging" techniques were suggested to overcome this difficulty. One suggestion is to modify the selection of $z_i^{N,k+1}$ by setting it to zero if $x_i^{N,k} \leq \epsilon$ and $r_i(x^{N,k}) > 0$, where $\epsilon > 0$ is some predetermined value, instead of the formula given by (14.61). Another device was proposed by McCormick [24]; in it z^{k+1} is computed on the basis of $\nabla f(x^{k-1})$ if α_k^* is not the unconstrained minimum of $f(x^{k-1} + \alpha_k z^k)$ along z^k.

If f is a linear function, the reduced-gradient method becomes the simplex method of linear programming by some easy modifications of the rules for changing bases. If no constraints are present, the reduced-gradient method becomes the steepest descent technique, and thus it can also be viewed as an extension of an unconstrained technique. McCormick [25] has suggested a modification of the reduced-gradient method that can be regarded as a certain extension of the DFP variable metric algorithm for linearly constrained problems. He proved convergence of his method to a Kuhn-Tucker point at a superlinear rate, provided that some mild conditions are satisfied.

Abadie and Carpentier [1] generalized the reduced-gradient method to problems with nonlinear equality constraints. Consider the bounded-variable problem

$$\text{(NEP)} \qquad \min f(x) \tag{14.84}$$

subject to

$$h_i(x) = 0, \qquad i = 1, \ldots, m \tag{14.85}$$

$$\beta \geq x \geq \alpha, \tag{14.86}$$

where the $f, h_1, h_2, \ldots, h_m$ are assumed to be continuously differentiable functions of $x \in R^n$ and $m \leq n$.

Every feasible vector x^0 is assumed to be nondegenerate, which means that x^0 can be partitioned, as before, into two subvectors, $x = (x^{B,0}, x^{N,0})^T$ where $x^{B,0} \in R^m$, $x^{N,0} \in R^{n-m}$. Similarly, $\alpha = (\alpha^{B,0}, \alpha^{N,0})^T$, $\beta = (\beta^{B,0}, \beta^{N,0})^T$ such that

$$\beta^{B,0} > x^{B,0} > \alpha^{B,0} \tag{14.87}$$

and the vectors

$$\nabla_{x^B} h_i(x^0) = \left[\frac{\partial h_i(x^0)}{\partial x_1^{B,0}}, \ldots, \frac{\partial h_i(x^0)}{\partial x_m^{B,0}}\right]^T, \qquad i = 1, \ldots, m \tag{14.88}$$

are linearly independent. That is, the $m \times m$ matrix $\Delta^B(x^0)$ whose columns are the vectors $\nabla_{x^B} h_i(x^0)$ is nonsingular. It can easily be verified that the Kuhn-Tucker necessary conditions of optimality at x^0 for problem (NEP) are the existence of vectors $\mu^0 \in R^m$, $\lambda^0 \in R^{n-m}$ such that

$$\nabla_{x^N} f(x^0) - \sum_{i=1}^{m} \mu_i^0 \nabla_{x^N} h_i(x^0) - \lambda^0 = 0 \tag{14.89}$$

$$\nabla_{x^B} f(x^0) - \sum_{i=1}^{m} \mu_i^0 \nabla_{x^B} h_i(x^0) = 0 \tag{14.90}$$

where

$$\lambda_j^0 \geq 0 \quad \text{if } x_j^{N,0} = \alpha_j^{N,0} \tag{14.91}$$

$$\lambda_j^0 = 0 \quad \text{if } \beta_j^{N,0} > x_j^{N,0} > \alpha_j^{N,0} \tag{14.92}$$

$$\lambda_j^0 \leq 0 \quad \text{if } x_j^{N,0} = \beta_j^{N,0}. \tag{14.93}$$

Hence

$$\mu^0 = [\Delta^B(x^0)]^{-1} \nabla_{x^B} f(x^0) \tag{14.94}$$

and

$$\lambda^0 = \nabla_{x^N} f(x^0) - \Delta^N(x^0)[\Delta^B(x^0)]^{-1} \nabla_{x^B} f(x^0), \tag{14.95}$$

where $\Delta^N(x^0)$ is the $(n - m) \times m$ matrix whose columns are the vectors $\nabla_{x^N} h_i(x^0)$. Note that λ^0 is the reduced gradient, analogous to (14.60), if the constraints are linear. It follows that, for every feasible x^0, we can compute a vector λ^0 by the last equation; and if λ^0 also satisfies (14.91) to (14.93), then the necessary optimality conditions for (NEP) hold at x^0. The **generalized reduced-gradient method** is based on iterative moving from a feasible x^0 to feasible x^1, x^2, and so on, until a point x^l is reached where a λ^l, computed from (14.95), satisfies (14.91) to (14.93). Suppose that we have a feasible x^k and the corresponding λ^k does not satisfy these three relations. Then we change the current value of the nonbasic vector by

$$x^{N,k+1} = x^{N,k} + \theta_{k+1} z^{k+1}, \qquad \theta_{k+1} \geq 0, \tag{14.96}$$

where $z_j^{k+1} = 0$ if

$$x_j^{N,k} = \alpha_j^{N,k} \quad \text{and} \quad \lambda_j^k > 0, \quad \text{or} \quad x_j^{N,k} = \beta_j^{N,k} \quad \text{and} \quad \lambda_j^k < 0. \tag{14.97}$$

Otherwise we set either $z_j^{k+1} = -\lambda_j^k$ or $z_j^{k+1} = 0$, according to rules that depend on the particular version of the method used.

1. In the GRG version, we set $z_j^{k+1} = -\lambda_j^k$ whenever (14.97) does not hold. Note that this choice of z_j^{k+1} is identical to the one used by Wolfe [38] if the constraints are linear, as we have seen above.

2. In the GRGS version, we first find an index s by

$$|\lambda_s^k| = \max_j |\lambda_j^k|, \tag{14.98}$$

where the indices j range over those for which (14.97) does not hold. Then we let

$$z_j^{k+1} = \begin{cases} 0 & j \neq s \\ -\lambda_s^k & j = s. \end{cases} \tag{14.99}$$

This version coincides with the simplex method if the optimization problem is actually a linear program.

3. In the GRGC version, we cyclically set $z_j^{k+1} = -\lambda_j^k$. A cycle consists of n iterations. For $k = 1, \ldots, n$, we set

$$z_j^{k+1} = \begin{cases} 0 & j \neq k \\ -\lambda_j^k & j = k \end{cases} \tag{14.100}$$

unless k is the index of a basic variable or (14.97) holds, in which case iteration k is omitted. After $k = n$ we return to $k = 1$ and so on.

Having determined a $z^{k+1} \in R^{n-m}$ by one of these versions, we take a trial value of the step length θ_{k+1}, say $\theta_{k+1}^1 > 0$, and define

$$\tilde{x}_j^{N,k} = \begin{cases} \beta_j^{N,k} & \text{if } x_j^{N,k} + \theta_{k+1}^1 z_j^{k+1} > \beta_j^{N,k} \\ \alpha_j^{N,k} & \text{if } x_j^{N,k} + \theta_{k+1}^1 z_j^{k+1} < \alpha_j^{N,k} \\ x_j^{N,k} + \theta_{k+1}^1 z_j^{k+1} & \text{otherwise.} \end{cases} \tag{14.101}$$

To maintain feasibility, we solve the set of m nonlinear equations

$$h_i(x^B, \tilde{x}^{N,k}) = 0, \qquad i = 1, \ldots, m \tag{14.102}$$

in m variables $x_1^B, \ldots, x_m^B$ by an iterative procedure, such as Newton's method. Then, at iteration l of Newton's method we have the formula

$$x^{B,l+1} = x^{B,l} - [\Delta^N(x^{B,l}, \tilde{x}^{N,k})]^{-1} h(x^{B,l}, \tilde{x}^{N,k}), \tag{14.103}$$

where $h = (h_1, \ldots, h_m)^T$ and $x^{B,l} = x^{B,k}$ for $l = 0$. Formula (14.103) is used until one of the following cases occurs:

1. The norm $\|h(x^{B,l}, \tilde{x}^{N,k})\|$ increases in a few successive iterations. In this case, we reduce θ_{k+1}^1 by some factor, recompute (14.101), and return to solve (14.102).
2. We obtain

$$f(x^{B,l}, \tilde{x}^{N,k}) > f(x^{B,l-1}, \tilde{x}^{N,k}). \tag{14.104}$$

The correction is the same as in case 1.

3. For some l, the point $(x^{B,l}, \tilde{x}^{N,k})$ is outside the bounds defined by (14.86). We find a point $\tilde{x}^{B,l}$ on the line segment joining $x^{B,l-1}$ and $x^{B,l}$ such that $\tilde{x}_r^{B,l} = \alpha_r^{B,l}$ or $\tilde{x}_r^{B,l} = \beta_r^{B,l}$ for some r and make a change in the current basis by replacing the variable $x_r^{B,l}$ by, say $x_s^{N,l}$, where the index s is determined by some rule, similar to the one used in the simplex method. For example, we can use the relation (omitting the iteration index)

$$|\Delta^N(\tilde{x})_s(\Delta^B(\tilde{x}))_r^{-1}|\, v_s = \max_p \{|\Delta^N(\tilde{x})_p(\Delta^B(\tilde{x}))_r^{-1}|\, v_p\} \tag{14.105}$$

where $\Delta^N(\tilde{x})_p$ is the row vector $(\partial h_1(\tilde{x})/\partial x_p^N, \ldots, \partial h_m(\tilde{x})/\partial x_p^N)$, evaluated at $\tilde{x} = (\tilde{x}^{B,l}, \tilde{x}^{N,k})$, $(\Delta^B(\tilde{x}))_r^{-1}$ is the rth column of the matrix $(\Delta^B(\tilde{x}))^{-1}$, and

$$v_p = \min \{(\tilde{x}_p^{N,k} - \alpha_p^{N,k}), (\beta_p^{N,k} - \tilde{x}_p^{N,k})\}. \tag{14.106}$$

The solution of (14.102) is now attempted with the new basis.

4. The iterations converge. That is,

$$\|h(x^{B,l^*}, \tilde{x}^{N,k})\| \leq \epsilon \tag{14.107}$$

for some l^*, where ϵ is a small positive number. If

$$\beta^{B,l^*} > x^{B,l^*} > \alpha^{B,l^*}, \tag{14.108}$$

we set $x^{k+1} = (x^{B,l^*}, \tilde{x}^{N,k})$ and find a new search direction. If some component r of x^{B,l^*} is at a lower or upper bound, we change basis as outlined in case 3. The computations terminate when a point that satisfies the necessary optimality conditions is found.

The actual way in which the computations are performed involves many important considerations, and as the reader has undoubtedly recognized, the implementation of this reduced-gradient method for nonlinear

constraints is not simple. It is interesting to note that, in a comparative study of nonlinear programming codes carried out by Colville [11], the generalized reduced-gradient method ranked among the best of the techniques tested.

14.3 CUTTING PLANE METHODS

So far we have been studying computational methods in which convexity and its generalizations do not play a major role. Nevertheless, many algorithms can be easier implemented for convex programs in which the limit points of the iterations usually satisfy necessary and sufficient conditions of optimality, an important feature not shared by every nonlinear program. The methods discussed here apply to nonlinear programs that possess some convexity properties. The underlying principle of cutting plane methods is to approximate the feasible set of a nonlinear program by a finite set of closed half spaces and to solve a sequence of approximating linear programs.

Consider the problem

$$\text{(I)} \qquad \min \hat{g}_0(x) \tag{14.109}$$

subject to

$$g_i(x) \geq 0, \qquad i = 1, \ldots, l, \tag{14.110}$$

where $\hat{g}_0$ is a closed proper convex function and the g_i, $i = 1, \ldots, l$, are closed proper concave functions. This nonlinear program can be easily modified so that we obtain an equivalent program with a linear objective function. Defining a new variable $x_0 \in R$, we write

$$\text{(II)} \qquad \min x_0 \tag{14.111}$$

subject to

$$g_i(x) \geq 0, \qquad i = 0, 1, \ldots, l, \tag{14.112}$$

where

$$g_0(x) = x_0 - \hat{g}_0(x) \tag{14.113}$$

is a closed proper concave function. Thus by the addition of an extra variable and an extra constraint, we transformed program (I) into (II), such that x^* solves (I) if and only if (x_0^*, x^*) solves (II), with $x_0^* = \hat{g}_0(x^*)$. Without loss of generality, we can therefore consider the convex programming problem

(CPP) $$\min f(x) = c^T x \tag{14.114}$$

subject to

$$g_i(x) \geq 0, \qquad i = 1, \ldots, m, \tag{14.115}$$

where $x = (x_1, x_2, \ldots, x_n)^T$ and $g_1, \ldots, g_m$ are closed proper concave functions on R^n. Let

$$G = \{x : g_i(x) \geq 0, i = 1, \ldots, m\} \tag{14.116}$$

and assume that G is contained in a compact set $T \subset R^n$, defined by a finite set of linear inequalities

$$T = \{x : x \in R^n, Ax \geq b\}. \tag{14.117}$$

For $x \in T$, let $\partial g_i(x) \subset R^n$ be the set of subgradients, or subdifferential of g_i (see Section 4.3); that is, $\xi \in \partial g_i(x)$ implies

$$g_i(y) \leq g_i(x) + \xi^T(y - x) \tag{14.118}$$

for every $y \in R^n$. If g_i is differentiable at x, then $\xi = \nabla g_i(x)$. We shall assume that

$$\partial g_i(x) \neq \emptyset, \qquad i = 1, \ldots, m \tag{14.119}$$

for $x \in T$ and that the sets

$$\bigcup_{x \in T} \partial g_i(x), \qquad i = 1, \ldots, m \tag{14.120}$$

are bounded. This assumption is satisfied if, for example, the g_i are continuously differentiable. We can now state a **cutting plane algorithm** for solving (CPP), derived by Kelley [20] and Cheney and Goldstein [10], which we shall call the KCG algorithm:

1. Solve the linear program of minimizing $f(x) = c^T x$, subject to $x \in T$, and let x^0 be the optimal solution. If x^0 is contained in the set

$$G(\epsilon) = \{x : x \in T, g_i(x) \geq -\epsilon, i = 1, \ldots, m\}, \tag{14.121}$$

where ϵ is a small positive number, stop; an optimum of (CPP) has been reached. Otherwise let $k = 0$ and go to step 2.

2. Given an $x^k \in T$, such that $x^k \notin G(\epsilon)$, find the index s_k by

$$g_{s_k}(x^k) = \min\{g_i(x^k), i = 1, \ldots, m\} < 0 \tag{14.122}$$

and select a $\xi^k \in \partial g_{s_k}(x^k)$. Solve the linear program

$$\min c^T x \tag{14.123}$$

$$\tilde{g}_{s_h}(x, x^h) = g_{s_h}(x^h) + (\xi^h)^T(x - x^h) \geq 0, \qquad h = 0, 1, \ldots, k \tag{14.124}$$

$$x \in T. \tag{14.125}$$

3. Let x^{k+1} be the optimal solution of the preceding linear program. If $x^{k+1} \in G(\epsilon)$, stop. Otherwise set $k = k + 1$ and return to step 2.

Note that (14.118), which follows from the concavity of g_i, and constraints (14.124) explain the name cutting plane. Suppose that $x^k \notin G(\epsilon)$. Then there is at least one constraint such that $g_s(x^k) < -\epsilon$ for some index s. The half space defined by

$$g_s(x^k) + \xi^T(x - x^k) \geq 0, \tag{14.126}$$

where $\xi \in \partial g_s(x^k)$, cuts off the point x^k, which does not satisfy (14.126), from the feasible set G that is contained in the half space defined by (14.126). Denote by S_k the feasible set of the linear program solved in step 2 of iteration k. These sets are nested—that is,

$$S_k \subset S_{k-1} \subset \cdots \subset S_0. \tag{14.127}$$

At each iteration, a new linear inequality constraint is added to the set of constraints already present from the previous iteration. This addition of a new constraint at each iteration suggests using a dual simplex method [12] for solving the linear programming problems. Let us state and prove the convergence of the KCG method for convex programs.

Theorem 14.1

Let $g_1, \ldots, g_m$ be closed concave functions on the compact convex set $T \subset R^n$ such that at every point $x \in T$ the sets of subgradients $\partial g_i(x)$ are nonempty for $i = 1, \ldots, m$ and there exists a K such that

$$\sup\{\|\xi^i\| : \xi^i \in \partial g_i(x),\ i = 1, \ldots, m;\ x \in T\} \leq K. \tag{14.128}$$

Further assume that G, the feasible set of (CPP), *is nonempty and contained in T. Let*

$$S_k = S_{k-1} \cap \{x : \tilde{g}_{s_k}(x, x^k) \geq 0\}, \tag{14.129}$$

where $S_0 = T$. If $x^{k+1} \in S_k$ is such that

$$f(x^{k+1}) = c^T x^{k+1} = \min\{c^T x : x \in S_k\}, \tag{14.130}$$

then the sequence $\{x^k\}$ *contains a subsequence that converges to an optimal solution of* (CPP).

Proof. First we observe from (14.127) that $\{f(x^k)\}$ is monotone increasing. Hence if $\{x^k\}$ contains a subsequence that converges to a point $x^* \in G$, then $\{f(x^k)\}$ converges to $f(x^*)$ and x^* solves (CPP). Suppose now that $\{x^k\}$ does not have a subsequence converging to a point in G. Then there exists an $\alpha > 0$ such that

$$g_{s_h}(x^h) = \min_i \{g_i(x^h), i = 1, \ldots, m\} \leq -\alpha \tag{14.131}$$

for $h = 0,1, \ldots, k$. If x^{k+1} minimizes $c^T x$ on S_k, then $x^{k+1} \in T$ and

$$g_{s_h}(x^h) + (\xi^h)^T(x^{k+1} - x^h) \geq 0, \qquad h = 0, 1, \ldots, k. \tag{14.132}$$

From the last two relations and the Schwarz inequality, it follows that

$$\alpha \leq -g_{s_h}(x^h) \leq (\xi^h)^T(x^{k+1} - x^h) \leq K\|x^{k+1} - x^h\|. \tag{14.133}$$

Hence for every subsequence $\{k_p\}$ of indices we have

$$\|x^{k_p} - x^{k_q}\| \geq \frac{\alpha}{K}, \qquad q < p \tag{14.134}$$

that is, $\{x^k\}$ does not have a Cauchy subsequence, which contradicts that $\{x^k\} \subset T$ is bounded. ■

A typical trajectory of points of the sequence $\{x^k\}$ found by successive construction of cutting planes and solution of approximating linear programs is shown in Figure 14.3.

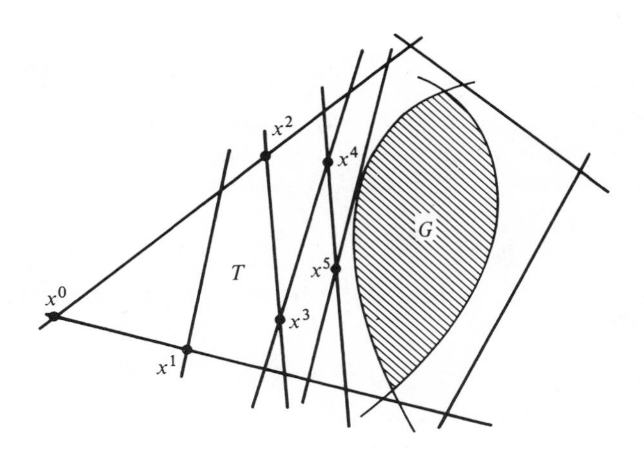

Fig. 14.3 The KCG cutting plane algorithm.

Example 14.3.1

Consider the convex programming problem

$$\min f(x) = 4x_1 + 5x_2 \tag{14.135}$$

subject to

$$g_1(x) = -(x_1)^2 - 2x_1x_2 - 2(x_2)^2 + 4 \geq 0 \tag{14.136}$$

$$g_2(x) = -(x_1)^2 - (x_2)^2 + 4x_1 - 3 \geq 0. \tag{14.137}$$

Let

$$T = \{x : x \in R^2, 4 \geq x_1 \geq 0, 4 \geq x_2 \geq -4\}. \tag{14.138}$$

Applying the Kelley-Cheney-Goldstein cutting plane algorithm, we first solve the linear program of minimizing $f(x)$ subject to $x \in T$. The optimal solution is at $x^0 = (0, -4)$, where $f(x^0) = -20$, $g_1(x^0) = -28$, $g_2(x^0) = -19$, and $s_0 = 1$. Since constraint (14.136) is the most violated at x^0, we construct the linear constraint

$$g_1(x^0) + (x - x^0)^T \nabla g_1(x^0) \geq 0 \tag{14.139}$$

or

$$2x_1 + 4x_2 + 9 \geq 0. \tag{14.140}$$

Solving the linear program

$$\min f(x) = 4x_1 + 5x_2 \tag{14.141}$$

subject to

$$2x_1 + 4x_2 + 9 \geq 0 \tag{14.142}$$

$$4 \geq x_1 \geq 0 \tag{14.143}$$

$$4 \geq x_2 \geq -4, \tag{14.144}$$

we find the optimal solution at $x^1 = (\frac{7}{2}, -4)$, where $f(x^1) = -16$, $g_1(x^1) = -\frac{49}{4}$, $g_2(x^1) = -\frac{69}{4}$, and $s_1 = 2$. Consequently, we add the constraint

$$g_2(x^1) + (x - x^1)^T \nabla g_2(x^1) \geq 0 \tag{14.145}$$

or

$$-12x_1 + 32x_2 + 101 \geq 0 \tag{14.146}$$

to the linear program (14.141) through (14.144) and find a new optimum at $x^2 = (\frac{29}{28}, -\frac{155}{56})$, where $f(x^0) = -9.7$. ∎

The reader will be asked to plot the foregoing problem and observe the rather slow progress of the algorithm. The rate of convergence of the KCG cutting plane method has been investigated by Levitin and Polyak [22] and Wolfe [39] for certain special types of convex programs. The results obtained are not encouraging for it was found that generally a convergence rate faster than linear cannot be expected. Nevertheless, combined with an efficient dual simplex algorithm, the method of cutting planes can be a useful technique for solving convex programs of moderate size. In the case of a nonconvex program, this method is quite useless, since the linear constraints added at each iteration may cut off portions of the feasible set, including the optimum.

It is interesting to note that the dual of the linear program to be solved at each iteration of the cutting plane method is itself the linear subproblem to be solved by the "generalized programming" (decomposition) algorithm of Dantzig and Wolfe [12, 39].

Perhaps the most important step that influences the efficiency of the method just described is the selection of a cutting plane. Several modifications of the basic KCG method have been proposed. Veinott [32] extended the cutting plane method to problems having quasiconcave constraint functions that, as we have seen in Chapter 6, define a convex feasible set. The construction of cutting planes is carried out by actually finding supporting hyperplanes to the feasible set. Consequently, Veinott's algorithm is called the **supporting hyperplane method** and is applied to solving problem (CPP), given by (14.114) and (14.115), where the $g_1, \ldots, g_m$ are assumed to be continuously differentiable quasiconcave functions. The set G defined by (14.116) is convex and is assumed to be contained in the compact set T given by (14.117). It is also assumed that there exists a point y such that

$$g_i(y) > 0, \qquad i = 1, \ldots, m. \tag{14.147}$$

The method consists of the following steps:

1. Step 1 of the supporting hyperplane method is identical to the corresponding step of the KCG algorithm.

2. Given an $x^k \in T$, such that $x^k \notin G(\epsilon)$, and a y satisfying (14.147), find a number θ^k such that the point z^k, given by

$$z^k = (1 - \theta^k)y + \theta^k x^k, \tag{14.148}$$

is feasible and satisfies

$$g_{s_k}(z^k) = 0 \tag{14.149}$$

for some s_k. Solve the linear program

$$\min c^T x \tag{14.150}$$

$$(x - z^h)\nabla g_{s_h}(z^h) \geq 0, \qquad h = 0, 1, \ldots, k \tag{14.151}$$

$$x \in T. \tag{14.152}$$

3. Let x^{k+1} be the optimal solution of the linear program in the preceding step. If $x^{k+1} \in G(\epsilon)$, stop. Otherwise set $k = k + 1$ and return to step 2.

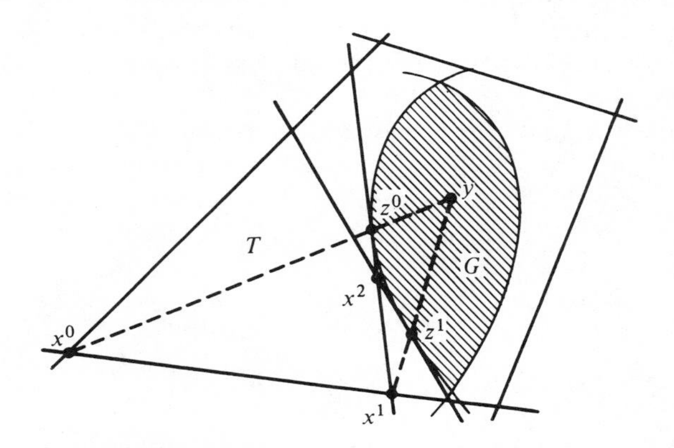

Fig. 14.4 Supporting hyperplane algorithm.

A typical situation is shown in Figure 14.4. Note that although extra computations are needed to find a point on the boundary of the feasible set, we can construct much deeper cuts than in the KCG method. Convergence of the supporting hyperplane method can be proven under the assumptions stated above.

One clear disadvantage of the cutting plane methods discussed so far is that the size of the linear programming subproblems increases from iteration to iteration, since cutting plane constraints are always added to the existing set of constraints but are never deleted. Some later works on cutting plane methods center on the question of dropping inactive constraints or, more generally, approximating the feasible set of the nonlinear program by linear constraints without nesting. Topkis [31] has derived such a method and proved convergence under assumptions similar to those stated earlier in this section. Eaves and Zangwill [15] have studied cutting plane methods in a more general framework, which contains the KCG, Veinott, Topkis, and several other versions of cutting plane algorithms as special cases.

14.4 COMPLEMENTARY CONVEX PROGRAMMING

In this section we seek the minimum of a linear function over a set in R^n that is an intersection of certain convex sets with complements of convex sets. Such a problem is usually a nonconvex program, and it is called a

complementary convex program or a "reverse" convex program. The method of solution, derived by Rosen [29] and Avriel and Williams [4], consists of successively approximating the "reverse" constraints by linear ones and thereby obtaining a sequence of convex programs to be solved by one of the methods discussed earlier, such as penalty functions or cutting planes. In order to prove convergence of the method described below, we rely on a fairly general theory of convergence that is due to Zangwill [40, 41] and that can also be used to prove the convergence of many other algorithms presented in this book.

Consider the nonlinear programming problem

$$\text{(CCP)} \qquad \min f(x) = c^T x \tag{14.153}$$

subject to

$$g_i(x) \geq 0, \qquad i = 1, \ldots, m \tag{14.154}$$

$$h_l(x) \geq 0, \qquad l = 1, \ldots, p \tag{14.155}$$

where $c \in R^n$ is a given nonzero vector and the g_i and h_l are real-valued continuously differentiable convex and concave functions respectively. Define

$$T = \{x : x \in R^n, g_i(x) \geq 0, i = 1, \ldots, m\} \tag{14.156}$$

$$U = \{x : x \in R^n, h_l(x) \geq 0, l = 1, \ldots, p\} \tag{14.157}$$

and let $X = T \cap U$. Note that (CCP) is generally not a convex program, since the set T is not convex if the g_i are nonlinear. Actually, T is the intersection of the **complements** of m open convex sets. We are specifically interested in the case of g_i convex and nonlinear. Since we are dealing with a nonconvex program, convergence of the algorithm to the global constrained minimum of f cannot be ensured. However, we shall show that, except for certain degenerate situations, the algorithm converges to a local minimum. It will be assumed that the following regularity conditions hold for (CCP):

1. The feasible set X is compact and program (CCP) is strongly consistent. That is, there exists at least one $x \in X$ such that (14.154) and (14.155) are satisfied as strict inequalities.
2. For $\tilde{x} \in X$, define the set of active constraints by

$$I(\tilde{x}) = \{i : g_i(\tilde{x}) = 0\} \tag{14.158}$$

$$L(\tilde{x}) = \{l : h_l(\tilde{x}) = 0\}. \tag{14.159}$$

Then for all numbers r_i, $i \in I(\tilde{x})$, and q_l, $l \in L(\tilde{x})$, such that

$$r_i \geq 0, \quad q_l \geq 0, \quad \sum_{i \in I(\tilde{x})} r_i + \sum_{l \in L(\tilde{x})} q_l > 0, \tag{14.160}$$

we must have

$$\sum_{i \in I(\tilde{x})} r_i \frac{\partial g_i(\tilde{x})}{\partial x_j} + \sum_{l \in L(\tilde{x})} q_l \frac{\partial h_l(\tilde{x})}{\partial x_j} \neq 0 \tag{14.161}$$

for some j, $j = 1, \ldots, n$.

3. Let ϕ be a real function of $x \in R^n$. An inequality $\phi(x) \geq 0$ is said to be nontrivial if there exists a vector $\hat{x}$ such that $\phi(\hat{x}) < 0$. It is assumed that each of the inequalities in (14.154) and (14.155) is nontrivial.

A feasible point satisfying (3.72) to (3.74) in the Kuhn-Tucker necessary conditions for optimality will be called a **K-T point**. It can be shown that the regularity conditions stated above imply the existence of Kuhn-Tucker multipliers at every solution of (CCP). Moreover, every K-T point x^* must satisfy

$$I(x^*) \cup L(x^*) \neq \emptyset. \tag{14.162}$$

The latter result follows from (3.72) and (3.73), since if the preceding set is empty, then all Kuhn-Tucker multipliers λ_i^*, $i = 1, \ldots, m$, and π_l^*, $l = 1, \ldots, p$, corresponding to constraints (14.154) and (14.155), respectively, must vanish, thus contradicting that $\nabla f(x^*) = c$ is a nonzero vector.

It can also be verified that local (or global) maxima of (CCP) cannot qualify as K-T points. For suppose that x^* is a K-T point. Then it must satisfy (14.162). Hence for some j,

$$c_j = \sum_{i=1}^{m} \lambda_i^* \frac{\partial g_i(x^*)}{\partial x_j} + \sum_{l=1}^{p} \pi_l^* \frac{\partial h_l(x^*)}{\partial x_j} \neq 0, \tag{14.163}$$

and the necessary conditions for maximizing $c^T x$ over X cannot hold at x^*.

The algorithm to solve problem (CCP) is described next. Starting from some feasible point $x^0 \in X$, a sequence of feasible points $\{x^k\}$ is generated as follows: Given a point $x^k \in X$, we replace the functions g_i in (14.154) by their first-order Taylor approximation about x^k. In other words, we replace $g_i(x) \geq 0$ by the linear constraints

$$\tilde{g}_i(x, x^k) = g_i(x^k) + (x - x^k)^T \nabla g_i(x^k) \geq 0, \qquad i = 1, \ldots, m. \tag{14.164}$$

If we let

$$T(x^k) = \{x : \tilde{g}_i(x, x^k) \geq 0, i = 1, \ldots, m\}, \tag{14.165}$$

we obtain a new program

$$(\text{CCP}_k) \qquad \min c^T x \tag{14.166}$$

subject to

$$x \in X(x^k) = T(x^k) \cap U, \tag{14.167}$$

which is a convex program. The next point, x^{k+1}, is taken to be any optimal solution of (CCP$_k$). That is, denoting the set of optimal solutions to (CCP$_k$) by A^k, we choose x^{k+1} from A^k. We shall show that if x^0 was feasible for (CCP), then so is each member of the sequence $\{x^k\}$. Moreover, we shall show that any convergent subsequence of $\{x^k\}$ converges to a K-T point of (CCP).

Before turning to questions of convergence, however, we demonstrate the proposed algorithm by a simple example.

Example 14.4.1

Consider the program

$$\min f(x) = x_1 \tag{14.168}$$

subject to

$$g_1(x) = (x_1 - 3)^2 + (x_2 - 2)^2 - 13 \geq 0 \tag{14.169}$$

$$h_1(x) = -(x_1 - 4)^2 - (x_2)^2 + 16 \geq 0. \tag{14.170}$$

The shaded area in Figure 14.5 represents the feasible set, which is clearly nonconvex. There are three K-T points in this problem. Two local minima are located at (0, 0) and at (6.4, 3.2), respectively, and a K-T point at $(3 + \sqrt{13}, 2)$ which is not a local minimum. Suppose that we start at $x^0 = (7, 0)$. Then we replace g_1 by its first-order Taylor approximation about x^0:

$$\tilde{g}_1(x, x^0) = 8x_1 - 4x_2 - 49 \tag{14.171}$$

and solve the convex program

$$(\text{CCP}_0) \qquad \min f(x) = x_1 \tag{14.172}$$

subject to the constraints

$$\tilde{g}_1(x, x^0) = 8x_1 - 4x_2 - 49 \geq 0 \tag{14.173}$$

$$h_1(x) = -(x_1 - 4)^2 - (x_2)^2 + 16 \geq 0. \tag{14.174}$$

Solution of this program yields

$$A^0 = \{(4.126, -3.998)\}. \tag{14.175}$$

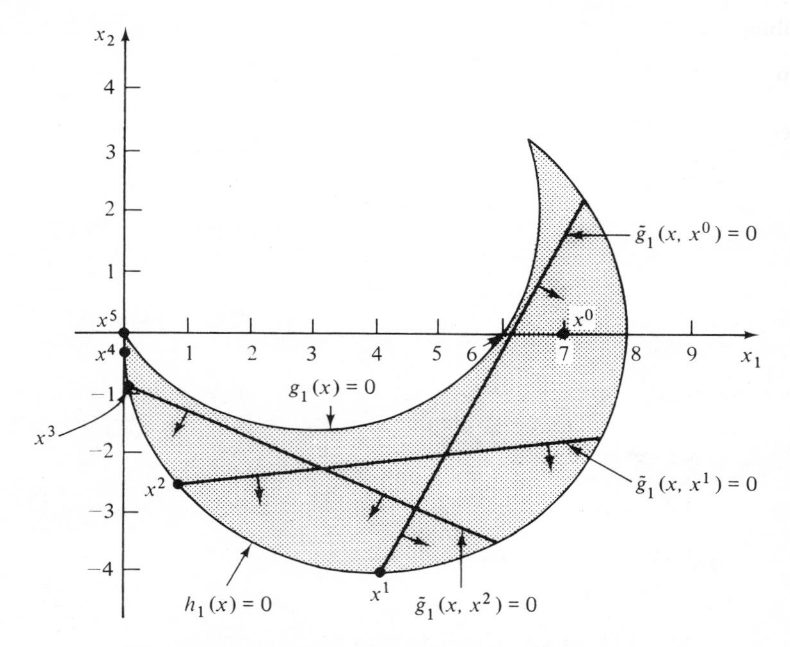

Fig. 14.5 Complementary convex programming.

Hence we choose $x^1 = (4.126, -3.998)$ and expand g_1 at x^1, solve the resulting convex program, and so forth.

Table 14.1

Iteration	x_1	x_2
0	7.000	0.000
1	4.126	−3.998
2	0.939	−2.575
3	0.094	−0.864
4	0.002	−0.134
5	0.000	−0.004

Table 14.1 lists the optimal solutions of the convex programs as generated by the iterative procedure of the algorithm. It can be seen that, in five iterations, the optimal solutions of the convex programs converged reasonably close to one of the local minima.

The choice of the starting point has an important effect on the location of the K-T point to which the solutions of the convex programs converge. To illustrate this effect, suppose that the starting point for the same example

is taken as $x^0 = (7.0, 2.1)$. In this case, the algorithm converges to the other local minimum in three iterations, as Table 14.2 shows.

Table 14.2

Iteration	x_1	x_2
0	7.000	2.100
1	6.600	3.040
2	6.411	3.191
3	6.400	3.200

In both cases, the optimal sets A^k consisted of a single point. To illustrate that an optimal set A^k may consist of more than one point, we choose $x^0 = (7, 2)$. Now we get

$$\tilde{g}_1(x, x^0) = 8x_1 - 53 \geq 0 \tag{14.176}$$

and the optimal set A^0 is given by

$$A^0 = \left\{(x_1, x_2) : x_1 = \frac{53}{8}, |x_2| \leq \sqrt{\frac{15{,}943}{8}}\right\} \tag{14.177}$$

The next point, x^1, is given by any point in A^0. If we choose $x^1 = (\frac{53}{8}, 2)$ and solve (CCP_1), we get a new set A^1, which is a line segment, and again we can choose $x_2^2 = 2$. By choosing $x_2^k = 2$ in all subsequent iterations, we finally converge to the third K-T point at $(3 + \sqrt{13}, 2)$, which is not a local minimum. ∎

Let us examine the convergence of the complementary convex programming algorithm. First we note that the algorithm can be described in terms of a point-to-set mapping F, whose semicontinuity properties play a central role in the subsequent theorems. We remind the reader that F is a point-to-set mapping if, for each $x \in X \subset R^n$, a subset $F(x)$ in $Y \subset R^q$ is uniquely defined. Alternatively, if $\sigma(Y)$ denotes the set of subsets of Y, we can say that F maps X into $\sigma(Y)$. We use the notation $F: X \longrightarrow \sigma(Y)$. Of course, X and Y may be the same sets. An important and relevant property of such maps is that of upper semicontinuity [9]. A point-to-set mapping $F: X \longrightarrow \sigma(Y)$ is said to be **upper semicontinuous** (usc) at a point $\bar{x} \in X$ if $\{x^k\} \longrightarrow \bar{x}$, $\{y^k\} \longrightarrow \bar{y}$ (that is, the sequences $\{x^k\}$ and $\{y^k\}$ converge to $\bar{x}$ and $\bar{y}$, respectively), $x^k \in X$, and $y^k \in F(x^k) \subset Y$ imply $\bar{y} \in F(\bar{x})$. We say that F is upper semicontinuous if it is usc at each point $x \in X$. The reader is invited to compare upper semicontinuity, as defined here, with lower semicontinuity, which was introduced in Chapter 6.

As an immediate consequence of this definition, we derive

Lemma 14.2

Let the point-to-set map $F: X \longrightarrow \sigma(Y)$ be upper semicontinuous at $\bar{x}$. If $\{x^k\} \longrightarrow \bar{x}$ and $y^k \in F(x^k)$, then every cluster point y of $\{y^k\}$ is in the set $F(\bar{x})$.

Proof. If y is a cluster point of $\{y^k\}$, then there exists a subsequence $\{y^{k_i}\}$ that converges to y. Since $y^{k_i} \in F(x^{k_i})$ and the sequence $\{x^{k_i}\}$ is a subsequence of $\{x^k\}$, hence $\{x^{k_i}\} \longrightarrow \bar{x}$. By the upper semicontinuity of F, it follows that $y \in F(\bar{x})$. ■

Next we state and prove a slightly modified version of a theorem due to Zangwill [40].

Theorem 14.3

Let a point-to-set map $A : X \longrightarrow \sigma(X)$ define a mapping from X to subsets of X. Let X be compact.

Suppose that A satisfies the following conditions:

(i) *If $x \in X$ is not a K-T point of* (CCP), *then $y \in A(x)$ implies $f(y) < f(x)$. If $x \in X$ is a K-T point of* (CCP), *then $y \in A(x)$ implies $f(y) = f(x)$.*

(ii) *The map A is upper semicontinuous for all $x \in X$.*

Then any sequence $\{x^k\}$ with $x^0 \in X$, $x^{k+1} \in A(x^k)$ will contain a subsequence that converges to a K-T point of (CCP).

Proof. The sequence $\{f(x^k)\}$ is monotone nonincreasing and bounded below (since X is compact), and so $\lim \{f(x^k)\}$ exists. Let it be denoted by $\hat{f}$.

Again, by the compactness of X, the sequence $\{x^k\}$ must contain a convergent subsequence, say $\{x^{k_i}\}$, with limit x^* in X. For this subsequence also, $\lim \{f(x^{k_i})\} = \hat{f}$. Now consider the sequence $\{x^{k_i+1}\}$, for which, again, $\lim \{f(x^{k_i+1})\} = \hat{f}$, since this sequence is still a subsequence of $\{x^k\}$. Now $\{x^{k_i}\} \longrightarrow x^*$, $x^{k_i+1} \in A(x^{k_i})$, and A is upper semicontinuous. By Lemma 14.2, therefore, every cluster point x^{**} of $\{x^{k_i+1}\}$ is in $A(x^*)$. Property (i) would require that if x^* were not a K-T point, then $f(x^{**}) < f(x^*)$. However, since f is continuous and $\lim \{f(x^k)\} = \hat{f}$, we have $f(x^{**}) = \hat{f}$ and also $f(x^*) = \hat{f}$. Thus the point x^*, must be a K-T point. ■

To apply the theorem to the proposed algorithm, we must show that the algorithm is, in fact, a point-to-set map with the properties described in the theorem.

For each $x^k \in X$, the approximation of the constraints $g_i(x) \geq 0$ by the linear constraints $\tilde{g}_i(x, x^k) \geq 0$ can be viewed abstractly as a mapping

A_1 of x^k into the coefficient set $\{\nabla g_i(x^k), (x^k)^T\nabla g_i(x^k) - g_i(x^k)\}$. Since the g_i are continuously differentiable, the (single-valued) map A_1 is continuous and thus upper semicontinuous [9]. The set $A(x^k)$ of optimal solutions to the convex program (CCP$_k$) may then be regarded as a function of the coefficient set. Abstractly, we may define A_2 as a point-to-set mapping from the coefficient set into a set in R^n. Thus $A = A_2A_1$ is a composite point-to-set mapping from X into subsets of R^n. We shall now show that it is, in fact, from X into subsets of X. Moreover, we shall show that A is upper semicontinuous and that a point $x^* \in X$ is a K-T point for (CCP) if and only if $x^* \in A(x^*)$.

Theorem 14.4

The mapping A is from X into $\sigma(X)$.

Proof. For $x^k \in X$, the set $A(x^k)$ is the set of optimal solutions to (CCP$_k$). For all $\hat{x} \in A(x^k)$, we have $\hat{x} \in X(x^k)$. That is,

$$g_i(x^k) + (\hat{x} - x^k)^T\nabla g_i(x^k) \geq 0, \qquad i = 1, \ldots, m \qquad (14.178)$$

$$h_l(\hat{x}) \geq 0, \qquad l = 1, \ldots, p. \qquad (14.179)$$

Since the g_i are convex,

$$g_i(\hat{x}) \geq g_i(x^k) + (\hat{x} - x^k)^T\nabla g_i(x^k) \qquad (14.180)$$

and $\hat{x}$ satisfies $g_i(\hat{x}) \geq 0$. Hence $\hat{x} \in X$—that is, $A(x^k) \subset X$. ■

The next theorem establishes property (i) of Theorem. 14.3.

Theorem 14.5

The vector x^* is a K-T point for (CCP) *if and only if $x^* \in A(x^*)$.*

Proof. Suppose that x^* is a K-T point. Then, by Theorem 3.8, there exist multipliers λ^*, π^* that satisfy (3.72) to (3.74). The existence of these multipliers is equivalent to a necessary and sufficient condition for x^* to be an optimal solution of the convex program (CCP$_*$) obtained by linearizing (14.154) around x^*; that is, $x^* \in A(x^*)$. Conversely, let $x^* \in A(x^*)$. Then again there exist multipliers that satisfy (3.72) to (3.74). Thus x^* is a K-T point for (CCP). ■

After establishing property (i) of the convergence theorem, we turn now to property (ii)—that is, to verifying the upper semicontinuity of A. Here we use the following result of Dantzig, Folkman, and Shapiro [13] and Williams [35], reworded to fit our case.

Theorem 14.6

Suppose that for any $\tilde{x} \in X$ *none of the linear constraints* $\tilde{g}_i(x, \tilde{x}) \geq 0$ *is trivial and there exists an* $\hat{x}$ *such that*

$$\tilde{g}_i(\hat{x}, \tilde{x}) > 0, \qquad i = 1, \ldots, m \tag{14.181}$$

$$h_l(\hat{x}) > 0, \qquad i = 1, \ldots, p. \tag{14.182}$$

Then the point-to-set mapping $A_2(\tilde{x})$ *is upper semicontinuous.*

Next we state the following well-known result [9].

Theorem 14.7

If A_1 *is a usc mapping of* X *into* Y *and* A_2 *is a usc mapping of* Y *into* Z, *then the composite mapping,* $A = A_2 A_1$, *is a usc mapping of* X *into* Z.

Since the mapping A_1 described above is usc, it is sufficient to show that the hypotheses of Theorem 14.6 hold for the mapping A_2, in order to verify property (ii) of the convergence theorem for the composite map $A = A_2 A_1$.

Theorem 14.8

The linear constraints $\tilde{g}_i(x, \tilde{x}) \geq 0$ *are nontrivial.*

Proof. By assumption (3) on page 485, the constraints $g_i(x) \geq 0$ are nontrivial. That is, there exists a vector $w \in R^n$ such that $g_i(w) < 0$. It then follows from the convexity of the g_i that

$$\tilde{g}_i(w, \tilde{x}) \leq g_i(w) < 0, \qquad i = 1, \ldots, m, \tag{14.183}$$

so that the $\tilde{g}_i(x, \tilde{x}) \geq 0$ are nontrivial. ■

Now we must show that there exists an $\hat{x}$ satisfying (14.181) and (14.182). Here Theorem 4.19 will prove useful.

Theorem 14.9

Let $\tilde{x}$ *be in* X. *There exists no* $\hat{x}$ *satisfying* (14.181) *and* (14.182) *if and only if there exist numbers* $r_1, \ldots, r_m, q_1, \ldots, q_p$ *satisfying*

$$r_i \geq 0, \qquad i = 1, \ldots, m \quad \textit{and} \quad q_l \geq 0. \qquad l = 1, \ldots, p \tag{14.184}$$

$$\sum_{i \in I(\tilde{x})} r_i + \sum_{l \in L(\tilde{x})} q_l > 0 \tag{14.185}$$

such that

$$\sum_{i \in I(\tilde{x})} r_i \frac{\partial g_i(\tilde{x})}{\partial x_j} + \sum_{l \in L(\tilde{x})} q_l \frac{\partial h_l(\tilde{x})}{\partial x_j} = 0, \qquad j = 1, \ldots, n \tag{14.186}$$

$$r_i g_i(\tilde{x}) = 0, \qquad i = 1, \ldots, m \tag{14.187}$$

$$q_l h_l(\tilde{x}) = 0, \qquad l = 1, \ldots, p. \tag{14.188}$$

Proof. Suppose that there is no $\hat{x}$ with the required property. Then, by Theorem 4.19, there exist nonnegative vectors r and q such that

$$\sum_{i=1}^{m} r_i + \sum_{l=1}^{p} q_l > 0 \tag{14.189}$$

and

$$\sum_{i=1}^{m} r_i \left[g_i(\tilde{x}) + \sum_{j=1}^{n} \frac{\partial g_i(\tilde{x})}{\partial x_j}(x_j - \tilde{x}_j) \right] + \sum_{l=1}^{p} q_l h_l(x) \leq 0 \tag{14.190}$$

for all $x \in R^n$.

Setting $x = \tilde{x}$, substituting into (14.190), and rearranging, we get

$$\sum_{i \in I(\tilde{x})} r_i g_i(\tilde{x}) + \sum_{l \in L(\tilde{x})} q_l h_l(\tilde{x}) + \sum_{i \notin I(\tilde{x})} r_i g_i(\tilde{x}) + \sum_{l \notin L(\tilde{x})} q_l h_l(\tilde{x}) \leq 0. \tag{14.191}$$

The first two terms in (14.191) clearly vanish. For $i \notin I(\tilde{x})$ and $l \notin L(\tilde{x})$, we have $g_i(\tilde{x}) > 0$ and $h_l(\tilde{x}) > 0$, respectively. Thus $r_i = 0$ and $q_l = 0$ for $i \notin I(\tilde{x})$ and $l \notin L(\tilde{x})$. Consequently, (14.187) and (14.188) are satisfied, and (14.188) reduces to

$$\sum_{j=1}^{n} \left[\sum_{i \in I(\tilde{x})} r_i \frac{\partial g_i(\tilde{x})}{\partial x_j} \right](x_j - \tilde{x}_j) + \sum_{l \in L(\tilde{x})} q_l h_l(x) \leq 0 \tag{14.192}$$

for all $x \in R^n$. Clearly, the function on the left-hand side of this inequality attains its unconstrained maximum at $\tilde{x}$. Hence its first partial derivatives vanish at $\tilde{x}$:

$$\sum_{i \in I(\tilde{x})} r_i \frac{\partial g_i(\tilde{x})}{\partial x_j} + \sum_{l \in L(\tilde{x})} q_l \frac{\partial h_l(\tilde{x})}{\partial x_j} = 0, \qquad j = 1, \ldots, n. \tag{14.193}$$

Thus (14.186) also holds.

Now suppose there exist vectors $p \geq 0, q \geq 0$ such that (14.184) through (14.188) hold and suppose that there is an $\hat{x}$ satisfying (14.181) and (14.182). Since the h_l are concave, (14.182) can be replaced by

$$h_l(\tilde{x}) + \sum_{j=1}^{n} \frac{\partial h_l(\tilde{x})}{\partial x_j}(\hat{x}_j - \tilde{x}_j) > 0, \qquad l = 1, \ldots, p, \tag{14.194}$$

Multiplying each of the inequalities in (14.181) and (14.194) by the corresponding r_i and q_l, respectively, we obtain

$$\sum_{j=1}^{n}\left[\sum_{i=1}^{m} r_i \frac{\partial g_i(\tilde{x})}{\partial x_j} + \sum_{l=1}^{p} \frac{\partial h_l(\tilde{x})}{\partial x_j}\right](\hat{x}_j - \tilde{x}_j) + \sum_{i=1}^{m} r_i g_i(\tilde{x}) + \sum_{l=1}^{p} q_l h_l(\tilde{x}) > 0, \tag{14.195}$$

contradicting the hypothesis. ■

Comparing the conditions of Theorem 14.9 with assumption (2) on pages 484–485, we conclude that there is an $\hat{x}$ with the required property. Theorems 14.7 to 14.9 thus establish the upper semicontinuity of the mapping A, and so property (ii) of Theorem 14.3 is satisfied. This completes the proof of convergence of the proposed algorithm for the solution of complementary convex programs.

It is clear that this method can be extended to programs with generalized convex and concave functions by some easy modifications. Another extension can be developed by constructing supporting hyperplanes to the constraints (14.154), similar to Veinott's method as described in the preceding section. In that case, of course, the g_i can be quasiconvex and not necessarily convex.

Perhaps the most important application of the complementary convex programming algorithm just described is in solving signomial or complementary geometric programming problems, already mentioned in Chapter 7.

Consider the problem

$$\text{(CGP)} \qquad \min x_0 \tag{14.196}$$

subject to

$$P_i(x) - Q_i(x) \leq 1, \qquad i = 1, \ldots, m \tag{14.197}$$

where $x = (x_0, x_1, \ldots, x_n)^T$ and P_i, Q_i are posynomials. It can be shown that such a program can be transformed into a "complementary (log, log)-convex" program (see Chapters 6 and 7) in which (log, log)-convex and (log, log)-concave functions appear in the constraints. Such a program can, of course, also be transformed into an equivalent complementary convex program.

For details of the conceptual method, the reader is referred to Avriel [2] and Avriel and Williams [5, 6]. Dembo [14] has successfully implemented the method. He used a cutting plane algorithm for solving the convex or (log, log)-convex subproblems. The complete algorithm is described in Avriel, Dembo, and Passy [3]. In Exercise 14.M, we shall see that more general problems, involving certain functions of posynomials other than their

differences as in (14.197), can also be handled by the method of complementary convex programming.

EXERCISES

14.A. Continue Example 14.1.1 by performing two more iterations of the Griffith-Stewart method and plotting the constraints of the subproblems. Compute analytically the optimal solution of the original problem and observe converge to it.

14.B. Draw a flow diagram of Fletcher's method of hypercubes and illustrate it on the problem

$$\min f(x) = \left(x_1 + \frac{1}{x_2}\right)^{1/2} + \frac{4}{(x_1 - x_2)^2} \tag{14.198}$$

subject to

$$x_1 + x_2 \geq 2 \tag{14.199}$$

$$x_1 \geq 0, x_2 \geq 0 \tag{14.200}$$

by performing at least two iterations. Choose your own parameters as needed in the algorithm.

14.C. Let us see now how to construct test problems for nonlinear programming algorithms with optimal solutions known in advance. We shall use a construction suggested by Rosen and Suzuki [30]. Suppose that we are interested in solving the problem

$$\min f(x) = \phi(x) + c^T x, \qquad x \in R^n \tag{14.201}$$

subject to

$$g_i(x) = \psi_i(x) + b_i \geq 0, \qquad i = 1, \ldots, m, \tag{14.202}$$

where ϕ, and $\psi_1, \ldots, \psi_m$ are, respectively, given real differentiable, strictly convex and concave functions. Choose any $x^* \in R^n$ to be the predetermined optimum and $\lambda^* \in R^m$, $\lambda^* \geq 0$, as the corresponding Kuhn-Tucker multipliers. Choose b_i, $i = 1, \ldots, m$, so that

$$g_i(x^*) = 0 \quad \text{if} \quad \lambda_i^* > 0 \tag{14.203}$$

$$g_i(x^*) \geq 0 \quad \text{if} \quad \lambda_i^* = 0. \tag{14.204}$$

Compute the vector c by

$$c = \sum_{i=1}^{m} \lambda_i^* \nabla \psi_i(x^*) - \nabla \phi(x^*). \tag{14.205}$$

Show that this choice of the parameters b and c indeed ensures that x^* is the desired optimum. Indicate how the convexity of ϕ and the concavity of the ψ could be replaced by other conditions so that the construction method would still be usable.

14.D. Use the method of the preceding exercise to construct a nonquadratic test problem with linear constraints and perform a few iterations of the reduced-gradient method on it.

14.E. Verify that (14.89) and (14.90) are the Kuhn-Tucker necessary conditions of optimality for problem (NEP).

14.F. Draw a flow diagram of the GRG method. Construct a nonlinearly constrained test problem in two variables by the Rosen-Suzuki method of Exercise 14.C. Apply the GRG method to your problem and plot the trajectory of the optimization.

14.G. Suppose that we are seeking the solution of the nonlinear equation $f(x) = 0$ in the single variable x, where f is a strictly increasing, differentiable concave function. Transform this problem into a nonlinear program suitable for solution by cutting plane methods. Find a relationship between the points generated by, say, the KCG cutting plane method and those of Newton's algorithm, applied directly to the equation-solving problem.

14.H. Plot the nonlinear program of Example 14.3.1 and find the optimal solution by inspection. Plot the cutting planes added and note the progress of the algorithm.

14.I. Apply the supporting hyperplane algorithm to the problem of Example 14.3.1. Carry out at least two plane-adding steps.

14.J. Derive a "cutting surface" method for (h, ϕ)-convex programs. Illustrate your method on the solution of geometric programming problems that are known to be (log, log)-convex.

14.K. The complementary convex programming method of Section 14.4 can be viewed as a "cutting plane" method (why?). Modify the method proposed in the text to a kind of "supporting hyperplane" method. Try out your method on the problem of Example 14.4.1.

14.L. Formulate the complementary geometric program (CGP) as a complementary convex programming problem by making the necessary transformations on the functions and variables involved and by adding extra variables and constraints as needed. Write the problem of Example 14.4.1 as a complementary geometric program first and then use your derivations to write it as a complementary convex program. Observe that the approximating programs are ordinary geometric (posynomial) programs in the original variables.

14.M. Given the nonlinear program

(AP)
$$\min \sum_{i \in I_0} w_{i0} \left| \frac{P_{i0}(x) - Q_{i0}(x)}{R_{i0}(x)} \right|^{a_{i0}} \tag{14.206}$$

subject to

$$\sum_{i \in I_k} w_{ik} \left| \frac{P_{ik}(x) - Q_{ik}(x)}{R_{ik}(x)} \right|^{a_{ik}} \leq 1, \qquad k = 1, \ldots, m, \qquad (14.207)$$

where P_{ik}, Q_{ik}, R_{ik} are posynomials and w_{ik}, a_{ik} are given positive numbers for $i \in I_k$, $k = 0, \ldots, m$, show that (AP) can be transformed into an equivalent complementary geometric program by adding extra variables and constraints. Formulate the approximating geometric programs and then try to eliminate the extra variables added.

REFERENCES

1. Abadie, J., and J. Carpentier, "Generalization of the Wolfe Reduced Gradient Method to the Case of Nonlinear Constraints," in *Optimization*, R. Fletcher (Ed.), Academic Press, London, 1969.
2. Avriel, M., "Methods for Solving Signomial and Reverse Convex Programming Problems," in *Optimization and Design*, M. Avriel, M. J. Rijckaert, and D. J. Wilde (Eds.), Prentice-Hall, Englewood Cliffs, N.J., 1973.
3. Avriel, M., R. S. Dembo, and U. Passy, "Solution of Generalized Geometric Programs," *Int. J. Numer. Methods in Engineering*, **9**, 149–168 (1975).
4. Avriel, M., and A. C. Williams, "Complementary Convex Programming," Mobil R & D Corp. Central Research Div. Progress Memorandum, Princeton, N.J., May 1968.
5. Avriel, M., and A. C. Williams, "Complementary Geometric Programming," *SIAM J. Appl. Math.*, **19**, 125–141 (1970).
6. Avriel, M., and A. C. Williams, "An Extension of Geometric Programming with Applications in Engineering Optimization," *J. Eng. Math.*, **5**, 187–194 (1971).
7. Beale, E. M. L., "Numerical Methods," in *Nonlinear Programming*, J. Abadie (Ed.), North-Holland Publishing Co., Amsterdam, 1967.
8. Beale, E. M. L., "Nonlinear Optimization by Simplex-like Methods," in *Optimization*, R. Fletcher (Ed.), Academic Press, London, 1969.
9. Berge, C., *Topological Spaces*, Oliver & Boyd Ltd., Edinburgh, 1963.
10. Cheney, E. W., and A. A. Goldstein, "Newton's Method for Convex Programming and Tchebycheff Approximation," *Numer. Math.*, **1**, 253–268 (1959).
11. Colville, A. R., "A Comparative Study on Nonlinear Programming Codes," *Proceedings of the Princeton Symposium on Mathematical Programming*, H. W. Kuhn (Ed.), Princeton University Press, Princeton, N.J., 1970.
12. Dantzig, G. B., *Linear Programming and Extensions*, Princeton University Press, Princeton, N.J., 1963.

13. Dantzig, G. B., J. Folkman, and N. Z. Shapiro, "On the Continuity of the Minimum Set of a Continuous Function," *J. Math. Anal. & Appl.*, **17**, 519–548 (1967).

14. Dembo, R. S., "Solution of Complementary Geometric Programming Problems," M.Sc. thesis, Technion, Israel Institute of Technology, Haifa, 1972.

15. Eaves, B. C., and W. I. Zangwill, "Generalized Cutting Plane Algorithms," *SIAM J. Control*, **9**, 529–542 (1971).

16. Fletcher, R., "A General Quadratic Programming Algorithm, *J. Inst. Math. Appl.*, **7**, 76–91 (1971).

17. Fletcher, R., "An Algorithm for Solving Linearly Constrained Optimization Problems," *Math. Prog.*, **2**, 133–165 (1972).

18. Goldfeld, S. M., R. E. Quandt, and H. F. Trotter, "Maximization by Quadratic Hill-Climbing," *Econometrica*, **34**, 541–551 (1966).

19. Griffith, R. E., and R. A. Stewart, "A Nonlinear Programming Technique for the Optimization of Continuous Processing Systems," *Management Science*, **7**, 379–392 (1961).

20. Kelley, J. E., "The Cutting-Plane Method for Solving Convex Programs," *J. SIAM*, **8**, 703–712 (1960).

21. Levenberg, K. A., "A Method for the Solution of Certain Nonlinear Problems in Least Squares," *Quart. Appl. Math.*, **2**, 164–168 (1944).

22. Levitin, E. S., and B. T. Polyak, "Constrained Minimization Methods," *USSR Comp. Math. & Math. Phys.*, **6**, 5, 1–50 (1966).

23. Marquardt, D. W., "An Algorithm for Least-Squares Estimation of Nonlinear Parameters," *J. SIAM*, **11**, 431–441 (1963).

24. McCormick, G. P., "Anti-zig-zagging by Bending," *Management Science*, **15**, 315–320 (1969).

25. McCormick, G. P., "The Variable Reduction Method for Nonlinear Programming," *Management Science*, **17**, 146–160 (1970).

26. Pearson, J. D., "Variable Metric Methods of Minimisation," *Computer J.*, **12**, 171–178 (1969).

27. Powell, M. J. D., "A New Algorithm for Unconstrained Minimization," in *Nonlinear Programming*, J. B. Rosen, O. L. Mangasarian, and K. Ritter (Eds.), Academic Press, New York, 1970.

28. Powell, M. J. D., "Convergence Properties of a Class of Minimization Algorithms," Report C.S.S. 8, A.E.R.E. Harwell, United Kingdom, April 1974.

29. Rosen, J. B., "Iterative Solution of Nonlinear Optimal Control Problems," *SIAM J. Control*, **4**, 223–244 (1966).

30. Rosen, J. B., and S. Suzuki, "Construction of Nonlinear Programming Test Problems," *Comm. of the ACM*, **8**, 113 (1965).

31. TOPKIS, D. M., "Cutting-Plane Methods without Nested Constraint Sets," *Operations Research*, **18**, 404–413 (1970).

32. VEINOTT, A. F., "The Supporting Hyperplane Method for Unimodal Programming," *Operations Research*, **15**, 147–152 (1967).

33. WILDE, D. J., "Jacobians in Constrained Nonlinear Optimization," *Operations Research*, **13**, 848–856 (1965).

34. WILDE, D. J., and C. S. BEIGHTLER, *Foundations of Optimization*, Prentice-Hall, Englewood Cliffs, N.J., 1967.

35. WILLIAMS, A. C., "Marginal Values in Linear Programming," *J. SIAM*, **11**, 82–94 (1963).

36. WILSON, R. B., "A Simplicial Algorithm for Concave Programming," Doctoral dissertation, Harvard University, Cambridge, Mass., 1963.

37. WOLFE, P., "Methods of Nonlinear Programming," in *Recent Advances in Mathematical Programming*, R. L. Graves and P. Wolfe (Eds.), McGraw-Hill Book Co., New York, 1963.

38. WOLFE, P., "Methods of Nonlinear Programming," in *Nonlinear Programming*, J. Abadie (Ed.), North-Holland Publishing Co., Amsterdam, 1967.

39. WOLFE, P., "Convergence Theory in Nonlinear Programming," in *Integer and Nonlinear Programming*, J. Abadie (Ed.), North-Holland Publishing Co., Amsterdam, 1970.

40. ZANGWILL, W. I., "Convergence Conditions for Nonlinear Programming Algorithms," *Management Science*, **16**, 1–13 (1969).

41. ZANGWILL, W. I., *Nonlinear Programming: A Unified Approach*, Prentice-Hall, Englewood Cliffs, N.J., 1969.

AUTHOR INDEX

Each chapter of this book is accompanied by a bibliography of works cited in the chapter. In this index, numbers in *italics* refer to the end-of-chapter reference pages on which the particular references are *first* listed.

C

D

E

F

G

N

O

P

Q

R

SUBJECT INDEX

D

N

O

P

D1222961